Biocomposites: Design and Mechanical Performance

Related titles

Biofiber reinforcement in composite materials
(ISBN 978-1-78242-122-1)

Natural fibre composites: Materials, processes and properties
(ISBN 978-0-85709-524-4)

Environmentally friendly polymer nanocomposites: Types, processing and properties
(ISBN 978-0-85709-777-4)

Woodhead Publishing Series in Composites Science and Engineering: Number 61

Biocomposites: Design and Mechanical Performance

Edited by

Manjusri Misra
Jitendra K. Pandey
Amar K. Mohanty

ELSEVIER

AMSTERDAM • BOSTON • CAMBRIDGE • HEIDELBERG
LONDON • NEW YORK • OXFORD • PARIS • SAN DIEGO
SAN FRANCISCO • SINGAPORE • SYDNEY • TOKYO
Woodhead Publishing is an imprint of Elsevier

WP
WOODHEAD PUBLISHING

Woodhead Publishing is an imprint of Elsevier
80 High Street, Sawston, Cambridge, CB22 3HJ, UK
225 Wyman Street, Waltham, MA 02451, USA
Langford Lane, Kidlington, OX5 1GB, UK

ISBN: 978-1-78242-373-7 (print)
ISBN: 978-1-78242-394-2 (online)

British Library Cataloguing-in-Publication Data
A catalogue record for this book is available from the British Library

Library of Congress Control Number: 2015935091

For information on all Woodhead Publishing publications visit our website at http://store.elsevier.com/

Contents

Contributors

H.P.S. Abdul Khalil Universiti Sains Malaysia, Penang, Malaysia

D. Acierno Dipartimento di Ingegneria Chimica, dei Materiali e della Produzione Industriale, Università di Napoli Federico II, Naples, Italy

S. Aprilia Syiah Kuala University, Banda Aceh, Indonesia

M.I. Aranguren Institute of Research in Materials Science and Technology (INTEMA), National University of Mar del Plata, Mar del Plata, Argentina

A.A. Astimar Malaysian Palm Oil Board, Kuala Lumpur, Malaysia

L. Avérous Université de Strasbourg, Strasbourg, France

C. Baley Université de Bretagne Sud, Lorient Cedex, France

B.I. Biddappa Clemson University, Greenville, SC, USA

A. Bourmaud Université de Bretagne Sud, Lorient Cedex, France

A. Cloutier Université Laval, Quebec City, QC, Canada

R.N. Darie-Niţă "P. Poni" Institute of Macromolecular Chemistry, Iasi, Romania

Y. Davoudpour Universiti Sains Malaysia, Penang, Malaysia

H.N. Dhakal University of Portsmouth, Portsmouth, UK

R. Dungani Universiti Sains Malaysia, Penang, Malaysia, and School of Life Sciences and Technology, Institut Teknologi Bandung, Bandung, Indonesia

R.F. El-Hajjar University of Wisconsin, Milwaukee, WI, USA

G. Filippone Dipartimento di Ingegneria Chimica, dei Materiali e della Produzione Industriale, Università di Napoli Federico II, Naples, Italy

P.J. Halley The University of Queensland, Brisbane, QLD, Australia

M.S. Hossain Universiti Sains Malaysia, Penang, Malaysia

L. Jiang North Dakota State University, Fargo, ND, USA

A. Kaboorani Université Laval, Québec, QC, Canada

A.R. Kakroodi Université Laval, Quebec City, QC, Canada

Y. Kazemi Université Laval, Quebec City, QC, Canada

A.B. Kousaalya Clemson University, Greenville, SC, USA

A. Le Duigou Université de Bretagne Sud, Lorient Cedex, France

H.H. León-Santiesteban Department of Biotechnology, Universidad Autónoma Metropolitana-Iztapalapa, Mexico, Mexico, D.F.

P. Liu Guangzhou University, Guangzhou, China

N.E. Marcovich Institute of Research in Materials Science and Technology (INTEMA), National University of Mar del Plata, Mar del Plata, Argentina

T.H. Mekonnen University of Guelph, Guelph, ON, Canada

M. Misra University of Guelph, Guelph, ON, Canada

A.K. Mohanty University of Guelph, Guelph, ON, Canada

M.A. Mosiewicki Institute of Research in Materials Science and Technology (INTEMA), National University of Mar del Plata, Mar del Plata, Argentina

R. Muthuraj University of Guelph, Guelph, ON, Canada

V. Nagarajan University of Guelph, Guelph, ON, Canada

Z. Nahrul Hayawin Malaysian Palm Oil Board, Kuala Lumpur, Malaysia

J.K. Pandey University of Petroleum and Energy Studies, Dehradun, India

E. Părpăriţă "P. Poni" Institute of Macromolecular Chemistry, Iasi, Romania

S. Pilla Clemson University, Greenville, SC, USA

I.I. Qamhia University of Wisconsin, Milwaukee, WI, USA

S. Rai Clemson University, Greenville, SC, USA

B. Riedl Université Laval, Québec, QC, Canada

D. Rodrigue Université Laval, Quebec City, QC, Canada

P. Russo Institute for Polymers, Composites and Biomaterials, National Research Council, Pozzuoli (Naples), Italy

N.L.M. Suraya Universiti Sains Malaysia, Penang, Malaysia

E. Ten Washington State University, Pullman, WA, USA

A. Tomasini Department of Biotechnology, Universidad Autónoma Metropolitana-Iztapalapa, Mexico, Mexico, D.F.

L.-S. Turng University of Wisconsin–Madison, Madison, WI, USA

C. Vasile "P. Poni" Institute of Macromolecular Chemistry, Iasi, Romania

M.P. Wolcott Washington State University, Pullman, WA, USA

F. Xie The University of Queensland, Brisbane, QLD, Australia

J. Zhang Washington State University, Pullman, WA, USA

H. Zhao Shandong University, Jinan, China

Woodhead Publishing Series in Composites Science and Engineering

1 **Thermoplastic aromatic polymer composites**
F. N. Cogswell
2 **Design and manufacture of composite structures**
G. C. Eckold
3 **Handbook of polymer composites for engineers**
Edited by L. C. Hollaway
4 **Optimisation of composite structures design**
A. Miravete
5 **Short-fibre polymer composites**
Edited by S. K. De and J. R. White
6 **Flow-induced alignment in composite materials**
Edited by T. D. Papthanasiou and D. C. Guell
7 **Thermoset resins for composites**
Compiled by Technolex
8 **Microstructural characterisation of fibre-reinforced composites**
Edited by J. Summerscales
9 **Composite materials**
F. L. Matthews and R. D. Rawlings
10 **3-D textile reinforcements in composite materials**
Edited by A. Miravete
11 **Pultrusion for engineers**
Edited by T. Starr
12 **Impact behaviour of fibre-reinforced composite materials and structures**
Edited by S. R. Reid and G. Zhou
13 **Finite element modelling of composite materials and structures**
F. L. Matthews, G. A. O. Davies, D. Hitchings and C. Soutis
14 **Mechanical testing of advanced fibre composites**
Edited by G. M. Hodgkinson
15 **Integrated design and manufacture using fibre-reinforced polymeric composites**
Edited by M. J. Owen and I. A. Jones
16 **Fatigue in composites**
Edited by B. Harris
17 **Green composites**
Edited by C. Baillie
18 **Multi-scale modelling of composite material systems**
Edited by C. Soutis and P. W. R. Beaumont

19 **Lightweight ballistic composites**
Edited by A. Bhatnagar
20 **Polymer nanocomposites**
Y.-W. Mai and Z.-Z. Yu
21 **Properties and performance of natural-fibre composite**
Edited by K. Pickering
22 **Ageing of composites**
Edited by R. Martin
23 **Tribology of natural fiber polymer composites**
N. Chand and M. Fahim
24 **Wood-polymer composites**
Edited by K. O. Niska and M. Sain
25 **Delamination behaviour of composites**
Edited by S. Sridharan
26 **Science and engineering of short fibre reinforced polymer composites**
S-Y. Fu, B. Lauke and Y-M. Mai
27 **Failure analysis and fractography of polymer composites**
E. S. Greenhalgh
28 **Management, recycling and reuse of waste composites**
Edited by V. Goodship
29 **Materials, design and manufacturing for lightweight vehicles**
Edited by P. K. Mallick
30 **Fatigue life prediction of composites and composite structures**
Edited by A. P. Vassilopoulos
31 **Physical properties and applications of polymer nanocomposites**
Edited by S. C. Tjong and Y.-W. Mai
32 **Creep and fatigue in polymer matrix composites**
Edited by R. M. Guedes
33 **Interface engineering of natural fibre composites for maximum performance**
Edited by N. E. Zafeiropoulos
34 **Polymer-carbon nanotube composites**
Edited by T. McNally and P. Pötschke
35 **Non-crimp fabric composites: Manufacturing, properties and applications**
Edited by S. V. Lomov
36 **Composite reinforcements for optimum performance**
Edited by P. Boisse
37 **Polymer matrix composites and technology**
R. Wang, S. Zeng and Y. Zeng
38 **Composite joints and connections**
Edited by P. Camanho and L. Tong
39 **Machining technology for composite materials**
Edited by H. Hocheng
40 **Failure mechanisms in polymer matrix composites**
Edited by P. Robinson, E. S. Greenhalgh and S. Pinho
41 **Advances in polymer nanocomposites: Types and applications**
Edited by F. Gao

42 **Manufacturing techniques for polymer matrix composites (PMCs)**
Edited by S. Advani and K.-T. Hsiao
43 **Non-destructive evaluation (NDE) of polymer matrix composites: Techniques and applications**
Edited by V. M. Karbhari
44 **Environmentally friendly polymer nanocomposites: Types, processing and properties**
S. S. Ray
45 **Advances in ceramic matrix composites**
Edited by I. M. Low
46 **Ceramic nanocomposites**
Edited by R. Banerjee and I. Manna
47 **Natural fibre composites: Materials, processes and properties**
Edited by A. Hodzic and R. Shanks
48 **Residual stresses in composite materials**
Edited by M. Shokrieh
49 **Health and environmental safety of nanomaterials: Polymer nanocomposites and other materials containing nanoparticles**
Edited by J. Njuguna, K. Pielichowski and H. Zhu
50 **Polymer composites in the aerospace industry**
Edited by P. E. Irving and C. Soutis
51 **Biofiber reinforcement in composite materials**
Edited by O. Faruk and M. Sain
52 **Fatigue and fracture of adhesively-bonded composite joints: Behaviour, simulation and modelling**
Edited by A. P. Vassilopoulos
53 **Fatigue of textile composites**
Edited by V. Carvelli and S. V. Lomov
54 **Wood composites**
Edited by M. P. Ansell
55 **Toughening mechanisms in composite materials**
Edited by Q. Qin and J. Ye
56 **Advances in composites manufacturing and process design**
Edited by P. Boisse
57 **Structural integrity and durability of advanced composites: Innovative modelling methods and intelligent design**
Edited by P. W. R. Beaumont, C. Soutis and A. Hodzic
58 **Recent advances in smart self-healing polymers and composites**
Edited by G. Li and H. Meng
59 **Manufacturing of nanocomposites with engineering plastics**
Edited by V. Mittal
60 **Fillers and reinforcements for advanced nanocomposites**
Edited by Y. Dong, R. Umer and A. Kin-Tak Lau
61 **Biocomposites: Design and mechanical performance**
Edited by M. Misra, J. K. Pandey and A. K. Mohanty
62 **Numerical modelling of failure in advanced composite materials**
Edited by P.P. Camanho and S. R. Hallett

Preface

The well-known Green Revolution in the late 1960s was a groundbreaking agricultural initiative that resulted in high-yield crops, thereby satisfying the food demand of a growing population. History repeated in the 1990s, when agriculture—with other renewable resources—became the base for the growing biobased economy. In the developing bioeconomy era, the necessity for sustainable development has drawn a renewed interest for natural fiber composites in the manufacturing sector. Biocomposites and green-composites are gaining importance for their application in eco-friendly consumer products, rigid packaging, automotive parts, and housing structures.

The environmental awareness of our society has reached its highest level in decades. Most segments of society, particularly the burgeoning younger demographics, now realize the importance of alternative resources substituting the non-renewable and fast-depleting petroleum resources in different areas of application, such as energy, chemicals, and materials. The ever-increasing concern over climate change, growing greenhouse gas (GHG) emission, along with the growing population has necessitated the reinvention of sustainable development in the plastic-based manufacturing industries. Sustainable development can be realized with the overlap of its three foundational platforms: economic, social, and environmental benefits. Currently, we have to plan, design, and formulate new, alternative options from resources that are renewable, within acceptable timeframes.

Today, petroleum is the main feedstock for the production of polymeric materials. Almost 99% of the total global plastic production is from petroleum resources. Synthetic glass fiber has occupied the predominant place in the multi-billion-dollar composite industry. The long-standing R&D output in the area of natural fiber composites has created enough scientific data showing that several natural fibers can substitute certain types of glass fiber (like E-glass) in specific composite uses—with the added advantage of eco-friendliness. Many natural fibers are less expensive, are about half the density—thus lighter, have higher specific strength, require comparatively less energy to produce, have superior sound abatement features, and are biodegradable—as compared to E-glass fibers.

The material being lightweight, along with its competitive cost and performance attributes, is a driving force behind the use of natural fiber composites in automotive parts. The agro-based natural fibers, like flax, jute, kenaf, industrial hemp, sisal, and henequen, have been researched extensively in the biobased composites field. This class of long fibers has advantages in engineering fabric-based structural composites, alongside their use in injection molded short fiber composites. Forest resourced wood fiber has been used widely in wood-plastic composites. The purpose-grown perennial grasses, for example, switch grass and miscanthus, as well as agro-residues such as wheat and rice straws, corn stovers, soy stalks, etc., demonstrate strong potential in injection molded composites application. For the manufacturing sector, the constant supply chain of natural fibers, and the logistics of their collection and transportation are some of the challenges hindering the fibers' large-scale implementation.

The biobased composite materials from petro-based plastics and natural fibers have achieved some commercial success. The most used biocomposites are natural fiber reinforced polypropylene, under the thermoplastic composites category, and natural fiber reinforced unsaturated polyester resin, under the thermoset class. New composite materials from plant derived natural fibers and crop derived bioplastics are gaining more and more importance with the growing environmental need to have truly green composites. The hybrid biocomposites, derived from mixed sources, for example, both from bio-resources and petro-resources, are gaining more market and commercial attraction because of their cost-performance attributes. At present, society is ready to accept biobased alternatives; however, not with cost and performance penalties. For these materials, scientists, in collaboration with chemists, physicists, mechanical and chemical engineers, are seeking ways to provide knowledge, and to offer technologies that can overcome the obstacles. Policy makers are the first element needed to move this plan forward. Governmental mandates are necessary to promote the production and utilization of bio-products in different industrial sectors.

With the present book, *Biocomposites: Design and Mechanical Performance*, we hope to contribute to the awareness of our society about the environment; provide a scientific approach to overcome obstacles with substitutes for petro-based polymeric materials; and encourage the policy makers to keep driving the industry sectors, and academic research, towards a sustainable future. This book consists of 17 chapters written by leading researchers. It is a single source of information about a wide range of bioplastics and biocomposites.

We would like to acknowledge all authors who contributed, as well as the Woodhead Publishing staff who helped us immensely in making this book possible.

Finally, we would like to dedicate this book to the numerous natural fiber growers, the stalwarts of the biomass producing industry, biopolymer compounders and equipment manufacturers, policy makers, those looking for value-added solutions to co-product streams of the bioproducts industry, and to all those looking to create a sustainable future for mankind in an environmentally conscious society.

Manjusri Misra
Jitendra K. Pandey
Amar K. Mohanty

Foreword

Environmental concerns associated with the ever-expanding use of petro-based materials continue to grow, along with the rapid acceleration in global warming. Faster development of sustainable materials and green processes for mainstream markets is critically important for tackling both these issues. This is particularly true for composites and other petro-based products that combine two dissimilar materials, because, as of now, no environmentally friendly end-of-life solutions are available for them. Unfortunately, most composites end up in landfills. Replacing petro-based materials with their bio-based counterparts has great potential to limit and even reduce the accumulation of new fossil carbon in the atmosphere. It is no wonder that, over the past few years, an increasing number of scientific publications, reports, patents, international conferences, and symposia have been covering the exploration, development, and use of renewable resource- or plant-based materials as alternatives to petro-based materials in broad and diverse applications. Companies with long-term vision are already moving away from petro-based materials and chemicals, and at the same time, governments are enacting new regulations to increase the use of bio-based products, for a better tomorrow.

Natural fiber composites, developed from traditional and engineering plastics, are being widely researched, and these materials have begun to find their way into diverse industrial and sports applications. On the other hand, researchers are increasingly exploring the design and engineering of *green* biocomposites derived from bio-based plastics and natural fibers, because most of these products tend to be environmentally benign. More and more research initiatives are also aimed at realizing the full potential of such bio-based and biodegradable plastics and composite materials. Improvements in the performance, efficiency, durability, and cost effectiveness of these materials, through further research, can improve their competitiveness in the current market. Current bio-based materials research is focused on developing and evaluating a rich array of bioproducts. As we make an effort to substitute the traditional materials with bio-based products, we must learn to better assess these sustainable materials to ensure their quality and long-term performance, while enhancing their functionalities.

The success of these materials in every possible field of application requires a full understanding of the material properties and manufacturing technologies. For this reason, this book focuses on detailing the design and mechanical performance of biocomposites. The chapters in this book provide a comprehensive scientific and technological appraisal of the current research and perspectives related to biocomposites made from traditional and engineering plastics, bio-based plastics, and their blends reinforced with natural fibers. This book should appeal to a broad range of research scientists, scholars, academics, industrialists, and government officials across various related fields.

Professor Anil Netravali
Jean and Douglas McLean Professor in
Fiber Science and Apparel Design
Cornell University, USA

Commercial potential and competitiveness of natural fiber composites

1

J.K. Pandey, V. Nagarajan†, A.K. Mohanty†, M. Misra†*
*University of Petroleum and Energy Studies, Dehradun, India; †University of Guelph, Guelph, ON, Canada

1.1 Introduction

Use of renewable resource-based materials is increasing in importance with environmental concerns mounting worldwide over the use of nonrenewable resources such as petroleum. Carbon dioxide emissions resulting from human activities have been the constant topic of discussion in the scientific community. In recent years, many industries across various manufacturing sectors have acknowledged the need to adapt environmentally friendly manufacturing technologies and products. Multiple industrial adaptations and explorations of innovative materials are underway to meet the environmental concerns of end users. The scope for composite materials and technologies based on renewable resources is vast; extensive research and laboratory demonstrations have repeatedly proved the technical, environmental, and economical benefits of such innovative materials.

Composite material is a heterogeneous combination of two or more different phase constituents (matrix, load bearing or reinforcement elements, fillers and compatibilizers) (Strong, 2008). Reinforcements and fillers can be long, short, continuous, discontinuous, or spherical particles (Strong, 2008). In this context, natural fibers are of definite interest to fabricate composites. Natural fibers offer numerous advantages; they are discussed in subsequent sections of this chapter. Synergistic combination of natural fibers and plastics can result in a material having favorable level of performance, quality, and cost, and therefore possesses immense potential and prospects for serving different industries and applications. A desirable level of mechanical performance and durability of natural fiber composites (NFCs) with cost competitiveness is the key driver for commercializing these materials for high-volume applications. For the past two decades, comprehensive research has been conducted on NFCs and numerous research articles have been published addressing various challenges in the fabrication of composites where the final goal has always been achieving NFCs with desired level of mechanical performance and cost attributes for certain specific industrial uses. A number of research articles published on natural fibers has reached a new high of 34,385 articles as of March 1, 2015, based on Scifinder resources (http://www.cas.org/products/scifindr/index.html).

Generally, synthetic polymer matrixes are the choice to develop natural fiber-based composites for multiple applications such as automotive, packaging,

Biocomposites: Design and Mechanical Performance. http://dx.doi.org/10.1016/B978-1-78242-373-7.00001-9

flexible electronics, and others. Because petroleum is a depleting resource with fluctuating cost, adding natural fibers to petroleum-based plastics can provide considerable environmental and economic advantages. Natural fibers are available in abundance and are relatively inexpensive. NFCs based on polyolefins such as polypropylene (PP) and polyethylene have been widely researched and are successful in finding applications in the development of interior parts for the automotive industry (Zampaloni et al., 2007; Mohanty et al., 2002; Ellison and McNaught, 2000; Suddell and Rosemaud, 2009). Recently, natural fibers have been explored as reinforcements for engineering thermoplastics such as polycarbonate (PC) in an aim to widen the adaptability window of natural fibers to suit high-performance engineering applications. Mechanical performance of NFCs containing polyolefins and PC as matrix material are discussed in detail in Chapters 9–11.

The environmental aspect concerned with the disposal of composites based on synthetic resins still requires quantitative research. To develop a material with complete environmental compatibility, it is important that every phase of the material possesses the ability to undergo biological degradation. Certain renewable resource-based biopolymers have the potential to be an important component of a continuously emerging new line of low carbon footprint materials. Biopolymers refer to either biodegradable materials, derived from both nonrenewable and renewable resources, or nonbiodegradable materials derived from renewable biological resources (Mohanty et al., 2000). In other words, for a polymer to be called a biopolymer, it has to be either biodegradable, biobased, or both. Biopolymers are often also called bioplastics in many published studies. Combining bioplastics with natural fibers is a strategy to produce advanced composite materials. This strategy enables research scientists to design and engineer materials according to the requirements of the end user while keeping sustainability, economic feasibility, and environmental considerations under the design framework. Chapters 2–7 in this book are dedicated to discussing the mechanical performance of NFCs containing various biopolymers as matrix materials.

1.2 Classification and composition of natural fibers

Natural fibers are mainly classified as vegetable fibers, animal fibers, and mineral fibers (e.g., asbestos) depending on the origin (Mohanty et al., 2000). Vegetable fibers are further classified as wood (soft and hard) fibers and nonwood fibers. Five basic types of nonwood fibers with few representative examples are

- Bast: hemp, flax, jute, kenaf, and ramie
- Leaf: sisal, pineapple, abaca
- Seed/fruit: coir, kapok, coconut
- Straw: wheat, rice, soy, and corn straw
- Grass/reed: switchgrass, miscanthus

Generally, bast fibers are preferred for applications requiring high tensile strength, stiffness, and modulus and they are available in chopped, nonwoven, and mat forms. Bast and leaf fibers find their application in particle boards, fiberboards, automotive components, and products for housing and infrastructure (Anandjiwala and Blouw,

2007). Leaf fibers and coir are employed as cordage materials, cotton in apparel, and jute in carpeting and sacking. Other lignocellulosic residues such as sugarcane bagasse, coffee chaff, and hulls and residues resulting from other agricultural and industrial processing also represent a major fiber source for the development of renewable resource-based biocomposites for industrial applications. Distillers dried grains with solubles (DDGS), a coproduct of the corn ethanol industry, and lignin, a coproduct of the bioethanol and pulp and paper industries are now gaining increased interest for creating new value-added biocomoposite materials (Lamontagne, 2013). Figure 1.1 shows digital photographs of different classes of natural fibers, agricultural residues, and industrial coproducts/byproducts that are being used and researched in the design and engineering of biobased composite materials.

The universe of natural fiber is broad and, even so, the basic chemical structure of many of these fibers is similar in terms of the constituents and varieties in their composition, depending upon the type and origin of the fiber. Natural fiber itself is a

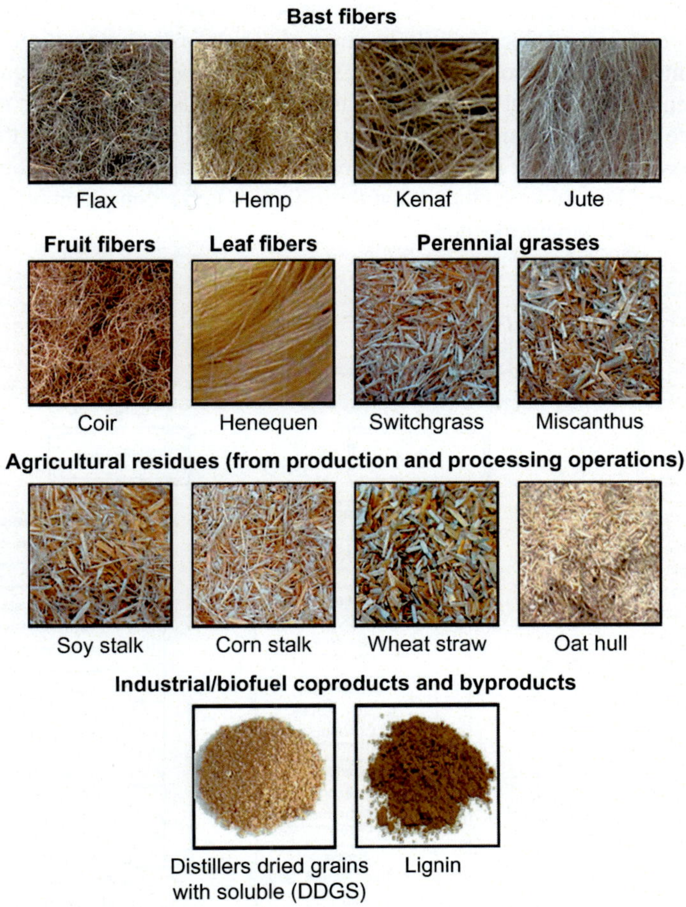

Figure 1.1 Photographs of different classes of natural fibers, agricultural residues, and industrial coproducts/byproducts.

composite by nature as it contains rigid, crystalline cellulose microfibrils in a matrix of lignin and hemicellulose that is amorphous. It also contains pectin, waxes, and some water-soluble compounds. Cellulose is the most abundant organic compound available in earth. Plant cell walls get their strength, stiffness, and stability from this main structural component, cellulose. The amount of cellulose present in a fiber has a significant influence on the properties and performance of the resulting composite materials produced for various applications.

The compositions of various natural fibers have been listed in detail in several review articles and book chapters (Faruk et al., 2012; Calistor et al., 2014). General properties and composition of natural fiber are additionally reviewed in Chapter 5. Figure 1.2 shows the general classification of lignocellulosic natural fibers, surface treatment methods, and challenges, which are discussed in detail in later sections of this chapter. Research and development is directed more toward in situ compatibilization and reactive extrusion techniques for injection molding types of biocomposites.

The mechanical properties of resulting composite materials are highly dependent on the source of natural fiber, mainly because of variation in composition. For example, two composites, made of jute and flax, with similar host matrix and equal concentrations even after identical treatments will not essentially show similar mechanical performance (Mohanty et al., 2000). Nagarajan et al. (2013) investigated the effect of fiber composition and fiber length on biocomposites containing perennial grasses and agricultural residues and concluded that these factors had a complementary effect on the performance of biocomposites.

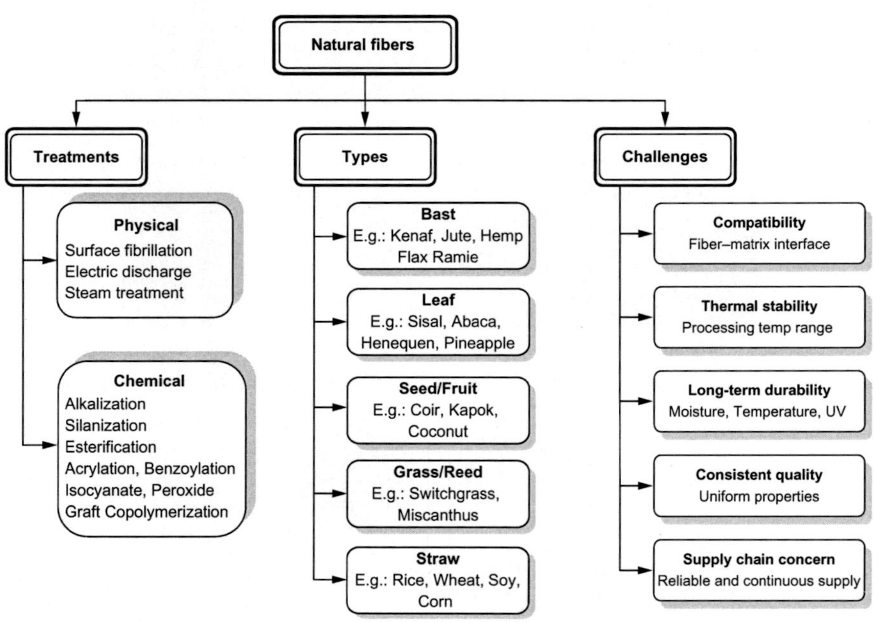

Figure 1.2 Natural fiber types, surface treatment options, and challenges.

1.3 Advantages and attributes of natural fibers

Principal advantages of using natural fibers in fabrication of composites are listed below (Mohanty et al., 2000; Faruk et al., 2012):

- Renewability, biodegradability, and CO_2 neutrality nature of natural fibers makes it environmentally friendly. Carbon sequestration attribute helps in reduction of GHG emissions and, hence, reducing the implications on climate change.
- Reinforcing traditional and biobased polymers with natural fibers can reduce the dependence on petro-based materials, as typically up to 50 wt.% of the matrix can be replaced with natural fibers for several injection-molded applications. Weight percentages of natural fiber in compression-molded composites could be higher than 50% depending on the matrix system and compatibilizers used for certain target applications.
- Natural fibers are available at a lower cost; therefore, they can help in offsetting the cost of matrix materials. Final cost of the parts ends up being cheaper or on par with the synthetic counterparts.
- Natural fibers can be an effective substitute for synthetic fibers such as E-glass fibers in certain applications. This advantage will be well received by industries because several countries have imposed severe restrictions on disposal of products made with glass fibers and the use of asbestos has also been banned in several parts of the world.
- When composites with natural fibers are incinerated for energy recovery process, mainly harmless residues are released without any trace of harmful heavy metals.
- Use of agricultural, lignocellulosic residues for high-volume applications can bring in new source of income for farmers while extending the value chain of the crops. This can help in job creation in rural farming communities.
- Natural fibers have environmentally benign production process and disposal options compared to synthetic fibers. They can be processed in conventional processing equipment; unlike glass fibers, natural fibers are nonabrasive to processing machinery used for manufacturing the composite materials, resulting in overall better efficiency.
- Natural fibers in the bast and leaf fibers category like jute, flax, kenaf, and others have high specific strength and modulus, and low density compared to E-glass fibers. NFCs are lightweight materials compared to glass fiber-filled composites, provided the fiber loading level remains the same; for automotive applications, this translates to a weight saving advantage and improved fuel efficiency.
- Natural fibers exhibit good acoustic and sound abatement properties. They have a relatively safe manufacturing process with reduced dermal and respiratory irritation.

1.4 Challenges encountered in adapting natural fibers for composite applications

Several challenges are encountered while developing composites from natural fibers and they all need to be effectively addressed to elevate the performance and market acceptability of NFCs. The challenges listed below are interconnected, and good control over these factors is important to create composites meeting the desired level of performance set by the industries. Knowledge on physical characteristics of fibers such as fiber architecture, strength, variability, and crystallinity are necessary to design composite formulations with desired performance attributes.

Processing temperature and moisture sensitivity: Most of the traditional polymers are processed between 180 and 200 °C and there are reports on time-dependent degradation of natural fiber at this temperature range. Literature on thermal resistance during processing of certain kind of fibers seems contradictory. For example, processing temperature of 170 °C was reported to affect the thermal stability of jute and flax fibers (Mohanty et al., 2002) whereas, in another study, high temperatures (170–180 °C) (Joshi et al., 2004) did not have any significant effect on the composite tensile properties. However, major damage to flax fiber has been reported at a temperature above 240 °C (Yu et al., 2006). Natural fibers are hydrophilic in nature, which can have adverse effects on the properties of the resulting composites. The moisture content is usually dependent on the composition of noncrystalline parts and the presence of voids on the fiber surface. Appropriate drying methods have to be established before processing the material with polymer matrix. In general, NFCs have higher susceptibility to moisture absorption.

Compatibility: Unfortunately, in most cases the NFC does not possess the same level of performance as glass fiber composites, owing to the incompatibility between hydrophobic polymer matrix and hydrophilic natural fiber. Although this issue can be solved by adequate physical and/or chemical modification, the risk of chain degradation and cost increment is a major thrust area for research. A significant part of Chapter 5 has been dedicated to discussing the effects of the compatibilization techniques explored so far in NFCs based on biodegradable blends as matrix material.

1.5 Supply chain management

Reliable and continuous availability of natural fibers/biomass is extremely important for any successful commercial development of composite materials. Farmers producing biomass for composite applications want to see solid markets and grantees; while on the other hand, industries want to have a guarantee of sustainable biomass supply. Therefore, it is imperative to strike a balance between the needs of farmers and industries. Performance of the supply chain largely depends on the degree of integration and coordination between the individual entities. Seasonal high volumes of agricultural residues available for use in composite applications, could affect the year round production of biocomposite materials. Abundant quantities of biomass are available; however, there is a need to control the fluctuating supply throughout the year. Cost, characteristics, and quality of fibers could vary substantially depending on the harvest conditions. Production patterns affecting the properties of the fibers have to be studied and surveyed over a significant period of time to acquire an understanding of producing fibers with consistent qualities.

One important aspect in the fiber supply chain is logistics. Several studies (Akgul et al., 2010; Yu and Tao, 2009; Sharma et al., 2013) have used mathematical models to assess the complete energy cost involved in the use of biomass along the entire supply chain (development, establishment, production, harvest, storage, and transport to facilities manufacturing the end use products). On the basis of these studies, numerous cost-reduction strategies have been proposed. Reduction in logistics cost could have a major impact on regional and domestic competitive positioning. Therefore, the presence of an industrial biocomposite manufacturing facility in the vicinity of farmland and agricultural production areas can have great economic advantages. This has been one of the main reasons for the development of small and localized biorefineries demonstrating the concept of regional sustainable manufacturing (Waldron, 2014). This ensures that biomass is locally sourced, as close as possible, to avoid high transportation costs and to keep the feedstock cost to a minimum while increasing the sustainability of the manufacturing process (Waldron, 2014). Another strategy that can effectively address fiber supply concerns is the hybridization of two or more different fibers which can ease the demand on just one fiber type. Hybridization of banana fibers with kenaf (Thiruchitrambalam et al., 2009) and hemp (Idicula et al., 2009) has been researched, and hybridization of agricultural residues (Nyambo et al., 2010) and perennial grasses (Nanda et al., 2012) is being actively explored. Hybridization of natural fibers in combination with coproduct/by-product residues from various industrial operations (Lignin, DDGS) (Sahoo et al., 2004; Vivekanandhan et al., 2013) also offer new avenues.

1.6 Commercial competitiveness, market development, and growth scenario

1.6.1 Factors promoting commercial competiveness

Specific characteristics of some natural fibers are helpful in innovation leading to commercialization of biobased materials. Novel fiber formatting processes can be used to create new biomaterials. NFCs have improved commercial competitiveness because of two major reasons: facilitating mitigation of price fluctuations in the crude oil market and the environmental implications of using nonrenewable materials and products that by definition have a finite supply chain. The annual average (nominal) price for crude oil was in the range of US$15–18 per barrel for most of late 1980s, reaching a maximum of US$20 in 1990 (http://inflationdata.com/Inflation/Inflation_Rate/Historical_Oil_Prices_Table.asp). Prices remained more or less the same: a low of US$12 in 1998 and a high of US$20.50 in 1996. The beginning and early 2000s saw crude prices in the range of US$50/barrel. However, during the past decade, crude prices have steadily increased with a maximum of US$92 in 2008 and 2013 (http://inflationdata.com/Inflation/Inflation_Rate/Historical_Oil_Prices_Table.asp). Although currently a weakening is observed, and they stand at US$89 as of November 2014 (http://inflationdata.com/Inflation/Inflation_Rate/Historical_Oil_Prices_Table.asp), the prices still remain high in terms of historical prices recorded for crude oil. The price of PP has been observed to be tracking the rising prices of crude oil. In the 1980s, the average price of PP was reported to be $803/mt, and $771/mt in the 1990s (FAO Consultation on Natural Fibres, 2007). However, in more recent years the prices for PP seem to have almost doubled with a high of $1367/mt as of September 2013 (Global Petrochemical Prices, 2013); this was when average petroleum prices increased by 2.3% between September 2012 and September 2013 (Global Petrochemical Prices,

2013). A 25% plunge in PP price to $869/mt was recorded as of December 2014 (http://www.platts.com/news-feature/2014/petrochemicals/pgpi/propylene). Such highly volatile prices for traditional plastics are creating scenarios that are favoring the use of alternative materials such as natural fibers that reduce the dependence on fossil fuel based polymers and facilitate steady price offers for longer time periods. The need to spread the environmental advantages of using natural fibers became imperative with fluctuating crude oil prices and oil's depleting nature. However, at the time, the early 1990s, no solid scientific research was being conducted on natural fibers with a view to possible commercialization. That was when the United Nations Conference on Trade and Development and International Jute Organization started addressing the positive environmental attributes of different bast fibers (United Nations Conference on Trade, 1996). Quantitative results demonstrating the environmental advantages of using natural fibers were significantly lacking. van Dam and Bos (2004a) have investigated the environmental impact of natural fibers in industrial applications and have mentioned in their report that "natural fiber production requires less than 10% of the energy originally consumed for the production of PP fibers (around 90 GJ/ton)." In another review, production of China reed fibers was reported to show around 15% lower energy consumption compared to the production of glass fibers (Joshi et al., 2004). The inherent advantages of natural fibers and their characteristics like biodegradability and a low carbon footprint add to the sustainability of the production, distribution, consumption, and disposal of biodegradable/compostable composites for certain specific applications, besides being biobased materials. However, ecological sustainability needs to be addressed in an unambiguous manner through independent studies involving the entire life cycle of natural fibers. Only since the early 2000s have concerted efforts been taken in demonstrating the true environmental benefits through life cycle analysis (LCA). Overall better efficiency is expected with natural fibers compared to synthetic fibers, as revealed through LCA (van Dam and Bos, 2004b). This work laid the groundwork for follow-up actions both in terms of research as well as applied technology. This life cycle perspective could also offer a significant competitive edge for marketing biocomposites based on natural fibers and agricultural residues. LCA can be used to effectively quantify the environmental impacts of natural fibers starting from raw material to processing and to final product disposal; in essence covering the whole life cycle of the fibers (Patel et al., 2005). In this regard, composites containing natural fibers have a distinct advantage in relation to energy consumption, toxicity, emission of effluent, ease of disposal options, and so forth (Patel et al., 2005). Therefore, industries are bound to latch onto NFCs as a promising option to substitute for certain nonrenewable counterparts.

There are also several other factors that contribute toward enhancing the market development and opportunities for natural fibers and biobased products. Production/manufacturing processes, and use of specific consumer goods that are developed with intentions to reduce the environmental damages and cause less hazard are being backed by legislative provisions. Norway is considering tax incentives for biobased polymers and the imposition of new taxes on petro-based CO_2 content in polymers with an aim to increase the market demand for biobased products (Nova-newsletter, 2015). This initiative proposed by Norway is the first of its kind, and is expected to be followed by

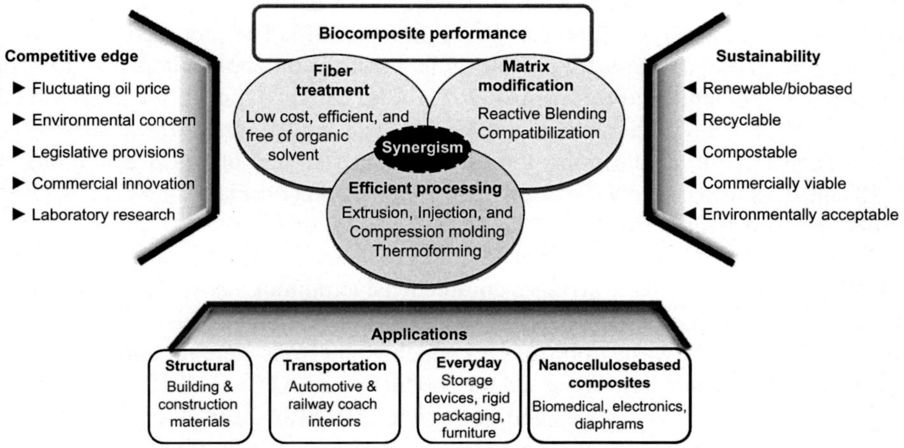

Figure 1.3 Performance, competitiveness, sustainability balance, and applications of biocomposites.

other countries of the world (Nova-newsletter, 2015). The environmental awareness in general has increased tremendously; the importance of natural fibers have been recognized and acknowledged by observing 2009 as the year of natural fibers by the United Nations (www.naturalfibres2009.org/en/). There are now more international platforms available for discussion of current status and market issues being faced in industries based on bioproducts. Examples of a few such organizations include Food and Agriculture Organization of the United Nations (FAO), European Bioplastics, Biotechnology Industry Organization (BIO), Cluster Industrielle Biotechnologie (CLIB) 2021. Aside from all these abovementioned factors, one important player determining the future prospects of NFCs is the research institutions and centers located around the world. These are dedicated to finding the most promising areas of commercial applicability, thereby elevating the status of natural fibers to a new level while considering the balance between performance, economics, and sustainability. Figure 1.3 provides a schematic of a synergistic approach requiring the coactions of efficient natural fiber and matrix treatment, and a suitable processing technique to ensure design and production of biocomposites with the desired level of performance for various applications. Factors influencing competitiveness and sustainability of biocomposites are also represented in the schematic in Figure 1.3.

1.6.2 Market development and growth scenario

According to a report published by Lucintel, TX, on the "Natural Fiber Composites Market Trend and Forecast 2011–2016: Trend, Forecast and Opportunity Analysis" in 2010, total global NFC market reached 430.7 million pounds, valued at US$289.3 million. The predictions forecast that the market will grow to US$531.3 million in 2016 with 11% compound annual growth rate (CAGR) for 2011–2016 (http://www.lucintel.com/reports_details.aspx?RepId=RPT1072). A plastics compounding market report released by BCC Research, MA, in November 2013 has provided North American market estimates for NFCs (Forman, 2013) through 2018. The proposed CAGR percentage for 2013–2018 is 9.3% for NFCs in construction applications and 15% in nonconstruction applications (Forman, 2013). The market estimate for construction applications in US $ millions was at 2530 and 3924 in 2013 and 2018, respectively. For nonconstruction applications, the estimate was 330 and 664 (US $ million) in 2013 and 2018, respectively (Forman, 2013). This market estimate for NFCs includes wood-fiber composites as well.

Nova Institute, Germany, published "Market study on wood plastic composites (WPC) and NFC" in 2014, giving a comprehensive outlook on the WPC and NFC market in the European Union (Carus et al., 2014). According to this report, production of NFCs in 2012 was 92,000 tons, of which 90,000 tons were for automotive applications. This report has forecasted the production of NFCs in 2020 to go as high as 350,000 tons for automotives and more than 20,000 tons for other applications, provided the strong incentives offered for bioproducts are taken into account; otherwise, the production is forecasted to be 130,000 tons in 2020 (Carus et al., 2014). The most dominant use of natural fibers has been found to be in the automotive industries, for interior trims in doors and dashboards of high-end cars; application in the consumer sector is said to be still at a very early stage (Carus et al., 2014). In 2011, 15.7 million passenger cars and another 2 million motor vehicles such as trucks and motorbikes were manufactured in the EU. Assuming 30,000 tons of natural fibers and 30,000 tons of wood fibers were used in all passenger cars manufactured in 2011, about 1.9 kg of natural fibers and 1.9 kg of wood fibers would be present in every car in the EU (Carus et al., 2014). Technically, however, vehicles containing about 20 kg of natural and wood fibers have also been manufactured (Carus et al., 2014). Leão et al. (2006) estimated the potential use of natural fiber in Brazil to be around 23 kg per automobile, making its use 40,000 tons per annum for the automotive industry alone.

Automotive interior parts applications are currently dominated by NFCs based on synthetic plastics like PP and polyurethane (John and Thomas, 2008; Koronis et al., 2013). However, with rapid developments of biobased and biodegradable polymers, completely biobased/biodegradable NFCs are expected to become prevalent. The global biobased polymer market in terms of production capacity reached 4.67 million metric tons (value of US$13.86 billion) in 2014 and is expected to further rise to 9.77 million metric tons (value of US$23.39 billion) by 2019 with CAGR of 15.91% (Global biobased polymers market, 2015). NFCs based on a combination of polylactic acid/kenaf fibers (Koronis et al., 2013) is being used for making spare tire covers; another such application is tailgate trim, where NFCs based on polybutylene succinate/bamboo fibers are being used (http://www.plastemart.com/upload/Literature/Biopolymers-in-Automotive-Interiors.asp).

There are some promising technological breakthroughs concerning new uses of perennial grasses in biocomposite materials for niche market application. Switchgrass, native to North America (NA), and miscanthus, introduced to NA toward the end of nineteenth century, are the most popular perennial grasses. They can provide high yields under poor, low-input conditions and are used for forage production, soil conservation, and as ornamental crops. During the last 10 years switchgrass and miscanthus have been developed for energy and fiber applications in North America and, more recently, in Europe (Girouard et al., 1995; Huisman, 1999). These perennial grass fibers may be an interesting reinforcing and filling agent for thermoplastic composites because they provide relatively good quality fibers that can be produced at low cost compared to other agricultural bast fibers, especially for injection molding type applications. Energy use and environmental impacts of growing switchgrass in Ontario have been evaluated through LCA by Kalita (2012). Bioproducts Discovery and Development Centre (BDDC), a research center at the University of Guelph, has been successful in producing NFCs containing perennial grasses (switchgrass and miscanthus) and agricultural residues. The formulations developed at the research center, with the help of collaborating industries, are now being used to make storage bins sold in major Canadian hardware stores (Lamontagne, 2013). The storage bins are said to contain around 25% of perennial grasses. The center has also commercialized NFC formulations that are now being used for manufacturing flowerpots from biobased resins. They contain about 25% miscanthus and are available for purchase from department stores such as Kroger (US) and Lowe's (Canada) (http://www.guelphmercury.com/news-story/4619563-an-eco-friendly-pot-for-your-posies/). Competitive Green Technologies, a biocomposite compounding facility in Ontario, Canada, is the licensed manufacturer for these biobased resins. These NFCs now available on the market have capitalized on the importance of having a local supply of natural fibers to avoid transportation of bulk fibers over long distances, in a way promoting the local economy by acknowledging the supply chain issues.

Another such product that has made it to the market is NCell®, a natural fiber-reinforced composite from GreenCore Composite Inc. The composite contains about 40 wt.% of cellulosic microfibers in PP matrix. This technology has been developed by the University of Toronto and has been exclusively licensed to GreenCore Composites (Lamontagne, 2013). The materials are provided as pellets to be molded as products for desired applications. UPM, a new forest industry company based in Finland has also been successful in producing cellulose fiber-reinforced PP composites. The materials are commercialized under the name UPM ForMi and are available in different grades, in granule forms. A maximum of 50% renewable content has been achieved in certain grades and they are said to offer smooth, reliable, and odorless composite products for a wide range of applications (http://www.upm.com/formi/Pages/default.aspx).

1.7 Future prospects and developments

As discussed throughout this chapter, natural fibers are the suitable material of choice to be used in combination with several polymer matrixes for the development of composites for myriad applications such as automobiles, packaging, flexible electronics, construction, and so on. Abundant availability of natural fibers combined with other advantageous characteristics, such as being nonabrasive to processing machines, easy processability, and biodegradability, are favoring innovations in designing and manufacturing of NFCs. Rapidly fluctuating price of petroleum-based products combined with strong political and social support to develop ecofriendly materials are expected to drive the growth of NFCs to a greater extent in the future. NFCs are applicable in almost every dimension of life (Suddell and Rosemaud, 2009), ranging from commodity to engineering applications. Because natural fibers possess low density compared to glass fibers, it is possible to achieve weight reduction in automotive parts by adapting natural fibers, provided the substitution is at same volume percentage. According to an estimate, achieving 25% reduction in weight of an automobile could save 250 million barrels of crude oil, which in turn could result in reduction of CO_2 emissions to the tune of 220 billion pounds per year (Mair, 2000; Kamath et al., 2005). It has been demonstrated that a reduction in CO_2 emission, about 3 t CO_2/t of the material, is possible when glass fibers are substituted with hemp fibers at same volume percentage, resulting in saving 1.16 m^3 of crude oil (Pervaiz and Sain, 2003).

Qualities, such as the ability to achieve specific orientation under magnetic field, and possibilities of desirable surface modifications, have directed intense research efforts in the area of cellulose nanofiber-based composites to develop optical devices, magnetic strips, and biosensors (Pandey and Takagi, 2011). Their application in the development of conducting composites with future of flexible and biodegradable electronics is another impressive area for natural fiber material. Cellulose nanofibers may also have a decisive role for development of lightweight, high-strength composites as they are optically active, such property can be utilized in fabricating colored films (Nishio et al., 1998). Such films can find application in bank notes, electoral cards, security papers and certificates, passports, and visas. Cellulose whiskers and semiconducting polymers have also been used recently to produce conductive composites (van den Berg et al., 2007).

Further, NFCs developed for structural applications such as infrastructure and housing applications, including decking, windows, doors, fencing, and construction of bridges, will have significant market share in the future. It is important to mention that although complete substitution of traditional, synthetic fibers with natural fibers are not anticipated or intended, NFCs will continue to find their own success in niche markets. Mass uses of NFCs are also envisaged for high-volume applications such as automobiles, provided that such products are commercially competitive through improved efficiency of processes and techniques.

Acknowledgments

The authors would like to acknowledge the funding support from (i) Ontario Ministry of Agriculture, Food and Rural Affairs, (OMAFRA)–University of Guelph Bioeconomy–Industrial Uses Theme; (ii) OMAFRA–New Directions Research program; (iii) the Ontario Ministry of Economic Development and Innovation (MEDI), Ontario Research Fund, Research Excellence Round 4 program (ORF–RE04); (iv) the Natural Sciences and Engineering Research Council (NSERC) Canada Discovery grant (Mohanty) and Network of Centres of Excellence (NCE)

AUTO21 program; (v) the Canadian foundation for Innovation's Leaders Opportunity Fund (CFI–LOF); (vi) Ontario Agri-Food Technologies (OAFT); (vii) Federal Economic Development Agency (FedDev #509260), Southern Ontario; (viii) Grain Farmers Ontario (GFO); and (ix) Hannam Soybean Utilization Fund (HSUF).

References

Akgul, O., Zamboni, A., Bezzo, F., Shah, N., Papageorgiou, L.G., 2010. Optimization-based approaches for bioethanol supply chains. Ind. Eng. Chem. Res. 50 (9), 4927–4938.

Anandjiwala, R.D., Blouw, S., 2007. Composites from bast fibres—prospects and potential in the changing market environment. J. Nat. Fibers 4 (2), 91–109.

Calistor, N., Nagarajan, V., Mohanty, A.K., Misra, M., 2014. Natural fiber composites from agricultural by-products: an overview. In: Netravali, A.N., Pastore, C.M. (Eds.), Sustainable Composites: Fibers, Resins and Applications. DEStech Publications, Inc. Lancaster, Pennsylvania, USA.

Carus, M., Eder, A., Dammer, L., Korte, H., Scholz, L., Essel, R., Breitmayer, E., 2014. Wood–plastic composites (WPC) and natural fibre composites (NFC): European and Global Markets 2012 and future trends. Nova Institute, Germany. www.nova-institut.de/download/market_study_wpc_nfc_short_version (last accessed March, 2015).

Ellison, G.C., McNaught, R., 2000. The use of natural fibres in nonwoven structures for applications as automotive component substrates. Research and development report, Ministry of Agriculture Fisheries and Food Agri-Industrial Materials, UK. www.maff.gov.uk/farm/acu/acu.htm, February.

FAO Consultation on Natural Fibres, 2007. The co-movement of jute and hard fibres prices with the prices of polypropylene and crude oil. FAO Consultation on Natural Fibres, Rome, 31 January–1 February.

Faruk, O., Bledzki, A.K., Fink, H.P., Sain, M., 2012. Biocomposites reinforced with natural fibers: 2000–2010. Prog. Polym. Sci. 37 (11), 1552–1596.

Forman, C., 2013. The plastics compounding market. Report by BCC research, November. http://www.bccresearch.com/market-research/plastics/plastics-compounding-market-pls018d.html (last accessed March, 2015).

Girouard, P., Henning, J.C., Samson, R., 1995. Economic assessment of short rotation forestry and switchgrass plantations for energy production in central Canada. In: The Canadian Energy Plantation Workshop, Ottawa, Canada. pp. 11–16.

Global biobased polymers market, 2015–2019. Market report by Technavia analysis.

Global Petrochemical Prices, September 2013. http://www.platts.com/news-feature/2013/petrochemicals/pgpi/index (last accessed March, 2015).

Huisman, W., 1999. Harvesting and handling of *Miscanthus giganteus*, *Phalaris arundinacea* and *Arundo donax* in Europe. In: Proceedings of the Fourth Biomass Conference of the Americas. Elsevier Science, Oxford, pp. 327–333.

Idicula, M., Sreekumar, P.A., Joseph, K., Thomas, S., 2009. Natural fiber hybrid composites – a comparison between compression molding and resin transfer molding. Polym. Compos. 30, 1417–1425.

John, M.J., Thomas, S., 2008. Biofibres and biocomposites. Carbohydr. Polym. 71 (3), 343–364.

Joshi, S.V., Drzal, L.T., Mohanty, A.K., Arora, S., 2004. Are natural fiber composites environmentally superior to glass fiber reinforced composites? Compos. Part A: Appl. Sci. Manuf. 35 (3), 371–376.

Kalita, B., 2012. Life cycle assessment of switchgrass (*Panicum virgatum* L.) biomass production in Ontario. Master of Science, University of Guelph, Guelph, Canada.

Kamath, M.G., Bhat, G.S., Parikh, D.V., Mueller, D., 2005. Cotton fiber nonwovens for automotive composites. Int. Nonwovens J. 14 (1), 34–40.

Koronis, G., Silva, A., Fontul, M., 2013. Green composites: a review of adequate materials for automotive applications. Compos. Part B: Eng. 44 (1), 120–127.

Lamontagne, N.D., 2013. New natural-fiber composites find their roles. Plastics Engineering (June). http://www.nxtbook.com/nxtbooks/wiley/pe_201306/index.php?startid=22 (last accessed March, 2015).

Leão, A., Sartor, S.M., Caraschi, J.C., 2006. Natural fibers based composites – technical and social issues. Mol. Cryst. Liq. Cryst. 448 (1), 161–763.

Mair, R.L., 2000. Tomorrow's plastic cars. ATSE focus no. 113, July/August.

Mohanty, A.K., Misra, M., Hinrichsen, G., 2000. Biofibres, biodegradable polymers and biocomposites: an overview. Macromol. Mater. Eng. 276–277, 1–24.

Mohanty, A.K., Misra, M., Drzal, L.T., 2002. Sustainable biocomposites from renewable resources: opportunities and challenges in the green materials world. J. Polym. Environ. 10, 19–26.

Nagarajan, V., Mohanty, A.K., Misra, M., 2013. Sustainable green composites: value addition to agricultural residues and perennial grasses. ACS Sustain. Chem. Eng. 1 (3), 325–333.

Nanda, M.R., Misra, M., Mohanty, A.K., 2012. Performance evaluation of biofibers and their hybrids as reinforcements in bioplastic composites. Macromol. Mater. Eng. 298, 779–788.

Nishio, Y., Kai, T., Kimura, N., Oshima, K., Suzuki, H., 1998. Controlling the selective light reflection of a cholesteric liquid crystal of (hydroxypropyl) cellulose by electrical stimulation. Macromolecules 31 (7), 2384–2386.

Nova-newsletter, 2015. From http://news.bio-based.eu/ (last accessed March, 2015).

Nyambo, C., Mohanty, A.K., Misra, M., 2010. Polylactide-based renewable green composites from agricultural residues and their hybrids. Biomacromolecules 11 (6), 1654–1660.

Pandey, J.K., Takagi, H., 2011. Self healing potential of green nanocomposites from crystalline cellulose. Int. J. Mod. Phys. B 25 (31), 4216–4219.

Patel, M., Bastioli, C., Marini, L., Würdinger, E., 2005. Life-cycle assessment of bio-based polymers and natural fiber composites. Biopolymers Online.

Pervaiz, M., Sain, M.M., 2003. Carbon storage potential in natural fiber composites. Resour. Conserv. Recycl. 39 (4), 325–340.

Sahoo, S., Misra, M., Mohanty, A.K., 2004. Biocomposites from switchgrass and lignin hybrid and poly(butylene succinate) bioplastic: studies on reactive compatibilization and performance evaluation. Macromol. Mater. Eng. 299 (2), 178–189.

Sharma, B., Ingalls, R.G., Jones, C.L., Khanchi, A., 2013. Biomass supply chain design and analysis: basis, overview, modeling, challenges, and future. Renew. Sustain. Energy Rev. 24, 608–627.

Strong, A.B., 2008. Fundamentals of Composites Manufacturing: Materials, Methods and Applications. Society of Manufacturing Engineers, Dearborn, MI.

Suddell, B.C., Rosemaud, A., 2009. Industrial fibres: recent and current developments. In: Proceedings of the Symposium on Natural Fibres, pp. 71–82. ftp://ftp.fao.org/docrep/fao/011/i0709e/i0709e10.pdf (last accessed March, 2015).

Thiruchitrambalam, M., Alavudeen, A., Athijayamani, A., Venkateshwaran, N., Perumal, A.E., 2009. Improving mechanical properties of banana/kenaf polyester hybrid composites using sodium lauryl sulfate treatment. Mater. Phys. Mech. 8, 165–173.

United Nations Conference on Trade and Development, 1996. Jute and hard fibres: overview of major current issues. http://www.unctad.info/upload/Infocomm/Docs/Jute/docs/unctadcom71.en.pdf (last accessed March, 2015).

van Dam, J.E.G., Bos, H.L., 2004. Consultation on natural fibres: the environmental impact of hard fibres and jute in non-textile industrial applications. ESC-fibres consultation no. 04/4, Rome, 15–16 December.

van Dam, J.E., Bos, H.L., 2004. The environmental impact of fibre crops in industrial applications. Hintergrundpapier zu, van Dam, JEG.

van den Berg, O., Schroeter, M., Capadona, J.R., Weder, C., 2007. Nanocomposites based on cellulose whiskers and (semi) conducting conjugated polymers. J. Mater. Chem. 17 (26), 2746–2753.

Vivekanandhan, S., Zarrinbakhsh, N., Misra, M., Mohanty, A.K., 2013. Coproducts of biofuel industries in value-added biomaterials uses: a move towards a sustainable bioeconomy. In: Fang, Z. (Ed.), Liquid, Gaseous and Solid Biofuels – Conversion Techniques. INTECH Open Access Publisher, Rijeka, Croatia.

Waldron, K.W., 2014. Advances in Biorefineries: Biomass and Waste Supply Chain Exploitation. Woodhead Publishing, Swaston, Cambridge, UK.

Yu, S., Tao, J., 2009. Economic, energy and environmental evaluations of biomass-based fuel ethanol projects based on life cycle assessment and simulation. Appl. Energy 86, S178–S188.

Yu, L., Dean, K., Li, L., 2006. Polymer blends and composites from renewable resources. Prog. Polym. Sci. 31 (6), 576–602.

Zampaloni, M., Pourboghrat, F., Yankovich, S.A., Rodgers, B.N., Moore, J., Drzal, L.T., Mohanty, A.K., Misra, M., 2007. Kenaf natural fiber reinforced polypropylene composites: a discussion on manufacturing problems and solutions. Compos. Part A 38, 1569–1580.

Mechanical performance of polylactic based formulations

2

P. Russo, D. Acierno†, G. Filippone†*
*Institute for Polymers, Composites and Biomaterials, National Research Council, Pozzuoli (Naples), Italy; †Dipartimento di Ingegneria Chimica, dei Materiali e della Produzione Industriale, Università di Napoli Federico II, Naples, Italy

2.1 Introduction

A wide variety of petroleum-based synthetic polymers are produced worldwide each year, and remarkable amounts of such plastic materials are introduced into the ecosystem as industrial waste products (Shah et al., 2008). Polymers have been designed in the past to resist degradation. Nowadays, public concerns about the environment, climate change, and limited fossil fuel resources are driving governments, companies, and scientists to find alternatives to crude oil. Biopolymers, meaning polymers that are biodegradable and/or derived from biological resources, can form the basis for an environmentally preferable, sustainable alternative to current materials based exclusively on petroleum feedstock. One such polymer is poly(lactic acid) (PLA), a biodegradable, renewably derived thermoplastic polyester extensively investigated over the last several decades (Rasal et al., 2010). PLA exhibits the stiffness and tensile strength of polyethylene terephthalate, which makes it potentially suitable for a variety of commodity applications, ranging from packaging to fibers, electronic and electrical devices, mechanical and automotive parts, and so forth (Anderson et al., 2008).

From a chemical point of view, PLA resins are linear aliphatic thermoplastic polyester of the lactic acid (2-hydroxypropanoic acid) monomers (Figure 2.1) obtained either from bacterial fermentation of sugar feed stocks, corn, wheat and other starch-rich products, or (less frequently) by chemical synthesis.

The fermentation involves species of the *Lactobacillus* genus, a pH range from 5.4 to 6.4, a temperature range from 38 to 42 °C, and a low oxygen concentration. In addition, in order to have high efficiency and maintain cell growth and production, neutralization and extraction of the produced acid are usually performed by adding calcium hydroxide or calcium carbonate to get soluble calcium lactate solutions. This broth is filtered to remove the cell biomass and other insolubles, then evaporated, recrystallized, and acidified with sulfuric acid to yield the crude lactic acid. The insoluble calcium sulfate (gypsum) is filtered off and discarded with up to 1 ton of gypsum per ton of lactic acid produced (Figure 2.2).

PLA resins are synthesized by either direct polycondensation or ring-opening polymerization (ROP) of lactides. The former technology includes solution and melt polycondensation, depending on whether a solvent is used or not to dissolve the PLA. The solution polycondensation, developed by Carothers et al. (1932) at DuPont and still used by Mitsui Chemicals, essentially consists in heating lactic acid monomers

Biocomposites: Design and Mechanical Performance. http://dx.doi.org/10.1016/B978-1-78242-373-7.00002-0

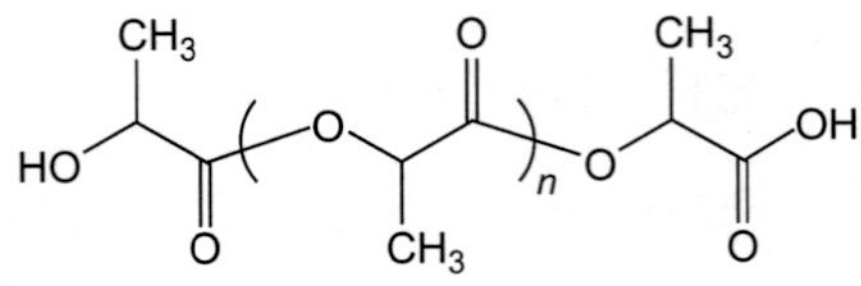

Figure 2.1 Poly(lactic acid).

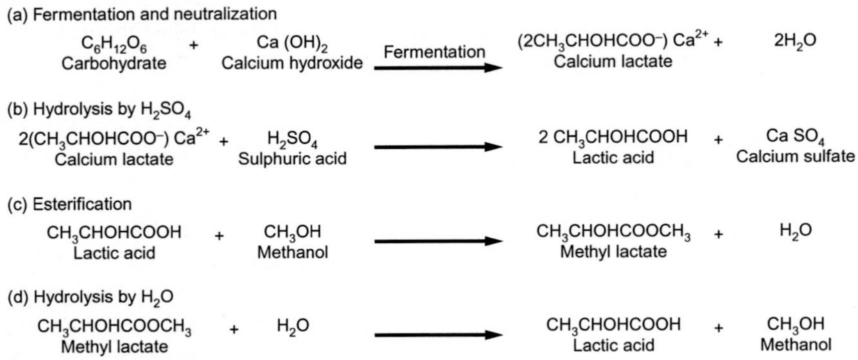

Figure 2.2 Conventional fermentation-based lactic acid production scheme.

under vacuum while removing the condensed water. This procedure yields PLA grades with an average molecular weight of over 200,000 Da.

The melt polycondensation, instead, is a one-step procedure, relatively economical and easy to control, but the equilibrium reactions are affected by numerous parameters, such as temperature, reaction time, catalysts, pressure, and so on. These factors can strongly influence the molecular weight of products. Moreover, the generated water can reduce the molecular weight by depolymerization phenomena. Thus, melt polycondensation usually leads to a low molecular weight, brittle and glassy resins, and is unusable for many industrial applications.

An approach to solve this drawback makes use of chain extender additives, namely bifunctional compounds carrying highly reactive functional groups. In this case, good results have been achieved with hexamethylene diisocyanate (Zhong et al., 1999; Gu et al., 2008). Although reactions involving chain-extending agents are economically feasible, as they can be done in the melt without the subsequent need to separate the added chain extender, disadvantages arise from the fact that the unreacted additive is generally not biodegradable and, in addition, it can be toxic. For example, the drawbacks related to the use of isocyanates as chain extenders are the toxicity and sensitivity effects related with the isocyanate monomers, and their subsequent toxic amine hydrolysis products (Hartmann, 1998).

ROP is the process mainly employed to produce PLA resins. This process was first demonstrated by Carothers et al. (1932), but high molecular weight products were not obtained until improved lactide purification techniques were developed by Lowe (1954) at DuPont.

Owing to the accurate control of residence time and temperatures, possible in the case of ROP, the desired ratio and sequence of isomer units in the final polymer can be designed, provided that adequate catalysts and concentration of reactants are selected.

Concerning the catalyst, currently tin(II)-bis(2-ethylhexanoate) ($SnOct_2$), approved as a food additive by FDA, is the most widely used, even if numerous novel efficient metal-free catalytic systems are emerging as viable alternatives, especially for biomedical applications (Albertsson and Varma, 2003).

For a commercial PLA resin the ratio of L- and D-enantiomers influences the mechanical properties, crystallinity, and even the degradation characteristics. In particular, the two isomers give rise to four distinct materials: hemicrystalline poly(L-lactic acid) (PLLA); crystalline poly(D-lactic acid) (PDLA), both with a regular chain structure and a melting temperature $T_m \approx 180\,°C$; amorphous poly(D,L-lactic acid) (PDLLA) with a glass transition temperature $T_g \approx 50$–$60\,°C$; and meso-PLA obtained by the polymerization of meso-lactide with T_g of 46 °C.

When compared to other biomaterials, such as polycaprolactone (PCL), polyhydroxyalkanoates (PHAs), and poly(ethylene glycol) (PEG), PLA is readily biodegradable via hydrolytic actions and can be processed by conventional technologies, such as film casting, extrusion, blow molding, and fiber spinning. In addition, PLA exhibits properties comparable to many petroleum-based plastics and, consequently, it has been proposed for use in service ware, grocery bags, waste-composting bags, films, and controlled release systems. Furthermore, PLA is biocompatible, energy saving, and approved by the US Food and Drug Administration for direct contact with biological fluids since 1970.

In terms of mechanical properties, PLA polymers behave as a glassy and quite brittle material, typically exhibiting tensile modulus $E = 2 \div 4\,GPa$, tensile strength $\sigma_R = 30 \div 50\,MPa$, and deformation at break $\varepsilon_R = 1 \div 7\%$. It is important to stress that the mechanical properties of PLA strongly depend on molecular weight, enantiomeric purity, and crystallinity content.

Despite the abovementioned potentialities, applications of PLA resins are still limited by weaknesses strictly related to their intrinsic chemical structures. The aim of the following sections is to provide an overview of the main experimental approaches proposed to overcome PLA limitations and broaden their areas of application. Attention is focused on the benefits gained in terms of mechanical properties of PLA-based products. In particular, the use of reinforcing nano- and microsized fillers and fibers and the modification via physical and chemical approaches will be discussed. Finally, a section will be devoted to the strategies recently proposed to overcome the poor mechanical resistance of commercial-grade PLA resins above their T_g, which is limiting the use of this biopolymer in many relevant industrial sectors where a certain resistance is required at relatively high temperature.

2.2 Challenges in the application of PLA

PLA can be considered an eco-friendly biomaterial with satisfying properties. However, it has many drawbacks. Among others, it is worth noting the following limitations:

- slow hydrolytic degradation of the backbone ester groups limiting applications in biomedical and packaging fields;
- low crystallization ability and degree, typical of polyester resins, limiting its industrial implementation in different applications, particularly in durable applications like automotive and electronics;

- high brittleness, with less than 10% of elongation at break, and low impact strength, making PLA not suitable for demanding structural applications;
- low solvent resistance (e.g., against water);
- relatively low heat distortion temperature;
- poor thermal stability;
- strong hydrophobicity and inflammatory responses from the tissue of living hosts, related to its low affinity with cells when used as a tissue engineering material;
- limited gas barrier properties preventing its complete access to the packaging sector;
- high cost.

2.3 Current approaches to improve PLA mechanical properties

2.3.1 PLA-based composites

In recent decades, several PLA-based composites containing micro- and nanofillers have been considered with the aims to improve some specific structural properties, such as rigidity, dimensional stability, heat deflection temperature (HDT), processability, and so on (Pluta et al., 2002; Huang et al., 2009). Various techniques can be used to disperse solid particulate inside the host matrix, such as in situ polymerization (Paul et al., 2005; Li et al., 2011a), solution compounding (Krikorian and Pochan, 2003; Chen et al., 2011; Oliveira et al., 2013) and melt compounding (Pluta et al., 2002; Molnar et al., 2009; Najafi et al., 2012; Yu and Qiu, 2011; Anderson et al., 2003). The latter method has been found to be the most practical, as no solvent is required and conventional industrial processing technology can be used.

With reference to microcomposites, in the past PLA resins were filled with different kinds of filler, such as hydroxyl apatite (Fambri et al., 2003; Russias et al., 2006), calcium phosphate, calcium carbonate (Urayama et al., 2003), and talc (Jain et al., 2010). Typically, the particle loadings were between 5% and 25% by weight. In particular, taking into account that during the purification of the lactic acid monomer considerable amounts of calcium sulfate (gypsum) is formed as a by-product, a reasonable interest has been dedicated to formulations containing calcium sulfate properly coated to improve interactions at the interface with the matrix (Murariu et al., 2007). This modification decreases the cost of the matrix and makes it competitive with traditional petrochemical polymers, although studies are still in progress to determine stricture–property correlations, deformation and failure behavior, and performance and potential application areas of these classes of composites. Molnar et al. (2009) demonstrated that tensile strength of PLA/$CaSO_4$ microcomposites decreases by increasing the filler content. The effects were more pronounced in presence of microparticles coated with stearic acid due to the achievement of both higher interactions at the interface and better dispersion in composites containing modified calcium sulfate. This effect has been also ascribed to the plasticizing action of stearic acid, which, dissolving in PLA, can lead to local plastic deformation.

Regarding PLA-based nanocomposites, various types of nanosized particles have been considered as reinforcing agents in order to enhance thermomechanical properties of the matrix as well as to provide additional functionalities (e.g., fire resistance (Fukushima et al., 2010; Zhan et al., 2009; Li et al., 2009; Wang et al., 2010; Gonzalez et al., 2012)). Among the most used fillers, it is worth mentioning organomodified layered silicates (OMLS) (Krikorian and Pochan, 2003; Najafi et al., 2012; Kubies et al., 2006; Dorigato et al., 2012; Bordes et al., 2009; Ray et al., 2002, 2003; Russo et al., 2014a), halloysites (Liu et al., 2013; Murariu et al., 2012), sepiolites (Liu et al., 2012), carbon nanotubes (CNTs) (Chrissafis et al., 2010; Chiu et al., 2008; Wu et al., 2010; Wu and Liao, 2007; Kuan et al., 2008), graphene (Pinto et al., 2013; Cao et al., 2010; Kim and Jeong, 2010), and polyhedral oligomeric silsesquioxanes (POSS) (Yu and Qiu, 2011; Pan and Qiu, 2010; Qiu and Pan, 2010; Zhang et al., 2011; Lee and Jeong, 2010).

Ray et al. (2003) observed a significant increase in flexural modulus and also in PLLA crystallinity only for isothermal intercalated PLLA nanocomposites with relatively high MMT loading (e.g., 4, 5, and 7 wt.%). The authors also reported that the 4 wt.% loading of organomodified MMT seems to be the best compromise to achieve a high value of flexural strength while preserving ductility.

Investigating the inclusion of commercial halloysite nanotubes into a PLA matrix via melt-compounding technology, Murariu et al. (2012) demonstrated that both rigidity and maximum tensile strength are gradually improved by increasing the nanofiller content up to 6% by weight, suggesting that the PLA–HNT composites can be potentially interesting in engineering applications. Although the rigidity remains at a high level, no further increase in tensile strength was noticed at higher particle loadings. This was ascribed to a detrimental effect of the inevitable aggregation of the nanofiller. The same authors also highlighted some improvement for the Izod impact strength of nanocomposites with respect to the neat matrix. Moreover, the reinforcement factor, evaluated as the ratio of the storage modulus of composite on the storage modulus of the matrix at the same temperature, E'_c/E'_m, was found to grow upon the addition of halloysites, not only in the glassy state but also above the glass transition temperature of the matrix. This consideration allows us to conclude that the inclusion of HNTs may favor the use of PLA in applications requiring rigidity at high temperatures of utilization.

Kuan et al. (2008) studied the effect of CNTs in two types of PLA with a low and a high degree of crystallinity. The authors reported that the nanocomposites exhibit enhanced tensile strength and flexural strength with respect to the matrices, especially in the case of low crystallinity and in the presence of a suitable coupling agent. This behavior was explained assuming a better dispersion of CNTs in the latter matrix.

Pinto and coworkers incorporated graphene oxide (GO) and graphene nanoplatelets (GNP) in PLA thin films using a solvent cast method (Pinto et al., 2013). They observed that the inclusion of relatively low amounts of these fillers (0.2–0.6 wt.%) significantly improve the mechanical properties and reduce the gas permeability to oxygen and nitrogen of the investigated thin films. Regarding the mechanical behavior, an optimum loading was identified around 0.4 wt.% for both fillers, with increases of tensile strength and Young's modulus of approximately 15% and 75%, respectively, and decreases of the same parameters for higher contents probably related to filler agglomerations phenomena.

Analogously, Cao et al. (2010) increased Young's modulus of PLA by 18% with the addition of only 0.2% by weight of reduced GO.

Zhang et al. (2011) studied the properties of hybrid films prepared incorporating Octa(3-ChloroPropylSilsesquioxane) (OCPS) into a commercial PLLA matrix by the solution blending method. They observed, among other things, a significant alteration in the tensile behavior of films, with a transition from brittle to ductile behavior by varying the filler content in the range 0–15% by weight. The strain at break was greatly improved as the result of the plasticizer effects that POSS cages take on to increase the free volume of the PLLA matrix.

Luo et al. (2009) added grafted titania (g-TiO_2) nanoparticles to a PLA matrix via melt processing, finding that both the matrix crystallinity and mechanical properties were improved. In particular, although the tensile strength of the composite with 1 wt.% of g-TiO_2 only slightly increased, the strain at break had a tenfold increase with respect to the neat PLA matrix.

Jiang et al. (2007) investigated melt-extruded PLA nanocomposites containing nanosized precipitated calcium carbonate (NPCC) and organically modified montmorillonite (MMT), showing significantly different effects on the strength, modulus, and elongation at break depending on the filler type. In particular, the authors revealed that for contents of fillers up to 7.5% by weight, the presence of NPCC induces evident increases of the strain at break up to 13%, with respect to the neat matrix, with just slight benefits in terms of tensile modulus accompanied by decreases of the yield stress. For systems filled with MMT, instead, a dramatic reduction of the strain at break with significant increases of the tensile modulus were observed. These results, suggesting different reinforcing and toughening effects on the PLA matrix of NPCC and MMT, were explained in terms of interfacial adhesion and aspect ratio aspects.

The easy debonding of NPCC particles due to the stearic acid coating, limiting their interaction with the matrix, and their cubic shape, causing a high stress concentration around the same, may justify the reduction of the yield strength and the significant increases of the strain at break by massive crazing. In contrast, the large aspect ratio (about 40) and surface characteristics of MMT increased the interfacial stress transfer resulting in high tensile strength of the PLA nanocomposites.

Yan et al. (2007) reported on the surface modification of silica nanoparticles with a L-lactic acid oligomer by direct grafting of silanol groups onto the surface of the silica nanoparticles. The authors found that the toughness and the tensile strength of the PLLA matrix were greatly improved because of the good dispersion of silica nanofiller.

2.3.2 PLA–natural fiber composites

Fine particulate apart, several authors investigated the possibility of using PLA as a matrix for the production of natural fiber reinforced composites (Oksman et al., 2003; Mukherjee and Kao, 2011). In this case, many natural fibers have been taken into account, such as kenaf (Ochi, 2008), hemp (Hu and Lim, 2007; Masirek et al., 2007), lyocell (Shibata et al., 2004), jute (Plackett et al., 2003), flax (Bax and Mussig, 2008), silk (Cheung et al., 2008), sisal (Li et al., 2011b), wood pulp (Huda et al., 2006), cellulose fibers (Huda et al., 2005; Graupner et al., 2009), and so on, always paying special attention to the optimization of the fiber–matrix interaction. The latter is typically improved by means of suitable pretreatments of the reinforcement and/or of the matrix (Sujaritjun et al., 2013; Hassan et al., 2012).

At this regard, Huda et al. (2008) analyzed the effects of alkali treatment, silane treatment, and the combination of these two approaches of kenaf fibers on mechanical, dynamic–mechanical and thermomechanical properties of PLA-based laminates. Composite laminates, produced by compression molding using the typical film-stacking route and reinforced with about 40% by weight of fibers, were characterized taking materials reinforced with untreated fibers and neat PLA systems as references. Results highlighted the superior performances of laminates containing surface-treated reinforcements not only compared to neat PLA-based samples but also with respect to ones filled with as-received kenaf fibers. In particular, improvements in strength approximately equal to 30%, 50%, and 60% compared to composite samples of untreated fibers were detected for the alkali, silane, and combined treatments, respectively. However, regardless of the surface treatment, the flexural strength of the composites was lower than that of the unreinforced matrix material. Moreover, although the composites of untreated fibers was stiffer than neat PLA, this mechanical parameter was further increased by 48%, 69%, and 80%, respectively, using alkali treatment, silane treatment, and the combined approach. The authors supported results by SEM morphological observations, which demonstrated that detected benefits can be attributed to an improved adhesion at the fiber–matrix interface.

Goriparthi et al. (2012) investigated polylactide composites containing surface-treated jute fibers. They considered the effects of alkalinization, permanganate, peroxide, and silane treatments on mechanical and abrasive wear performances of unidirectional PLA composites prepared by hot pressing of solvent impregnated prepregs. Results showed that fiber treatments significantly influence strength and moduli of products evaluated by both tensile and flexural tests. Dynamic–mechanical measurements carried out on the temperature range between 25 and 120 °C evidenced a relevant increase of the storage modulus for surface-modified jute fiber composites with respect to systems containing untreated fibers. Moreover, abrasive wear analysis, in terms of variation of weight loss with respect to sliding distance at a constant sliding velocity of 0.418 m/s, demonstrated that surface-modified jute fiber systems have higher wear resistance when compared to composites with untreated fibers. The authors explained the higher weight loss of these latter systems to lower stiffness and poor interface between matrix and reinforcing fibers.

Nowadays, some of these products are already commercialized. For example, Jakob Winter (Satzung, Germany) produces biodegradable urns from flax and PLA by compression molding (Reinhardt et al., 2013); Nek Corporation and UNITIKA Ltd have announced joint development of new composite materials for mobile phone shells consisting of PLA and 15–20% of kenaf fibers (http://www.nec.co.jp/press/en/0603/2001.html).

2.3.3 PLA-based polymer blends

As an alternative to filling with solid particles, attempts have been made to improve flexibility of PLA by blending it with other polymers (Hiljanen-Vainio et al., 1996; Shibata et al., 2006) or incorporating in it various plasticizers (Jacobsen and Fritz, 1999; Ren et al., 2006). For this purpose, it is worth mentioning blends of PLA with petrochemical-derived polymers, such as PEG (Sungsanit et al., 2012), poly(propylene glycol) (Piorkowska et al., 2006), polyoxyethylene (Nijenhuis et al., 1996), and PCL (Odent et al., 2013), or with biobased resins, such as oligomers of lactic acid, citrate esters (Labrecque et al., 1997; Ljungberg and Wesslen, 2003), and epoxidized oils (Giita Silverajah et al., 2012; Fathilah et al., 2009; Xu and Qu, 2009). The aim was to lower the glass transition temperature and to increase the PLA chains mobility so as to favor the phenomenon of cold crystallization (i.e., crystallization upon heating) and, at the same time, to increase ductility and softness of the PLAs. In general, the choice of polymers or plasticizers is driven by the requirements of the specific application of products. For example, in cases of packaging and hygiene application, only nontoxic substances approved for contact and personal care can be considered as plasticizing agents. Moreover, the plasticizer should be miscible with PLA, thus creating a homogeneous blend. In addition, it should not be too volatile, because this would cause evaporation to occur at the elevated temperature used during processing. Finally, a good plasticizer should not be prone to migration, because this would cause contamination of the materials in contact with the plasticized PLA and regaining of brittleness of pure matrix.

Ali and coworkers prepared and characterized PLA/epoxidized soybean oil (ESO) blends (Ali et al., 2009). The authors found that ESO is partially miscible with PLA, and blends containing up to 30% by weight of epoxidized oil exhibited a ductile behavior with yielding and a subsequent plastic deformation. PLA/ESO specimens showed decreases of tensile modulus, yield stress, and tensile strength coupled with a noticeable increase of about 40% of the elongation at break with respect to the neat matrix. Despite the relatively high T_g of the investigated systems (about 60 °C), such a result was ascribed to local plasticization phenomena induced by microdroplets of low molecular weight ESO dispersed in the PLA matrix.

2.3.4 Physical modifications

Physical modification of PLA is another possible way to improve its mechanical properties. Among other strategies, annealing is commonly used to favor PLA crystallization and, consequently, to improve the performances of PLA products (Tabi et al., 2010; Yu et al., 2008).

When a semicrystalline polymer is processed, its crystallinity strongly depends on the cooling rate. Because of the particularly low crystallization rate of PLA, products prepared by means of typical industrial processing methods, such as blow molding, injection molding, and thermoforming, are usually totally/predominantly amorphous. A postprocessing annealing step can represent a cost-effective way to improve final properties of PLA-based products.

Srithep et al. (2013) evaluated the degree of crystallinity, heat resistance, and mechanical properties of injection-molded PLA specimens annealed under different times and temperature conditions. Experimental data showed that the evolution of PLA crystallinity as a function of annealing time and temperature closely follow the time–temperature superposition relationship and, consequently, master curves based on the well-known Williams–Landel–Ferry equation may be built and used as an easy tool to tailor the thermal and mechanical properties of injection-molded PLA parts. Results confirmed that tensile strength and modulus of the annealed PLA items were higher with respect to the molded ones with temperature effects essentially ascribed to the difference in the achieved crystalline morphology.

Increases of the crystallinity degrees may result in an embrittlement of the polymer under quasi-static load, while the material results toughened under impact load. Park and coworkers investigated the physical properties of PLA plates prepared by compression molding at 180 °C and then quenched to room temperature for 10 min (Park et al., 2004). Some of the quenched plates were then annealed for 3 h at 100 °C to crystallize them. Specimens derived from the amorphous and crystallized plates were tested at a quasi-static rate of 1 mm/min using a servohydraulic testing machine and at an impact rate of 1 m/s using an instrumental drop weight machine. The authors noticed that impact parameters of the quenched (amorphous) plates were higher than those of the annealed ones at quasi-static loading rate, but this situation was reversed at high impact rate. This behavior was explained by morphological investigation, demonstrating that the formation in the quenched plates of multiple crazes in the vicinity of the crack tip is responsible for high impact parameters at quasi-static loading conditions. The suppression of craze formation under high impact loading, clearly highlighted by polarized optical microscopy (POM) observations, has allowed us to explain the reduction of the toughness. For annealed samples, instead, the formation of microcracks through spherulites and at the interfaces with few or no crazes might be responsible for low values of impact parameters under quasi-static conditions and increased fracture toughness at high impact rates.

Russo et al. (2014a) studied the effect of barrier properties of PLA films upon the addition of nanoclays and subsequent thermal annealing at 80 °C. The authors showed that both water vapor transmission rate and permeability values of neat PLA are reduced by approximately 10% after the thermal treatment as a consequence of increased crystallinity and reduced free volume (Figure 2.3).

2.3.5 Chemical modifications

Chemical modifications can be also performed aiming at improving the mechanical properties of PLA products. These types of treatments mainly involve copolymerization and cross-linking (Rasal et al., 2010). The former process can be carried out through polycondensation with lactone-type monomers like ε-caprolactone and glycolic acid, leading to low molecular weight copolymers or, alternatively, through the ring-opening copolymerization of lactide with other cyclic monomers like ethylene oxide to produce high molecular weight copolymers.

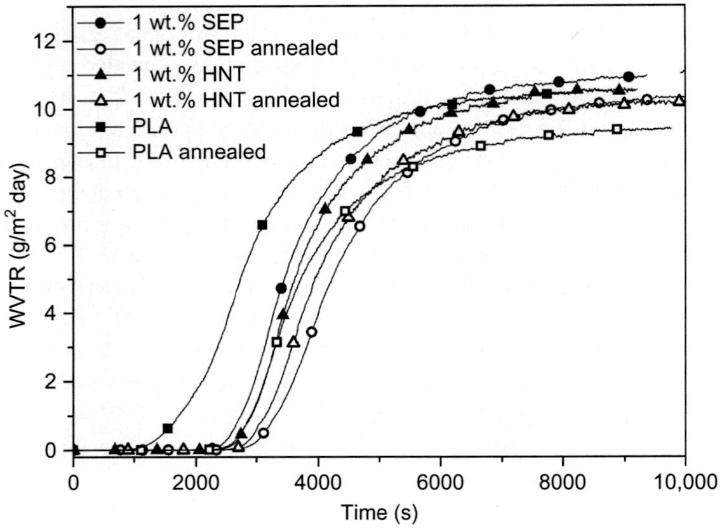

Figure 2.3 WVTR of untreated and annealed neat PLA and nanocomposites.
Image taken from Russo et al. (2014a).

Cross-linking of PLA can be instead performed either by irradiation or through chemical reactions. In this regard, γ-irradiation and electron beam irradiation were widely applied to cross-link PLA in the presence of a small amount of a cross-linking agent like triallyl isocyanurate (Nagasawa et al., 2005a; Jin et al., 2002; Quynh et al., 2007). The double bonds in TAIC, whose structural formula is given in Figure 2.4, can be easily broken to produce monomer radicals, which combine with the polymer radicals to form the cross-linking network.

Alternatively, improvements of thermal and mechanical properties of PLA products can be obtained by exposing PLA items to high-energy electron beam irradiation as long as appropriate cross-linking agents are used (Ragoubi et al., 2012). Rytlewski et al. (2010) compared the influence of two cross-linking additives, namely trimethylopropane trimethacylate (TMPTA) and triallyl isocyanurate (TAIC), on the thermal and mechanical properties of electron beam irradiated PLA. It was found that TMTPA favors degradation over cross-linking reactions, contrary to the TAIC, which favors cross-linking of the amorphous PLA phase. This different behavior has been ascribed to the better miscibility of polylactide with TAIC with respect to TMTPA.

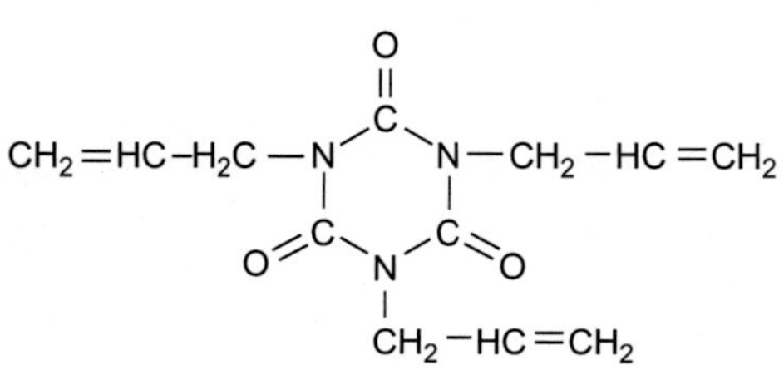

Figure 2.4 Structural formula of triallyl isocyanurate (TAIC).

Similarly, the effectiveness of the TAIC has also been demonstrated by Mitomo et al. (2005), who found that PLLA containing 3 wt.% of TAIC and irradiated at 50 kGy of γ-rays reached a 84% gel fraction with improved heat stability. In the same category of modifications, it is worth citing other techniques such as corona discharge (Zenkiewicz et al., 2009, 2010; Amirou et al., 2013), laser (Mitomo et al., 2005; Rytlewski et al., 2012), and plasma (Boselli et al., 2012; Jacobs et al., 2013; Chaiwong et al., 2010) treatments, essentially considered to achieve desirable surface properties. In particular, the latter approach has the ability to control the surface structure, energy, and charge, and to uniformly modify the surface without impacting on bulk PLA properties. For example, Nagasawa and coworkers proved that PLLA filled with 3 wt.% TAIC exhibits excellent heat stability after electron beam irradiation (Nagasawa et al., 2005b).

However, the cross-linking of PLA by means of irradiation is limited due to the high cost of irradiation equipment. In addition, it has to be stressed that PLA samples must be processed as thin plates to absorb enough energy to undergo a satisfactory level of cross-linking.

About cross-linking through chemical reactions (Yang et al., 2008), the use of dicumyl peroxide (DCP) has shown good results in terms of thermal stability even if strong effects may cause, expectedly, a steep drop of the tensile strength.

2.4 Mechanical properties of PLA at high temperature

One of the main drawbacks of PLA is its low mechanical resistance at high temperature, which is limiting the expansion and diversification of its possible applications. In particular, the slow crystallization rate of commercial-grade PLA resins generally results in amorphous products that lose their mechanical properties as soon as the glass transition temperature is approached. This means that the employment of commercial-grade PLA is prevented if the operating temperature is expected to exceed ~50 °C. Different approaches have been pursued in recent years to overcome this shortcoming, which can be essentially divided into two main groups: (1) promoting PLA crystallization and (2) blending PLA with heat-resistant polymer matrices.

Crystallinity improves the mechanical properties of polymers above the glass transition temperature as the crystallites act as constraints that limit macroscopic flows. Annealing PLA at a temperature slightly above its T_g favors a slow crystallization (Perego et al., 1996). The partial thermal degradation occurring during the annealing results in an increase of free volume that allows the polymer chains to rearrange and fold up into crystallites. Although postprocessing annealing steps may be unaffordable in industrial contexts, such an approach has been sometimes proposed to overcome the scarce mechanical performances of PLA-based products above the T_g. Alternatively, the use of different kinds of nucleating agents has been proposed to accelerate crystallization. Talc and ethylene bis-stearamide (Harris and Lee, 2008), various compounds having hydrazide groups (Kawamoto et al., 2007), titanium dioxide (Pluta et al., 2007), calcium sulfate (Zhu et al., 2011), sodium stearate and calcium lactate (Li and Huneault, 2007), organoclays (Shina Ray et al., 2002, 2003; Shina Ray and Okamoto, 2003), are some examples of additives that can assist crystallization by acting as nucleating agents (Figure 2.5). Actually, more than the nature of the nucleating agent, chain mobility seems to be the key factor for a fast crystallization of PLA. In fact, the combination of nucleating agent and plasticizer, or the adoption of targeted processing conditions, such as mild postprocessing annealing steps or heated molds, are generally needed to promote fast crystallization from the melt upon cooling.

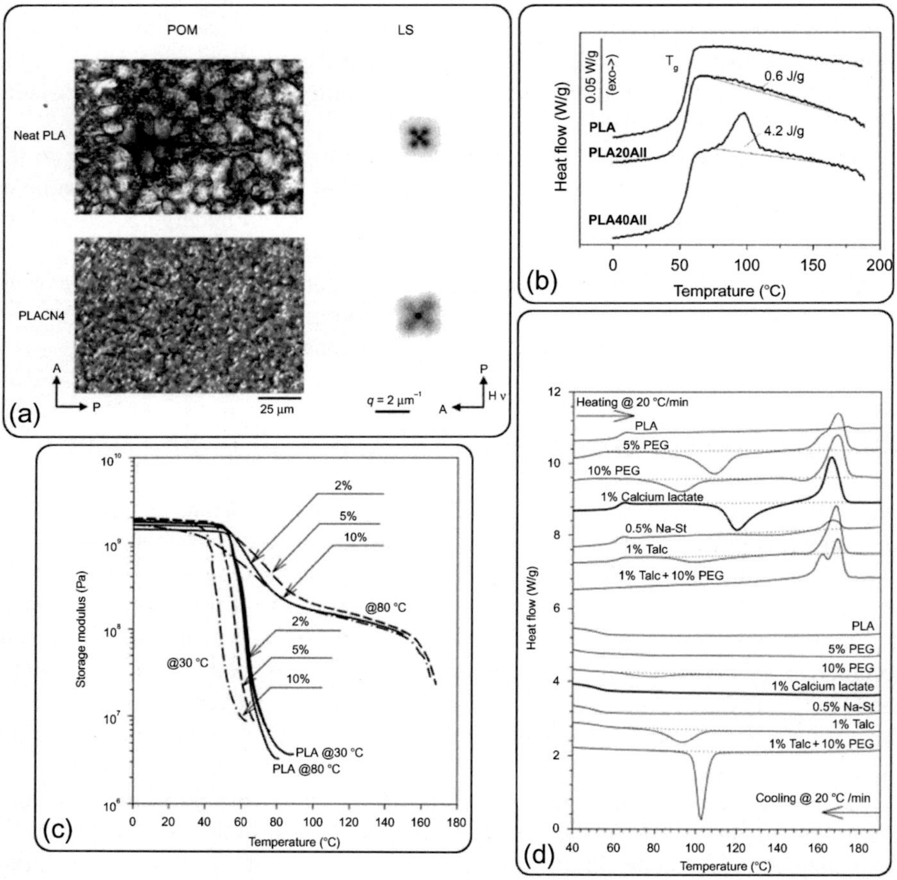

Figure 2.5 Examples of improved crystallinity and high temperature mechanical properties of PLA. (a) Polarized optical micrographs and corresponding Hv-light scattering patterns for unfilled and organoclay-filled (PLACN4) PLA isothermally crystallized at 110 °C for 1.5 h (image taken from Shina Ray et al. (2003)). (b) DSC thermograms for PLA and calcium sulfate-filled composites (PLA20AII and PLA40AII) during cooling from the melt at rate of 3 °C/min (image taken from Pluta et al. (2007)). (c) Effect of temperature on the storage modulus for PLA control and for formulations comprising 1% talc with 2%, 5%, and 10% PEG, molded at 30 and 80 °C (image taken from Li and Huneault (2007)). (d) DSC thermograms for PLA and PLA formulations cooled at 20 °C/min and subsequently heated at 20 °C/min (image taken from Li and Huneault (2007)).

Blending PLA with a second biodegradable organic phase has been also investigated as a possible way to improve the ability of PLA to crystallize upon cooling. Tsuji et al. (2009) compared various biodegradable polyesters as crystallization-accelerating agents for PLA, identifying different action mechanisms, such as nucleation-assisting effect and spherulite growth-accelerating effect. Due to a partial miscibility with PLA, PCL was found to enhance to some extent the PLA crystallization rate both from the melt or the glassy state, most likely because of the increase in nucleation rate (López-Rodríguez et al., 2006; Dell'Erba et al., 2001; Russo et al., 2014b). In particular, Sakai et al. (2009) showed that nucleation of the PLA occurs even 40 °C below its glass transition temperature in the presence of small amounts of PCL, which causes a relevant local depression of the PLA glass transition at the interface with the PCL domains, thus enhancing PLA nucleation.

Rather than acting on the PLA crystallizability, one can take advantage of the inherent self-supporting capacity of a secondary polymer phase to remedy the softness of PLA above its T_g. Most polymer pairs are thermodynamically immiscible because of the typically low mixing entropy of macromolecular materials. The final properties of immiscible blends will strictly depend on their microstructure; that is, on the space arrangement of the polymer phases achieved during the whole thermomechanical history experienced in the course of mixing. For low amounts of either of the phases, immiscibility results in a "drop-matrix morphology," with spherical domains of the minor phase suspended in the majority component. The properties of such a blend are essentially governed by those of the matrix, but in many cases the presence of the dispersed phase can have a drastic detrimental effect. As the content of the minor component is raised up to a critical value, the blend microstructure turns from drop-matrix into "co-continuous." The interpenetration of the phases that characterizes this particular morphology may result in a synergistic combination of the properties of the blend constituents, providing improved mechanical properties, heat resistance, and electrical and thermal conductivity (Pötschke and Paul, 2003).

Hashima et al. (2010) prepared a blend at 40 wt.% of PLA with good toughness and HDT of about 90 °C. Ductility was provided by hydrogenated styrene–butadiene–styrene block copolymer, whereas the superior mechanical properties of polycarbonate were exploited to get high temperature resistance. Tuning the composition, Zhang et al. (2012) designed fully biorenewable and biodegradable ternary blends with balanced properties. The authors found that the formulation at 30 wt.% of PLA exhibits the best compromise between stiffness/toughness and thermal resistance, with HDT ≈72 °C. Stoclet et al. (2011) have recently blended PLA with polyamide 11 (PA11), a not biodegradable yet biobased polymer. Due to its semicrystalline feature, PA11 exhibits high temperature resistance, which makes it suitable to correct the PLA softness above its T_g. In addition, the relatively low processing temperature of PA11 allows for melt blending with PLA, minimizing the thermomechanical degradation of the latter during processing. To fully exploit the superior high temperature mechanical performances of PA11, however, it must represent a continuous phase in the blend. Stoclet and coworkers reached this goal by increasing the PA11 content up to 40 wt.%. The goal, however, is to maximize the content of PLA, cheaper and much more "eco-sustainable" than PA11, while preserving the continuity of the latter phase. A possible strategy envisages the

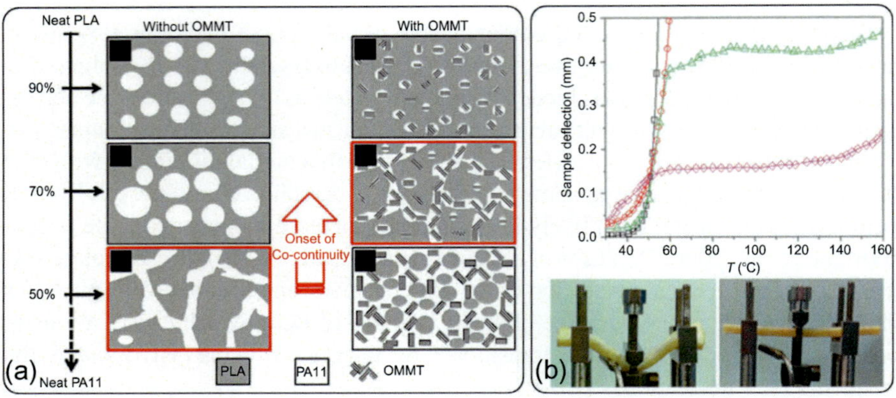

Figure 2.6 (a) Schematic of the morphology evolution of the unfilled and organoclay-filled PLA/PA11 blends as a function of composition. (b) Sample deflection recorded during temperature ramp at 2 °C/min under a load of 0.455 MPa for the sample PLA (squares), PA11 (diamonds), PLA70 (circles), and PLA70-C3 (triangles). The pictures show the unfilled (left) and organoclay-filled (right) blend at 30 wt.% of PA11 at the end of the test, that is after the temperature had reached ≈160 °C.
Images taken from Nuzzo et al. (2014).

employment of nanoparticles. When incorporated in multicomponent systems, nanosized fillers are usually inclined to enrich specific regions (e.g., one of the phases or the interface). The clever control of the nonuniform particle distribution may lead to novel opportunities, such as preferential reinforcement of one phase, targeted reinforcement of the interphase, compatibilization, and filler-induced changes in phase morphology. In particular, nanoparticles have revealed to be effective in promoting cocontinuity in immiscible binary blends at low amounts of either of the polymer phases (Li and Shimizu, 2005; Zou et al., 2007). Exploiting this phenomenon, Filippone and coworkers added organoclay to induce cocontinuity in the same PLA/PA11 blend as that by Stoclet et al. The filled blend preserved structural integrity up to ≈160 °C, while the unfilled counterpart, having drop-matrix morphology, softened as soon as the T_g of the PLA matrix phase was reached (Nuzzo et al., 2014) (Figure 2.6).

The preferential positioning of the filler inside the minor PA11 phase is responsible for a drastic slowing down of the melt state relaxation dynamics, eventually preserving the interpenetrated morphology attained during melt mixing. Further analyses are currently in progress to assess the dominant mechanisms behind the phenomenon of nanoparticle-induced cocontinuity.

References

Albertsson, A.-C., Varma, I.K., 2003. Recent developments in ring opening polymerization of lactones for biomedical applications. Biomacromolecules 4, 1466.

Ali, F., Chang, Y.-W., Kang, S.C., Yoon, Y.G., 2009. Thermal, mechanical and rheological properties of poly(lactic acid)/epoxidized soybean oil blends. Polym. Bull. 62, 91.

Amirou, S., Zerizer, A., Haddadou, I., Merlin, A., 2013. Effects of corona discharge treatment on the mechanical properties of biocomposites from polylactic acid and Algerian date palm fibres. Sci. Res. Essays 8, 946.

Anderson, K.S., Lim, S.H., Hillmyer, M.A., 2003. Toughening of polylactide by melt blending with linear low-density polyethylene. J. Appl. Polym. Sci. 89, 3757.

Anderson, K.S., Schreck, K.M., Hillmyer, M.A., 2008. Toughening polylactide. Polym. Rev. 48, 85.

Bax, B., Mussig, J., 2008. Impact and tensile properties of PLA/Cordenka and PLA/flax composites. Compos. Sci. Technol. 68, 1601.

Bordes, P., Pollet, E., Averous, L., 2009. Nano-biocomposites: biodegradable polyester/nanoclay systems. Prog. Polym. Sci. 34, 125.

Boselli, M., Colombo, V., De Angelis, M.G., Ghedini, E., Gherardi, M., Laurita, R., Liguori, A., MInelli, M., Rotundo, F., Sanibondi, P., Stancampiano, A., 2012. Comparing the effect of different atmospheric pressure non-equilibrium plasma sources on PLA oxygen permeability. J. Phys. Conf. Ser. 406, 012038.

Cao, Y., Feng, J., Wu, P., 2010. Preparation of organically dispersible graphene nanosheet powders through a lyophilication method and their poly(lactic acid) composites. Carbon 48, 3834.

Carothers, W.H., Dorough, G.L., Van Natta, F.J., 1932. Studies of polymerization and ring formation. X. The reversible polymerization of six-membered cyclic esters. J. Am. Chem. Soc. 54, 761.

Chaiwong, C., Rachtanapun, P., Wongchaiya, P., Auras, R., Boonyawan, D., 2010. Effect of plasma treatment on hydrophobicity and barrier property of polylactic acid. Surf. Coat. Technol. 204, 2933.

Chen, Y., Yao, X., Zhou, X., Pan, Z., Gu, Q., 2011. Poly(lactic acid)/graphene nanocomposites prepared via solution blending using chloroform as a mutual solvent. J. Nanosci. Nanotechnol. 11, 7813.

Cheung, H.-Y., Lau, K.-T., Tao, X.-M., Hui, D., 2008. A potential material for tissue engineering: silkworm silk/PLA biocomposite. Compos. Part B: Eng. 39, 1026.

Chiu, W.M., Chang, Y.A., Kuo, H.Y., Lin, M.H., Wen, H.C., 2008. A study of carbon nanotubes/biodegradable plastic polylactic acid composites. J. Appl. Polym. Sci. 108, 3024.

Chrissafis, K., Paraskevopolous, K.M., Jannakoudakis, A., Beslikas, T., Bikiaris, D., 2010. Oxidized multiwalled carbon nanotubes as effective reinforcement and thermal stability agents of poly(lactic acid) ligaments. J. Appl. Polym. Sci. 118, 2712.

Dell'Erba, R., Groeninckx, G., Maglio, G., Malinconico, M., Migliozzi, A., 2001. Immiscible polymer blends of semicrystalline biocompatible components: thermal properties and phase morphology analysis of PLLA/PCL blends. Polymer 42, 7831.

Dorigato, A., Sebastiani, M., Pegoretti, A., Fambri, L., 2012. Effect of silica nanoparticles on the mechanical performances of poly(lacic acid). J. Polym. Environ. 20, 713.

Fambri, L., Kesenci, K., Migliaresi, C., 2003. Characterization of modulus and glass transition phenomena in poly(L-lactide)/hydroxyapatite composites. Polym. Compos. 24, 100.

Fathilah, A., Chang, Y.W., Kang, S.C., Yoon, J.Y., 2009. Thermal, mechanical and rheological properties of poly(lactic acid)/epoxidized soybean oil blends. Polym. Bull. 62, 91.

Fukushima, K., Murariu, M., Camino, G., Dubois, P., 2010. Effect of expanded graphite/layered-silicate clay on thermal, mechanical and fire retardant properties of poly(lactic acid). Polym. Degrad. Stab. 95, 1063.

Giita Silverajah, V.S., Ibrahim, N.A., Zainuddin, N., Wan Yunus, W.M.Z., Abu Hassan, H., 2012. Mechanical, thermal and morphological properties of poly(lactic acid)/epoxidized palm olein blend. Molecules 17, 11729.

Gonzalez, A., Dasari, A., Herrero, B., Plancher, E., Santaren, J., Esteban, A., Lim, S.H., 2012. Fire retardancy behavior of PLA based nanocomposites. Polym. Degrad. Stab. 97, 248.

Goriparthi, B.K., Suman, K.N.S., Rao, N.M., 2012. Effect of fiber surface treatments on mechanical and abrasive wear performance of polylactide/jute composites. Compos. Part A 43, 1800.

Graupner, N., Hermann, A.S., Mussig, J., 2009. Natural and man-made cellulose fibre-reinforced poly(lactic acid) (PLA) composites: an overview about mechanical characteristics and application areas. Compos. Part A 40, 810.

Gu, S., Yang, M., Yu, T., Ren, T., Ren, J., 2008. Synthesis and characterization of biodegradable lactic acid-based polymers by chain extension. Polym. Int. 57, 982.

Harris, M., Lee, E.C., 2008. Improving mechanical performance of injection molded PLA by controlling crystallinity. J. Appl. Polym. Sci. 107, 2246.

Hartmann, M.H., 1998. High molecular weight polylactic acid polymers. In: Kaplan, D.L. (Ed.), Biopolymers from Renewable Resources. Springer-Verlag, Berlin, pp. 367–411.

Hashima, K., Nishitsuji, S., Inoue, T., 2010. Structure-properties of super-tough PLA alloy with excellent heat resistance. Polymer 51, 3934.

Hassan, E., Wei, Y., Jiao, H., Huo, Y.M., 2012. Plant fibers reinforced poly(lactic acid) (PLA) as a green composites: review. Int. J. Eng. Sci. Technol. 4, 4429.

Hiljanen-Vainio, M., Varpomaa, P., Seppala, J., Tormala, P., 1996. Modification of poly(L-lactides) by blending: mechanical and hydrolitic behavior. Macromol. Chem. Phys. 197, 1503.

Hu, R., Lim, J.-K., 2007. Fabrication and mechanical properties of completely biodegradable hemp fiber reinforced polylactic acid composites. J. Compos. Mater. 41, 1655.

Huang, J.W., Hung, Y.C., Wen, Y.L., Kang, C.C., Yeh, M.Y., 2009. Polylactide/nano and microscale silica composite films. 1. Preparation and characterization. J. Appl. Polym. Sci. 112, 1688.

Huda, M.S., Mohanty, A.K., Drzal, L.T., Schut, E., Misra, M., 2005. "Green" composites from recycled cellulose and poly(lactic acid): physic-mechanical and morphological properties evaluation. J. Mater. Sci. 40, 4221.

Huda, M.S., Drzal, L.T., Misra, M., Mohanty, A.K., 2006. Wood-fiber-reinforced poly(lactic acid) composites: evaluation of the physicomechanical and morphological properties. J. Appl. Polym. Sci. 102, 4856.

Huda, M.S., Drzal, L.T., Mohanty, A.K., Misra, M., 2008. Effect of fiber surface-treatments on the properties of laminated biocomposites from poly(lactic acid) (PLA) and kenaf fibers. Compos. Sci. Technol. 68, 424.

Jacobs, T., Declercq, H., De Geyter, N., Cornelissen, R., Dubruel, P., Leys, C., Beaurain, A., Payen, E., Morent, R., 2013. Plasma surface modification of polylactic acid to promote interaction with fibroblast. J. Mater. Sci. Mater. Med. 24, 469.

Jacobsen, S., Fritz, H.G., 1999. Plasticizing polylactide—the effect of different plasticizers on the mechanical properties. Polym. Eng. Sci. 39, 1303.

Jain, S., Reddy, M.M., Mohanty, A.K., Misra, A.M., Ghosh, A.K., 2010. A new biodegradable flexible composite sheet from poly(lactic acid)/poly(epsiloncaprolactone) blends and micro-talc. Macromol. Mater. Eng. 295, 750.

Jiang, L., Zhang, J., Wolcott, M.P., 2007. Comparison of polylactide/nano-sized calcium carbonate and polylactide/montmorillonite composites: reinforcing effects and toughening mechanisms. Polymer 48, 7632.

Jin, F.Z., Hyon, S.H., Iwata, H., 2002. Tsutsumi, crosslinking of poly(L-lactide) by γ-irradiation. Macromol. Rapid. Commun. 23, 909.

Kawamoto, N., Sakai, A., Horikoshi, T., Urushihara, T., Tobita, E., 2007. Nucleating agent for poly (L-lactic acid) – an optimization of chemical structure of hydrazide compound for advanced nucleation ability. J. Appl. Polym. Sci. 103, 244.

Kim, I.H., Jeong, Y.G., 2010. Polylactide/exfoliated graphite nanocomposites with enhanced thermal stability, mechanical modulus, and electrical conductivity. J. Polym. Sci. B Polym. Phys. 48, 850.

Krikorian, V., Pochan, D.J., 2003. Poly(L-lactic acid)/layered silicate nanocomposite: fabrication, characterization, and properties. Chem. Mater. 15, 4317.

Kuan, C.F., Kuan, H.C., Ma, C.C.M., Chen, C.H., 2008. Mechanical and electrical properties of multi-wall carbon nanotubes/poly(lactic acid) composites. J. Phys. Chem. Solids 69, 1395.

Kubies, D., Scudla, J., Puffr, R., Sikora, A., Baldrian, J., Kovarova, J., 2006. Structure and mechanical properties of poly(L-lactide)/layered silicate nanocomposites. Eur. Polym. J. 42, 888.

Labrecque, L.V., Kumar, R.A., Dave, V., Gross, R.A., Mecarthy, S.P., 1997. Citrate esters as plasticizers for poly(lactic acid). J. Appl. Polym. Sci. 66, 1507.

Lee, J.H., Jeong, Y.G., 2010. Preparation and characterization of nanocomposites based on polylactides tethered with polyhedral oligomeric silsesquioxane. J. Appl. Polym. Sci. 115, 1039.

Li, H., Huneault, M.A., 2007. Effect of nucleation and plasticization on the crystallization of poly(lactic acid). Polymer 48, 6855.

Li, Y., Shimizu, H., 2005. Co-continuous polyamide 6 (PA6)/acrylonitrile-butadiene-styrene (ABS) nanocomposites. Macromol. Rapid Commun. 26, 710.

Li, S., Yuan, H., Yu, T., Yuan, W., Ren, J., 2009. Flame-retardancy and anti-dripping effects of intumescent flame retardant incorporating montmorillonite on poly(lactic acid). Adv. Polym. Technol. 20, 1114.

Li, Y., Chen, C., Li, J., Sun, X.S., 2011a. Synthesis and characterization of bionancomposites of poly(lactic acid) and TiO_2 nanowires by in situ polymerization. Polymer 52, 2367.

Li, Z., Zhou, X., Pei, C., 2011b. Effect of sisal fiber surface treatment on properties of sisal fiber reinforced polilactide composites. Int. J. Polym. Sci. 7. http://dx.doi.org/10.1155/2011/803428.

Liu, M., Pu, M., Ma, H., 2012. Preparation, structure and thermal properties of polylactide/sepiolite nanocomposites with and without organic modifiers. Compos. Sci. Technol. 72, 1508.

Liu, M., Zhang, Y., Zhou, C., 2013. Nanocomposites of halloysite and polylactide. Appl. Clay Sci. 75, 52.

Ljungberg, N., Wesslen, B., 2003. Tributyl citrate oligomers as plasticizers for poly(lactic acid): thermomechanical film properties and aging. Polymers 44, 7679.

López-Rodríguez, N., López-Arraiza, A., Meaurio, E., Sarasua, J.R., 2006. Crystallization, morphology, and mechanical behavior of polylactide/poly (ε-caprolactone) blends. Polym. Eng. Sci. 46, 1299.

Lowe, C.E., 1954. Preparation of High Molecular Weight Polyhydroxyacetic Ester. U.S. Patent 2,668,162.

Luo, Y.B., Li, W.D., Wang, X.L., Xu, D.Y., Wang, Y.Z., 2009. Preparation and properties of nanocomposites based on poly(lactic acid) and functionalized TiO_2. Acta Mater. 57, 3182.

Masirek, R., Kulinski, Z., Chionna, D., Piorkowska, E., Pracella, M., 2007. Composites of poly(L-lactide) with hemp fibers: morphology and thermal and mechanical properties. J. Appl. Polym. Sci. 105, 255.

Mitomo, H., Kaneda, A., Quynh, T.M., Nagasawa, N., Yoshii, F., 2005. Improvement of heat stability of poly(L-lactic acid) by radiation-induced crosslinking. Polymer 46, 4695.

Molnar, K., Moczo, J., Murariu, M., Dubois, P., Pukanszky, B., 2009. Factors affecting the properties of PLA/$CaSO_4$ composites: homogeneity and interactions. Express Polym. Lett. 3, 49.

Mukherjee, T., Kao, N., 2011. PLA based biopolymer reinforced with natural fibre: a review. J. Polym. Environ. 19, 714.

Murariu, M., Ferreira, A.D., Degee, P., Alexandre, M., Dubois, P., 2007. Polylactide compositions. Part 1: effect of filler content and size on mechanical properties of PLA/calcium sulfate composites. Polymer 48, 2613.

Murariu, M., Dechief, A.L., Paint, Y., Peeterbroeck, S., Bonnaud, L., Dubois, P., 2012. Polylactide (PLA)-halloysite nanocomposites: production, morphology and key-properties. J. Polym. Environ. 20, 932.

Nagasawa, N., Kaneda, A., Kanazawa, S., Yagi, T., Mitomo, H., Yoshii, F., Tamada, M., 2005a. Application of poly(lactic acid) modified by radiation crosslinking. Nucl. Instrum. Meth. Phys. Res. B 236, 611.

Nagasawa, N., Kaneda, S., Kanazawa, S., Yagi, T., Mitomo, H., Yoshii, F., Tamada, M., 2005b. Nucl. Instrum. Meth. Phys. Res., Sect. A-Beam Interact. Mater. Atoms 236, 611.

Najafi, N., Heuzey, M.C., Carreau, P.J., 2012. Polylactide (PLA)-clay nanocomposites prepared by melt compounding in the presence of a chain extender. Compos. Sci. Technol. 72, 608.

Nijenhuis, A.J., Colstee, E., Grijpma, D.W., Pennings, A.J., 1996. High molecular weight poly(L-lactide) and poly(ethylene oxide) blends: thermal characterization and physical properties. Polymer 37, 5849.

Nuzzo, A., Coiai, S., Carroccio, S.C., Dintcheva, N.Tz., Gambarotti, C., Filippone, G., 2014. Heat-resistant fully bio-based nanocomposite blends based on poly(lactic acid). Macromol. Mater. Eng. 299, 31.

Ochi, S., 2008. Mechanical properties of kenaf fibers and kenaf/PLA composites. Mech. Mater. 40, 446.

Odent, J., Habibi, Y., Raquez, J.M., Dubois, P., 2013. Ultra-tough polylactide-based materials synergistically designed in the presence of rubbery ε-caprolactone-based copolyester and silica nanoparticles. Compos. Sci. Technol. 84, 86.

Oksman, K., Skrifvars, M., Selin, J.-F., 2003. Natural fibers as reinforcement in polylactic acid (PLA) composites. Compos. Sci. Technol. 63, 1317.

Oliveira, J.E., Moraes, E.A., Marconcini, J.M., Mattoso, L.H.C., Glenn, G.M., Medeiros, E.S., 2013. Properties of poly(lactic acid) and poly(ethylene oxide) solvent polymer mixtures and nanofibers made by solution blown spinning. J. Appl. Polym. Sci. 129, 3672.

Pan, H., Qiu, Z., 2010. Biodegradable poly(l-lactide)/polyhedral oligomeric silsesquioxanes nanocomposites: enhanced crystallization, mechanical properties, and hydrolytic degradation. Macromolecules 43, 1499.

Park, S.-D., Todo, M., Arakawa, K., 2004. Effect of annealing on the fracture toughness of poly(lactic acid). J. Mater. Sci. 39, 1113.

Paul, M.A., Delcourt, C., Alexandre, M., Degée, P., Monteverde, F., Rulmont, A., Dubois, P., 2005. (Plasticized) polylactide/(organo-)clay nanocomposites by in situ intercalative polymerization. Macromol. Chem. Phys. 206, 484.

Perego, G., Cella, G.D., Bastioli, C., 1996. Effect of molecular weight and crystallinity on poly (lactic acid) mechanical properties. J. Appl. Polym. Sci. 59, 37.

Pinto, A.M., Cabral, J., Pacheco Tanaka, D.A., Mendes, A.M., Magalhaes, F.D., 2013. Effect of incorporation of graphene oxide and graphene nanoplatelets on mechanical and gas permeability properties of poly(lactic acid) films. Polym. Int. 62, 33.

Piorkowska, E., Kulinski, Z., Galeski, A., Masirek, R., 2006. Plasticization of semicrystalline poly(L-lactide) with poly(propylene glycol). Polymer 47, 7178.

Plackett, D., Andersen, T.L., Pedersen, W.B., Nielsen, L., 2003. Biodegradable composites based on L-polylactide and jute fibres. Compos. Sci. Technol. 63, 1287.

Pluta, M., Galeski, A., Alexandre, M., Paul, M.A., Dubois, P., 2002. Polylactide/montmorillonite nanocomposites and microcomposites prepared by melt blending: structure and some physical properties. J. Appl. Polym. Sci. 86, 1497.

Pluta, M., Murariu, M., Da Silva Ferreira, A., Alexandre, M., Galeski, A., Dubois, P., 2007. Polylactide compositions. II. Correlation between morphology and main properties of PLA/calcium sulfate composites. J. Polym. Sci. B Polym. Phys. 45, 2770.

Pötschke, P., Paul, D.R., 2003. Formation of co-continuous structures in melt-mixed immiscible polymer blends. J. Macromol. Sci. C Polym. Rev. 43, 87.

Qiu, Z., Pan, H., 2010. Preparation, crystallization and hydrolytic degradation of biodegradable of poly(l-lactide)/polyhedral oligomeric silsesquioxanes nanocomposite. Compos. Sci. Technol. 70, 1089.

Quynh, T.M., Mitomo, H., Nagasawa, N., Wada, Y., Yoshii, F., Tamada, M., 2007. Properties of crosslinked polylactides (PLLA & PDLA) by radiation and its biodegradability. Eur. Polym. J. 43, 1779.

Ragoubi, M., George, B., Molina, S., Bienaime, D., Merlin, A., Hiver, J.-M., Dahoun, A., 2012. Effect of corona discharge treatment on mechanical and thermal properties of composites based on miscanthus fibres and polylactic acid or polypropylene matrix. Compos. A: Appl. Sci. Manuf. 43, 675.

Rasal, R.M., Janorkar, A.V., Hirt, D.E., 2010. Poly (lactic acid) modifications. Prog. Polym. Sci. 35, 338.

Ray, S.S., Yamada, K., Okamoto, M., Ueda, K., 2002. Polylactide-layered silicate nanocomposites: a novel biodegradable material. Nano Lett. 2, 1093.

Ray, S.S., Yamada, K., Okamoto, M., Ueda, K., 2003. New polylactide-layered silicate nanocomposites. 2. Concurrent improvements of material properties, biodegradability and melt rheology. Polymer 44, 857.

Reinhardt, M., Kaufmann, J., Kausch, M., Kroll, L., 2013. PLA–viscose-composites with continuous fiber reinforcement for structural applications. Proc. Mater. Sci. 2, 137.

Ren, Z., Dong, L., Yang, Y., 2006. Dynamic mechanical and thermal properties of plasticized poly(lactic acid). J. Appl. Polym. Sci. 101, 183.

Russias, J., Saiz, E., Nalla, R.K., Gryn, K., Richtie, R.O., Tomsia, A.P., 2006. Fabrication and mechanical properties of PLA/HA composites: a study of in vitro degradation. Mater. Sci. Eng. C 26, 1289.

Russo, P., Cammarano, S., Bilotti, E., Peijs, T., Cerruti, P., Acierno, D., 2014a. Physical properties of poly lactic acid/clay nanocomposite films: effect of filler content and annealing treatment. J. Appl. Polym. Sci. 131, 397.

Russo, P., Acierno, D., Vignali, A., Lavorgna, M., 2014b. Poly(lactic acid) based systems filled with talc microparticles: thermal, structural and morphological issues. Polym. Compos. 35, 1093–1103. http://dx.doi.org/10.1002/pc.22757.

Rytlewski, P., Malinowski, R., Moraczewski, K., Zenkiewicz, M., 2010. Influence of some crosslinking agents on thermal and mechanical properties of electron beam irradiated polylactide. Radiat. Phys. Chem. 79, 1052.

Rytlewski, P., Mroz, W., Zenkiewicz, M., Czwartos, J., Budner, B., 2012. Laser induced surface modification of polylactide. J. Mater. Process. Technol. 212, 1700.

Sakai, F., Nishikawa, K., Inoue, Y., Yazawa, K., 2009. Nucleation enhancement effect in poly (l-lactide)(PLLA)/poly (ϵ-caprolactone)(PCL) blend induced by locally activated chain mobility resulting from limited miscibility. Macromolecules 42, 8335.

Shah, A.A., Hasan, F., Hameed, A., Ahmed, S., 2008. Biological degradation of plastics: a comprehensive review. Biotechnol. Adv. 26, 246.

Shibata, M., Oyamada, S., Kobayashi, S., Yaginuma, D., 2004. Mechanical properties and biodegradability of green composites based on biodegradable polyesters and lyocell fabric. J. Appl. Polym. Sci. 92, 3857.

Shibata, M., Inoue, Y., Miyoshi, M., 2006. Mechanical properties, morphology, and crystallization behavior of blends of poly(L-lactide) with poly(butylene succinate-co-L-lactate) and poly(butylenes succinate). Polymer 47, 3557.

Shina Ray, S., Okamoto, M., 2003. Biodegradable polylactide and its nanocomposites: opening a new dimension for plastics and composite. Macromol. Rapid Commun. 24, 815.

Shina Ray, S., Maiti, P., Okamoto, M., Yamada, K., Ueda, K., 2002. New polylactide/layered silicate nanocomposites. 1. Preparation, characterization, and properties. Macromolecules 35, 3104.

Shina Ray, S., Yamada, K., Okamoto, M., Ogami, A., Ueda, K., 2003. New polylactide/layered silicate nanocomposites. 3. High-performance biodegradable materials. Chem. Mater. 15, 1456.

Srithep, Y., Nealey, P., Turng, L.-S., 2013. Effects of annealing time and temperature on the crystallinity and heat resistance behavior of injection molded poly(lactic acid). Polym. Eng. Sci. 53, 580.

Stoclet, G., Seguela, R.J., Lefebvre, M., 2011. Morphology, thermal behavior and mechanical properties of binary blends of compatible biosourced polymers: polylactide/polyamide11. Polymer 52, 1417.

Sujaritjun, W., Uawongsuwan, P., Pivsa-Art, W., Hamada, H., 2013. Mechanical property of surface modified natural fiber reinforced PLA biocomposites. Energy Procedia 34, 664.

Sungsanit, K., Kao, N., Bhattacharya, S.N., 2012. Properties of linear poly(lactic acid)/polyethylene glycol blends. Polym. Eng. Sci. 52, 108.

Tabi, T., Sajo, E., Szabo, F., Luyt, A.S., Kovacs, J.G., 2010. Crystalline structure of annealed polylactic acid and its relation to processing. Express Polym. Lett. 4, 659.

Tsuji, H., Sawada, M., Bouapao, L., 2009. Biodegradable polyesters as crystallization-accelerating agents of poly(L-lactide). ACS Appl. Mater. Interfaces 1, 1719.

Urayama, H., Ma, C., Kimura, Y., 2003. Mechanical and thermal properties of poly(L-lactide) incorporating various inorganic fillers with particle and whisker shapes. Macromol. Mater. Eng. 288, 562.

Wang, D.Y., Leuteritz, A., Wang, Y.Z., Wagenknecht, U., Heinrich, G., 2010. Preparation and burning behaviors of flame retarding biodegradable poly(lactic acid) nanocomposite based on zinc aluminum layered double hydroxide. Polym. Degrad. Stab. 95, 2474.

Wu, C.S., Liao, H.T., 2007. Study on the preparation and characterization of biodegradable polylactide/multi-walled carbon nanotubes nanocomposites. Polymer 48, 4449.

Wu, D., Wu, L., Zhou, W., Sun, Y., Zhang, M., 2010. Relations between the aspect ratio of carbon nanotubes and the formation of percolation networks in biodegradable polylactide/carbon nanotubes composites. J. Polym. Sci. B Polym. Phys. 48, 479.

Xu, Y.Q., Qu, J.P., 2009. Mechanical and rheological properties of epoxidized soybean oil plasticized poly(lactic acid). J. Appl. Polym. Sci. 112, 3185.

Yan, S.F., Yin, J.B., Yang, Y., Dai, Z.Z., Ma, J., Chen, X.S., 2007. Surface-grafted silica linked with L-lactic acid oligomer: a novel nanofiller to improve the performance of biodegradable poly(L-lactide). Polymer 48, 1688.

Yang, S.-L., Wu, Z.-H., Yang, W., Yang, M.-B., 2008. Thermal and mechanical properties of chemical crosslinked polylactide (PLA). Polym. Test. 27, 957.

Yu, J., Qiu, Z., 2011. Preparation and properties of biodegradable poly(L-lactide)/octamethyl-polyhedral oligomeric silsesquioxanes nanocomposites with enhanced crystallization rate via simple melt compounding. ACS Appl. Mater. Interfaces 3, 890.

Yu, L., Liu, H., Xie, F., Chen, L., Li, X., 2008. Effect of annealing and orientation on microstructures and mechanical properties of polylactic acid. Polym. Eng. Sci. 48, 634.

Zenkiewicz, M., Richert, J., Rytlewski, P., Moraczewski, K., 2009. Some effects of corona plasma treatment of polylactide/montmorillonite nanocomposite films. Plasma Process. Polym. 6, S387.

Zenkiewicz, M., Rytlewski, P., Malinowski, R., 2010. Compositional, physical and chemical modification of polylactide. J. Achiev. Mater. Manufact. Eng. 43, 192.

Zhan, J., Song, L., Nie, S., Hu, Y., 2009. Combustion properties and thermal degradation behavior of polylactide with an effective intumescent flame retardant. Polym. Degrad. Stab. 94, 291.

Zhang, X., Sun, J., Fang, S., Han, X., Li, Y., Zhang, C., 2011. Thermal, crystalline and mechanical properties of octa(3-chloropropylsilsesquioxane)/poly(L-lactic acid) hybrid films. J. Appl. Polym. Sci. 122, 296.

Zhang, K., Mohanty, A.K., Misra, M., 2012. Fully biodegradable and biorenewable ternary blends from polylactide, poly(3-hydroxybutyrate-co-hydroxyvalerate) and poly(butylene succinate) with balanced properties. ACS Appl. Mater. Interfaces 4, 3091.

Zhong, W., Ge, J., Gu, Z., Li, W., Chen, X., Zang, Y., Yang, Y., 1999. Study on biodegradable polymer materials based on poly(lactic acid). I. Chain extending of low molecular weight poly(lactic acid) with methylenediphenyl diisocyanate. J. Appl. Polym. Sci. 74, 2546.

Zhu, Y., Buonocore, G.G., Lavorgna, M., Ambrosio, L., 2011. Poly(lactic acid)/titanium dioxide nanocomposite films: influence of processing procedure on dispersion of titanium dioxide and photocatalytic activity. Polym. Compos. 32, 519.

Zou, H., Ning, N., Su, R., Zhang, Q., Fu, Q., 2007. Manipulating the phase morphology in PPS/PA66 blends using clay. J. Appl. Polym. Sci. 106, 2238.

Mechanical performance of polyhydroxyalkanoate (PHA)-based biocomposites

3

E. Ten, L. Jiang†, J. Zhang*, M.P. Wolcott**
*Washington State University, Pullman, WA, USA; †North Dakota State University, Fargo, ND, USA

3.1 Introduction

Polyhydroxyalkanoates (PHAs) are biodegradable polyesters that are naturally produced by microorganisms. A wide variety of microorganisms (De Koning, 1993; Madison and Huisman, 1999; Santhanam and Sasidharan, 2010) accumulate PHA as storage polymers for carbon and energy sources. Depending on the carbon substrates of the microorganisms, different monomers, and therefore copolymers, can be obtained (Santhanam and Sasidharan, 2010). The main variant of the PHA family is the polyhydroxybutyrate (PHB) homopolymer. As a naturally produced biopolymer, PHB not only can be produced without a petrochemical feedstock, but it can also rapidly biodegrade in aerobic, anaerobic, and saline environments (Lee, 1996).

PHB is a highly crystalline polyester with properties similar to conventional plastics such as polypropylene (Madison and Huisman, 1999; Barham and Organ, 1994; Harding et al., 2007). Melting temperature of PHB is about 173–180 °C and glass transition temperature is around 5 °C. The homopolymer has the disadvantages of being brittle and having a narrow processability window compared with petroleum-based plastics due to the relative proximity of its thermal decomposition and melt temperatures. The brittleness of PHB is primarily attributed to large spherulite formation and secondary crystallization. Although PHB has a high degree of crystallinity, it demonstrates low nucleation density and consequently forms large spherulites, which are known to impair the mechanical properties of the polymer. To overcome these drawbacks, various copolymers such as poly(hydroxybutyrate-co-hydroxyvalerate) (PHBV), poly(hydroxybutyrate-co-hydroxyhexanoate) (PHBHx), poly(hydroxybutyrate-co-hydroxyoctanoate) (PHBO), and poly(hydroxybutyrate-cohydroxyoctadecanoate) (PHBOd) have been biosynthesized (Reddy et al., 2003).

PHBV is a common representative of the PHA family. Compared to PHB, PHBV possesses higher ductility and can be processed at lower temperatures. This change in processability alleviates the problem of thermal degradation; however, PHBV exhibits a slow crystallization rate, low crystallinity, and low-impact strength (Qian et al., 2007). These properties become prominent with increasing 3-HV comonomer content. By adding nucleating agents, the crystallization rates of PHB and PHBV can be significantly accelerated. Nucleating agents introduce extra sites to initiate heterogeneous

Biocomposites: Design and Mechanical Performance. http://dx.doi.org/10.1016/B978-1-78242-373-7.00008-1

nucleation, which results in small and more homogeneous spherulite sizes, enhancing the overall mechanical properties of the material. Qian et al. (2007) reported using thymine and melamine as nucleating agents for both PHB and different types of PHBVs, and the nucleating effects and crystallization kinetics were compared with those of a commonly used nucleating agent, boron nitride (BN). It has been shown that BN and thymine greatly increased the nucleation ability of PHBV8, PHBV12, and PHBV21. These and other nucleating agents evaluated resulted in fine crystalline structures with similar perfection, and the nucleated PHBVs demonstrated higher melting temperatures than neat PHBVs (Qian et al., 2007).

Due to their biocompatibility and biodegradability, PHB and PHBV can potentially replace petroleum-based plastics and present long lasting benefits for sustainable development, waste management, and environmental pollution issues. However, commercialization of PHAs is still limited due to the high recovery and production costs of these biopolymers. There are several ways to improve the economic feasibility of these processes, such as changing carbon source and bacterial strains. In addition to altering production variables to lower production costs, introducing superior properties to the bioplastics such as improved mechanical properties and biodegradability can improve the value of the polymer, especially for specialty applications. There is an increasing interest in the study of improving mechanical properties of PHB and PHBV for potential applications in the field of biomedical, packaging, and agricultural plastics.

3.2 Mechanical properties of PHB—biodegradable polymer composites

Blending with various biodegradable polymers, for example, starch (Parulekar and Mohanty, 2007; Zhang and Thomas, 2010), corn gluten meal (Corradini et al., 2011), soy protein isolate (Ma and Cheng, 2010), poly(butylene adipate-co-terephthalate) (PBAT) (Javadi et al., 2010), poly(propylene carbonate) (PPC) (Corre et al., 2013), and dried distillers grains with solubles (DDGS) (Zarrinbakhsh et al., 2011; Madbouly et al., 2014), is a common approach to significantly reduce production costs, while potentially improving properties of PHAs. Polymer blends have attracted more attention in the last few decades because physical blending is an effective and simple way to prepare biocomposites with varying morphologies and physical characteristics.

Blending PHBV with biodegradable polymers of greater toughness, such as PBAT, can improve impact strength of PHBV. PBAT is a biodegradable aliphatic aromatic copolyester that has a tensile strength of 32–36 MPa and elongation at failure of ~700% (Raquez et al., 2008). Javadi et al. reported that blending PHBV with PBAT improved the specific toughness up to two orders of magnitude compared to the neat PHBV. Also, the PHBV blends with 70 wt.% PBAT exhibited a more than 500% increment in strain-at-failure. However, the dramatic increase in toughness and strain to failure were accompanied by a coincident reduction in strength and modulus compared with the pure PHBV (Javadi et al., 2010).

According to Corre et al., the low-impact strength of pure PHBV8 can be directly improved through the addition of PPC. PPC is a biodegradable aliphatic polycarbonate that is synthesized by combining propylene oxide and carbon dioxide. PPC exhibited very ductile behavior with a low Young's modulus of 319 MPa and strain at failure close to 400%, while PHBV featured a fragile behavior with a high modulus close to 4 GPa and a strain at failure lower than 1%. Similar to the Javadi et al. (2010) work, a large content of tough polymer is required to significantly improve strain at failure of PHBV. Strain at failure was found to be doubled when the PPC ratio was 40 wt.% and quintupled for a ratio of 60 wt.% (Corre et al., 2013).

Biodegradation is one of the intrinsic properties of PHBV and there are multiple factors influencing the biodegradation rates (moisture, pH, temperature, as well as composition, crystallinity, surface area, and additives). Therefore, it is important to investigate the change in mechanical properties with biodegradation. Madbouly et al. measured thermomechanical properties of PHB/PHBV blended with DDGS. This co-product of the grain ethanol production process is a fibrous material that contains proteins, carbohydrates, and oils (Madbouly et al., 2014). The biodegradation rate was found to have a significant influence on the thermomechanical and viscoelastic properties of the PHA/DDGS biocomposites. It was evident that storage modulus (E') decreased systematically with increasing biodegradation time (with intervals up to 20 weeks) over a wide range of temperatures. The authors speculated that this process could be related to the typical decrease in the molecular weight of the biocomposites during biodegradation. Adding 10 wt.% DDGS can decrease material costs, increase the biodegradation rate, and enhance the viscoelastic and thermomecanical properties of PHA. The blend containing 10 wt.% DDGS was reported to be nontoxic to plants before and after 24 weeks of biodegradation in soil, implying the potential use of the composites in agricultural applications (Zarrinbakhsh et al., 2011).

3.3 Mechanical properties of PHB, PHBV/natural fiber-reinforced composites

Low-impact resistance and brittleness of PHBV can also be improved by reinforcing with natural fibers. Besides being renewable and biodegradable, natural fibers offer a number of benefits including abundant availability, low cost, low density, and high mechanical performance (John and Thomas, 2008).

Mechanical properties of PHB and PHBV were improved when reinforced with flax fiber (Wong et al., 2002; Barkoula et al., 2010); jute fiber (Khan et al., 1999) and yarn (Mohanty et al., 2000); wheat (Avella et al., 2000) and rice straw (Buzarovska et al., 2008); cellulose (Bhardwaj et al., 2006) and wood fiber (WF) (Reinsch and Kelley, 1997; Anderson et al., 2013); and wood flour (Singh and Mohanty, 2007), pineapple (Luo and Netravalli, 1999), kenaf (Persico et al., 2011), and bamboo fibers (Singh et al., 2008; Jiang et al., 2008a, 2010). The PHA composites in these various research initiatives were typically processes by melt compounding. For comparison, Table 3.1 presents tensile properties (strength and modulus) of PHA/natural fiber-reinforced composites.

Table 3.1 Tensile properties of selected PHA/natural fiber composites

Fiber (wt.% fraction in PHA composite)	Tensile properties		References
	Strength σ (MPa)	**Modulus E (GPa)**	
Wood flour (PHBV/40%)	16.75	2.73	Singh and Mohanty (2007)
Recycled cellulose fibers (PHBV/40%)	~24	3.2	Bhardwaj et al. (2006)
Flax (PHB/40%)	40	~8.5	Barkoula et al. (2010)
Pineapple fiber (PHBV/30%)	~60	2.3	Luo and Netravalli (1999)
Kenaf (PHBV/40%)	Flexural 16.7	Flexural 3.6	Persico et al. (2011)
Rice straw (PHBV/30%)	5.24	1.17	Buzarovska et al. (2008)
Wood fiber (PHBV/40%)	16.75	2.73	Singh et al. (2008)
Bamboo fiber (PHBV/40%)	16.74	2.8	Singh et al. (2008)
Bamboo pulp fiber (PHBV/20%)	39	4.6	Jiang et al (2008a)

In general, these biocomposites possess increased Young's modulus, tensile strength, and thermal properties, for example, glass transition temperature (T_g) and heat deflection temperature (HDT), with increasing filler content. The composites possess varying degrees of mechanical enhancement when reinforced with either fine wood flour or wood/natural fibers. For example, the tensile modulus of PHBV-based biocomposites reinforced with 40 wt.% of wood flour was improved by ~167% (2.73 ± 0.22 GPa) when compared with neat PHBV (1.02 ± 0.09 GPa) (Singh and Mohanty, 2007). In contrast, the tensile strength continuously decreased from 21.42 to 16.75 MPa with increasing wood flour contents ranging from 0 to 40 wt.%. The heat deflection temperature (HDT) was increased by 21%, while the coefficient of linear thermal expansion (CLTE) was reduced by 18% for PHBV with 40 wt.% of wood flour content (Singh and Mohanty, 2007).

Bhardwaj et al. reported a similar effect for PHBV reinforced with recycled cellulose fibers (RCF) prepared by extrusion and injection molding processes. The tensile and storage moduli of the PHBV-based composites were improved by 220% and 190%, respectively, when the reinforcement was at 40 wt.%. However, unlike the case with wood flour mentioned above, the tensile strength showed an interesting trend. Initially, the tensile strength decreased from 26 (neat PHBV) to 19 MPa with the addition of 15 wt.% RCF. However, the addition of 30 and 40 wt.% RCF resulted in increases. The authors speculated that, at low fiber volume, the matrix was not constrained resulting in relatively low stresses causing fiber–matrix debonding. When fiber volume was above a critical value, the authors speculated that the matrix was restrained, resulting in lower strain at comparatively high stresses. The CLTE value for the PHBV was also reduced significantly upon reinforcement with RCF (Bhardwaj et al., 2006).

Reduced tensile strength with increasing filler amount was also reported for PHBV/rice straw biocomposites (Buzarovska et al., 2008). This behavior was ascribed to the poor distribution of filler within the matrix and weak interfacial adhesion. This subsequently resulted in a decrease of tensile strength. The results were confirmed by SEM micrographs where it could be seen that the rice straw husks were completely pulled out of the polymer (Buzarovska et al., 2008).

Flexural properties of PHBV can be improved by the addition of natural fibers in a similar manner to tension. PHBV composites with 40% kenaf fibers demonstrated increase in flexural modulus from 1280 MPa for neat to 3605 MPa and flexural strength from 28.2 to 16.7 MPa (Persico et al., 2011). The authors interpreted these results to suggest that there was a considerable interfacial interaction between the fibers and PHBV matrix, while fibers entanglements acted as resistant junction points against deflection forces.

As it is evident from the examples above, weak interfacial bonding restricts the enhancement of tensile and flexural properties of PHB and PHBV biocomposites. As with other polyesters, PHB and PHBV are highly hydrophobic and have very different surface energies compared to the highly hydrophilic surfaces of natural fibers. Therefore, insufficient interfacial adhesion can lead to inadequate stress transfer between fiber and matrix and commensurately reduced tensile and impact strengths compared to those of neat polymers. It has been shown that improving interfacial adhesion by coupling agent or reactive compatibilizers is the most effective way to achieve good tensile and flexural mechanical properties of natural fiber-reinforced composites (Anderson et al., 2013). Natural fibers possess abundant hydroxyl groups, which can be readily reacted with various chemical compounds to alter surface energy or polymer interaction.

Anderson et al. (2013) investigated the effect of four interfacial modifiers on the mechanical properties of PHB with WF composites. The modifiers included maleated PHB (PHB-g-MA), a low molecular weight epoxy, a low molecular weight polyester, and poly(methylene-diphenyl diisocyante) (pMDI). Results showed that all modifiers resulted in improvements in tensile strength and modulus with pMDI showing the highest enhancement (Figure 3.1). Adding 4% pMDI to PHB/WF composites produced a 17% increase in Young's modulus and a 53% increase in tensile strength. Because the isocyanate functional groups were highly reactive to hydroxy functionality, pMDI may simply react with the hydroxyls of WFs, altering the surface energy of the fiber to improve wetting. Alternatively, though less likely, the difunctional diisocyante may link the matrix and fiber by bridging adjacent hydroxyl groups. Study of the fracture surfaces showed signs of improved interfacial adhesion through increased propensity of fiber breakage (Anderson et al., 2013).

Introducing plasticizers to the biocomposites was also reported to improve interfacial bonding between PHB and natural fibers (Wong et al., 2002). The plasticizers chosen in this study were glycerol triacetate (GTA), tributyl citrate (TBC), and a low molecular weight poly-(ethylene glycol) (PEG). PHB reinforced with flax fibers composites were prepared by chloroform solution casting onto glass plate. The plasticizers TBC and GTA were also chosen for their potential for renewable sourcing. The presence of plasticizer and its migration from the fibers to the PHB matrix modified the crystallization, degree of crystallinity, and glass transition temperature of the PHB. Incorporation of GTA and PEG in the PHB composites resulted in higher storage

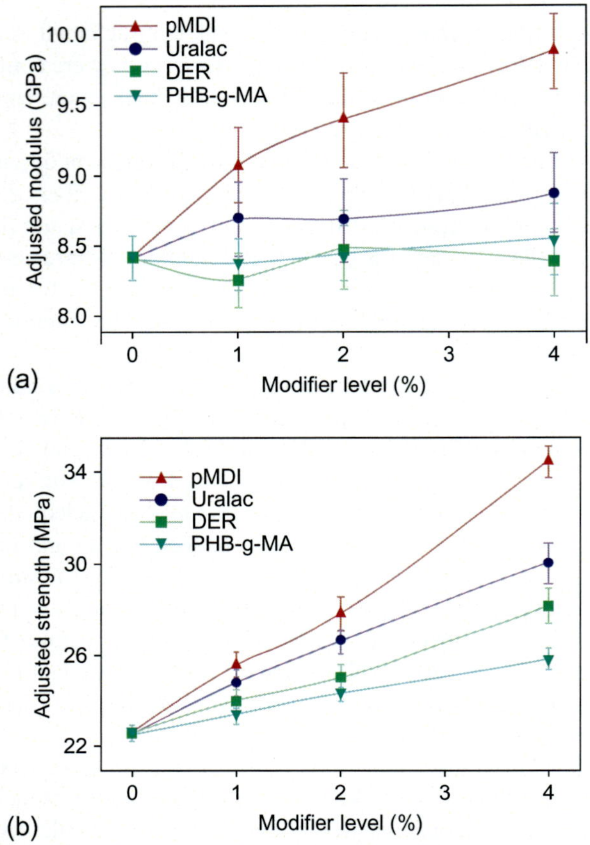

Figure 3.1 Effect of interfacial modifiers on tensile properties of PHB composites. The moduli (a) and strengths (b) are normalized against density.
Reproduced with permission from Anderson et al. (2013). Copyright 2013 Springer.

modulus values than the untreated PHB composite. This was attributed to a better interfacial bonding between the matrix and the fibers caused by a change in surface energy from the plasticizers. A high degree of fiber dispersion within the PHB matrix was also exhibited by the composites with GTA and PEG (Wong et al., 2002). Indeed, the dispersion of the fibers plays an important role in the resulting composite properties. It is commonly known that if the fibers are well dispersed throughout the polymer matrix, the composites have higher mechanical properties due to effective stress transfer.

Unlike property relationships for many other short natural fiber-reinforced polymer composites, the strength, modulus, elongation, and impact strength of PHBV8 reinforced with long, discontinuous bamboo pulp fiber (BPF) were all simultaneously improved (Jiang et al., 2008a, 2010). The tensile modulus of PHBV8/BPF (30 wt.%) and the flexural modulus of PHBV8/BPF (20 wt.%) were 142 and 111% higher than those of PHBV8, respectively. The maximum increase in strength was reported to occur at 20 wt.% of BPF: the tensile and flexural strengths were 42% and 91% higher than those of the neat PHBV8, respectively. With the fiber content higher than 20 wt.%, both tensile and flexural strength tended to decrease slightly. Tensile strain revealed little change before it began to decrease above 20 wt.% of BPF, while flexural strain was increased by 57% at the same fiber content. Typically, the strain and strength of natural fiber biocomposites are lower than that of neat polymers. However, the unique properties of BPF that facilitated stress transfer, including large aspect ratio (*L*/*D*), high strength, high modulus, large elongation, and good interfacial bonding, contributed to the superior reinforcing effects. The BPF carried a large amount of load through interfacial shear stress, which was supported by the fact that under load, the fibers extended with the polymer matrix until the matrix fractured.

The authors also extensively investigated the effect of BN as a nucleation agent and the effectiveness of maleic anhydride (Jiang et al., 2008a) and pMDI (Jiang et al., 2010) as a compatibilizer for PHBV8 composites. As demonstrated earlier with short fibers, by using appropriate compatibilizers and therefore changing matrix/fiber interaction, one can tailor the strength, modulus, and toughness of the composites. For example, maleic anhydride compatibilized PHBV8 composites demonstrated an increase in their tensile and flexural strengths at the expense of the ductility and impact strength of the composites. Even though the tensile and flexural strengths increased 32.2% and 14.4%, respectively, with 5% content of MA, the strain at failure in both tensile and flexural testing decreased. This suggests that while interfacial interaction was improved, the ductility or energy to failure decreased. In addition, the impact strength of the composites was found to monotonously decrease with the MA-PHBV8 concentration.

Similar results were reported for pMDI compatibilized PHBV8/BPF composites. Tensile and flexural strengths increased 28.8% and 13.1% at 3% pMDI, respectively (Jiang et al., 2010). As the improved interfacial adhesion enables greater stress transfer between the fibers and PHBV8, it reduces energy dissipation through fiber–matrix debonding. The enhanced stress transfer into the fibers enhanced the modulus and strength of the composites. In contrast, the improved interfacial adhesion hindered fiber pullout, which is a major source of energy dissipation for increasing toughness of the composites. Therefore, the impact strength was reduced with the addition of MA-PHBV8 and the decreased fiber pullout was confirmed by SEM micrographs (Jiang et al., 2008a).

Addition of 1 wt.% of BN led to improvements in both strength and elongation at failure but decreased the modulus. This was attributed to the refined crystalline structures, reduced spherulite size, and increased spherulite density. The impact strength of the polymer and the composites were also improved by BN.

3.4 Mechanical properties of PHB and PHBV nanocomposites

Nanofillers have been effectively used in biocomposites as a reinforcing agent due to their high surface to volume ratio and high aspect ratio (for nonspherical nanoparticles). Polysaccharide nanocrystals including cellulose, chitosan, and chitin enjoy preference as reinforcing agent for biocomposites due to their low cost and density, biocompatibility, nontoxicity, and high surface reactivity.

Chitin is a natural polysaccharide that exists in considerable amounts as the exoskeleton of arthropods and fungi (Muzzarelli, 1977) and is biosynthesized by a large number of organisms including crustaceans such as crab, lobster, and shrimp, as well as insects. The form of chitin with a high degree of deacetylation is called chitosan. Recently, chitin (Wang et al., 2012, 2013; Ikejima and Inoue, 2000; Sankar et al., 2012) and chitosan (Ikejima and Inoue, 2000; Tong et al., 2012; Hu et al., 2003, 2004; Yu et al., 2006; Veleirinho et al., 2011, 2012; Shih et al., 2007; Cao et al., 2005; Chen et al., 2005) have gained considerable scientific attention for their reinforcing potential, biodegradability, and biocompatibility. Because chitin and chitosan have an abundance of modifiable functional groups, such as hydroxyls, amines, and amides, PHB- and PHBV-based composites can be chemically modified when reinforced with chitin or chitosan. The application of these unique biomaterials include biomedical applications such as tissue engineering scaffolds, wound dressing, and drug delivery (Sankar et al., 2012; Tong et al., 2012; Hu et al., 2003, 2004; Yu et al., 2006; Veleirinho et al., 2011, 2012; Shih et al., 2007; Cao et al., 2005).

The reinforcing effect of chitin nanofibrils on PHBV was evaluated by Wang et al., who found that the addition of both acetylated and native chitin nanofibrils significantly enhanced the ultimate tensile strength and Young's modulus of the PHBV matrix. Compared with neat PHBV, the incorporation of 5.0 wt.% acetylated chitin nanofibrils improved the ultimate tensile strength and Young's modulus of PHBV by 44% and 67%, respectively, while the addition of 5.0 wt.% native chitin nanofibrils enhanced the same parameters by only 24% and 43%. The difference in these behaviors was attributed to improved dispersion in the weakly polar PHBV matrix that was achieved by incorporating acetylated chitin nanofibrils compared to the original chitin nanofibrils (Wang et al., 2013).

Cellulose nanowhiskers (CNWs) or cellulose nanocrystals (CNCs) are among the most studied polysaccharide nanocrystals. CNWs are needle-shaped highly crystalline fibrils that are native in many plants (e.g., wood, hemp, sugar beet, sisal, cotton, and ramie), bacteria, and sea animals (e.g., tunicin). Nano- and microstructures can be isolated from the hierarchical composition of cellulose using mechanical or chemical processes. These natural nanoparticles are promising candidates for composites because of their relative abundance, low cost, high specific modulus, stiffness (100–143 GPa) (Sakurada et al., 1962; Matsuo et al., 1990; Stucova et al., 2005), and high aspect ratio. As such, CNWs have found success as a reinforcement for PHBV as well (Jiang et al., 2008b; Ten et al., 2010, 2012a,b, 2013; Yu et al., 2011, 2012, 2013; Yu and Qin, 2014).

Jiang et al. (2008b) first compared the mechanical properties of PHBV/CNW nanocomposites prepared by solution casting and melt blending. Solution casting produced PHBV/CNW composites with homogeneous particle dispersion, resulting in PHBV with increased tensile strength and modulus but lower elongation. In this work, the strength was increased from 14.1 MPa for neat PHBV to 26.1 MPa at 5% CNWs and the corresponding strain at failure was reduced from 12.4% to 7.8%. Modulus was enhanced from 0.82 GPa for neat polymer to 1.76 GPa for PHBV/5% CNW.

Later, Ten et al. (2010) further investigated thermal and mechanical properties of PHBV/CNW composites prepared by solution casting. Significant improvements in tensile strength, Young's modulus, toughness, and dynamic modulus were observed. Compared to the neat PHBV, the moduli of the nanocomposite with 5% CNWs were increased by 77% (tensile test) and 91% (bulge test), respectively. This modulus increase was attributed to the restrained chain movement during the deformation due to the existence of CNWs. The tensile strength of the PHBV/CNW composites also increased with the CNW content, indicating strong interfacial bonding between the nanowhiskers and the PHBV matrix. The tensile strength of PHBV/5% CNWs was enhanced by 35.5% compared to that of the neat PHBV. Moreover, the tensile toughness was also shown to increase by up to 41% for the nanocomposite with 5 wt.% CNWs. In a following study, thermal, mechanical, dielectric, and dynamic–mechanical properties for the PHBV/CNW nanocomposites exhibited abrupt transitions at 2.3–2.9 wt.% CNW concentrations due to the percolation of CNWs (Ten et al., 2012b). CNWs also contributed to the substantial changes in crystallization kinetics observed in the PHBV (Ten et al., 2010, 2012a,b).

The ability to control fiber orientation is highly desirable to alter mechanical properties of the composites. Highly reinforced composites can be manufactured by stacking layers of unidirectional reinforced composite lamina at alternating orientations. A recent study by Ten et al. (2013) demonstrated successful orientation of CNWs in the PHBV matrix using an electric field, where the CNWs were aligned in the direction of the applied electric field. TEM and DMA results showed that CNW concentration strongly influenced the degree of CNW alignment. High CNW concentration (>4 wt.%) caused more pronounced fiber–fiber and fiber–polymer interactions, which led to high viscosity of the suspension that consequently restricted the CNW mobility (Figure 3.2). Therefore, the electric field became ineffective in aligning the CNW at 5% concentration where the composites were largely isotropic. On the other hand, the aligned PHBV/CNWs composites showed substantial mechanical anisotropy under low CNW concentrations (<4 wt.%).

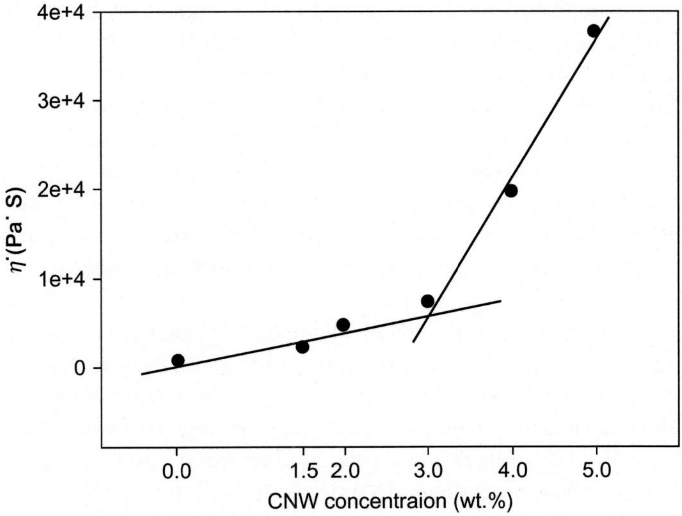

Figure 3.2 The effect of CNW concentration on complex viscosity η^* of PHBV/CNW composites at angular frequency $\omega = 1.5$ rad/s.
Reproduced with permission from Ten et al. (2013). Copyright 2013 Springer.

3.5 Concluding remarks and future trends

PHA biocomposites have shown their versatility and tremendous potential as lightweight sustainable composites. The variety of PHA representatives and fillers, the number of material combinations, environmental conditions, processing technologies, and potential applications offer extensive opportunities; however, there are many challenges that must be overcome during the development of these biocomposites. Key activities to reach this goal will be the identification of economically and environmentally sound production of PHA biocomposites. By changing carbon source and bacterial strains in the production of the PHAs it may be feasible to reduce high production costs and broaden the application areas of such biopolymers. The next generation of PHA composites can be developed through advanced processing and nanotechnology along with utilization of waste and by-products.

It is clear from the work summarized in this chapter that the use of PHA in composites is continuously growing and producing promising results. Blending with different biodegradable polymers, natural particles, and fibers including nanosized fillers such as chitin, chitosan, and CNWs have already been investigated as reinforcement for PHAs. An abundance of research addresses the synthesis and characterization of the polymers, the selection of the reinforcement/filler, as well as the characterization techniques for the material structure and performance (thermal, mechanical, impact, and fracture), including specific ones (crystallization and biodegradation kinetics, hardness, thermal and dielectrical conductivity, etc.). Different manufacturing processes have also been utilized, clearly the most common being solution casting and injection molding. The greatest challenge to commercial application of natural fiber-reinforced plastic composites arises from their large variation in properties and characteristics.

In most cases, compatibility is cited as an issue leading to deterioration of mechanical properties when combining natural fibers with PHA. Poor filler–matrix compatibility can be improved by using compatibilizing agents or filler modifications. Efforts in this area are needed to facilitate further mechanical performance improvement for better load transfer. Although tensile, flexural, and impact properties are the most commonly investigated mechanical properties for PHA biocomposites, impact strength is a weakness for the mechanical performance of these materials. To advance performance to the desired level, research is needed in the areas of fiber processing, fiber dispersion, and composite manufacturing with optimized processing parameters. In addition, research is needed to include the chemical, physical, and biological treatment routes to tailor PHAs and the wide variety of natural fiber types that can be utilized.

From a composite engineering point of view, the production of highly oriented fiber-reinforced composites is a natural progression toward manufacturing composites with superior mechanical properties. Ability to align fibers within PHA matrix using shear stresses, electrical and magnetic fields opens up new platforms of discovery and innovation for PHA biocomposites. If continued research thrusts are taken, it should not be long before PHA biocomposites are able to become an accepted and reliable material for a wide variety of applications.

In summary, the authors foresee developments in the area of sustainable design of PHA biocomposites. Careful life cycle analysis is anticipated to establish the balance between optimum processing conditions and performance of the composites. Further research is required to overcome PHA drawbacks such as brittleness and difficulty in controlling biodegradation. Nanotechnology offers multiple opportunities to enhance overall properties of the biocomposites including mechanical properties and degradation rates. The use of CNCs is being researched for a wide range of applications due to their high aspect ratio and modulus. PHA composites reinforced with CNCs could provide high value, advanced performance, and a fully sustainable design.

References

Anderson, S., Zhang, J., Wolcott, M.P., 2013. Effect of interfacial modifiers on mechanical and physical properties of the PHB composite with high wood flour content. J. Polym. Environ. 21, 631–639. http://dx.doi.org/10.1007/s10924-013-0586-y.

Avella, M., La Rota, G., Martuscelli, E., Raimo, M., 2000. Poly(3-hydroxybutyrate-co-3-hydroxyvalerate) and wheat straw fibre composites: thermal mechanical properties and biodegradation behavior. J. Mater. Sci. 35, 829–836.

Barham, P.J., Organ, S.J., 1994. Mechanical properties of polyhydroxybutyrate–hyd roxybutyratehydroxyvalerate copolymer blends. J. Mater. Sci. 29, 1676–1679.

Barkoula, N.M., Garkhail, S.K., Peijs, T., 2010. Biodegradable composites based on flax/polyhydroxybutyrate and its copolymer with hydroxyvalerate. Ind. Crop. Prod. 31, 34–42.

Bhardwaj, R., Mohanty, A.K., Drzal, L.T., Pourboghrat, F., Misra, M., 2006. Renewable resource-based green composites from recycled cellulose fiber and poly(3-hydroxybutyrate-co-3-hydroxyvalerate) bioplastic. Biomacromolecules 7, 2044–2051.

Buzarovska, A., Bogoeva-Gaceva, G., Grozdanov, A., Avella, M., Gentile, G., Errico, M., 2008. Potential use of rice straw as filler in eco-composite materials. Aust. J. Crop. Sci. 1, 37–42.

Cao, W., Wang, A., Jing, D., Gong, Y., Zhao, N., Zhang, X., 2005. Novel biodegradable films and scaffolds of chitosan blended with poly(3-hydroxybutyrate). J. Biomater. Sci. Polym. Ed. 16 (11), 1379–1394. http://dx.doi.org/10.1163/156856205774472308.

Chen, C., Zhou, X., Zhuang, Y., Dong, L., 2005. Thermal behavior and intermolecular interactions in blends of poly(3-hydroxybutyrate) and maleated poly(3-hydroxybutyrate) with chitosan. J. Polym. Sci. B Polym. Phys. 43, 35–47.

Corradini, E., Marconcini, J.M., Agnelli, J.A.M., Mattoso, L.H.C., 2011. Thermoplastic blends of corn gluten meal/starch (CGM/starch) and corn gluten meal/polyvinyl alcohol and corn gluten meal/poly (hydroxybutyrate-co-hydroxyvalerate) (CGM/PHB-V). Carbohydr. Polym. 83, 959–965.

Corre, Y.M., Bruzaud, S., Grohens, Y., 2013. Poly(3-hydroxybutyrate-co-3-hydroxyvalerate) and poly(propylene carbonate) blends: an efficient method to finely adjust properties of functional materials. Macromol. Mater. Eng. 298, 1176–1183.

De Koning, G.J.M., 1993. Prospects of Bacterial Poly[(R)-3-Hydroxyalkanoates]. Center for Polymers and Composites (CPC), Eindhoven University of Technology, Eindhoven.

Harding, K.G., Dennis, J.S., von Blottnitz, H., Harrison, S.T.L., 2007. Environmental analysis of plastic production processes: comparing petroleum-based polypropylene and polyethylene with biologically-based poly-β-hydroxybutyric acid using life cycle analysis. J. Biotechnol. 130, 57–66.

Hu, S.-G., Jou, C.-H., Yang, M.-C., 2003. Antibacterial and biodegradable properties of polyhydroxyalkanoates grafted with chitosan and chitooligosaccharides via ozone treatment. J. Appl. Polym. Sci. 88, 2797–2803.

Hu, S.-G., Jou, C.-H., Yang, M.-C., 2004. Biocompatibility and antibacterial activity of chitosan and collagen immobilized poly(3-hydroxybutyric acid-co-3-hydroxyvaleric acid). Carbohydr. Polym. 58, 173–179.

Ikejima, T., Inoue, Y., 2000. Crystallization behavior and environmental biodegradability of the blend films of poly(3-hydroxybutyric acid) with chitin and chitosan. Carbohydr. Polym. 41, 351–356.

Javadi, A., Kramschuster, A.J., Pilla, S., Lee, J., Gong, S., Turng, L., 2010. Processing and characterization of microcellular PHBV/PBAT blends. Polym. Eng. Sci. 50, 1440–1448.

Jiang, L., Huang, J., Qian, J., Chen, F., Zhang, J., Wolcott, M.P., Zhu, Y., 2008a. Study of poly(3-hydroxybutyrate-co-3-hydroxyvalerate) (PHBV)/bamboo pulp fiber composites: effects of nucleation agent and compatibilizer. J. Polym. Environ. 16, 83–93.

Jiang, L., Morelius, E., Zhang, J., Wolcott, M., Holbery, J., 2008b. Study of the poly(3-hydroxybutyrate-co-3-hydroxyvalerate)/cellulose nanowhisker composites prepared by solution casting and melt processing. J. Compos. Mater. 42 (24), 2629–2645. http://dx.doi.org/10.1177/0021998308096327.

Jiang, L., Chen, F., Qian, J., Huang, J., Wolcott, M.P., Liu, L., Zhang, J., 2010. Reinforcing and toughening effects of bamboo pulp fiber on poly(3-hydroxybutyrate-co-3 hydroxyvalerate) fiber composites. Ind. Eng. Chem. Res. 49, 572–577.

John, M.J., Thomas, S., 2008. Biofibres and biocomposites. Carbohydr. Polym. 71, 343–364.

Khan, M.A., Ali, K.M.I., Hinrichsen, G., Kopp, C., Kropke, S., 1999. Study on physical and mechanical properties of biopol-jute composite. Polym.-Plast. Technol. Eng. 38, 99–112. http://dx.doi.org/10.1080/03602559909351562.

Lee, S.Y., 1996. Bacterial polyhydroxyalkanoates. Biotechnol. Bioeng. 49 (1), 1–14.

Luo, S., Netravalli, A.N., 1999. Interfacial and mechanical properties of environment-friendly "green" composites made from pineapple fibers and poly(hydroxybutyrate-co-valerate) resin. J. Mater. Sci. 34, 3709–3719.

Ma, X., Cheng, G., 2010. Property of poly(hydroxybutyrate) complexed polymer with soy protein isolate using glycerol as a medium. e-Polymers, no. 076.

Madbouly, S.A., Schrader, J.A., Srinivasan, G., Liu, K., McCabe, K.G., Grewell, D., Graves, W.R., Kessler, M.R., 2014. Biodegradation behavior of bacterial-based polyhydroxyalkanoate (PHA) and DDGS composites. Green Chem. 16, 1911–1920. http://dx.doi.org/10.1039/C3GC41503A.

Madison, L.L., Huisman, G.W., 1999. Metabolic engineering of poly(3-hydroxyalkanoates): from DNA to plastic. Microbiol. Mol. Biol. Rev. 63, 21–53.

Matsuo, M., Sawatari, C., Iwai, Y., Ozaki, F., 1990. Effect of orientation distribution and crystallinity on the measurement by X-ray diffraction of the crystal lattice moduli of cellulose I and II. Macromolecules 23, 3266–3275.

Mohanty, A.K., Khan, M.A., Sahoo, S., Hinrichsen, G., 2000. Effect of chemical modification on the performance of biodegradable jute yarn-Biopol® composites. J. Mater. Sci. 35, 2589–2595.

Muzzarelli, R.A.A., 1977. Chitin. Pergamon, Oxford.

Parulekar, Y., Mohanty, A.K., 2007. Extruded biodegradable cast films from polyhydroxyalkanoate and thermoplastic starch blends: fabrication and characterization. Macromol. Mater. Eng. 292, 1218–1228.

Persico, P., Acierno, D., Carfagna, C., Cimino, F., 2011. Mechanical and thermal behaviour of ecofriendly composites reinforced by kenaf and caroà fibers. Int. J. Polym. Sci. 2011, 1–7. http://dx.doi.org/10.1155/2011/841812, Article ID 841812.

Qian, J., Zhu, L., Zhang, J., Whitehouse, R.S., 2007. Comparison of different nucleating agents on crystallization of poly(3-hydroxybutyrate-co-3-hydroxyvalerates). J. Polym. Sci. B Polym. Phys. 45, 1564–1577. http://dx.doi.org/10.1002/polb.

Raquez, J.M., Nabar, Y., Narayan, R., Dubois, P., 2008. Novel high-performance talc/poly[(butylene adipate)-co-terephthalate] hybrid materials. Macromol. Mater. Eng. 293, 310–320.

Reddy, C.S.K., Ghai, R., Rashmi, T., Kalia, V.C., 2003. Polyhydroxyalkanoates: an overview. Bioresour. Technol. 87, 137–146.

Reinsch, V., Kelley, S.S., 1997. Crystallization of poly (hydroxybutyrate-co-hydroxyvalerate) in wood fiber-reinforced composites. J. Appl. Polym. Sci. 64, 1785–1796.

Sakurada, I., Nukushina, Y., Ito, T., 1962. Experimental determination of the elastic modulus of crystalline regions in oriented polymers. J. Polym. Sci. 57, 651–660.

Sankar, D., Chennazhi, K.P., Nair, S.V., Jayakumar, R., 2012. Fabrication of chitin/poly(3-hydroxybutyrate-co-3-hydroxyvalerate) hydrogel scaffold. Carbohydr. Polym. 90, 725–729.

Santhanam, A., Sasidharan, S., 2010. Microbial production of polyhydroxy alkanotes (PHA) from *Alcaligens* spp. and *Pseudomonas oleovorans* using different carbon sources. Afr. J. Biotechnol. 9 (21), 3144–3150.

Shih, W.-J., Chen, Y.-H., Shih, C.-J., Hon, M.-H., Wang, M.-C., 2007. Structural and morphological studies on poly(3-hydroxybutyrate acid) (PHB)/chitosan drug releasing microspheres prepared by both single and double emulsion processes. J. Alloys Compd. 434–435, 826–829.

Singh, S., Mohanty, A.K., 2007. Wood fiber reinforced bacterial bioplastic composites: fabrication and performance evaluation. Compos. Sci. Technol. 67, 1753–1763.

Singh, S., Mohanty, A.K., Sugie, T., Takai, Y., Hamada, H., 2008. Renewable resource based biocomposites from natural fiber and polyhydroxybutyrate-co-valerate (PHBV) bioplastics. Compos. Part A 39, 875–886.

Stucova, A., Davies, G., Eichhorn, S.J., 2005. The elastic modulus and stress-transfer properties of tunicate cellulose whiskers. Biomacromolecules 6, 1055–1061.

Ten, E., Turtle, J., Bahr, D., Jiang, L., Wolcott, M.P., 2010. Thermal and mechanical properties of poly(3-hydroxybutyrate-*co*-3-hydroxyvalerate)/cellulose nanowhiskers composites. Polymer 51, 2652–2660.

Ten, E., Jiang, L., Wolcott, M.P., 2012a. Crystallization kinetics of Poly(3-hydroxybutyrate-*co*-3-hydroxyvalerate)/cellulose nanowhiskers composites. Carbohydr. Polym. 90, 541–550.

Ten, E., Jiang, L., Bahr, D.F., Li, B., Wolcott, M.P., 2012b. Effects of cellulose nanowhiskers on mechanical dielectric and rheological properties of poly(3-hydroxybutyrate-co-3-hydroxyvalerate) (PHBV)/cellulose nanowhiskers (CNW) composites. Ind. Eng. Chem. Res. 51, 2941–2951.

Ten, E., Jiang, L., Wolcott, M.P., 2013. Preparation and properties of aligned poly(3-hydroxybutyrate-*co*-3-hydroxyvalerate)/cellulose nanowhiskers composites. Carbohydr. Polym. 92, 206–213.

Tong, H.W., Zhang, X., Wang, M., 2012. A new nanofiber fabrication technique based on coaxial electrospinning. Mater. Lett. 66, 257–260.

Veleirinho, B., Ribeiro-do-Valle, R.M., Lopes-da-Silva, J.A., 2011. Processing conditions and characterization of novel electrospun poly(3-hydroxybutyrate-co-hydroxyvalerate)/chitosan blend fibers. Mater. Lett. 65, 2216–2219.

Veleirinho, B., Coelho, D.S., Dias, P.F., Maraschin, M., Ribeiro-do-Valle, R.M., Lopes-da-Silva, J.A., 2012. Nanofibrous poly(3-hydroxybutyrate-co-3-hydroxyvalerate)/chitosan scaffolds for skin regeneration. Int. J. Biol. Macromol. 51, 343–350.

Wang, J., Wang, Z., Li, J., Wang, B., Liu, J., Chen, P., Miao, M., Gu, Q., 2012. Chitin nanocrystals grafted with poly(3-hydroxybutyrate-co-3-hydroxyvalerate) and their effects on thermal behavior of PHBV. Carbohydr. Polym. 87, 784–789.

Wang, B., Li, J., Zhang, J., Li, H., Chen, P., Gu, Q., Wang, Z., 2013. Thermo-mechanical properties of the composite made of poly(3-hydroxybutyrate-co-3-hydroxyvalerate) and acetylated chitin nanocrystals. Carbohydr. Polym. 95, 100–106.

Wong, S., Shanks, R., Hodzic, A., 2002. Properties of poly(3-hydroxybutyric acid) composites with flax fibres modified by plasticiser absorption. Macromol. Mater. Eng. 287, 647–655.

Yu, H.-Y., Qin, Z.-Y., 2014. Surface grafting of cellulose nanocrystals with poly(3-hydroxybutyrate-co-3-hydroxyvalerate). Carbohydr. Polym. 101, 471–478.

Yu, D.-G., Lin, W.-C., Lin, C.-H., Yang, M.-C., 2006. Cytocompatibility and antibacterial activity of a PHBV membrane with surface-immobilized water-soluble chitosan and chondroitin-6-sulfate. Macromol. Biosci. 6, 348–357.

Yu, H.-Y., Qin, Z.-Y., Zhou, Z., 2011. Cellulose nanocrystals as green fillers to improve crystallization and hydrophilic property of poly(3-hydroxybutyrate-co-3-hydroxyvalerate). Prog. Nat. Sci.: Mater. Int. 21, 478–484.

Yu, H.-Y., Qin, Z.-Y., Liu, Y.-N., Chen, L., Liu, N., Zhou, Z., 2012. Simultaneous improvement of mechanical properties and thermal stability of bacterial polyester by cellulose nanocrystals. Carbohydr. Polym. 89, 971–978.

Yu, H.-Y., Qin, Z.-Y., Liu, L., Yang, X.-G., Zhou, Y., Yao, J.-M., 2013. Comparison of the reinforcing effects for cellulose nanocrystals obtained by sulfuric and hydrochloric acid hydrolysis on the mechanical and thermal properties of bacterial polyester. Compos. Sci. Technol. 87, 22–28.

Zarrinbakhsh, N., Misra, M., Mohanty, A.K., 2011. Biodegradable green composites from distiller's dried grains with solubles (DDGS) and a polyhydroxy(butyrate-*co*-valerate) (PHBV)-based bioplastic. Macromol. Mater. Eng. 296, 1035–1045.

Zhang, M., Thomas, N.L., 2010. Preparation and properties of polyhydroxybutyrate blended with different types of starch. J. Appl. Polym. Sci. 116, 688–694.

Mechanical performance of starch-based biocomposites

4

F. Xie, L. Avérous*[†]*, P.J. Halley*, P. Liu*[‡]

*The University of Queensland, Brisbane, QLD, Australia; [†]Université de Strasbourg, Strasbourg, France; [‡]Guangzhou University, Guangzhou, China

4.1 Introduction

The increasing environmental concerns and the expected shortage of fossil resources (mainly petroleum) have greatly stimulated the use of biobased compounds extracted from the biomass. Following this mainstream, starch, as a natural polymer directly extracted from renewable resources (plants), has experienced considerable development in the past decades in material applications (Avérous and Halley, 2014). Over many traditional synthetic polymers, starch has strong advantages such as renewability, low cost, wide availability, and total compostability without toxic residues. However, neat starch-based materials have low mechanical properties compared to most traditional synthetic polymers, which greatly limits its application. This has led to the strong development, up to an industrial level, of multiphase systems obtained by the associations of starch with various compounds, such as starch-based composites, which have shown enhanced mechanical properties compared with the neat matrix.

In the following sections, the fundamental knowledge of starch (including its structure, phase transition, and processing strategies) is first introduced, considering that starch is a biopolymer (i.e. polymer directly extracted from biomass) that is unique to other conventional polymers. Following that, there is general discussion of the mechanical properties of starch-based materials and macrobiocomposites. As nanocomposite (i.e. composite based on a nanofiller with dimensions of less than around 100 nm) has now been a strong focus in the polymer materials area (Alexandre and Dubois, 2000; Crosby and Lee, 2007; Paul and Robeson, 2008; Pavlidou and Papaspyrides, 2008; Sinha Ray and Bousmina, 2005; Sinha Ray and Okamoto, 2003; Spitalsky et al., 2010), this chapter subsequently focuses on starch-based nanobiocomposites. Thus, the various nanofillers for starch-based nanobiocomposites that has been reported in the literature are generally described, followed by a discussion of the mechanical properties of starch-based nanobiocomposites resulting from the use of these different nanofillers.

4.2 Structures of native starch

The starch granule provides the main way of storing carbon and energy over long periods in green plants (Pérez et al., 2009). Starch granules are mainly found in seeds, roots, and tubers, and display different shapes (spherical, lenticular, and polygonal) depending on their origins such as maize (corn), wheat, potato, and rice. An overview of starch granule structure is shown in Figure 4.1, which presents a concentric three-dimensional

Biocomposites: Design and Mechanical Performance. http://dx.doi.org/10.1016/B978-1-78242-373-7.00011-1

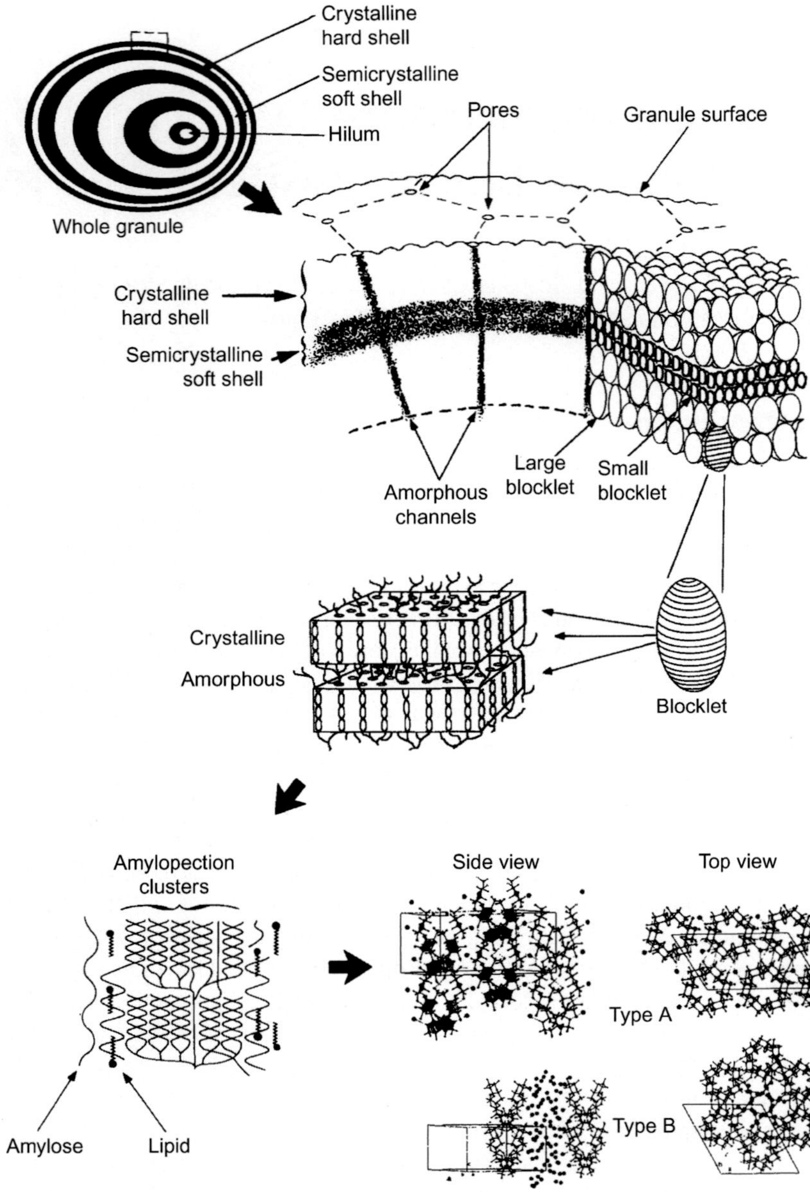

Figure 4.1 Overview of starch granule structure. At the lowest level of granule organization (upper left), alternating crystalline (hard) and semicrystalline (soft) shells are shown (dark and light colours, respectively). Shells are thinner toward the granule exterior (due to increasing surface area to be added to by constant growth rate) and the hilum is shown off centre. At a higher level of structure, the blocklet structure is shown in association with amorphous radial channels. Blocklet size is smaller in the semicrystalline shells than in the crystalline shells. At the next highest level of structure, one blocklet is shown containing several amorphous crystalline lamellae. In the next diagram, amylopectin is shown in the lamellae. The next image is a reminder of the importance of amylose–lipid (and protein) components in the organization of amylopectin chains. At the highest level of order, crystal structures of the starch polymers are shown. Reprinted from Gallant et al. (1997), Copyright (1997), with permission from Elsevier. Redrawn in Gallant et al. (1997) from Imberty et al. (1988) and Imberty and Perez (1988).

architecture from the hilum, with crystallinity varying from 15% to 45% depending on the botanical source (Zobel, 1988). The starch granule is well known to possess multi-level structures from macro- to molecular scales; that is, starch granules (<1 – 100 μm), alternating amorphous and semicrystalline shells (growth rings) (100 – 400 nm), crystalline and amorphous lamellae (periodicity) (9 – 10 nm), and macromolecular chains (~nm) (Buléon et al., 1998; Jane, 2009; Pérez et al., 2009; Pérez and Bertoft, 2010).

In molecular level of structure, starch is a polysaccharide consisting of D-glucose units, referred to as homoglucan or glucopyranose. Figure 4.2 presents the two major biomacromolecules of starch (i.e. amylose and amylopectin). Amylose is a sparsely branched carbohydrate mainly based on α(1–4) bonds with a molar mass of 10^5–10^6 and can have a degree of polymerisation (DP) as high as 600 (Pérez et al., 2009). The number of macromolecular configurations based on α(1–6) links is directly proportional to the amylose molar mass (Tako and Hizukuri, 2002). The chains show spiral-shaped single- or double-helices with a rotation on the α(1–4) link and with six glucoses per turn, where the hydroxyl groups are mainly located toward the exterior of the helices. As a result, the core of the helix is mainly hydrophobic. On the other hand, amylopectin is a highly multiple-branched polymer with a high molar mass of 10^7–10^9. It is one of the largest natural polymers on earth (Pérez et al., 2009). Amylopectin is based on α(1–4) (around 95%) and α(1–6) (around 5%) links, with constituting branching points localised every 22–70 glucose units, generating a highly branched structure with a lot of pending chains of DP ≈ 15, which are mainly responsible for the material crystallinity (between 20% and 45%). This specific structure has a profound effect on the physical and biological properties (Pérez et al., 2009; Pérez and Bertoft, 2010). Depending on the source, the amylose/amylopectin ratio of starch can be varied from <1% to 83 % (Tan et al., 2007).

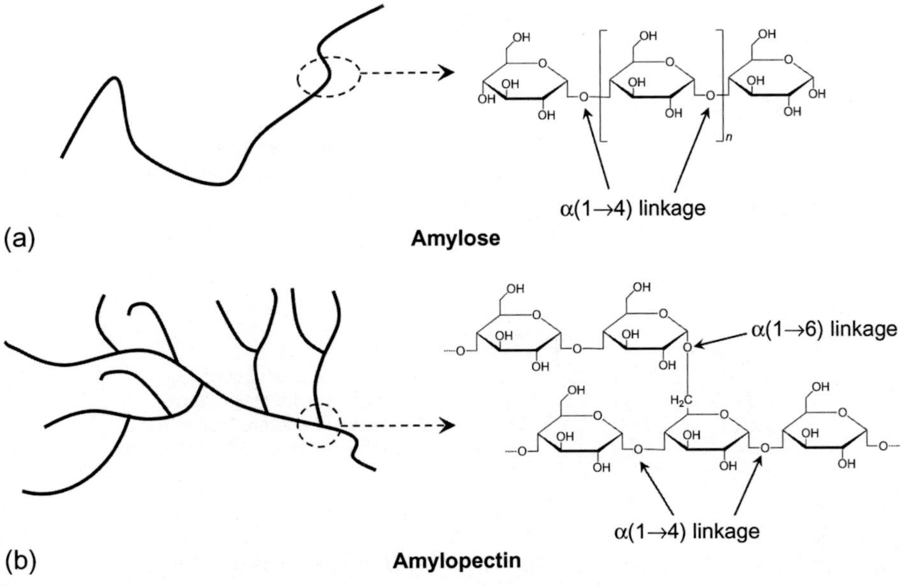

Figure 4.2 Chemical structures of amylose (a) and amylopectin (b) macromolecules of starch.

Besides, very small amounts of proteins, lipids, and phosphorus are also found in starch granules depending on the botanical resource (Jane, 2009; Pérez and Bertoft, 2010). These components can interact with the carbohydrate chains during processing (e.g. Maillard reaction) and then modify the behaviour of the starch-based materials.

4.3 From native starch to plasticised starch

In order to convert native starch into plasticised starch or thermoplastic starch (TPS), it is important to understand gelatinisation, which is the key process to transform starch into a processable material. This term describes the phase transition of starch from its original ordered granule structure into a disordered state, as native starch granules are heated in water (Atwell et al., 1988; Lelievre, 1974; Ratnayake et al., 2008). Gelatinisation is an irreversible process for starch that includes, in a broad sense and in time/temperature sequence, granule swelling, melting of native crystalline structure (loss of birefringence), and solubilisation of molecules (Russo et al., 2009), as shown in Figure 4.3. Full gelatinisation of starch under shearless condition requires excess

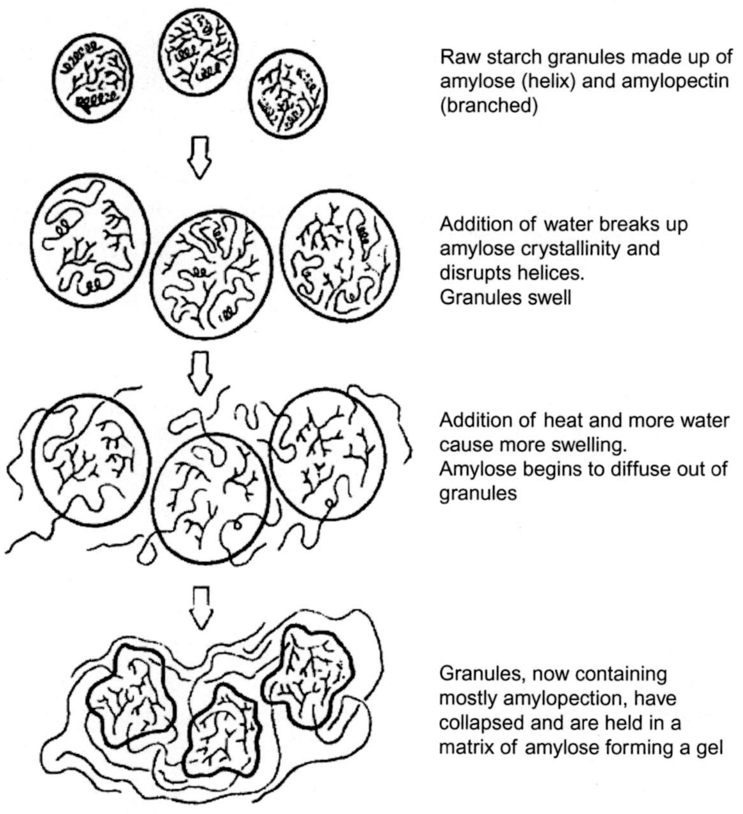

Figure 4.3 Mechanism of starch gelatinisation.
Copyright © 1978 Wiley Periodicals, Inc. Reproduced from Remsen and Clark (1978) with permission.

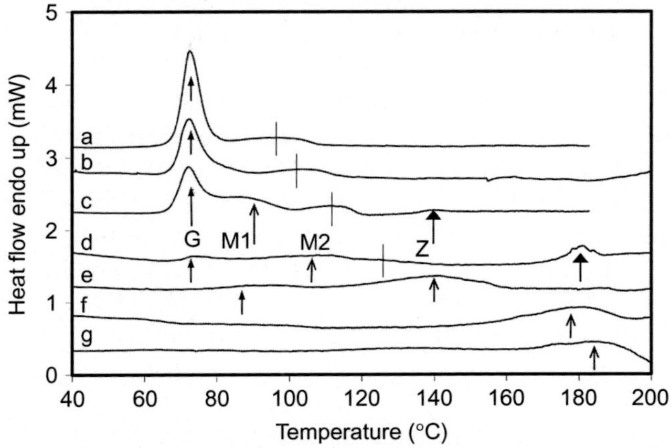

Figure 4.4 Thermal endotherms of regular maize starch with different water contents (a–g: 74.57%, 65.3%, 51.8%, 40%, 29.9%, 16.15%, and 9%) measured by differential scanning calorimetry.
Reprinted from Liu et al. (2006), Copyright 2006, with permission from Elsevier.

water (>63% for waxy maize starch, for example, Wang et al., 1991). While with abundant water the crystallites in starch might be pulled apart primarily by swelling by water, this swelling force will be much less significant if the water content is limited, and in this case, higher temperature is required to facilitate the mobility of starch molecules and the destructuration of the crystalline regions (Donovan, 1979; Lai and Kokini, 1991; Liu et al., 2005, 2006). In the area of starch-based materials, the process of gelatinisation with low water content could more accurately be defined as the "melting" of starch (Lai and Kokini, 1991). The gelatinisation/melting behaviours of regular maize starch with different moisture contents are shown in Figure 4.4.

Figure 4.4 shows that by reducing the moisture content, the melting temperature (T_m) of starch would progressively increase, and that of dry starch is often larger than its decomposition temperature (T_d), as extrapolated by Flory's law (Liu et al., 2008; Russell, 1987).

4.4 Processing for starch-based materials

To produce plasticised starch-based materials, gelatinisation/melting needs to occur. This can be achieved using conventional polymer processing techniques such as extrusion techniques and solution casting (Liu et al., 2009). In practical processing, water functions by lowering the T_m and plasticising the starch polymers, acting as a "plasticiser". However, the volatility of water could result in unstable processing or undesirable foaming (e.g. at extrusion exit). Besides, the final products based on starch containing only water usually have poor mechanical properties especially due to the brittleness, because its final temperature is usually lower than its glass transition temperature (T_g) (Chaudhary et al., 2011; Enrione et al., 2010; Forssell et al., 1997; Liu et al., 2010, 2011c;

Lourdin et al., 1997), and/or resulting from the densification (happening below T_g) or retrogradation (also recrystallisation, happening above T_g) (Atwell et al., 1988; Bulkin et al., 1987; Gudmundsson, 1994; Liu and Thompson, 1998). To overcome these issues, nonvolatile (at the processing temperature) plasticisers such as polyols (glycerol, sorbitol, polyethylene glycol, etc.) (Da Róz et al., 2006; Enrione et al., 2010; Forssell et al., 1997; Gaudin et al., 1999; Kirby et al., 1993; Lourdin et al., 1997; Mathew and Dufresne, 2002b; Qiao et al., 2011; Shi et al., 2007b; van Soest et al., 1994; Yan et al., 2012), nitrogen-based compounds (urea, ammonium derived, amines, etc.) (Dai et al., 2009a, 2010a,b; Ma and Yu, 2004a,b,c; Ma et al., 2004, 2005, 2006; Shogren et al., 1992; Yang et al., 2006a,b,c; Zhang et al., 2008; Zheng et al., 2009a), or citric acid (CA) (Shi et al., 2007a; Wang et al., 2007; Yu et al., 2005) are utilised.

4.5 Mechanical properties of starch-based materials

The mechanical properties of neat starch-based materials can be influenced by several factors: (a) polymer entanglement and interactions; (b) moisture content and plasticiser type and content; (c) crystallinity; and (d) T_g value. It has been reported that starch with higher amylose content has stronger mechanical properties (Li et al., 2011a), as shown in Table 4.1. However, Follain et al. (2005) have pointed out that both amylose and amylopectin cannot develop long-distance interactions (e.g. entanglements) to present good ultimate mechanical properties, and that starch-based systems are not cohesive enough. This suggests that the effect of amylose content on mechanical properties might be more appropriately associated with the easier and faster recrystallisation of amylose (Rindlav-Westling and Gatenholm, 2002). Besides, it is important to note the presence of three hydroxyl groups (one primary and two secondary) per glucose sugar unit of starch, which makes starch highly hydrophilic. This makes the moisture content in starch-based materials closely rely on the ambient relative humidity (RH), which leads to a large variation of physical properties (Follain et al., 2005). In addition, the moisture content and plasticiser content of starch can greatly affect the recrystallisation of starch-based materials (Forssell et al., 1999; Mathew and Dufresne, 2002b; van Soest and Borger, 1997; van Soest and Knooren, 1997). In particular, Mathew and Dufresne (2002b) found that the crystallisation of starch is favoured by high molecular weight plasticisers in dry atmosphere and by low molecular weight plasticisers in moist atmosphere. And T_g can be influenced by the type (molecular weight) and content of plasticiser (Lourdin et al., 1997; Mathew and Dufresne, 2002b). The moisture content, the plasticiser type and content, the crystallinity, and T_g all greatly impact on mechanical properties in a complex way.

Blending starch with another polymer might be a good way to improve the mechanical properties. Starch is often blended with other polymers (mainly biodegradable) such as poly(lactic acid) (PLA), polycaprolactone (PCL), poly(butylene succinate adipate) (PBSA), poly(butylene adipate-co-terephthalate), and poly(vinyl alcohol) (PVA), as extensively reviewed in several papers (Avérous, 2004; Wang et al., 2003; Yu et al., 2006). However, it is noteworthy that the mechanical properties of the blends largely depend on the interfacial compatibility between the phases. Starch normally

Table 4.1 Mechanical properties of extruded starch films

Sample	Thickness (mm)	Tensile testing			Impact testing		
		Young's modulus (MPa)	Tensile strength (MPa)	Elongation at break (%)	Peak load (kN)	Deflection at peak load (mm)	Energy to peak load (J)
Waxy	0.23 ± 0.02	730.7 ± 120.0	12.68 ± 2.41	2.28 ± 0.38	0.0573 ± 0.0085	2.76 ± 0.15	0.088 ± 0.012
Maize	0.23 ± 0.02	866.6 ± 124.9	15.66 ± 2.14	4.23 ± 0.35	0.1427 ± 0.0015	6.95 ± 0.13	0.515 ± 0.010
Gelose 50	0.23 ± 0.04	1139.9 ± 382.6	20.22 ± 1.65	4.67 ± 0.95	0.3886 ± 0.0077	11.05 ± 0.22	2.201 ± 0.044
Gelose 80	0.23 ± 0.04	1217.2 ± 334.7	23.82 ± 1.85	4.95 ± 0.80	0.4148 ± 0.0082	11.52 ± 0.23	2.262 ± 0.045

Reprinted from Li et al. (2011a), Copyright 2011, with permission from Elsevier.

is not compatible with other polymers, resulting in heterogeneous multiphase systems displaying poor mechanical properties, equal to or even weaker than the neat starch reference (Follain et al., 2005). So it is important to choose polymers that are relatively compatible with starch (e.g. PVA) to realise enhancement of mechanical properties (Follain et al., 2005). Otherwise, the use of a compatibiliser is useful. For example, for starch blended with a polyester (PLA, PCL, or PBSA), methylenediphenyl diisocyanate (MDI) could reactively compatibilise the two components (Xie et al., 2006, 2007). Nonetheless, the toxicity of many effective compatibilisers like MDI has forced people to seek more friendly strategies. Huneault and Li (2007) claimed that the interface of starch and PLA could be improved by using PLA grafted with maleic anhydride (MA) for the interfacial modification. Besides, it has been demonstrated very recently that chemically modified plant oils (MA-grafted tung oils (Xiong et al., 2013a) and epoxidised soybean oil (Xiong et al., 2013b)) are also effective in increasing the compatibility between starch (MA-grafted) and PLA.

4.6 Mechanical properties of starch-based macrobiocomposites

Starch biocomposites can be produced with the reinforcement by cellulose fibres (Alvarez et al., 2004; Averous, 2007; Avérous, 2002; Averous and Boquillon, 2004; Avérous et al., 2001b; Belhassen et al., 2009; Carvalho et al., 2003; Curvelo et al., 2001; Dufresne et al., 2000; Funke et al., 1998; Gironès et al., 2012; Lawton et al., 2004; Matsui et al., 2004; Romhány et al., 2003; Wollerdorfer and Bader, 1998) and mineral fillers such as kaolinite (de Carvalho et al., 2001), halloysite (He et al., 2012; Xie et al., 2011), and organomodified hectorite (Bentone 109) (Chen and Evans, 2005). The incorporation of these fillers with macrosize can normally result in improved mechanical performance compared to the neat matrix (Avérous, 2004; Avérous and Halley, 2009; Kalambur and Rizvi, 2006; Tang and Alavi, 2011; Wang et al., 2003; Yu et al., 2006)—but the improvement is quite limited compared with that from nanocomposites based on nanosized fillers (see the following sections). In some studies, the improvement has been linked to a classical reinforcing effect, and most of the results follow a simple mixing law (Averous, 2007; Avérous et al., 2001a,b; Curvelo et al., 2001). Some others are linked to the strong interactions between the matrix and the filler (Averous, 2007; Averous and Boquillon, 2004; Avérous et al., 2001b). It has also been shown that cellulose fibres may also facilitate recrystallisation of starch matrix, accounting for improved mechanical properties (Müller et al., 2009).

4.7 Nanofillers for starch-based nanobiocomposites

4.7.1 General description

During this last decade, along with the exponential momentum of the development in polymer nanocomposites (Alexandre and Dubois, 2000; Crosby and Lee, 2007; Paul and Robeson, 2008; Pavlidou and Papaspyrides, 2008; Sinha Ray and Bousmina,

2005; Sinha Ray and Okamoto, 2003; Spitalsky et al., 2010), much attention has been focused on the use of nanosized fillers in producing starch-based nanobiocomposite materials. When the dispersed fillers come in nanoscale, conventional polymer composites become nanocomposites that are expected to exhibit drastic modification in their properties because the global materials properties are mainly driven by the huge filler–matrix interface (2-D surface) or interphase (3-D volume), including further improvement in the mechanical properties (Alexandre and Dubois, 2000).

Various nanofillers have been examined with plasticised starch, including phyllosilicates, polysaccharide nanofillers, carbonaceous nanofillers, and many more. They have different geometry (size and shape) and surface chemistry. Regarding their shapes, three distinct types of nanofillers can be observed: nanoparticles, nanotubes, and nanolayers (Kumar et al., 2009). Nevertheless, the term "nanoparticles" is also frequently used in a broad sense to describe a nanofiller regardless of its shape.

The different nanofillers that have been examined to be used in starch-based nanobiocomposites are summarised in Table 4.2. They have different geometry (size and shape) and surface chemistry; therefore, different kinds of interactions between the nanofiller and the starch matrix and variation in the reinforcing capability are expected.

Among different nanofillers for starch, phyllosilicates (layered silicates), especially montmorillonite (MMT), are most frequently utilised due to their appealing advantages such as wide availability, low cost, and low toxicity. They have different structure, texture, and/or morphology (interestingly, some phyllosilicates such as sepiolite and halloysite do not display a normal layered structure). Cationic exchange is the most common technique for chemical modification of the phyllosilicate surface, which increases the interlayer spacing (d_{001}) (Lagaly, 1986) and the compatibility with the matrix, through some specific chemical groups brought by the organomodifier. By this technique, various organomodified MMTs (OMMTs) and hectorites used for starch-based nanobiocomposites (see Table 4.3) can be produced, which mainly differ in the countercation nature and the cationic exchange capacity (CEC). Most conventional organomodifiers increase the hydrophobicity of the phyllosilicate, resulting in reduced compatibility with the hydrophilic starch matrix. This is one of the key points to consider for developing starch-based nanobiocomposites.

Polysaccharide nanofiller represent the second popular group of nanofiller for starch due to their similar chemical structure to starch. Normally, they can be produced from cellulose (Angles and Dufresne, 2000; Azizi Samir et al., 2004, 2005; Bondeson et al., 2006; Cao et al., 2008a,b; Chen et al., 2009a,b; Dong et al., 1996; Lu et al., 2006; Mathew and Dufresne, 2002a; Teixeira et al., 2009), starch (Angellier et al., 2004; García et al., 2009; Putaux et al., 2003; Viguié et al., 2007; Zeng et al., 2011), and chitin (Gopalan Nair et al., 2003; Morin and Dufresne, 2002; Paillet and Dufresne, 2001; Saito et al., 1997; Sriupayo et al., 2005) through acid hydrolysis with a concentrated mineral acid (typically sulphuric acid) under strictly controlled conditions of time and temperature. While other methods for producing polysaccharide nanofillers have also been reported (Xie et al., 2013), acid action can result in a decrease of the amorphous parts and, thus, material with high crystallinity is obtained (and, thus, high E is expected (Kumar et al., 2009)), which is beneficial to mechanical property improvement of the resulting nanobiocomposites.

Table 4.2 **Overview of common nanofillers used for starch-based nanocomposites**

Category/source	Nanofiller	Dimensional type	Remark
Phyllosilicates			
Smectite group	Montmorillonite	Nanolayer	Most expandable; most frequently used; natural sodium montmorillonite being a preferable choice due to the matching of polarity with starch
	Hectorite	Nanolayer	Only natural hectorite expandable
Sepiolite group	Sepiolite	Nanolayer (actually in needle shape)	–
Synthetic clays	Somasif™ ME 100 fluorohectorite/fluoromica	Nanolayer	Montmorillonite- or hectorite-type synthetic clay
	Laponite® B, Laponite® RD	Nanolayer	Hectorite-type synthetic clay
Polysaccharides			
Cellulose	Nanowhiskers/nanofibrils/nanofibres from plant cellulose (flax, hemp, ramie, wood, pea hull, cassava bagasse, wheat straw) and microcrystalline cellulose (commercially available)	Nanotube/ nanoparticle	Tend to form aggregates; size, shape, and crystallinity largely dependent on the preparation method and conditions
	Nanorods from tunicin (animal cellulose)	Nanotube	
	Nanofibrils/nanofibres from bacterial cellulose	Nanotube	High crystallinity
Starch	Nanoparticles/nanocrystals from starch (waxy maize and regular maize)	Nanoparticle	Tend to form aggregates
Chitin	Nanoparticles from chitin	Nanoparticle	
Chitosan	Nanoparticles from chitosan	Nanoparticle	

Carbonaceous materials			
	Carbon nanotubes	Nanotube	–
	Graphite oxide	Nanolayer	–
	Carbon black	Nanoparticle	–
Metalloid oxides			
	Silicon dioxide (also silica) (SiO_2)	Nanoparticle	–
	Antimony trioxide (Sb_2O_3)	Nanoparticles	–
Metal oxides and chalcogenides			
	Zinc oxide (ZnO)	Nanoparticle	–
	Hydrous zirconium dioxide (also zirconia) ($ZrO_2 \cdot nH_2O$)	Nanoparticle	–
	Titanium dioxide (also titania) (TiO_2)	Nanoparticle	–
	Cadmium sulphide (CdS)	Nanoparticle	–
	Cadmium selenide (CdSe)	Nanoparticle	–
Metal phosphates			
	α-Zirconium phosphate ($Zr(HPO_4) \cdot H_2O$)	Nanolayer	–

Somasif™ is a trademark of CBC Co., Ltd. Japan; Laponite® is a registered trademark of Laporte Ind. Ltd. (SCP).

Table 4.3 Unmodified and organomodified smectite group clays used for starch-based nanocomposites with the corresponding chemical structures of counterions and their commercial trade names

Clay type	Countercation	Name
Montmorillonite (MMT)	Na^+	Natural sodium MMT; MMT–Na^+; Cloisite® Na^+; Dellite® LVF; Dellite® HPS; Nanofil® 757; BH Natural
	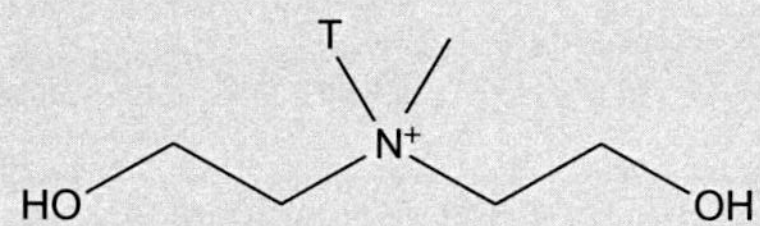 Methyl-tallow-bis-2-hydroxyethyl ammonium	Cloisite® 30B
	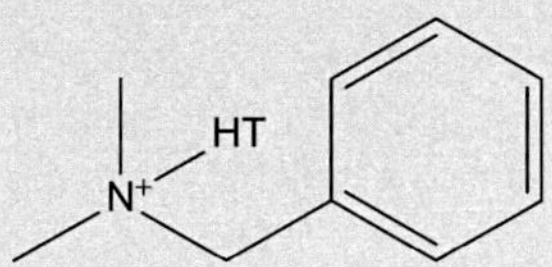 Dimethyl-benzyl-hydrogenated tallow ammonium	Cloisite® 10A; Bentone® 111; Dellite® 43B
	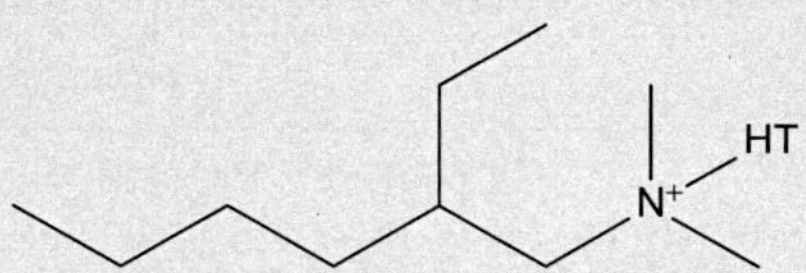 Dimethyl-hydrogenated tallow-2-ethylhexyl ammonium	Cloisite® 25A
	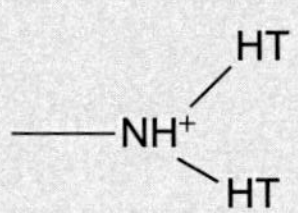 Methyl-dihydrogenated tallow ammonium	Cloisite® 93A

	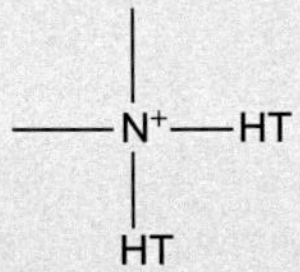 Dimethyl-dihydrogenated-tallow ammonium	Cloisite® 20A; Cloisite® 15A; Cloisite® 6A; Dellite® 67G; Dellite® 72T
	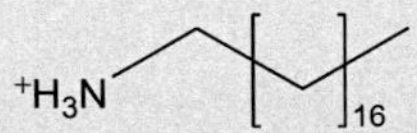 Octadecyl ammonium	Nanomer® I.30E
	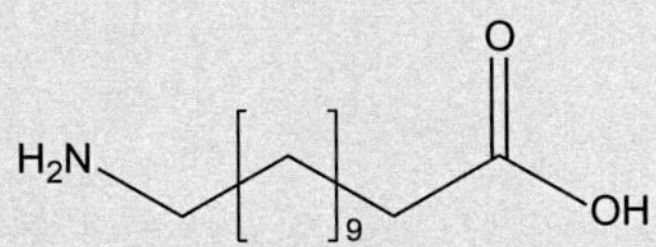 Aminododecanoic acid	Nanofil® 784
	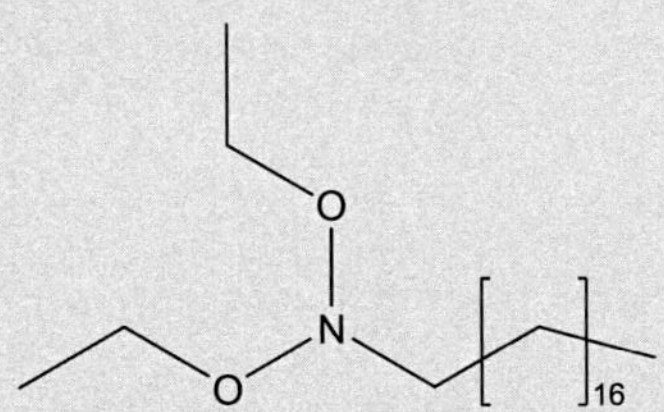 Stearyl dihydroxyethyl ammonium	Nanofil® 804
	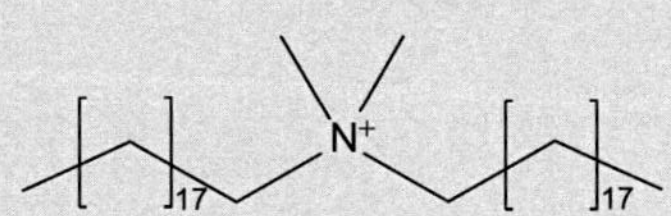 Distearyl dimethyl ammonium chloride	Nanofil® 948

Continued

Table 4.3 **Continued**

Clay type	Countercation	Name
	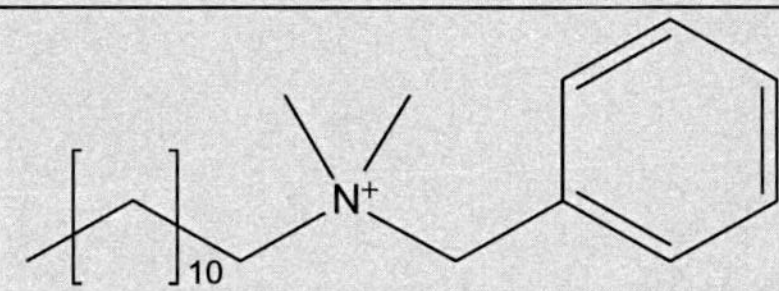 Dodecyl benzyl dimethyl ammonium bromide	–
	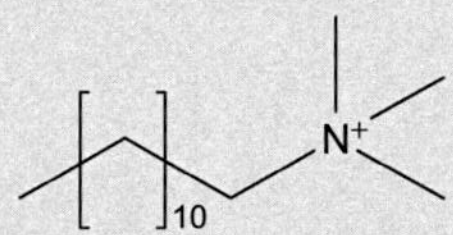 Dodecyl trimethyl ammonium	–
	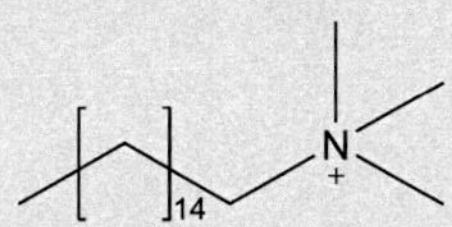 Hexodecyl trimethyl ammonium	–
	Cationic starch	–
	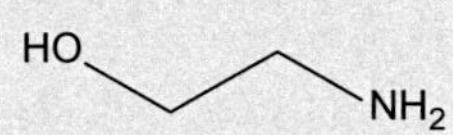 Ethanolamine	–

	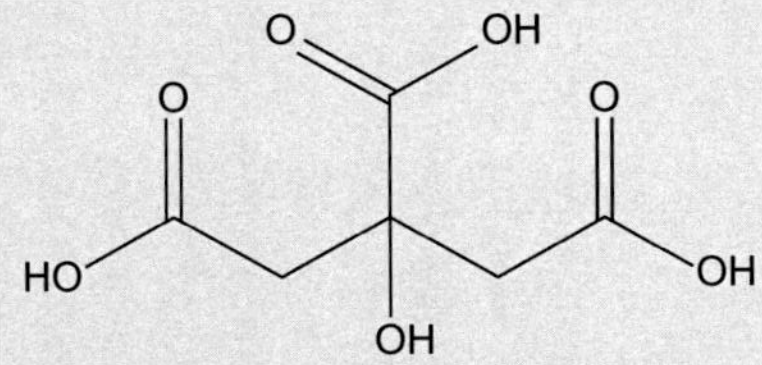 Citric acid	–
Hectorite	Ca^{2+}	Natural calcium hectorite; Bentone EA-163
	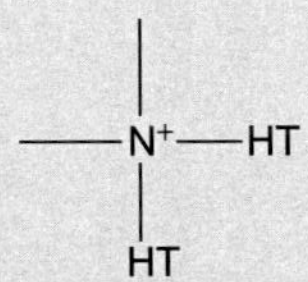 Dimethyl-dihydrogenated-tallow ammonium	Bentone 109

Cloisite® is a trademark of South Clay Products, Inc. (USA); Nanomer® is a trademark of Nanocor, Inc.; Dellite® is a trademark of Laviosa Chimica Mineraria, S.p.A. (Italy); Nanofil® is a trademark of Süd Chemie AG (Germany); BH Natural is a product from Blackhill Bentonite LLC (USA); Bentone® is a trademark of Elementis Specialties (USA); Nanomer® is a trademark of Nanocor, Inc. (USA).

The surface hydrophobicity of Cloisite clays: $Na^+ < 30B < 10A < 25A < 93A < 20A < 15A$.

T = tallow (≈65% C18, ≈30% C16, ≈5% C14), HT = hydrogenated tallow.

4.7.2 Key factors determining the nanofiller efficacy

Theoretically, when the filler for starch-based materials goes from macro- to nanoscale, the material becomes nanocomposites. The nanosize of nanofillers can generate huge areas at the matrix–filler interface, which largely control the global macrostructures and associated global properties of the corresponding materials.

However, whether a nanocomposite is realised not only relies on the size of the nanofiller, but more often depends on the dispersion state of the nanofiller (Schaefer and Justice, 2007). In particular, of layer-type nanofillers such as phyllosilicates, the dispersion state in a polymer determines the structure of the resulting composite, which can be either phase-separated composite (microcomposite), intercalated nanocomposite, or exfoliated nanocomposite (Alexandre and Dubois, 2000; Avérous and Halley, 2009; Paul and Robeson, 2008; Sinha Ray and Bousmina, 2005; Sinha Ray et al., 2003; Sinha Ray and Okamoto, 2003).

In order to achieve nanodispersion, the nanofiller and the matrix need to be chemically compatible. In particular for starch-based nanocomposites, the nanofiller needs to match the hydrophilicity of, and to achieve strong affinity with, starch for easier dispersion in the starch matrix. Regarding the compatibility between the two phases, one of the advantages in using polysaccharide nanofiller is the similar polysaccharide chemical structure of the nanofiller and the matrix, which benefits the nanofiller–matrix interactions. While natural phyllosilicates may be relatively hydrophilic to interact with starch, the layer spacing (d_{001}) is small and less easily penetrated by starch molecules. As mentioned above, while organomodification of phyllosilicates may increase d_{001} potentially for better dispersion (which is a common method for conventional polymers), the increased hydrophobicity could also reduce the compatibility between the nanofiller and starch, making the dispersion even worse. This will be further discussed later.

4.7.3 Strategies for enhancing the nanofiller efficacy

The enhancement of properties of nanocomposites depends on the nature, geometry, surface area, and surface chemistry of the nanofiller (Sinha Ray and Okamoto, 2003). In order to increase the nanofiller–starch compatibility, mainly two strategies have been used, including (a) surface modification of nanofillers, and (b) in-situ synthesis of nanofillers in a polysaccharide matrix.

Nanofiller modification can occur in diverse ways. Chemical modification is a typical way. For example, graphite oxide (GO) instead of graphite is more suitable as it has various oxygen functional groups including hydroxyls, epoxides, carbonyls, and carboxyls, which makes it hydrophilic and can form strong interactions and bondings with starch (Li et al., 2011b). As another example, to obtain strong interactions with starch, carbon nanotubes (CNTs) can be treated by carboxylation with strong acids (e.g. sulphuric acid and nitric acid) (Cao et al., 2007; Liu et al., 2011d), or be grafted with polysaccharides such as chitosan (Wu et al., 2007), cellulose acetate (Ke, 2010), and starch (Yan et al., 2011). However, it is noteworthy that covalent sidewall functionalisation usually destroys the extended networks on the CNT surfaces, diminishing their mechanical properties (Famá et al., 2011).

CNTs can also be modified in noncovalent ways, although this may result in less effective interactions with starch. This involves treatment by a surfactant; for example, sodium dodecyl sulphate (Ma et al., 2008a) and sodium dodecyl benzene sulphonate (Bonnet et al., 2007). In addition, wrapping of CNTs by an aqueous solution of a starch–iodine complex (Famá et al., 2011) is also interesting.

The strategy of in-situ synthesis in a polysaccharide is mainly for nanoparticles of metalloid oxides, metal oxides, and chalcogenides, which involve two methods. In one method, the nanoparticles, such as ZnO (Ma et al., 2009; Yu et al., 2009) or Sb_2O_3 (Chang et al., 2009; Zheng et al., 2009b), are first synthesised with a stabilising template, for example, native starch (Rodriguez et al., 2008), soluble starch (Chairam et al., 2009; Li et al., 2007; Ma et al., 2009; Radhakrishnan et al., 2007; Vigneshwaran et al., 2006; Wei et al., 2004), or carboxymethyl cellulose (Chang et al., 2009; Liu et al., 2011a; Yu et al., 2009; Zheng et al., 2009b), and then the encapsulated nanoparticles are incorporated into a starch dispersion. In another method, the nanoparticles (SiO_2 (Tang et al., 2008a; Yao et al., 2011), TiO_2 (Liao and Wu, 2008), CdS (Radhakrishnan et al., 2007), and CdSe (Zhao et al., 2008)) are directly synthesised in a starch dispersion, which acts not only as the stabiliser but also as the matrix (sol–gel method).

4.8 Mechanical properties of starch-based nanobiocomposites reinforced by phyllosilicates

4.8.1 Effect of phyllosilicate addition

Starch-based systems reinforced by phyllosilicates normally exhibit increases in tensile strength (σ) (Chivrac et al., 2008a, 2010a; Chung et al., 2010; Cyras et al., 2008; Dai et al., 2009b; de Carvalho et al., 2001; Gao et al., 2012; Hassan Nejad et al., 2011; He et al., 2012; Huang et al., 2004, 2005a,b, 2006; Huang and Yu, 2006; Ibrahim, 2011; Kvien et al., 2007; Lilichenko et al., 2008; Ma et al., 2007; Majdzadeh-Ardakani et al., 2010; Maksimov et al., 2009; Mondragón et al., 2008; Müller et al., 2011; Park et al., 2002, 2003; Qiao et al., 2005; Ren et al., 2009; Tang et al., 2008b,c; Wang et al., 2009, 2010; Xie et al., 2011), Young's modulus (E) (Chaudhary, 2008; Chen and Evans, 2005; Chivrac et al., 2009, 2010a,b; Chung et al., 2010; Cyras et al., 2008; de Carvalho et al., 2001; Hassan Nejad et al., 2011; He et al., 2012; Huang et al., 2004, 2005a,b; Huang and Yu, 2006; Lilichenko et al., 2008; Ma et al., 2007; Majdzadeh-Ardakani et al., 2010; Maksimov et al., 2009; Mondragón et al., 2008; Müller et al., 2011; Nejad et al., 2010; Pandey and Singh, 2005; Ren et al., 2009; Wang et al., 2010; Wilhelm et al., 2003b), and storage modulus (E') (measured by dynamic mechanical analysis) (Kvien et al., 2007; Park et al., 2003; Wilhelm et al., 2003a,b), even though the elongation at break (ε_b) was observed to be reduced in most studies (Chivrac et al., 2008a, 2010b; Chung et al., 2010; Cyras et al., 2008; Dai et al., 2009b; de Carvalho et al., 2001; Hassan Nejad et al., 2011; Huang et al., 2004, 2005a,b; Ibrahim, 2011; Ma et al., 2007; Majdzadeh-Ardakani et al., 2010; Maksimov et al., 2009; Pandey and Singh, 2005; Park et al., 2002, 2003; Qiao et al., 2005; Ren et al.,

2009; Tang et al., 2008b,c; Wang et al., 2009; Wilhelm et al., 2003b). The effectiveness of phyllosilicates relies on structural reasons such as (a) the homogeneous dispersion of silicate layers in the starch matrix (Chivrac et al., 2010a; Chung et al., 2010; de Carvalho et al., 2001; Huang et al., 2005a, 2006; Huang and Yu, 2006; Ibrahim, 2011; Kampeerapappun et al., 2007; Kvien et al., 2007; Ma et al., 2007; Majdzadeh-Ardakani et al., 2010; Nejad et al., 2010; Wang et al., 2009, 2010; Zhang et al., 2007), (b) the strong interactions (typically, hydrogen bonding) between the nanofiller and matrix (Chivrac et al., 2008a, 2010a; Cyras et al., 2008; Dai et al., 2009b; Huang et al., 2004; Ma et al., 2007; Majdzadeh-Ardakani et al., 2010; Park et al., 2003; Ren et al., 2009; Tang et al., 2008c; Wang et al., 2009, 2010), and (c) the high aspect (width-to-thickness) ratio and, thus, the vast exposed surface of the nanofiller (Cyras et al., 2008; Huang and Yu, 2006; Ibrahim, 2011). In particular, increases in E, E', and σ can be ascribed to (a) a good stress transfer between the matrix and the nanofiller (Chivrac et al., 2008b; Chung et al., 2010; Tang et al., 2008c), (b) the formation of a physical cross-linking network (Park et al., 2003; Ren et al., 2009), and (c) the stretching resistance of the oriented backbones of the chains in the interlayer spaces (Cyras et al., 2008). Besides, a decrease in ε_b can be attributed to the decreased flexibility of the starch matrix (Cyras et al., 2008; Ren et al., 2009; Tang et al., 2008c).

In spite of the general trends mentioned above, discrepancies have also existed in certain cases over ε_b (Chivrac et al., 2010a; Chung et al., 2010; Huang and Yu, 2006; Huang et al., 2006; Kvien et al., 2007; Majdzadeh-Ardakani et al., 2010; Pandey and Singh, 2005) and breaking energy (Chivrac et al., 2010b; Huang et al., 2004, 2005a,b; Huang and Yu, 2006; Ma et al., 2007; Wang et al., 2010). To address these, it might be useful to consider the crystalline structures in the materials as influenced by the nanofillers. Crystallinity and/or crystal size could affect the mechanical properties (Chivrac et al., 2010a,b; Chung et al., 2010; Majdzadeh-Ardakani et al., 2010). Most typically, higher crystallinity could embrittle materials with lower ε_b values (Chivrac et al., 2010a,b; Majdzadeh-Ardakani et al., 2010). Nevertheless, the literature shows no consistent trend of recrystallisation as influenced by the phyllosilicate addition restrained (Chung et al., 2010; Huang et al., 2004, 2005a; Liu et al., 2011b; Magalhães and Andrade, 2009, 2010; Ren et al., 2009; Wang et al., 2009; Zhang et al., 2007), unchanged (Zeppa et al., 2009), or enhanced (Chaudhary, 2008; Müller et al., 2011).

4.8.2 Effects of phyllosilicate type and content

Table 4.3 highlights various natural and OMMT nanofillers that have been used in starch-based nanobiocomposites. It is important to note that the surface hydrophobicity of Cloisite MMT nanofillers follows the following order: $Na^+ < 30B < 10A < 25A < 93A < 20A < 15A < 6A$. Although OMMT could provide higher d_{001} compared to natural sodium MMT (MMT–Na^+, or Cloisite Na^+) to possibly facilitate starch chains intercalation, the structure of the resulting composites has been demonstrated to be more dependent on the hydrophilicity of MMT (Zhang et al., 2007). It has been shown that incorporation of hydrophobic OMMT nanofillers such as Cloisite 15A, 6A, and 10A, Nanomer I.30E, and others, led to the formation of microcomposites (Chiou et al., 2005; Park et al., 2002; Tang et al., 2008c), as evidenced by the unchanged d_{001}. With Cloisite 30B, a more hydrophilic OMMT, higher d_{001} values were obtained, corresponding to higher dispersion and lower aggregation state (Chen and Evans, 2005; Chiou et al., 2005, 2006; Park et al., 2003). Exfoliated nanocomposites have also been produced with MMT–Na^+ due to the more hydrophilic character, which makes it more compatible with plasticised starch (Chen and Evans, 2005; Chen et al., 2005; Chiou et al., 2005, 2006, 2007; Cyras et al., 2008; Huang et al., 2004; Pandey and Singh, 2005; Park et al., 2002, 2003; Tang et al., 2008c; Zeppa et al., 2009; Zhang et al., 2007). Uniform dispersion of the MMT in the plasticised starch can be achieved in this case due to the polar interactions, especially hydrogen bonds formed between the –OH groups of the MMT and of the starch molecules (Huang et al., 2004; Park et al., 2002, 2003; Tang et al., 2008c).

The high compatibility of MMT–Na^+ with a starch matrix and the corresponding good dispersion could result in improved properties compared with those of other starch–OMMT hybrids. Typically, the tensile properties of starch-based biocomposites filled by Cloisite Na^+, 30B, 10A, and 6A examined by Park et al. (2002) are shown in Table 4.4.

In certain cases, MMT with proper organomodification may still display better reinforcing ability than MMT–Na^+. Qiao et al. (2005) showed that incorporation of OMMT based on trimethyl dodecyl ammonium into acetylated starch could result in a nanobiocomposite with higher σ and E' than the sample with MMT–Na^+. This was attributed to the higher dispersion and d_{001} of the OMMT in the acetylated starch matrix (Qiao et al., 2005). Chivrac et al. (2008a, 2009, 2010b,c,d) used cationic starch (CS) as a new organomodifier to better match the polarity of a starch matrix. Remarkably, morphological analyses showed that MMT organomodified by CS (OMMT–CS) allowed preparation of well-exfoliated nanobiocomposites, compared to natural MMT–Na^+, which only led to the formation of an intercalated structure (Chivrac et al., 2008a). Consequently, the OMMT–CS could lead to greater stiffness preserving the ε_b value, although with an increase in the OMMT–CS content a slight increase in ε_b was noticed (Chivrac et al., 2008a, 2009).

Although Tang et al. (2008c) reported that the MMT content did not have any significant impact on the occurrence of intercalation or exfoliation, it is worth noting that there can be an optimised level of MMT addition for the greatest improvement in the properties such as mechanical properties (Huang and Yu, 2006; Huang et al., 2006;

Table 4.4 Tensile properties of starch–clay nanocomposites as affected by the clay type and content

Clay type	Tensile strength (MPa)	Elongation at break ε_b (%)
Cloisite Na^+	2.79	48.9
Cloisite 30B	3.32	57.2
Cloisite 10A	3.20	52.0
Cloisite 6A	2.75	45.7
Blank (0%)	2.80	44.5

Kampeerapappun et al., 2007; Mondragón et al., 2008; Park et al., 2003; Tang et al., 2008c). A higher content of MMT might contribute to aggregates and stacks of MMT in a starch matrix (Chung et al., 2010; Huang et al., 2004; Kampeerapappun et al., 2007; Majdzadeh-Ardakani et al., 2010; Park et al., 2003) and also to lower plasticisation of the starch phase (Wang et al., 2009).

Other clays (natural or synthetic) such as hectorite (Chen and Evans, 2005; Kvien et al., 2007; Wilhelm et al., 2003a,b) and kaolinite (Chen and Evans, 2005; de Carvalho et al., 2001; Wilhelm et al., 2003a), and halloysite (He et al., 2012; Xie et al., 2011) have also been experimented into starch-based materials. However, due to the nonswelling nature of kaolinite (de Carvalho et al., 2001) and halloysite (Xie et al., 2011), only microcomposites can be formed from them and variations in the properties can be very limited. Natural hectorite might perform better than kaolinite for the formation of a nanocomposite structure (as shown by the increased d_{001}) (Chen and Evans, 2005; Wilhelm et al., 2003a); nevertheless, organomodified hectorite (Bentone 109) could only result in conventional composite just like kaolinite (Chen and Evans, 2005). Again, this can be explained by the dominant role of the phyllosilicate hydrophilicity in determining the biocomposite structure. In addition, a larger CEC could contribute to a stronger water retention property, which also influences the mechanical properties (Dean et al., 2007).

Chivrac et al. (2010a) initiated the use of sepiolite in starch-based materials, and found its reinforcing effect was even better than that of MMT, as evidenced by higher E, ε_b, and breaking tensile strength (σ_b), which was ascribed to the stronger nanofiller–matrix interactions.

4.8.3 Nanobiocomposites based on starch blends

In some studies (Arroyo et al., 2010; Avella et al., 2005; Bocchini et al., 2010; Dean et al., 2008; Majdzadeh-Ardakani and Nazari, 2010; McGlashan and Halley, 2003), the effects of different factors such as the second polymer content and the nanofiller type and content on the properties of nanocomposites were systematically investigated. These nanocomposites are expected to show improved characteristics (increases in the E' (Bocchini et al., 2010; Vertuccio et al., 2009), E (Arroyo et al., 2010; Avella et al., 2005; Dean et al., 2008, 2011; Ikeo et al., 2006; McGlashan and Halley, 2003; Mondragón et al., 2009; Pérez et al., 2007), and σ (Arroyo et al., 2010; Avella et al., 2005; Dean et al., 2008, 2011; Ikeo et al., 2006; McGlashan and Halley, 2003; Mondragón et al., 2009; Pérez et al., 2007)). However, it is quite significant to note that incorporation of another polymer (usually being relatively hydrophobic) would modify the hydrophilicity of the material. As a result, MMT–Na^+ might not match anymore the polarity of the blend. McGlashan and Halley (2003) proposed that the organic constituents (alcohols and hydrogenated tallow) of the OMMT could be more thermodynamically compatible with the polyester. Lee and coworkers (Lee and Hanna, 2008; Lee et al., 2007, 2008a,b; Lee and Hanna, 2009) compared addition of different MMTs in the melt processing of tapioca starch–PLA nanocomposite foams. Their results showed that Cloisite 30B, instead of MMT–Na^+, could result in the greatest extent of intercalation and also better functional properties (Lee et al.,

2007). This was caused by strong hydrogen bonding between the hydroxyl groups of the matrix and those of the Cloisite 30B organomodifier (Lee et al., 2007). A similar result was observed by Bocchini et al. (2010) who used a starch–PBSA matrix. Majdzadeh-Ardakani and Nazari (2010) showed that CA–modified MMT, compared with MMT–Na^+ and Cloisite 30B, led to better mechanical properties of starch–PVA nanocomposites because of strong interactions between the CA and the starch/PVA polymer chains (Majdzadeh-Ardakani and Nazari, 2010).

4.9 Mechanical properties of starch-based nanobiocomposites reinforced by cellulose nanowhiskers

4.9.1 Effect of cellulose nanowhiskers addition

The literature has shown that the mechanical properties (σ and E) generally show improvement with addition of CNWs to starch-based materials (Angles and Dufresne, 2000, 2001; Cao et al., 2008a,b; Chang et al., 2010c; Chen et al., 2009a,b; Grande et al., 2009; Kaushik et al., 2010; Kvien et al., 2007; Lu et al., 2006; Mathew and Dufresne, 2002a; Mathew et al., 2008; Svagan et al., 2009; Teixeira et al., 2009; Wan et al., 2009; Woehl et al., 2010), although a high level of nanofiller addition may result in aggregation, adversely affecting the properties (Chen et al., 2009a). The reinforcement in mechanical properties can be linked to not only the good nanofiller dispersion in the matrix, resulting from the chemical similarity, but also the strong nanofiller–matrix interactions by hydrogen bonding (Cao et al., 2008b; Chen et al., 2009b; Lu et al., 2006). In particular, improvement in mechanical properties (but usually at the expense of the ε_b) can be benefited by (a) the formation of a rigid network of the CNWs, the mutual entanglement between the nanofiller and the matrix, and the efficient stress transfer from the matrix to the nanofiller (Kaushik et al., 2010; Siqueira et al., 2008), and by (b) an increase in the overall crystallinity of the system resulting from the nucleating effect of the CNWs (Mathew and Dufresne, 2002a; Mathew et al., 2008). (The latter point has been in dispute ever since the hindrance of the lateral rearrangement of the starch chains, and hence that of the recrystallisation, was also observed (Kaushik et al., 2010).)

4.9.2 Effect of nanofiller preparation

Preparation can directly affect the quality of nanofiller, which then impacts the structure and performance of the resulting nanocomposites. Chen et al. (2009a,b) found that, while starch–CNWs nanobiocomposites (with CNWs prepared by acid hydrolysis of pea hull fibre for 4–24 h) generally exhibited higher σ, ε_b, and E (see Figure 4.5), than the microcomposites (with the native pea hull fibres without acid hydrolysis) (Chen et al., 2009a), 8 h of acid hydrolysis could result in most homogeneous dispersion of the CNWs within the starch matrix and, in this case, the CNWs were wrapped most tightly by the matrix (Chen et al., 2009b). It was suggested that this stronger

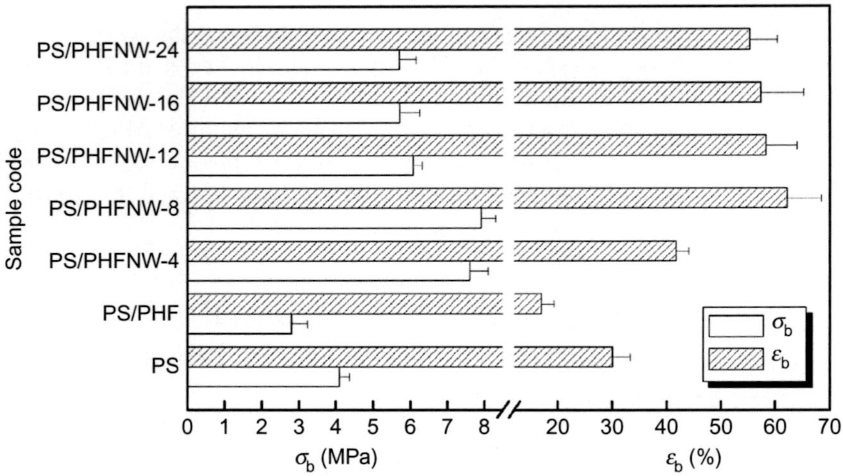

Figure 4.5 Mechanical properties of the pea starch (PS) film, the pea starch–pea hull fibre composite film (PS/PHF), and the pea starch–nanowhiskers nanocomposite films (PS/PHFNW-*t*, where *t* represents the acid hydrolysis time for the CNWs preparation). Reprinted from Chen et al. (2009b), Copyright 2009, with permission from Elsevier.

nanofiller–matrix adhesion resulted from the highest aspect ratio, and contributed to the superior properties of the nanobiocomposite such as σ and ε_b (see Figure 4.5) (Chen et al., 2009b). Moreover, Woehl et al. (2010) found that enzyme hydrolysis of bacterial cellulose for 60 min, compared with either a shorter or longer time, could generate CNWs with which the nanobiocomposites display the most improved mechanical properties. This is because 60 min is enough to disentangle the fibres and reduce the defects in the surface of the fibres without weakening the crystalline regions of the CNWs (Woehl et al., 2010).

4.9.3 Mechanical properties of starch-based nanobiocomposites reinforced by starch nanoparticles

With the addition of starch nanoparticles (SNPs), starch-based nanobiocomposites generally showed increased σ_b and E (Angellier et al., 2006; Ma et al., 2008b; Viguié et al., 2007) and decreased ε_b. This strong reinforcing effect of SNPs could be due to the same chemical structure, good interfacial interactions (nanofiller–nanofiller and nanofiller–matrix) by hydrogen bonding and, hence, a strong reinforcing effect of SNPs could be expected (Angellier et al., 2006; Ma et al., 2008b; Viguié et al., 2007). Besides, increased crystallinity in the starch matrix (especially at the interface) as a result of the nucleating effect of SNPs could also account for the changes in mechanical properties (Angellier et al., 2006; Viguié et al., 2007). The degree of crystallinity could be dependent on the content of plasticiser, which has the same affinity to both components (Angellier et al., 2006). These could explain a greater reinforcing effect of the SNPs when the plasticiser content (glycerol or sorbitol) was high (see Table 4.5) (Angellier et al., 2006; Viguié et al., 2007).

Table 4.5 **Mechanical properties (E, Young's modulus; σ_b; strength at break; ε_b, elongation at break) of glycerol-plasticised waxy maize starch-based films with different plasticiser (glycerol or sorbitol) and nanofiller (SNPs) contents**

Plasticiser type	Plasticiser content[a] (%)	Nanofiller content[a] (%)	Ageing[b]	E (MPa)	σ_b (MPa)	ε_b (%)
Glycerol	20	0	No	49 ± 12	2.4 ± 0.5	182 ± 52
		5	No	298 ± 54	13.2 ± 3.3	8.2 ± 1.5
		10	No	333 ± 54	13.6 ± 1.6	8.6 ± 1.5
		15	No	–	7.6 ± 2.9	4.5 ± 1.4
	25	0	No	11 ± 3.6	1.0 ± 0.1	297 ± 64
			Yes	59 ± 11	4.6 ± 0.7	61 ± 11
		2	No	75 ± 17	3.3 ± 0.4	122 ± 31
			Yes	147 ± 13	7.6 ± 0.7	57 ± 13
		5	No	80 ± 3.5	3.6 ± 0.4	97 ± 18
			Yes	154 ± 20	8.0 ± 0.7	52 ± 8.5
		10	No	82 ± 30	4.2 ± 1	57 ± 21
			Yes	173 ± 22	8.0 ± 0.4	33 ± 3.5
		15	No	241 ± 46	9.8 ± 1.4	20 ± 8
			Yes	259 ± ?	10.8 ± ?	9 ± ?
	30	0	No	0.46 ± 0.3	0.26 ± 0.1	551 ± 80
		5	No	3.4 ± 1.1	1.3 ± 0.5	236 ± 40
		10	No	25 ± 14	2.7 ± 0.5	236 ± 40
		15	No	44 ± 5	3.6 ± 0.3	82 ± 13

Continued

Table 4.5 Continued

Plasticiser type	Plasticiser content[a] (%)	Nanofiller content[a] (%)	Ageing[b]	E (MPa)	σ_b (MPa)	ε_b (%)
Sorbitol	20	0	No	46.8 ± 2.5	1.06 ± 0.05	23 ± 2
	25	0	No	17.2 ± 1.1	0.38 ± 0.02	63 ± 4
			Yes	116.9 ± 2.5	4.46 ± 0.27	35 ± 3
	25	5	No	36.6 ± 1.6	0.99 ± 0.05	57 ± 8
			Yes	120.7 ± 6.0	4.47 ± 0.22	28 ± 2
		10	No	38.3 ± 3.6	1.37 ± 0.95	58 ± 5
			Yes	108.6 ± 3.3	3.81 ± 0.26	18 ± 2
		15	No	46.2 ± 7.0	1.59 ± 0.14	41 ± 12
			Yes	110.0 ± 0.6	4.42 ± 0.15	23 ± 2
	35	0	No	6.02 ± 0.52	0.18 ± 0.05	107 ± 5
			Yes	51.1 ± 2.3	3.05 ± 0.05	42 ± 4
		5	No	9.93 ± 1.73	0.47 ± 0.03	92 ± 4
			Yes	69.1 ± 5.5	3.66 ± 0.11	26 ± 4
		10	No	18.8 ± 1.15	1.09 ± 0.02	64 ± 5
			Yes	24.9 ± 6.3	1.71 ± 0.35	29 ± 2
		15	No	42.5 ± 2.5	2.22 ± 0.14	45 ± 3
			Yes	58.4 ± 7.1	3.10 ± 0.10	28 ± 3
		25	No	103.0 ± 7.0	4.31 ± 0.21	31 ± 2

[a]Based on the total weight of starch and plasticiser.
[b]Carried out at 88% RH for 1 week in addition to conditioning at 43% RH for 1 week.
Adapted with permission from Angellier et al. (2006) Copyright 2006 American Chemical Society; Viguié et al. (2007) Copyright © 2002 Wiley Periodicals, Inc.

4.9.4 Mechanical properties of starch-based nanobiocomposites reinforced by chitin/chitosan nanoparticles

Chang et al. (2010a,b) found that chitin/chitosan nanoparticles could be uniformly dispersed in a starch matrix at low loading levels, resulting in improved properties (e.g. σ, E'; the mechanisms are similar to those for SNPs-reinforced nanobiocomposites); however, when the nanofiller addition was high (5 wt.% for chitin (Chang et al., 2010b), 6 wt.% for chitosan (Chang et al., 2010a)), conglomeration/aggregation occurred. Nevertheless, good interfacial interactions between the nanofiller and the matrix could be still observed (Chang et al., 2010b).

4.10 Mechanical properties of nanobiocomposites reinforced by CNTs

4.10.1 Effect of CNTs addition

Incorporation of multiwalled carbon nanotubes (MWCNTs) into starch-based materials generally increased ultimate tensile strength (σ_u) (Cao et al., 2007; Famá et al., 2011; Ma et al., 2008a), E (Cao et al., 2007; Famá et al., 2011; Ma et al., 2008a). This can be ascribed to (a) the formation of an isotropic, three-dimensional nanotube network, which inhibits the crack propagation (Ma et al., 2008a); (b) the effective stress transfer from the matrix to the nanofiller (Ma et al., 2008a); and (c) the increase in T_g, which contributes to an increase in stiffness (Famá et al., 2011, 2012). Meanwhile a decrease in toughness, or ε_b as usually observed, was attributed to the spatial restraint of the slippage movement of the starch molecules, yielding increasingly brittle samples (Ma et al., 2008a).

There could be an optimised level of nanofiller addition, which was, for example, 3.8% for surfactant-treated MWCNTs (Ma et al., 2008a) and 1.5% for carboxylated MWCNTs (Liu et al., 2011d).

4.10.2 Effect of nanofiller modification

Proper modification of MWCNTs could lead to homogeneous dispersion of the nanofiller in the matrix and strong nanofiller–matrix interactions, accounting for property enhancement (Cao et al., 2007; Famá et al., 2011, 2012; Liu et al., 2011d; Ma et al., 2008a). For the carboxylated MWCNTs, the acid-treatment process can incorporate various polar groups (carboxylic, carbonyl, and hydroxyl groups), which can improve the hydrophilicity and reduce the MWCNTs agglomeration (Cao et al., 2007; Liu et al., 2011d). As a result of the hydrogen bonding interactions and, thus, the high starch–MWCNTs compatibility, σ and E of the nanocomposites did not come at the expense of ε_b, which increased from 30% to 42% with the MWCNTs addition level up to 1% (Cao et al., 2007). Very recently, Famá et al. (2011, 2012) used a starch to wrap MWCNTs and the same starch to be as the matrix. The SEM results showed that the failure occurred within the matrix rather than between the MWCNTs and their starch coating, indicating strong nanofiller–matrix interactions (Famá et al., 2011).

Therefore, efficient load transfer was attained, resulting in increases in not only σ but also toughness and ε_b (Famá et al., 2011). This work is remarkable as very small content of MWCNTs (i.e. 0.055%) could result in great changes in the material performance (Famá et al., 2011).

4.11 Mechanical properties of starch-based nanobiocomposites reinforced by metalloid oxides, metal oxides, and metal chalcogenides

Uniform dispersion of a nanofiller (especially for polysaccharide-encapsulated metal nanoparticles) in a starch matrix and strong interfacial adhesion through hydrogen bonding between them could usually be achieved (due to the similar chemical structures of the stabiliser and the matrix) (Chang et al., 2009; Liu et al., 2011a; Ma et al., 2009; Yu et al., 2009; Zheng et al., 2009b). As a result, the resulting nanocomposites generally display increases in the mechanical properties (σ (Chang et al., 2009; Liu et al., 2011a; Ma et al., 2009; Wu et al., 2009b; Yu et al., 2009; Yun et al., 2011; Zheng et al., 2009b), E (Ma et al., 2009; Yu et al., 2009), and E' (Ma et al., 2009; Yu et al., 2009)).

A higher addition level of nanofiller may result in agglomeration of nanoparticles (Chang et al., 2009; Liu et al., 2011a; Ma et al., 2009; Tang et al., 2009; Yu et al., 2009; Zheng et al., 2009b), which is unfavourable for the performance improvement. For example, Wu et al. (2009b) found that greatest σ could be achieved when the nano-SiO_2 addition level was 4%. Tang et al. (2009) found that 3 wt.% was the optimised level of the nano-SiO_2 for greatest σ of starch–PVA-based nanocomposites

4.12 Mechanical properties of starch-based nanobiocomposites reinforced by other nanofillers

Besides what has been discussed above, other specific nanofillers have been used for starch, but to a lesser extent. Some of them only appear in some reports and their potential as nanofillers for reinforcing starch-based materials remains uncertain. For example, synthetic nanolayers like layered double hydroxides (LDH) and brucite can hardly be intercalated by starch chains (Wilhelm et al., 2003a), which greatly limited their usability.

Those nanofillers that are effective in reinforcing starch normally can form strong hydrogen bonding with starch, such as GO (Li et al., 2011b) and synthetic α-ZrP (i.e. $Zr(HPO_4)_2 \cdot H_2O$) (Wu et al., 2009a). Regarding the strong interaction between two phases, Li et al. (2011b) found that with increased GO loading level up to 2 wt.%, σ and E of resulting starch-based nanobiocomposites were continuously increased and ε_b were decreased (Li et al., 2011b); Wu et al. (2009a) found that, compared with the neat plasticised starch, the starch/α-zirconium phosphate nanobiocomposite films showed increases in σ and ε_b.

4.13 Summary

In order to enhance and widen the current applications of starch-based materials, starch-based biocomposites (starch-based materials incorporated with fillers) have experienced great development, which generally show improved mechanical properties compared with the neat starch matrix. The effect of fillers is much more evident when the dispersed fillers become nanoscaled. Regarding this, a wide range of nanofillers have been tested during the last two decades for starch-based materials, as shown in Table 4.2. In particular, phyllosilicates (especially MMT of the smectite group) have been mostly utilised due to their advantages such as wide availability, low cost, and high aspect ratio and, thus, vast interface area of around 700 m^2 per gram when fully exfoliated (and also the swelling nature). In addition, polysaccharide nanofillers represent the second most popular group due to their abundance in nature, the biological sources, and the chemical similarity to starch.

With the incorporation of the nanofiller, starch-based materials generally show improvement in mechanical properties (typically σ, E, and E'). This improvement can be fundamentally ascribed to the homogeneous dispersion of the nanofiller in the matrix and the strong interface adhesion, which can contribute to the formation of a rigid nanofiller network and influence the molecular and crystalline structures in the matrix. To realise these, the nanofiller–matrix compatibility is the key point to address, which mainly depends on the surface chemistry of the nanofiller and is usually achieved by hydrogen bonding. In addition, the aspect ratio/surface area and the mechanical properties of the nanofiller itself are highly important in determining the mechanical properties of the resulting starch-based nanobiocomposites.

4.14 Future trends

Starch is a special polymer that has many unique characteristics. This makes the development of starch-based biocomposites less straightforward than that of other polymer composites. In particular, many reports have shown the heterogeneous dispersion of the nanofiller in starch and the separation of two phases, which compromises the efficacy of the nanofiller. To maximise the compatibility between the nanofiller and starch is the key to maximise the mechanical performance of the resulting starch-based nanobiocomposites. This can be achieved by either choosing nanofillers that are relatively hydrophilic and/or have chemical affinity to starch, or by surface modification of nanofillers to achieve stronger interactions with starch. In this sense, it is worth testing more nanofillers for use in starch-based systems, and to explore the possibilities in chemistry. Moreover, it is worth it to carry out studies regarding the effectiveness of processing (e.g. in melt state with thermomechanical treatment) in nanofiller dispersion, while in most of the studies so far the starch-based biocomposites were prepared in solution.

The mechanism also needs to be further explored. Starch-based nanobiocomposites are not simple systems and the reinforcing effect of nanofillers can also be modified by the changes in, for example, crystallinity, T_g, and phase distribution (including more components such as the plasticiser), which greatly influence the mechanical properties. However, there is no consensus in the literature regarding how a nanofiller affects the crystallinity, T_g, and phase distribution. Of course, this can be different in different systems (different nanofiller, starch type and content, plasticiser type and content, processing technique conditions). But, it is definitely worth carrying out further investigation into how different factors can synergistically contribute to better mechanical properties.

These starch-based multiphase materials with such mechanical properties are and should be industrialised for a wide range of applications with short-term behaviour, such as compostable packaging, agriculture (planting pots, pickets, etc.), and leisure, hygiene, and personal care products.

Acknowledgements

The authors from The University of Queensland would like to acknowledge the Australian Research Council (ARC) for the research fund under the Discovery Project 120100344. P. Liu from Guangzhou University would like to thank the National Natural Science Foundation of China (NSFC) for the research fund under the project number 21106023, as well as the Ministry of Education Engineering Research Center of Starch and Vegetable Protein Processing (located at the South China University of Technology, Guangzhou, P. R. China) for providing financial support and lab resources under the project number 2012-ERC-01.

References

Alexandre, M., Dubois, P., 2000. Polymer-layered silicate nanocomposites: preparation, properties and uses of a new class of materials. Mater. Sci. Eng. R. Rep. 28, 1–63.

Alvarez, V.A., Fraga, A.N., Vázquez, A., 2004. Effects of the moisture and fiber content on the mechanical properties of biodegradable polymer–sisal fiber biocomposites. J. Appl. Polym. Sci. 91, 4007–4016.

Angellier, H., Choisnard, L., Molina-Boisseau, S., Ozil, P., Dufresne, A., 2004. Optimization of the preparation of aqueous suspensions of waxy maize starch nanocrystals using a response surface methodology. Biomacromolecules 5, 1545–1551.

Angellier, H., Molina-Boisseau, S., Dole, P., Dufresne, A., 2006. Thermoplastic starch-waxy maize starch nanocrystals nanocomposites. Biomacromolecules 7, 531–539.

Angles, M.N., Dufresne, A., 2000. Plasticized starch/tunicin whiskers nanocomposites. 1. Structural analysis. Macromolecules 33, 8344–8353.

Angles, M.N., Dufresne, A., 2001. Plasticized starch/tunicin whiskers nanocomposite materials. 2. Mechanical behavior. Macromolecules 34, 2921–2931.

Arroyo, O.H., Huneault, M.A., Favis, B.D., Bureau, M.N., 2010. Processing and properties of PLA/thermoplastic starch/montmorillonite nanocomposites. Polym. Compos. 31, 114–127.

Atwell, W.A., Hood, L.F., Lineback, D.R., Varrianomarston, E., Zobel, H.F., 1988. The terminology and methodology associated with basic starch phenomena. Cereal Foods World 33, 306–311.

Avella, M., De Vlieger, J.J., Errico, M.E., Fischer, S., Vacca, P., Volpe, M.G., 2005. Biodegradable starch/clay nanocomposite films for food packaging applications. Food Chem. 93, 467–474.

Avérous, L., 2002. Interactions between cellulose and plasticized wheat starch – properties of biodegradable multiphase systems. In: Renard, D., Della, G., Popineau, Y. (Eds.), Plant Biopolymer Science: Food and Non-Food Applications. Royal Society of Chemistry, London, UK, pp. 253–259.

Avérous, L., 2004. Biodegradable multiphase systems based on plasticized starch: a review. Polym. Rev. 44, 231–274.

Averous, L., 2007. Cellulose-based biocomposites: comparison of different multiphasic systems. Compos. Interfaces 14, 787–805.

Averous, L., Boquillon, N., 2004. Biocomposites based on plasticized starch: thermal and mechanical behaviours. Carbohydr. Polym. 56, 111–122.

Avérous, L., Halley, P.J., 2009. Biocomposites based on plasticized starch. Biofuels Bioprod. Biorefin. 3, 329–343.

Avérous, L.R., Halley, P.J., 2014. Starch polymers: from the field to industrial products. In: Halley, P., Avérous, L.R. (Eds.), Starch Polymers: From Genetic Engineering to Green Applications. Elsevier, Amsterdam, Boston, Heidelberg, London, New York, Oxford, Paris, San Diego, San Francisco, Singapore, Sydney, Tokyo, pp. 3–10.

Avérous, L., Fringant, C., Moro, L., 2001a. Starch-based biodegradable materials suitable for thermoforming packaging. Starch-Starke 53, 368–371.

Avérous, L., Fringant, C., Moro, L., 2001b. Plasticized starch–cellulose interactions in polysaccharide composites. Polymer 42, 6565–6572.

Azizi Samir, M.A.S., Alloin, F., Sanchez, J.-Y., El Kissi, N., Dufresne, A., 2004. Preparation of cellulose whiskers reinforced nanocomposites from an organic medium suspension. Macromolecules 37, 1386–1393.

Azizi Samir, M.A.S., Alloin, F., Dufresne, A., 2005. Review of recent research into cellulosic whiskers, their properties and their application in nanocomposite field. Biomacromolecules 6, 612–626.

Belhassen, R., Boufi, S., Vilaseca, F., et al., 2009. Biocomposites based on *Alfa* fibers and starch-based biopolymer. Polym. Adv. Technol. 20, 1068–1075.

Bocchini, S., Battegazzore, D., Frache, A., 2010. Poly (butylensuccinate co-adipate)-thermoplastic starch nanocomposite blends. Carbohydr. Polym. 82, 802–808.

Bondeson, D., Mathew, A., Oksman, K., 2006. Optimization of the isolation of nanocrystals from microcrystalline cellulose by acid hydrolysis. Cellulose 13, 171–180.

Bonnet, P., Albertini, D., Bizot, H., Bernard, A., Chauvet, O., 2007. Amylose/SWNT composites: from solution to film – synthesis, characterization and properties. Compos. Sci. Technol. 67, 817–821.

Buléon, A., Colonna, P., Planchot, V., Ball, S., 1998. Starch granules: structure and biosynthesis. Int. J. Biol. Macromol. 23, 85–112.

Bulkin, B.J., Kwak, Y., Dea, I.C.M., 1987. Retrogradation kinetics of waxy-corn and potato starches: a rapid Raman-spectroscopic study. Carbohydr. Res. 160, 95–112.

Cao, X., Chen, Y., Chang, P.R., Huneault, M.A., 2007. Preparation and properties of plasticized starch/multiwalled carbon nanotubes composites. J. Appl. Polym. Sci. 106, 1431–1437.

Cao, X., Chen, Y., Chang, P.R., Stumborg, M., Huneault, M.A., 2008a. Green composites reinforced with hemp nanocrystals in plasticized starch. J. Appl. Polym. Sci. 109, 3804–3810.

Cao, X., Chen, Y., Chang, P.R., Muir, A.D., Falk, G., 2008b. Starch-based nanocomposites reinforced with flax cellulose nanocrystals. Express Polym. Lett. 2, 502–510.

Carvalho, A.J.F., Zambon, M.D., Curvelo, A.A.S., Gandini, A., 2003. Size exclusion chromatography characterization of thermoplastic starch composites. 1. Influence of plasticizer and fibre content. Polym. Degrad. Stab. 79, 133–138.

Chairam, S., Poolperm, C., Somsook, E., 2009. Starch vermicelli template-assisted synthesis of size/shape-controlled nanoparticles. Carbohydr. Polym. 75, 694–704.

Chang, P.R., Yu, J., Ma, X., 2009. Fabrication and characterization of Sb_2O_3/carboxymethyl cellulose sodium and the properties of plasticized starch composite films. Macromol. Mater. Eng. 294, 762–767.

Chang, P.R., Jian, R., Yu, J., Ma, X., 2010a. Fabrication and characterisation of chitosan nanoparticles/plasticised-starch composites. Food Chem. 120, 736–740.

Chang, P.R., Jian, R., Yu, J., Ma, X., 2010b. Starch-based composites reinforced with novel chitin nanoparticles. Carbohydr. Polym. 80, 421–426.

Chang, P.R., Jian, R., Zheng, P., Yu, J., Ma, X., 2010c. Preparation and properties of glycerol plasticized-starch (GPS)/cellulose nanoparticle (CN) composites. Carbohydr. Polym. 79, 301–305.

Chaudhary, D.S., 2008. Understanding amylose crystallinity in starch–clay nanocomposites. J. Polym. Sci. B Polym. Phys. 46, 979–987.

Chaudhary, D.S., Adhikari, B.P., Kasapis, S., 2011. Glass-transition behaviour of plasticized starch biopolymer system – a modified Gordon–Taylor approach. Food Hydrocoll. 25, 114–121.

Chen, B., Evans, J.R.G., 2005. Thermoplastic starch-clay nanocomposites and their characteristics. Carbohydr. Polym. 61, 455–463.

Chen, M., Chen, B., Evans, J.R.G., 2005. Novel thermoplastic starch–clay nanocomposite foams. Nanotechnology 16, 2334.

Chen, Y., Liu, C., Chang, P.R., Anderson, D.P., Huneault, M.A., 2009a. Pea starch-based composite films with pea hull fibers and pea hull fiber-derived nanowhiskers. Polym. Eng. Sci. 49, 369–378.

Chen, Y., Liu, C., Chang, P.R., Cao, X., Anderson, D.P., 2009b. Bionanocomposites based on pea starch and cellulose nanowhiskers hydrolyzed from pea hull fibre: effect of hydrolysis time. Carbohydr. Polym. 76, 607–615.

Chiou, B.-S., Yee, E., Glenn, G.M., Orts, W.J., 2005. Rheology of starch–clay nanocomposites. Carbohydr. Polym. 59, 467–475.

Chiou, B.S., Yee, E., Wood, D., Shey, J., Glenn, G., Orts, W., 2006. Effects of processing conditions on nanoclay dispersion in starch–clay nanocomposites. Cereal Chem. 83, 300–305.

Chiou, B.-S., Wood, D., Yee, E., Imam, S.H., Glenn, G.M., Orts, W.J., 2007. Extruded starch–nanoclay nanocomposites: effects of glycerol and nanoclay concentration. Polym. Eng. Sci. 47, 1898–1904.

Chivrac, F., Pollet, E., Schmutz, M., Avérous, L., 2008a. New approach to elaborate exfoliated starch-based nanobiocomposites. Biomacromolecules 9, 896–900.

Chivrac, F., Gueguen, O., Pollet, E., Ahzi, S., Makradi, A., Averous, L., 2008b. Micromechanical modeling and characterization of the effective properties in starch-based nano-biocomposites. Acta Biomater. 4, 1707–1714.

Chivrac, F., Pollet, E., Avérous, L., 2009. Shear induced clay organo-modification: application to plasticized starch nano-biocomposites. Polym. Adv. Technol. 21, 578–583.

Chivrac, F., Pollet, E., Schmutz, M., Avérous, L., 2010a. Starch nano-biocomposites based on needle-like sepiolite clays. Carbohydr. Polym. 80, 145–153.

Chivrac, F., Pollet, E., Dole, P., Avérous, L., 2010b. Starch-based nano-biocomposites: plasticizer impact on the montmorillonite exfoliation process. Carbohydr. Polym. 79, 941–947.

Chivrac, F., Angellier-Coussy, H., Guillard, V., Pollet, E., Avérous, L., 2010c. How does water diffuse in starch/montmorillonite nano-biocomposite materials? Carbohydr. Polym. 82, 128–135.

Chivrac, F., Gueguen, O., Pollet, E., Avérous, L., Ahzi, S., Belouettar, S., 2010d. Micromechanically-based formulation of the cooperative model for the yield behavior of starch-based nano-biocomposites. J. Nanosci. Nanotechnol. 10, 2949–2955.
Chung, Y.-L., Ansari, S., Estevez, L., Hayrapetyan, S., Giannelis, E.P., Lai, H.-M., 2010. Preparation and properties of biodegradable starch–clay nanocomposites. Carbohydr. Polym. 79, 391–396.
Crosby, A.J., Lee, J.Y., 2007. Polymer nanocomposites: the "nano" effect on mechanical properties. Polym. Rev. 47, 217–229.
Curvelo, A.A.S., de Carvalho, A.J.F., Agnelli, J.A.M., 2001. Thermoplastic starch–cellulosic fibers composites: preliminary results. Carbohydr. Polym. 45, 183–188.
Cyras, V.P., Manfredi, L.B., Ton-That, M.-T., Vazquez, A., 2008. Physical and mechanical properties of thermoplastic starch/montmorillonite nanocomposite films. Carbohydr. Polym. 73, 55–63.
Da Róz, A.L., Carvalho, A.J.F., Gandini, A., Curvelo, A.A.S., 2006. The effect of plasticizers on thermoplastic starch compositions obtained by melt processing. Carbohydr. Polym. 63, 417–424.
Dai, H., Chang, P.R., Peng, F., Yu, J., Ma, X., 2009a. *N*-(2-hydroxyethyl)formamide as a new plasticizer for thermoplastic starch. J. Polym. Res. 16, 529–535.
Dai, H., Chang, P., Geng, F., Yu, J., Ma, X., 2009b. Preparation and properties of thermoplastic starch/montmorillonite nanocomposite using *N*-(2-hydroxyethyl)formamide as a new additive. J. Polym. Environ. 17, 225–232.
Dai, H., Chang, P.R., Yu, J., Geng, F., Ma, X., 2010a. *N*-(2-hydroxypropyl)formamide and *N*-(2-hydroxyethyl)-*N*-methylformamide as two new plasticizers for thermoplastic starch. Carbohydr. Polym. 80, 139–144.
Dai, H., Chang, P.R., Geng, F., Yu, J., Ma, X., 2010b. Preparation and properties of starch-based film using *N*, *N*-bis(2-hydroxyethyl)formamide as a new plasticizer. Carbohydr. Polym. 79, 306–311.
de Carvalho, A.J.F., Curvelo, A.A.S., Agnelli, J.A.M., 2001. A first insight on composites of thermoplastic starch and kaolin. Carbohydr. Polym. 45, 189–194.
Dean, K., Yu, L., Wu, D.Y., 2007. Preparation and characterization of melt-extruded thermoplastic starch/clay nanocomposites. Compos. Sci. Technol. 67, 413–421.
Dean, K.M., Do, M.D., Petinakis, E., Yu, L., 2008. Key interactions in biodegradable thermoplastic starch/poly(vinyl alcohol)/montmorillonite micro- and nanocomposites. Compos. Sci. Technol. 68, 1453–1462.
Dean, K.M., Petinakis, E., Goodall, L., Miller, T., Yu, L., Wright, N., 2011. Nanostabilization of thermally processed high amylose hydroxylpropylated starch films. Carbohydr. Polym. 86, 652–658.
Dong, X.M., Kimura, T., Revol, J.-F., Gray, D.G., 1996. Effects of ionic strength on the isotropic–chiral nematic phase transition of suspensions of cellulose crystallites. Langmuir 12, 2076–2082.
Donovan, J.W., 1979. Phase transitions of the starch–water system. Biopolymers 18, 263–275.
Dufresne, A., Dupeyre, D., Vignon, M.R., 2000. Cellulose microfibrils from potato tuber cells: processing and characterization of starch-cellulose microfibril composites. J. Appl. Polym. Sci. 76, 2080–2092.
Enrione, J., Osorio, F., Pedreschi, F., Hill, S., 2010. Prediction of the glass transition temperature on extruded waxy maize and rice starches in presence of glycerol. Food Bioprocess Technol. 3, 791–796.
Famá, L.M., Pettarin, V., Goyanes, S.N., Bernal, C.R., 2011. Starch/multi-walled carbon nanotubes composites with improved mechanical properties. Carbohydr. Polym. 83, 1226–1231.

Famá, L., Rojo, P.G., Bernal, C., Goyanes, S., 2012. Biodegradable starch based nanocomposites with low water vapor permeability and high storage modulus. Carbohydr. Polym. 87, 1989–1993.

Follain, N., Joly, C., Dole, P., Bliard, C., 2005. Mechanical properties of starch-based materials. I. Short review and complementary experimental analysis. J. Appl. Polym. Sci. 97, 1783–1794.

Forssell, P.M., Mikkilä, J.M., Moates, G.K., Parker, R., 1997. Phase and glass transition behaviour of concentrated barley starch-glycerol-water mixtures, a model for thermoplastic starch. Carbohydr. Polym. 34, 275–282.

Forssell, P.M., Hulleman, S.H.D., Myllärinen, P.J., Moates, G.K., Parker, R., 1999. Ageing of rubbery thermoplastic barley and oat starches. Carbohydr. Polym. 39, 43–51.

Funke, U., Bergthaller, W., Lindhauer, M.G., 1998. Processing and characterization of biodegradable products based on starch. Polym. Degrad. Stab. 59, 293–296.

Gallant, D.J., Bouchet, B., Baldwin, P.M., 1997. Microscopy of starch: evidence of a new level of granule organization. Carbohydr. Polym. 32, 177–191.

Gao, W., Dong, H., Hou, H., Zhang, H., 2012. Effects of clays with various hydrophilicities on properties of starch–clay nanocomposites by film blowing. Carbohydr. Polym. 88, 321–328.

García, N.L., Ribba, L., Dufresne, A., Aranguren, M.I., Goyanes, S., 2009. Physico-mechanical properties of biodegradable starch nanocomposites. Macromol. Mater. Eng. 294, 169–177.

Gaudin, S., Lourdin, D., Le Botlan, D., Ilari, J.L., Colonna, P., 1999. Plasticisation and mobility in starch-sorbitol films. J. Cereal Sci. 29, 273–284.

Gironès, J., López, J.P., Mutjé, P., Carvalho, A.J.F., Curvelo, A.A.S., Vilaseca, F., 2012. Natural fiber-reinforced thermoplastic starch composites obtained by melt processing. Compos. Sci. Technol. 72, 858–863.

Gopalan Nair, K., Dufresne, A., Gandini, A., Belgacem, M.N., 2003. Crab shell chitin whiskers reinforced natural rubber nanocomposites. 3. Effect of chemical modification of chitin whiskers. Biomacromolecules 4, 1835–1842.

Grande, C.J., Torres, F.G., Gomez, C.M., Troncoso, O.P., Canet-Ferrer, J., Martínez-Pastor, J., 2009. Development of self-assembled bacterial cellulose–starch nanocomposites. Mater. Sci. Eng. C 29, 1098–1104.

Gudmundsson, M., 1994. Retrogradation of starch and the role of its components. Thermochim. Acta 246, 329–341.

Hassan Nejad, M., Ganster, J., Bohn, A., Volkert, B., Lehmann, A., 2011. Nanocomposites of starch mixed esters and MMT: improved strength, stiffness, and toughness for starch propionate acetate laurate. Carbohydr. Polym. 84, 90–95.

He, Y., Kong, W., Wang, W., et al., 2012. Modified natural halloysite/potato starch composite films. Carbohydr. Polym. 87, 2706–2711.

Huang, M., Yu, J., 2006. Structure and properties of thermoplastic corn starch/montmorillonite biodegradable composites. J. Appl. Polym. Sci. 99, 170–176.

Huang, M.-F., Yu, J.-G., Ma, X.-F., 2004. Studies on the properties of montmorillonite-reinforced thermoplastic starch composites. Polymer 45, 7017–7023.

Huang, M.-F., Yu, J.-G., Ma, X.-F., Jin, P., 2005a. High performance biodegradable thermoplastic starch-EMMT nanoplastics. Polymer 46, 3157–3162.

Huang, M.F., Yu, J.G., Ma, X.F., 2005b. Preparation of the thermoplastic starch/montmorillonite nanocomposites by melt-intercalation. Chin. Chem. Lett. 16, 561–564.

Huang, M., Yu, J., Ma, X., 2006. High mechanical performance MMT-urea and formamide-plasticized thermoplastic cornstarch biodegradable nanocomposites. Carbohydr. Polym. 63, 393–399.

Huneault, M.A., Li, H., 2007. Morphology and properties of compatibilized polylactide/thermoplastic starch blends. Polymer 48, 270–280.

Ibrahim, S.M., 2011. Characterization, mechanical, and thermal properties of gamma irradiated starch films reinforced with mineral clay. J. Appl. Polym. Sci. 119, 685–692.

Ikeo, Y., Aoki, K., Kishi, H., Matsuda, S., Murakami, A., 2006. Nano clay reinforced biodegradable plastics of PCL starch blends. Polym. Adv. Technol. 17, 940–944.

Imberty, A., Perez, S., 1988. A revisit to the three-dimensional structure of B-type starch. Biopolymers 27, 1205–1221.

Imberty, A., Chanzy, H., Pérez, S., Bulèon, A., Tran, V., 1988. The double-helical nature of the crystalline part of A-starch. J. Mol. Biol. 201, 365–378.

Jane, J.-L., 2009. Structural features of starch granules II. In: James, B., Roy, W. (Eds.), Starch. third ed.. Academic Press, San Diego, pp. 193–236.

Kalambur, S., Rizvi, S.S.H., 2006. An overview of starch-based plastic blends from reactive extrusion. J. Plast. Film Sheeting 22, 39–58.

Kampeerapappun, P., Aht-ong, D., Pentrakoon, D., Srikulkit, K., 2007. Preparation of cassava starch/montmorillonite composite film. Carbohydr. Polym. 67, 155–163.

Kaushik, A., Singh, M., Verma, G., 2010. Green nanocomposites based on thermoplastic starch and steam exploded cellulose nanofibrils from wheat straw. Carbohydr. Polym. 82, 337–345.

Ke, G., 2010. A novel strategy to functionalize carbon nanotubes with cellulose acetate using triazines as intermediated functional groups. Carbohydr. Polym. 79, 775–782.

Kirby, A.R., Clark, S.A., Parker, R., Smith, A.C., 1993. The deformation and failure behaviour of wheat starch plasticized with water and polyols. J. Mater. Sci. 28, 5937–5942.

Kumar, A.P., Depan, D., Singh Tomer, N., Singh, R.P., 2009. Nanoscale particles for polymer degradation and stabilization—trends and future perspectives. Prog. Polym. Sci. 34, 479–515.

Kvien, I., Sugiyama, J., Votrubec, M., Oksman, K., 2007. Characterization of starch based nanocomposites. J. Mater. Sci. 42, 8163–8171.

Lagaly, G., 1986. Interaction of alkylamines with different types of layered compounds. Solid State Ionics 22, 43–51.

Lai, L.S., Kokini, J.L., 1991. Physicochemical changes and rheological properties of starch during extrusion. Biotechnol. Prog. 7, 251–266.

Lawton, J.W., Shogren, R.L., Tiefenbacher, K.F., 2004. Aspen fiber addition improves the mechanical properties of baked cornstarch foams. Ind. Crop. Prod. 19, 41–48.

Lee, S.-Y., Hanna, M.A., 2008. Preparation and characterization of tapioca starch-poly(lactic acid)-Cloisite Na^+ nanocomposite foams. J. Appl. Polym. Sci. 110, 2337–2344.

Lee, S.Y., Hanna, M.A., 2009. Tapioca starch-poly(lactic acid)-Cloisite 30B nanocomposite foams. Polym. Compos. 30, 665–672.

Lee, S.Y., Xu, Y.X., Hanna, M.A., 2007. Tapioca starch-poly (lactic acid)-based nanocomposite foams as affected by type of nanoclay. Int. Polym. Process. 22, 429–435.

Lee, S.Y., Hanna, M.A., Jones, D.D., 2008a. An adaptive neuro-fuzzy inference system for modeling mechanical properties of tapioca starch-poly(lactic acid) nanocomposite foams. Starch-Starke 60, 159–164.

Lee, S.Y., Chen, H., Hanna, M.A., 2008b. Preparation and characterization of tapioca starch-poly(lactic acid) nanocomposite foams by melt intercalation based on clay type. Ind. Crop. Prod. 28, 95–106.

Lelievre, J., 1974. Starch gelatinization. J. Appl. Polym. Sci. 18, 293–296.

Li, J.H., Ren, C.L., Liu, X., De Hu, Z., Xue, D.S., 2007. "Green" synthesis of starch capped CdSe nanoparticles at room temperature. Mater. Sci. Eng. A 458, 319–322.

Li, M., Liu, P., Zou, W., et al., 2011a. Extrusion processing and characterization of edible starch films with different amylose contents. J. Food Eng. 106, 95–101.

Li, R., Liu, C., Ma, J., 2011b. Studies on the properties of graphene oxide-reinforced starch biocomposites. Carbohydr. Polym. 84, 631–637.

Liao, H.-T., Wu, C.-S., 2008. New biodegradable blends prepared from polylactide, titanium tetraisopropylate, and starch. J. Appl. Polym. Sci. 108, 2280–2289.

Lilichenko, N., Maksimov, R., Zicans, J., Merijs Meri, R., Plume, E., 2008. A biodegradable polymer nanocomposite: mechanical and barrier properties. Mech. Compos. Mater. 44, 45–56.

Liu, Q., Thompson, D.B., 1998. Effects of moisture content and different gelatinization heating temperatures on retrogradation of waxy-type maize starches. Carbohydr. Res. 314, 221–235.

Liu, H., Xie, F., Chen, L., Yu, L., Dean, K., Bateman, S., 2005. Thermal behaviour of high amylose cornstarch studied by DSC. Int. J. Food Eng. 1, Article 3/1–6.

Liu, H., Yu, L., Xie, F., Chen, L., 2006. Gelatinization of cornstarch with different amylose/amylopectin content. Carbohydr. Polym. 65, 357–363.

Liu, X., Yu, L., Liu, H., Chen, L., Li, L., 2008. In situ thermal decomposition of starch with constant moisture in a sealed system. Polym. Degrad. Stab. 93, 260–262.

Liu, H., Xie, F., Yu, L., Chen, L., Li, L., 2009. Thermal processing of starch-based polymers. Prog. Polym. Sci. 34, 1348–1368.

Liu, P., Yu, L., Wang, X., Li, D., Chen, L., Li, X., 2010. Glass transition temperature of starches with different amylose/amylopectin ratios. J. Cereal Sci. 51, 388–391.

Liu, D., Chang, P.R., Deng, S., et al., 2011a. Fabrication and characterization of zirconium hydroxide-carboxymethyl cellulose sodium/plasticized *Trichosanthes Kirilowii* starch nanocomposites. Carbohydr. Polym. 86, 1699–1704.

Liu, H., Chaudhary, D., Yusa, S.-I., Tadé, M.O., 2011b. Glycerol/starch/Na^+-montmorillonite nanocomposites: a XRD, FTIR, DSC and ^1H NMR study. Carbohydr. Polym. 83, 1591–1597.

Liu, H., Chaudhary, D., Ingram, G., John, J., 2011c. Interactions of hydrophilic plasticizer molecules with amorphous starch biopolymer—an investigation into the glass transition and the water activity behavior. J. Polym. Sci. B Polym. Phys. 49, 1041–1049.

Liu, Z., Zhao, L., Chen, M., Yu, J., 2011d. Effect of carboxylate multi-walled carbon nanotubes on the performance of thermoplastic starch nanocomposites. Carbohydr. Polym. 83, 447–451.

Lourdin, D., Coignard, L., Bizot, H., Colonna, P., 1997. Influence of equilibrium relative humidity and plasticizer concentration on the water content and glass transition of starch materials. Polymer 38, 5401–5406.

Lu, Y., Weng, L., Cao, X., 2006. Morphological, thermal and mechanical properties of ramie crystallites—reinforced plasticized starch biocomposites. Carbohydr. Polym. 63, 198–204.

Ma, X., Yu, J., 2004a. Formamide as the plasticizer for thermoplastic starch. J. Appl. Polym. Sci. 93, 1769–1773.

Ma, X., Yu, J., 2004b. The plastcizers containing amide groups for thermoplastic starch. Carbohydr. Polym. 57, 197–203.

Ma, X., Yu, J., 2004c. The effects of plasticizers containing amide groups on the properties of thermoplastic starch. Starch-Starke 56, 545–551.

Ma, X., Yu, J., Feng, J., 2004. Urea and formamide as a mixed plasticizer for thermoplastic starch. Polym. Int. 53, 1780–1785.

Ma, X.F., Yu, J.G., Ma, Y.B., 2005. Urea and formamide as a mixed plasticizer for thermoplastic wheat flour. Carbohydr. Polym. 60, 111–116.

Ma, X.F., Yu, J.G., Wan, J.J., 2006. Urea and ethanolamine as a mixed plasticizer for thermoplastic starch. Carbohydr. Polym. 64, 267–273.

Ma, X., Yu, J., Wang, N., 2007. Production of thermoplastic starch/MMT-sorbitol nanocomposites by dual-melt extrusion processing. Macromol. Mater. Eng. 292, 723–728.

Ma, X., Yu, J., Wang, N., 2008a. Glycerol plasticized-starch/multiwall carbon nanotube composites for electroactive polymers. Compos. Sci. Technol. 68, 268–273.

Ma, X., Jian, R., Chang, P.R., Yu, J., 2008b. Fabrication and characterization of citric acid-modified starch nanoparticles/plasticized-starch composites. Biomacromolecules 9, 3314–3320.

Ma, X., Chang, P.R., Yang, J., Yu, J., 2009. Preparation and properties of glycerol plasticized-pea starch/zinc oxide-starch bionanocomposites. Carbohydr. Polym. 75, 472–478.

Magalhães, N.F., Andrade, C.T., 2009. Thermoplastic corn starch/clay hybrids: effect of clay type and content on physical properties. Carbohydr. Polym. 75, 712–718.

Magalhães, N.F., Andrade, C.T., 2010. Calcium bentonite as reinforcing nanofiller for thermoplastic starch. J. Braz. Chem. Soc. 21, 202–208.

Majdzadeh-Ardakani, K., Nazari, B., 2010. Improving the mechanical properties of thermoplastic starch/poly(vinyl alcohol)/clay nanocomposites. Compos. Sci. Technol. 70, 1557–1563.

Majdzadeh-Ardakani, K., Navarchian, A.H., Sadeghi, F., 2010. Optimization of mechanical properties of thermoplastic starch/clay nanocomposites. Carbohydr. Polym. 79, 547–554.

Maksimov, R.D., Lagzdins, A., Lilichenko, N., Plume, E., 2009. Mechanical properties and water vapor permeability of starch/montmorillonite nanocomposites. Polym. Eng. Sci. 49, 2421–2429.

Mathew, A.P., Dufresne, A., 2002a. Morphological investigation of nanocomposites from sorbitol plasticized starch and tunicin whiskers. Biomacromolecules 3, 609–617.

Mathew, A.P., Dufresne, A., 2002b. Plasticized waxy maize starch: effect of polyols and relative humidity on material properties. Biomacromolecules 3, 1101–1108.

Mathew, A.P., Thielemans, W., Dufresne, A., 2008. Mechanical properties of nanocomposites from sorbitol plasticized starch and tunicin whiskers. J. Appl. Polym. Sci. 109, 4065–4074.

Matsui, K.N., Larotonda, F.D.S., Paes, S.S., Luiz, D.B., Pires, A.T.N., Laurindo, J.B., 2004. Cassava bagasse–Kraft paper composites: analysis of influence of impregnation with starch acetate on tensile strength and water absorption properties. Carbohydr. Polym. 55, 237–243.

McGlashan, S.A., Halley, P.J., 2003. Preparation and characterisation of biodegradable starch-based nanocomposite materials. Polym. Int. 52, 1767–1773.

Mondragón, M., Mancilla, J.E., Rodríguez-González, F.J., 2008. Nanocomposites from plasticized high-amylopectin, normal and high-amylose maize starches. Polym. Eng. Sci. 48, 1261–1267.

Mondragón, M., Hernández, E.M., Rivera-Armenta, J.L., Rodríguez-González, F.J., 2009. Injection molded thermoplastic starch/natural rubber/clay nanocomposites: morphology and mechanical properties. Carbohydr. Polym. 77, 80–86.

Morin, A., Dufresne, A., 2002. Nanocomposites of chitin whiskers from riftia tubes and poly(caprolactone). Macromolecules 35, 2190–2199.

Müller, C.M.O., Laurindo, J.B., Yamashita, F., 2009. Effect of cellulose fibers on the crystallinity and mechanical properties of starch-based films at different relative humidity values. Carbohydr. Polym. 77, 293–299.

Müller, C.M.O., Laurindo, J.B., Yamashita, F., 2011. Effect of nanoclay incorporation method on mechanical and water vapor barrier properties of starch-based films. Ind. Crop. Prod. 33, 605–610.

Nejad, M.H., Ganster, J., Volkert, B., 2010. Starch esters with improved mechanical properties through melt compounding with nanoclays. J. Appl. Polym. Sci. 118, 503–510.

Paillet, M., Dufresne, A., 2001. Chitin whisker reinforced thermoplastic nanocomposites. Macromolecules 34, 6527–6530.

Pandey, J.K., Singh, R.P., 2005. Green nanocomposites from renewable resources: effect of plasticizer on the structure and material properties of clay-filled starch. Starch-Starke 57, 8–15.

Park, H.-M., Li, X., Jin, C.-Z., Park, C.-Y., Cho, W.-J., Ha, C.-S., 2002. Preparation and properties of biodegradable thermoplastic starch/clay hybrids. Macromol. Mater. Eng. 287, 553–558.

Park, H.-M., Lee, W.-K., Park, C.-Y., Cho, W.-J., Ha, C.-S., 2003. Environmentally friendly polymer hybrids. Part I. Mechanical, thermal, and barrier properties of thermoplastic starch/clay nanocomposites. J. Mater. Sci. 38, 909–915.

Paul, D.R., Robeson, L.M., 2008. Polymer nanotechnology: nanocomposites. Polymer 49, 3187–3204.

Pavlidou, S., Papaspyrides, C.D., 2008. A review on polymer-layered silicate nanocomposites. Prog. Polym. Sci. 33, 1119–1198.

Pérez, S., Bertoft, E., 2010. The molecular structures of starch components and their contribution to the architecture of starch granules: a comprehensive review. Starch-Starke 62, 389–420.

Pérez, C.J., Alvarez, V.A., Mondragón, I., Vázquez, A., 2007. Mechanical properties of layered silicate/starch polycaprolactone blend nanocomposites. Polym. Int. 56, 686–693.

Pérez, S., Baldwin, P.M., Gallant, D.J., 2009. Structural features of starch granules I. In: James, B., Roy, W. (Eds.), Starch, third ed. Academic Press, San Diego, pp. 149–192.

Putaux, J.-L., Molina-Boisseau, S., Momaur, T., Dufresne, A., 2003. Platelet nanocrystals resulting from the disruption of waxy maize starch granules by acid hydrolysis. Biomacromolecules 4, 1198–1202.

Qiao, X., Jiang, W., Sun, K., 2005. Reinforced thermoplastic acetylated starch with layered silicates. Starch-Starke 57, 581–586.

Qiao, X., Tang, Z., Sun, K., 2011. Plasticization of corn starch by polyol mixtures. Carbohydr. Polym. 83, 659–664.

Radhakrishnan, T., Georges, M.K., Sreekumari Nair, P., Luyt, A.S., Djokovic, V., 2007. Study of sago starch-CdS nanocomposite films: fabrication, structure, optical and thermal properties. J. Nanosci. Nanotechnol. 7, 986–993.

Ratnayake, W.S., Jackson, D.S., Steve, L.T., 2008. Starch gelatinization. Adv. Food Nutr. Res. 55, 221–268.

Remsen, C.H., Clark, J.P., 1978. A viscosity model for a cooking dough. J. Food Process Eng. 2, 39–64.

Ren, P., Shen, T., Wang, F., Wang, X., Zhang, Z., 2009. Study on biodegradable starch/OMMT nanocomposites for packaging applications. J. Polym. Environ. 17, 203–207.

Rindlav-Westling, A., Gatenholm, P., 2002. Surface composition and morphology of starch, amylose, and amylopectin films. Biomacromolecules 4, 166–172.

Rodriguez, P., Muñoz-Aguirre, N., San-Martín Martinez, E., González de la Cruz, G., Tomas, S.A., Zelaya Angel, O., 2008. Synthesis and spectral properties of starch capped CdS nanoparticles in aqueous solution. J. Cryst. Growth 310, 160–164.

Romhány, G., Karger-Kocsis, J., Czigány, T., 2003. Tensile fracture and failure behavior of thermoplastic starch with unidirectional and cross-ply flax fiber reinforcements. Macromol. Mater. Eng. 288, 699–707.

Russell, P.L., 1987. Gelatinisation of starches of different amylose/amylopectin content. A study by differential scanning calorimetry. J. Cereal Sci. 6, 133–145.

Russo, M.A.L., O'Sullivan, C., Rounsefell, B., Halley, P.J., Truss, R., Clarke, W.P., 2009. The anaerobic degradability of thermoplastic starch: polyvinyl alcohol blends: potential biodegradable food packaging materials. Bioresour. Technol. 100, 1705–1710.

Saito, Y., Putaux, J.L., Okano, T., Gaill, F., Chanzy, H., 1997. Structural aspects of the swelling of β chitin in HCl and its conversion into α chitin. Macromolecules 30, 3867–3873.

Schaefer, D.W., Justice, R.S., 2007. How nano are nanocomposites? Macromolecules 40, 8501–8517.

Shi, R., Zhang, Z., Liu, Q., et al., 2007a. Characterization of citric acid/glycerol co-plasticized thermoplastic starch prepared by melt blending. Carbohydr. Polym. 69, 748–755.

Shi, R., Liu, Q., Ding, T., et al., 2007b. Ageing of soft thermoplastic starch with high glycerol content. J. Appl. Polym. Sci. 103, 574–586.

Shogren, R.L., Swanson, C.L., Thompson, A.R., 1992. Extrudates of cornstarch with urea and glycols: structure/mechanical property relations. Starch-Starke 44, 335–338.

Sinha Ray, S., Bousmina, M., 2005. Biodegradable polymers and their layered silicate nanocomposites: in greening the 21st century materials world. Prog. Mater. Sci. 50, 962–1079.

Sinha Ray, S., Okamoto, M., 2003. Polymer/layered silicate nanocomposites: a review from preparation to processing. Prog. Polym. Sci. 28, 1539–1641.

Sinha Ray, S., Okamoto, K., Okamoto, M., 2003. Structure–property relationship in biodegradable poly(butylene succinate)/layered silicate nanocomposites. Macromolecules 36, 2355–2367.

Siqueira, G., Bras, J., Dufresne, A., 2008. Cellulose whiskers versus microfibrils: influence of the nature of the nanoparticle and its surface functionalization on the thermal and mechanical properties of nanocomposites. Biomacromolecules 10, 425–432.

Spitalsky, Z., Tasis, D., Papagelis, K., Galiotis, C., 2010. Carbon nanotube-polymer composites: chemistry, processing, mechanical and electrical properties. Prog. Polym. Sci. 35, 357–401.

Sriupayo, J., Supaphol, P., Blackwell, J., Rujiravanit, R., 2005. Preparation and characterization of α-chitin whisker-reinforced chitosan nanocomposite films with or without heat treatment. Carbohydr. Polym. 62, 130–136.

Svagan, A.J., Hedenqvist, M.S., Berglund, L., 2009. Reduced water vapour sorption in cellulose nanocomposites with starch matrix. Compos. Sci. Technol. 69, 500–506.

Tako, M., Hizukuri, S., 2002. Gelatinization mechanism of potato starch. Carbohydr. Polym. 48, 397–401.

Tan, I., Flanagan, B.M., Halley, P.J., Whittaker, A.K., Gidley, M.J., 2007. A method for estimating the nature and relative proportions of amorphous, single, and double-helical components in starch granules by ^{13}C CP/MAS NMR. Biomacromolecules 8, 885–891.

Tang, X., Alavi, S., 2011. Recent advances in starch, polyvinyl alcohol based polymer blends, nanocomposites and their biodegradability. Carbohydr. Polym. 85, 7–16.

Tang, S., Zou, P., Xiong, H., Tang, H., 2008a. Effect of nano-SiO_2 on the performance of starch/polyvinyl alcohol blend films. Carbohydr. Polym. 72, 521–526.

Tang, X., Alavi, S., Herald, T.J., 2008b. Effects of plasticizers on the structure and properties of starch-clay nanocomposite films. Carbohydr. Polym. 74, 552–558.

Tang, X., Alavi, S., Herald, T.J., 2008c. Barrier and mechanical properties of starch-clay nanocomposite films. Cereal Chem. 85, 433–439.

Tang, H., Xiong, H., Tang, S., Zou, P., 2009. A starch-based biodegradable film modified by nano silicon dioxide. J. Appl. Polym. Sci. 113, 34–40.

Teixeira, E.d.M, Pasquini, D., Curvelo, A.A.S., Corradini, E., Belgacem, M.N., Dufresne, A., 2009. Cassava bagasse cellulose nanofibrils reinforced thermoplastic cassava starch. Carbohydr. Polym. 78, 422–431.

van Soest, J.J.G., Borger, D.B., 1997. Structure and properties of compression-molded thermoplastic starch materials from normal and high-amylose maize starches. J. Appl. Polym. Sci. 64, 631–644.

van Soest, J.J.G., Knooren, N., 1997. Influence of glycerol and water content on the structure and properties of extruded starch plastic sheets during aging. J. Appl. Polym. Sci. 64, 1411–1422.

van Soest, J.J.G., de Wit, D., Tournois, H., Vliegenthart, J.F.G., 1994. The influence of glycerol on structural changes in waxy maize starch as studied by Fourier transform infra-red spectroscopy. Polymer 35, 4722–4727.

Vertuccio, L., Gorrasi, G., Sorrentino, A., Vittoria, V., 2009. Nano clay reinforced PCL/starch blends obtained by high energy ball milling. Carbohydr. Polym. 75, 172–179.

Vigneshwaran, N., Sampath, K., Kathe, A.A., Varadarajan, P.V., Prasad, V., 2006. Functional finishing of cotton fabrics using zinc oxide–soluble starch nanocomposites. Nanotechnology 17, 5087–5095.

Viguié, J., Molina-Boisseau, S., Dufresne, A., 2007. Processing and characterization of waxy maize starch films plasticized by sorbitol and reinforced with starch nanocrystals. Macromol. Biosci. 7, 1206–1216.

Wan, Y.Z., Luo, H., He, F., Liang, H., Huang, Y., Li, X.L., 2009. Mechanical, moisture absorption, and biodegradation behaviours of bacterial cellulose fibre-reinforced starch biocomposites. Compos. Sci. Technol. 69, 1212–1217.

Wang, S.S., Chiang, W.C., Zhao, B., Zheng, X.G., Kim, I.H., 1991. Experimental analysis and computer simulation of starch-water interactions during phase transition. J. Food Sci. 56, 121–124.

Wang, X.-L., Yang, K.-K., Wang, Y.-Z., 2003. Properties of starch blends with biodegradable polymers. Polym. Rev. 43, 385–409.

Wang, N., Yu, J., Chang, P.R., Ma, X., 2007. Influence of citric acid on the properties of glycerol-plasticized dry starch (DTPS) and DTPS/poly(lactic acid) blends. Starch-Starke 59, 409–417.

Wang, N., Zhang, X., Han, N., Bai, S., 2009. Effect of citric acid and processing on the performance of thermoplastic starch/montmorillonite nanocomposites. Carbohydr. Polym. 76, 68–73.

Wang, N., Zhang, X., Han, N., Liu, H., 2010. A facile method for preparation of thermoplastic starch/urea modified montmorillonite nanocomposites. J. Compos. Mater. 44, 27–39.

Wei, Q., Kang, S.-Z., Mu, J., 2004. "Green" synthesis of starch capped CdS nanoparticles. Colloids Surf. A Physicochem. Eng. Asp. 247, 125–127.

Wilhelm, H.-M., Sierakowski, M.-R., Souza, G.P., Wypych, F., 2003a. The influence of layered compounds on the properties of starch/layered compound composites. Polym. Int. 52, 1035–1044.

Wilhelm, H.M., Sierakowski, M.R., Souza, G.P., Wypych, F., 2003b. Starch films reinforced with mineral clay. Carbohydr. Polym. 52, 101–110.

Woehl, M.A., Canestraro, C.D., Mikowski, A., Sierakowski, M.R., Ramos, L.P., Wypych, F., 2010. Bionanocomposites of thermoplastic starch reinforced with bacterial cellulose nanofibres: effect of enzymatic treatment on mechanical properties. Carbohydr. Polym. 80, 866–873.

Wollerdorfer, M., Bader, H., 1998. Influence of natural fibres on the mechanical properties of biodegradable polymers. Ind. Crop. Prod. 8, 105–112.

Wu, Z., Feng, W., Feng, Y., et al., 2007. Preparation and characterization of chitosan-grafted multiwalled carbon nanotubes and their electrochemical properties. Carbon 45, 1212–1218.

Wu, H., Liu, C., Chen, J., Chang, P.R., Chen, Y., Anderson, D.P., 2009a. Structure and properties of starch/α-zirconium phosphate nanocomposite films. Carbohydr. Polym. 77, 358–364.

Wu, M., Wang, M., Ge, M., 2009b. Investigation into the performance and mechanism of SiO_2 nanoparticles and starch composite films. J. Text. Inst. 100, 254–259.

Xie, F., Yu, L., Liu, H., Dean, K., 2006. Effect of compatibilizer distribution on thermal and rheological properties of gelatinized starch/biodegradable polyesters blends. Int. Polym. Process. 21, 379–385.

Xie, F., Xue, T., Yu, L., Chen, L., Li, X., Zhang, X., 2007. Rheological properties of starch-based materials and starch/poly(lactic acid) blends. Macromol. Symp. 249–250, 529–534.

Xie, Y., Chang, P.R., Wang, S., Yu, J., Ma, X., 2011. Preparation and properties of halloysite nanotubes/plasticized *Dioscorea opposita* Thunb. starch composites. Carbohydr. Polym. 83, 186–191.

Xie, F., Pollet, E., Halley, P.J., Avérous, L., 2013. Starch-based nano-biocomposites. Prog. Polym. Sci. 38, 1590–1628.

Xiong, Z., Li, C., Ma, S., et al., 2013a. The properties of poly(lactic acid)/starch blends with a functionalized plant oil: tung oil anhydride. Carbohydr. Polym. 95, 77–84.

Xiong, Z., Yang, Y., Feng, J., et al., 2013b. Preparation and characterization of poly(lactic acid)/starch composites toughened with epoxidized soybean oil. Carbohydr. Polym. 92, 810–816.

Yan, L., Chang, P.R., Zheng, P., 2011. Preparation and characterization of starch-grafted multiwall carbon nanotube composites. Carbohydr. Polym. 84, 1378–1383.

Yan, Q., Hou, H., Guo, P., Dong, H., 2012. Effects of extrusion and glycerol content on properties of oxidized and acetylated corn starch-based films. Carbohydr. Polym. 87, 707–712.

Yang, J.-H., Yu, J.-G., Ma, X.-F., 2006a. Preparation of a novel thermoplastic starch (TPS) material using ethylenebisformamide as the plasticizer. Starch-Starke 58, 330–337.

Yang, J.-H., Yu, J.-G., Ma, X.-F., 2006b. Preparation and properties of ethylenebisformamide plasticized potato starch (EPTPS). Carbohydr. Polym. 63, 218–223.

Yang, J.-H., Yu, J.-G., Ma, X.-F., 2006c. Study on the properties of ethylenebisformamide and sorbitol plasticized corn starch (ESPTPS). Carbohydr. Polym. 66, 110–116.

Yao, K., Cai, J., Liu, M., et al., 2011. Structure and properties of starch/PVA/nano–SiO_2 hybrid films. Carbohydr. Polym. 86, 1784–1789.

Yu, J., Wang, N., Ma, X., 2005. The effects of citric acid on the properties of thermoplastic starch plasticized by glycerol. Starch-Starke 57, 494–504.

Yu, L., Dean, K., Li, L., 2006. Polymer blends and composites from renewable resources. Prog. Polym. Sci. 31, 576–602.

Yu, J., Yang, J., Liu, B., Ma, X., 2009. Preparation and characterization of glycerol plasticized-pea starch/ZnO-carboxymethylcellulose sodium nanocomposites. Bioresour. Technol. 100, 2832–2841.

Yun, Y.-H., Hwang, K.-J., Wee, Y.-J., Yoon, S.-D., 2011. Synthesis, physical properties, and characterization of starch-based blend films by adding nano-sized TiO_2/poly(methyl metacrylate-*co*-acrylamide). J. Appl. Polym. Sci. 120, 1850–1858.

Zeng, M., Huang, Y., Lu, L., Fan, L., Lourdin, D., 2011. Effects of filler-matrix morphology on mechanical properties of corn starch–zein thermo-moulded films. Carbohydr. Polym. 84, 323–328.

Zeppa, C., Gouanvé, F., Espuche, E., 2009. Effect of a plasticizer on the structure of biodegradable starch/clay nanocomposites: thermal, water-sorption, and oxygen-barrier properties. J. Appl. Polym. Sci. 112, 2044–2056.

Zhang, Q.-X., Yu, Z.-Z., Xie, X.-L., Naito, K., Kagawa, Y., 2007. Preparation and crystalline morphology of biodegradable starch/clay nanocomposites. Polymer 48, 7193–7200.

Zhang, J.-S., Chang, P.R., Wu, Y., Yu, J.-G., Ma, X.-F., 2008. Aliphatic amidediol and glycerol as a mixed plasticizer for the preparation of thermoplastic starch. Starch-Starke 60, 617–623.

Zhao, X., Wang, B., Li, J., 2008. Synthesis and electrorheological activity of a modified kaolinite/carboxymethyl starch hybrid nanocomposite. J. Appl. Polym. Sci. 108, 2833–2839.

Zheng, P., Chang, P.R., Yu, J., Ma, X., 2009a. Formamide and 2-hydroxy-*N*-[2-(2-hydroxypropionylamino)-ethyl] propionamide (HPEP) as a mixed plasticizer for thermoplastic starch. Carbohydr. Polym. 78, 296–301.

Zheng, P., Chang, P.R., Yu, J., Ma, X., 2009b. Preparation of Sb_2O_3-carboxymethyl cellulose sodium nanoparticles and their reinforcing action on plasticized starch. Starch-Starke 61, 665–668.

Zobel, H.F., 1988. Molecules to granules: a comprehensive starch review. Starch-Starke 40, 44–50.

Further Reading

Journal review papers

Avérous, L., 2004. Biodegradable multiphase systems based on plasticized starch: a review. Polym. Rev. 44, 231–274.

Avérous, L., Halley, P.J., 2009. Biocomposites based on plasticized starch. Biofuels Bioprod. Biorefin. 3, 329–343.

Chivrac, F., Pollet, E., Avérous, L., 2009. Progress in nano-biocomposites based on polysaccharides and nanoclays. Mater. Sci. Eng. R. Rep. 67, 1–17.

Follain, N., Joly, C., Dole, P., Bliard, C., 2005. Mechanical properties of starch-based materials. I. Short review and complementary experimental analysis. J. Appl. Polym. Sci. 97, 1783–1794.

Le Corre, D., Bras, J., Dufresne, A., 2010. Starch nanoparticles: a review. Biomacromolecules 11, 1139–1153.

Pandey, J.K., Pratheep Kumar, A., Misra, M., Mohanty, A.K., Drzal, L.T., Singh, R.P., 2005. Recent advances in biodegradable nanocomposites. J. Nanosci. Nanotechnol. 5, 497–526.

Pérez, S., Bertoft, E., 2010. The molecular structures of starch components and their contribution to the architecture of starch granules: a comprehensive review. Starch-Starke 62, 389–420.

Raquez, J.M., Nabar, Y., Narayan, R., Dubois, P., 2007. New developments in biodegradable starch-based nanocomposites. Int. Polym. Process. 22, 463–470.

Schaefer, D.W., Justice, R.S., 2007. How nano are nanocomposites? Macromolecules 40, 8501–8517.

Tang, X., Alavi, S., 2011. Recent advances in starch, polyvinyl alcohol based polymer blends, nanocomposites and their biodegradability. Carbohydr. Polym. 85, 7–16.

Xie, F., Halley, P.J., Avérous, L., 2012. Rheology to understand and optimize processibility, structures and properties of starch polymeric materials. Prog. Polym. Sci. 37, 595–623.

Xie, F., Pollet, E., Halley, P.J., Avérous, L., 2013. Starch-based nano-biocomposites. Prog. Polym. Sci. 38, 1590–1628.

Yu, L., Dean, K., Li, L., 2006. Polymer blends and composites from renewable resources. Prog. Polym. Sci. 31, 576–602.

Zhao, R., Torley, P., Halley, P., 2008. Emerging biodegradable materials: starch- and protein-based bio-nanocomposites. J. Mater. Sci. 43, 3058–3071.

Books

Avérous, L., Pollet, E., 2013. Macro-, micro-, and nanocomposites based on biodegradable polymers. In: Handbook of Biopolymer-Based Materials. Wiley-VCH Verlag GmbH & Co. KGaA, Weinheim, Germany, pp. 173–210.

Halley, P.J., Avérous, L.R., 2014. Starch Polymers: From Genetic Engineering to Green Applications. Elsevier, Amsterdam.

Studies on mechanical, thermal, and morphological characteristics of biocomposites from biodegradable polymer blends and natural fibers

5

R. Muthuraj, M. Misra, A.K. Mohanty
University of Guelph, Guelph, ON, Canada

5.1 Introduction

Bioplastics can contribute significantly to sustainable development in terms of the environment and ecological systems. Over the past two decades, the utilization of biobased and/or biodegradable polymeric materials has gained significant research attention from both academia and industry. Bioplastics are a group of plastics with a wide range of properties and applications. According to European Bioplastics, a bioplastic can be biobased, biodegradable, or a combination of both biobased and biodegradable. In 2010, the worldwide annual production of plastics was reported to be 265 Mt (Imre and Pukánszky, 2013), and reached 299 Mt in 2013 (http://www.plasticseurope.org). Meanwhile, bioplastics production was estimated to be 1.6 Mt in 2013 (http://en.european-bioplastics.org/market). A European bioplastics market report has predicted annual bioplastic production to be ~6.7 Mt in 2018 (http://en.european-bioplastics.org/market), which is a 22% hike compared to bioplastics production in 2012. Currently, bioplastics contribute 10–15% of the total plastic market share and by 2020 the market share is expected to be around 25–30%, mainly driven by the gradual increase in the number of bioplastics processing companies (http://www.hkc22.com/bioplastics.html). Based on overwhelming chemical and material demand, the U.S. Department of Energy aims to reach 10% renewable resource-based chemical production by 2020 and 50% by 2050 (Plant, 2020). The main reasons for biopolymers being a prime focus are reduction of CO_2 emissions, reduced waste problems, and establishment of sustainable alternatives to conventional materials.

Petroleum-based nonbiodegradable polymers can remain intact in landfills for decades, causing grave environmental concerns. Such nonbiodegradable polymers are not suited for applications requiring a short life span. Incineration of many nonbiodegradable plastics waste after their end use can produce harmful emissions. The U.S. Environmental Protection Agency (EPA) has reported that most of the synthetic plastic waste enters landfills, oceans, and lakes (DiGregorio, 2009). This problem can be overcome by using biodegradable polymers that can offer an effective solution to this

Biocomposites: Design and Mechanical Performance. http://dx.doi.org/10.1016/B978-1-78242-373-7.00014-7

scenario as they are degradable in the presence of naturally occurring microorganisms. Biodegradable polymers, whether obtained from renewable resources or petroleum resources, are potential substitutes for some nonbiodegradable polymers.

5.2 Biodegradable and compostable polymeric materials

All biodegradable plastics fall under the bioplastic category and for it to be biodegradable, it should possess the ability to break down into smaller molecules through the action of naturally occurring microbes (ASTM D6400-12, 2012). In order to assess the biodegradability of polymeric substances, standard experimental procedures are established by national and international organization. These standards are used to investigate the biodegradability of polymeric materials in specific environmental conditions. The ASTM D6400 standard defines compostable polymeric materials as "a plastic that undergoes biological degradation during composting to yield CO_2, H_2O, inorganic compounds and biomass at a rate consistent with other known compostable materials and leaves no visually distinguishable or toxic residues" (ASTM D6400-12, 2012). This (ASTM D6400) test method is similar to that given in ISO 17088. A detailed biodegradability and compostability mechanism of polymeric materials has been reviewed by Muniyasamy et al. (2013a). In the 1980s, synthetic biodegradable polymers were first introduced and were considered potential substitutes for nonbiodegradable polymers (Soroudi and Jakubowicz, 2013). Currently, there are many biodegradable polymers produced from petrochemical resources and renewable resources (John and Thomas, 2008). Biodegradability of the polymeric materials generally depends on the molecular formula and does not depend on the monomer origin (Witt et al., 1999, 2001). Therefore, biodegradable polymers can be derived from nonrenewable resources and/or renewable resources. We can sort these biodegradable polymers based on the source of the feedstock used to make them (Table 5.1).

5.3 Renewable resource-based biodegradable polymers: some examples

Renewable resource-based biodegradable polymers have their origin either from plant or biological resources. On the other hand, nonrenewable resources-based biodegradable polymers are generally made from fossil fuel-based monomers. A few examples of renewable resource-based biodegradable polymers are discussed below.

5.3.1 Poly(lactic acid)

Poly(lactic acid) (PLA) was first synthesized by DuPont scientist Wallace Hume Carothers in 1931 as a typical linear aliphatic thermoplastic polyester, derived from agricultural feedstock like corn, sugarcane, and sugar beet (Yu et al., 2006; Groot et al., 2011). PLA is synthesized from lactic acid (2-hydroxypropanoic acid) monomers

Table 5.1 Classification and molecular structure of some biodegradable polymers

	Biodegradable polymers	Chemical structures
Renewable resources	❖ Poly(lactic acid), PLA ❖ Polyhydroxybutyrate, PHB ❖ Poly(hydroxy butyrate-co-valerate), PHBV	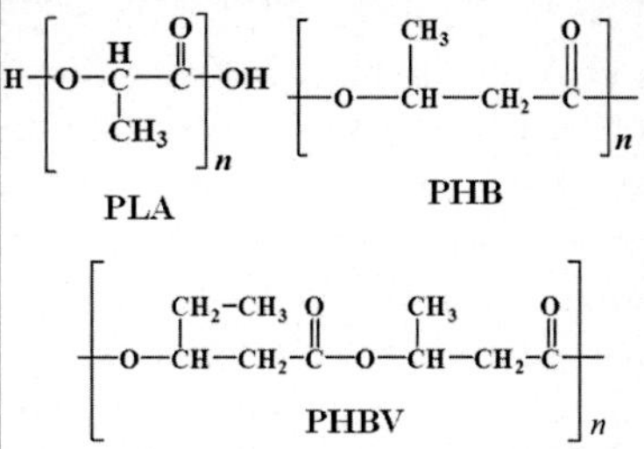
Fossil fuel resources	❖ Poly(butylene adipatee-co-terephthalate), PBAT ❖ Poly(butylene succinate), PBS ❖ Poly(caprolactone), PCL	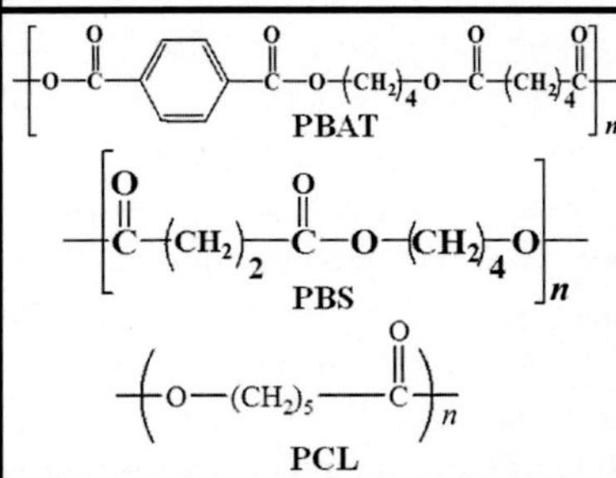

through the application of heat under vacuum conditions. PLA can be produced by various methods, such as direct condensation polymerization, ring-opening polymerization (ROP), and azeotropic dehydration (Auras et al., 2004). Low molecular weight PLA is not acceptable for some of the applications due to poor mechanical performance. High molecular weight PLA preparation by direct dehydration condensation is not feasible due to the equilibrium not favoring a high molecular weight polymer. Therefore, Cargill Dow LLC developed a new technology (ROP) to produce high molecular weight PLA through an economic route (Lim et al., 2008; Henton et al., 2005). This process involves three separate steps (Auras et al., 2004; Lim et al., 2008). First step involves polycondensation reaction, which converts lactic acid into PLA oligomers (1000–5000 Da). In the second step, PLA oligomers are catalytically converted into a mixture of lactide stereoisomers. After vacuum distillation of lactide stereoisomers, ROP is typically preferred to produce high molecular weight PLA due to its high degree of controllability. One of the biggest advantages of the ROP method is the elimination of water formation step, allowing high molecular weight structures to be attained (Garlotta, 2001). This was the first attempt to synthesize renewable resources based on economically and commercially viable biodegradable polymer (PLA) by melt stage rather than in solution (Madhavan Nampoothiri et al., 2010).

Stereochemistry of PLA plays a vital role in the synthesis of stereoisomers such as L (+), D (−), and meso (L, D) lactic acid by using an appropriate microorganism. Four different types of PLA can be produced using different kinds of lactide monomers: L-PLA (PLLA), D-PLA (PDLA), mixtures of L and D, and meso-PLA. The L-PLA and D-PLA are optically active, as opposed to mixtures of L and D and meso-PLA, which are optically inactive. These steriochemical structures determine the crystallinity of PLA. For instance, PLA with >93% L-lactic acid is semicrystalline in nature, whereas PLA containing 50–93% L-lactic acid becomes amorphous in nature (Auras et al., 2004). Due to its amorphous nature, PDLA does not show any crystallization or melting enthalpies under differential scanning calorimetry tests, while PLLA has melt crystallization, cold crystallization, and melting enthalpies due to formation and deformation of crystals. Other parameters affecting PLA properties include polymerization conditions, thermal history, purity, and molecular weight.

In terms of tensile strength, tensile modulus (Table 5.2; Huda et al., 2006), and gas (O_2 and CO_2) barrier properties (Vandewijngaarden et al., 2014), PLA is superior to polypropylene (PP); however, toughness and impact strength are lower than those of many commodity polymers (Lim et al., 2008). PLA can be processed in conventional processing equipment such as extrusion, injection, stretch blow molding, extrusion blow film, thermoforming, foaming, melt spinning, cast film, and sheet (Lim et al., 2008). Therefore, PLA has potential applications ranging from food packaging, compostable bags, household items, and textiles, to medical supplies. Due to these attractive properties of PLA, the global production of PLA is expected to reach \$5.2 billion by 2020 with a growth rate of 19.5% during 2013 to 2020 (http://www.reportsandintelligence.com/polylactic-acid-market). Unfortunately, PLA has some inherent drawbacks, including low heat deflection temperature (Zhang and Thomas, 2011), poor melt strength in comparison with polyolefins (Lunt, 1998), slow crystallization rate (Madhavan Nampoothiri et al., 2010), and brittleness. Moreover, PLA is not able to biodegrade at room temperature; that is, it is not home compostable (Zhang and Thomas, 2011). These properties prevent PLA from being applied in varied applications. Several attempts have been made to address these issues using plasticization, rubber toughening, blending with other polymers, and reinforcement with natural fibers (Ren, 2010; Kfoury et al., 2013; Terzopoulou et al., 2014). Recently, the toughness modification of PLA has been reviewed exhaustively by Kfoury et al. (2013).

5.3.2 Microbial polyesters–polyhydroxyalkanoates

Microbial polyester synthesis began in 1901 and detailed investigations were conducted in 1925 (Yu et al., 2006). In the late 1920s, Maurice Lemoigne first discovered polyhydroxyalkanoates (PHAs) using the *Bacillus megaterium* bacteria (Verlinden et al., 2007). Two different types of bacterial polyesters along with their chemical structure are shown in Table 5.1. PHAs are a family of linear thermoplastic polyesters, which can degrade by the microorganisms (Rosa and Lenz, 2013). PHAs can be found as homopolymers or copolymers. The performance of PHAs is dependent on their chemical composition. After the 1970s, researchers' curiosity and an impending energy crisis initiated the commercialization of poly(3-hydroxybutyrate), PHB. PHB is

Table 5.2 **Properties of some biodegradable polymers in comparison to non-biodegradable polymers**

Polymers	Density (g/cm^3)	Melting point (°C)	Tensile strength (MPa)	Tensile modulus (MPa)	Elongation at break (%)	References
PLA	1.21-1.25	150-162	21-60	350-3500	2.5-6	Koronis et al. (2013) and Van de Velde and Kiekens (2002)
PLLA	1.24-1.30	170-200	15.5-150	2700-4140	3-10	Koronis et al. (2013) and Van de Velde and Kiekens (2002)
PDLA	1.25-1.27	Amorphous	27.6-50	1000-3450	2-10	Van de Velde and Kiekens (2002)
PHB	1.18-1.26	168-182	40	3500-4000	5-8	Koronis et al. (2013) and Van de Velde and Kiekens (2002)
PHBV	1.23-1.25	144-172	20-25	500-1500	17.5-25	Koronis et al. (2013) and Mohanty et al. (2005)
PBS	1.26-1.32	96-114	19-36	324-647	200-807	Fujimaki (1998)
PCL	1.11-1.15	58-65	21-42	210-440	300-1000	Van de Velde and Kiekens (2002) and Bogoeva-Gaceva et al. (2007)
PBAT	1.25	110-115	36	80	820	Mohanty et al. (2005) and Audic and Chaufer (2012)
HDPE	0.92	110	10	177	700	Fujimaki (1998) and Bogoeva-Gaceva et al. (2007)
PP	0.9-1.16	161-170	30-40	1100-1600	20-400	Koronis et al. (2013)

the first microbial polyester synthesized through fermentation of sugar and starch with the help of a microorganism. PHB is a saturated linear homopolymer, which behaves like traditional thermoplastics. It can be processed in traditional processing equipment such as extrusion and injection and compression molding. PHB has properties that are very similar to those of a commodity polymer like PP (Avella et al., 2000a; Parulekar and Mohanty, 2006). However, poor thermal stability (Rosa and Lenz, 2013; Liu et al., 2009), embrittlement due to high crystallinity (Yu et al., 2006; Liu et al., 2009) and narrow processing window (Vroman and Tighzert, 2009) of PHB are hindering its wide range of applications. The brittleness/impact strength of PHB can be modified in two ways: copolymerization of hydroxybutyrate (HB) and hydroxyvalerate (HV) (Yu et al., 2006), and blending with tough polymers (Parulekar and Mohanty, 2006). For instance, poly(3-hydroxybutyrate-*co*-3-hydroxyvalerate) (PHBV) is more ductile and flexible compared to PHB (Liu et al., 2009).

In 1990, poly(3-hydroxybutyrate-*co*-3-valerate) (PHBV), was introduced into the market by Imperial Chemical Industries Inc. (ICI) (Yu et al., 2006). Currently, the major PHAs producers are Tianan (Enmat™), Kaneka (Nodax™), PHB Industrial (Biocycle®), Mitsubishi Gas Chemicals (Biogreen™), Biomer® biopolyester, and Metabolix (Mirel®). PHAs have desirable properties like higher melting point (Table 5.2) and higher crystallinity, comparable to commodity polymer (PP), with added advantages like good biodegradability, compostability, and biocompatibility. PHAs have potential applications in medical fields such as wound management, vascular system, drug delivery, orthopedics, and urology (Wu et al., 2009). However, the thermal degradation temperature range of PHBVs is 180–200 °C, which is very close to their melting temperature. In addition, the main drawback of PHB and low valerate content PHBV is their poor impact strength and low elongation at break caused by low nucleation density. All the above factors combined with high cost are hindering the application of PHBV in a wide range of fields (Grassie et al., 1984). Increasing the valerate unit up to 34 mol% can help in decreasing the brittleness of PHBV (Avella et al., 2000a). However, an increase in the HV content can decrease the melting temperature, tensile strength, modulus of elasticity, and glass transition temperature (Avella et al., 2000a). Most of the commercially produced PHBV contains HV content below 15 mol% (Liu et al., 2009). The melting temperature (T_m) of such PHBV is still high (e.g., HV = 11 mol%, T_m = 157 °C) and thermal degradation may occur during melt processing (Avella et al., 2000a). In addition, it possesses very low ductility, very similar to PHB. The synthesis of PHBV with higher valerate content requires a lot of energy consumption for production, sterilization, and purification. Therefore, intensive research needs to be directed toward reducing the production cost of PHBV, and in bringing the cost to be on par with traditional plastics like PP (Koronis et al., 2013). Extensive research has been conducted on polymers from the PHA family to overcome the above-discussed drawbacks (Avella et al., 2000a; Liu et al., 2009; Vroman and Tighzert, 2009; Vieira et al., 2011). Different approaches adopted to address the challenges of PHAs are discussed in the following blend and composites sections. In addition to PHBV-based polymer blends, natural fibers are a potential candidate to reduce the cost of PHBV-based materials while enhancing some of their mechanical properties (Terzopoulou et al., 2014).

5.4 Fossil fuel-based biodegradable polymers: some examples

Biodegradability of a polymer is basically based on chemical structure unlike from the source from which it is obtained. Thus, biodegradable polymers besides renewable-based sources (as discussed above) can also be made from fossil fuel sources. A few examples of such polymer are reviewed below.

5.4.1 *Poly(butylene succinate)*

Poly(butylene succinate) (PBS) is a promising aliphatic biodegradable thermoplastic polyester (it's chemical structure is shown in Table 5.1). It is synthesized by condensation polymerization of 1,4-butanediol (BDO) and succinic acid in the presence of a catalyst (Fujimaki, 1998). Low molecular weight PBS is weak and brittle in nature. Therefore, Showa Highpolymer Co. Ltd in Japan produced relatively high molecular weight PBS under the trade name Bionolle™ (Fujimaki, 1998). High molecular weight PBS was achieved through melt condensation polymerization followed by coupling reaction with a chain extender (e.g., hexamethylene diisocynate). The condensation polymerization can be performed using either direct melt polymerization or solution polymerization. Direct melt polycondensation is the most promising method for PBS synthesis as it can produce the high molecular weight PBS suitable for food contact applications (Xu and Guo, 2010). Currently, some of the companies producing PBS include Xinfu pharmaceutical (China), Hexing (China), Ire Chemical (Korea), Nippon Shokubai (Japan), Mitsubishi Chemical (Japan), Showa Highpolymer (Japan), and Jinfa Tech (China) (Kabasci and Bretz, 2011; Babu et al., 2013).

Interestingly, PBS can also be produced from either biobased, fossil fuel-based monomers or a combination of both biobased and fossil fuel-based monomers. There are several companies around the world producing pilot scale biobased succinic acid, including BioAmber and Myriant. Mitsubishi Chemical has developed a new technology to produce partially biobased PBS having 54% biobased content derived from biobased succinic acid (Zini and Scandola, 2011). It can be used in applications requiring food contact and applications besides nonfood contact uses while meeting global biodegradability standards. Traditionally, 1,4-BDO is produced from fossil fuel resources. Alternatively, the biobased 1,4-BDO is produced through a fermentation process using glucose monomers. Therefore, it is possibile to produce PBS from 100% renewable resource-based monomers (http://www.themoldingblog.com/2012/11/13/faurecia-bets-on-bio-pbs-for-auto-interiors/). PBS is a biodegradable polyester with balanced thermomechanical properties compared to other polymers such as PP, polyethylene, poly(hydroxybutyrate), and PLA (Ma et al., 2013a). Mechanical and thermal properties of PBS depend mainly on crystallinity, crystal structure, and molecular weight (Jacquel et al., 2011). Good thermal stability, processing properties, semicrystalline nature, and an ideal melting point range makes PBS a promising biodegradable polymer (Yoo and Im, 1999; Wang et al., 2009; Kim and Park, 1999). Notably, the strength of PBS is comparable to that of low-density polyethylene (LDPE) and PP, as shown in Table 5.2. PBS-based products are widely used for many applications, such as packaging film, agriculture mulch film, sheets, laminates, bags, hygiene products, nonwovens, split yarn, monofilament, and multifilament (Sinha Ray et al., 2003). However, the stiffness, water vapor barrier, melt viscosity, and impact strength of PBS are often not sufficient for many applications. These drawbacks can be effectively addressed by blending with other polymers (Muthuraj et al., 2014), fillers (Calabia et al., 2013; Risse et al., 2014), and additives (Wang et al., 2009). The stiffness of PBS has been improved by the addition of natural fibers (Calabia et al., 2013) and blending with brittle polymers like PHB (Ma et al., 2012). The melt viscosity and gas barrier properties of the PBS were improved with the help of reactive agents (Wang et al., 2009) and nanofillers (Risse et al., 2014), respectively.

Poly(butylene succinate-*co*-adipate), PBSA, is a random copolyester, made from 1,4-butanediol, succinic acid, and adipic acid by polycondensation reaction. It has excellent impact strength, elongation, and processability. Comonomer content dictates the physical properties of the copolymer of PBS, like PBSA. For instance, the adipate content must be lower than 15 mol% in order to maintain the melting point of PBSA above 100 °C (Tserki et al., 2006). In addition, adipate comonomer content has a greater influence on the biodegradability and mechanical properties of PBSA (Fujimaki, 1998). PBSA has higher impact strength and elongation as compared to PBS, whereas the tensile strength and melting point of the PBSA are lower than that of PBS.

5.4.2 Poly(butylene adipate-co-terephthalate)

Polymers that are solely aromatic polyesters such as poly(ethylene terephthalate) (PET) and poly(butylene terephthalate) (PBT) are resistant to microbial attack (Marten et al., 2005). However, copolymerization of aliphatic monomers with aromatic monomers results in a biodegradable polymer, like poly(butylene adipate-*co*-terephthalate) (PBAT). PBAT, a biodegradable aliphatic-aromatic copolyester, is derived from fossil fuel-based adipic acid, terephthalic acid, and butane diol by a polycondensation reaction (Gu et al., 2008). The chemical structure of PBAT is shown in Table 5.1. Due to its mixed chemical structure, PBAT is enzymatically hydrolyzed in the presence of microorganisms followed by mineralization, like aliphatic polyesters (Witt et al., 2001). The concentration of aromatic functionality plays a vital role in the performance of the PBAT. PBAT shows good thermal and mechanical properties with a terephthalic acid concentration above 35 mol% but good biodegradability was achieved with an aromatic moiety concentration lower than 55 mol% (Vroman and Tighzert, 2009). Therefore, PBAT with a range of about 35–55 mol% of terephthalic acid can offer an optimal performance (Witt et al., 1996).

In 1998, PBAT with 22.2 mol% of aromatic moiety was commercialized under the trade name Ecoflex®, by BASF (Witt et al., 2001). BASF studied the biodegradability of PBAT under composting conditions. More than 90% of PBAT was observed to be metabolized within 3 months (Siegenthaler et al., 2012). Witt and coworkers (2001) have studied the biodegradability of Ecoflex® under composting conditions according to the DIN-V-54900 standard and they have concluded that there is no environmental risk after composting Ecoflex®. Mechanical properties of PBAT are comparable to polyethylene (Table 5.2; Breulmann et al., 2000), but with water vapor permeability of

Table 5.3 **Ecoflex®-based masterbatches for different applications**

Masterbatches	Filler/additive types	Filler content	Proposed applications
Ecoflex® Batch AB1	Chalk	60%	Packaging films, compost bags, and agricultural films
Ecoflex® Batch SL 05	Erucamide ESA (lubricant)	5%	Packaging films, compost bags, and agricultural films
Ecoflex® Batch C Black	Carbon black	35%	Packaging films, compost bags, and agricultural films
Ecoflex® Batch C White	Titanium dioxide (TiO_2)	60%	Packaging films, compost bags, and agricultural films
Ecoflex® Batch SL 2	Lubricant (wax)	5%	Packaging films, compost bags, and agricultural films

http://worldaccount.basf.com/wa/plasticsAP~en_GB/portal/show/content/products/biodegradable_plastics/ecoflex_product_literature; http://www.plastixportal.co.za/basf_pdf/Ecoflex_Brochure.pdf

PBAT (240 $gm^{-2}d^{-1}$) being higher than LDPE (3 $gm^{-2}d^{-1}$) (Siegenthaler et al., 2012). In order to overcome this shortcoming, BASF is producing a wide range of biodegradable masterbatched Ecoflex® (mixed with wax, talc, carbon black, chalk, or silica, etc.) to fulfill the requirements set for different applications; details of the masterbatches are shown in Table 5.3 (http://worldaccount.basf.com/wa/plasticsAP~en_GB/portal/show/content/products/biodegradable_plastics/ecoflex_product_literature). Among them, wax-based masterbatch film has the potential to reduce the water vapor permeability up to 75% compared to virgin PBAT (Siegenthaler et al., 2012). PBAT possesses excellent toughness, biodegradability, and processability. Therefore, PBAT is widely used for compostable organic waste bags, agricultural mulch films, as well as lamination/coatings for starch-based products (Javadi et al., 2010a). Moreover, it can be used to tailor the properties of starch, PLA, and PHAs, thereby opening up new applications for PBAT-based materials. Currently, there are few companies producing PBAT: BASF (Germany), Ire Chemical (Korea), Shandong Fuwin New Material (China), and Xinfu Pharmaceutical (China).

5.4.3 Poly(caprolactone)

Poly(caprolactone) (PCL) is a linear aliphatic biodegradable polyester derived from caprolactone by ROP. PCL can also be produced by anionic, cationic, coordination, radical, and enzymatic catalyzed polymerization (Stridsberg et al., 2002; Uyama and Kobayashi, 2002). PCL is a typical semicrystalline synthetic thermoplastic with crystallinity ~50%, melting temperature of 58–65°C (depending on the crystallinity), and glass transition temperature of −60°C (Van de Velde and Kiekens, 2002). PCL has been most widely used in drug delivery, tissue engineering, and dentistry, and it is approved by the U.S. Food and Drug Administration (FDA). PCL is commercially available under different trade names: Capa™ (Perstorp, UK), Placcel™ (Daicel Chemical Indus, Japan), and Tone® (Dow, USA). A well-known PCL/starch biodegradable blend is commercially available in the market under the brand name Mater-Bi® Z grades, and is produced by Novamount, Italy (Bastioli, 1998). This blend is mainly used for biodegradable and/or compostable films and sheets. Inherent toughness and biocompatibility of PCL are desired attributes that makes it an ideal choice of blending partner. In this view, PCL has been blended with PLA, PHAs, and PBS for different biomedical applications (Sarazin et al., 2004; Hinüber et al., 2011; Casarin et al., 2011; Del Gaudio et al., 2012; Can et al., 2014).

5.5 Recyclability of biodegradable polymers

Recyclability of plastics is a process that extends their service lives before discarding. Plastic waste can be recycled in two different ways: mechanical and chemical recycling. Mechanical recycling involves a polymer being remelted and reprocessed into desired products through different processing techniques (http://www.plasticsrecyclers.eu/mechanical-recycling). Chemical recycling is a process (depolymerization) that converts polymers into monomers, which are then used as a feedstock for further polymerization process (Soroudi and Jakubowicz, 2013). Among these recycling methods, mechanical recycling is the most favored method for bioplastics recycling. Many studies have investigated the mechanical recyclability of biodegradable polymers, like PLA (Żenkiewicz et al., 2009; Pillin et al., 2008; Lopez et al., 2012; Zembouai et al., 2014), PCL (Moraczewski, 2014), PHBV (Zembouai et al., 2014; Zaverl et al., 2012), PLA/PHBV blend (Zembouai et al., 2014), and PBS (Kanemura et al., 2012). Multiple extrusion and injection molding of the PLA (3002D NatureWorks®, USA) has been studied by Żenkiewicz et al. (2009). After reextruding PLA for 10 times, the tensile strength, tensile strain, and impact strength were reduced by 5.2%, 2.4%, and 20.2%, respectively. On the other hand, melt flow rate, and oxygen and water vapor permeability rates were significantly increased. After the 10-times extrusion cycle of PLA, the oxygen and water vapor permeability rates were increased by 39% and 18%, respectively. Thermal properties, such as thermal stability, crystallization, and melting point, were slightly reduced while there was no change in glass transition temperature. Another study investigated the recyclability of PLLA (L9000, Biomer, Germany) up to seven-injection molding (Pillin et al., 2008). The authors found that the tensile modulus was not heavily affected, whereas rheological properties, molecular weight, hardness, tensile strength, and strain at break were reduced considerably. A recent study also proved that 8-times mechanical recycling of PLLA (L9000, Biomer, Germany) leads to considerable reduction in its mechanical properties as well as thermal stability (Lopez et al., 2012). Meanwhile, the melt flow index (MFI) of the 8-times recycled PLLA showed 15 times higher value than unprocessed PLLA. The reduced mechanical properties were attributed to thermomechanical degradation of PLLA during melt reprocessing. Consequently, the molecular weight, thermal stability, and viscosity of the PLLA decreased. Such reduction in properties can be overcome by using thermal stabilizers such as quinine and tropolone (Pillin et al., 2008). However, the study concluded that PLLA (L9000, Biomer) can be recycled up to 5 times without a significant change in the tensile properties and flexural modulus (Lopez et al., 2012). Zaverl and coworkers (2012) investigated thermal and mechanical properties of PHBV being recycled 7 times. Mechanical properties such as impact, tensile, and flexural properties were not affected after recycling up to 4 times; however, there was a slight decrease after a fourth cycle. According to this study (Zaverl et al., 2012), there was no change observed in the molecular weight of PHBV after 2 cycles. A considerable molecular weight reduction was observed after the third (8.7%), fourth (13.5%), and fifth cycles (16.6%). Even after the fifth round of recycling, melting temperature and thermal stability of the PHBV were not affected significantly. The chemical structure of the PHBV was not changed during melt reprocessing, which was confirmed by Fourier transform spectroscopy. On the contrary, PHBV is more susceptible to thermomechanical degradation compared to PLA (Zembouai et al., 2014). Therefore, blending of PHBV with PLA can minimize the degradation of PHBV during recycling. Recyclability of PCL has been investigated by Moraczewski (2014). After 8 cycles, there were no changes observed in tensile strength and elongation at break. However, MFI, thermal stability, and Charpy impact strength of PCL were dependent on the number of cycles. This study suggests that the recycled PCL can be blended with virgin PCL. According to Kanemura et al. (2012), PBS (Bionolle® 1020) can be recycled without affecting its bending strength and modulus when reprocessed up to 3 times at 140 °C using a compression molding method. In addition, the molecular weight of the PBS was not heavily affected after 3 times of reprocessing at 140 °C. Therefore, this study concluded that PBS is a suitable candidate for material recycling.

5.6 Durability of biodegradable polymers

Generally, durability is very important for polymers as it increases their suitability for a wide range of applications. The durability of polymeric materials can be investigated by exposure to simulated environmental variables such as temperature and humidity. Only a few studies have examined the durability of biodegradable polymeric materials (Kim and Kim, 2008; Kim et al., 2014a,b; Harris and Lee, 2010, 2013). These literature sources report that biodegradable polymers are very sensitive to elevated temperature and humidity. Recently, Harris and Lee (2010, 2013) have assessed the durability of PLA at elevated temperature and humidity with respect to time. Their experimental findings suggest that after 28 days of exposure at 70 °C temperature with 90% relative humidity, PLA materials experienced severe loss in mechanical properties. Therefore, commercial PLA cannot fulfill long-term durability requirements set by the automotive industry. This drawback can be addressed either by reducing the hydrolysis reaction through the addition of scavengers/sacrificial compounds or by blending/alloying with other durable polymers (Harris and Lee, 2010). The degradation of PHBV has been examined in distilled water and marine environments at different temperatures (Deroiné et al., 2014a,b). The mechanical property performance of the PHBV gradually deteriorated with increasing immersion time. This effect is attributed to the hydrolytic degradation of ester bonds in the PHBV backbone. Usually, the durability of the polyesters depends on the chemical structure. Among PLA, PBS, PBAT, and PBSA, PLA and PBAT are more durable compared to PBS and PBSA (Kim and Kim, 2008). However, the durability of PBS and its composites has been noticed to improve with the addition of trimethylolpropane triacrylate (TMPTA) as an antihydrolysis agent (Kim and Kim, 2008). The durability of PBS, PBAT, and their blend has been investigated by exposure to 50 °C and 90% relative humidity for a duration of up to 30 days (Muthuraj et al., 2015). This study concluded that the biodegradable polymers (PBS, PBAT, and PBS/PBAT) could readily undergo hydrolytic degradation in the presence of elevated temperature and humidity. The durability of biodegradable polymers needs to be improved under high humidity and temperature to expand their applications.

5.7 Polymer blends: some examples

Use of some neat polymers for commercial applications cannot fulfill all product requirements. One of the alternatives for modifying the properties of polymers is to blend with other polymer (one or more), which have complementary properties. Blending two or more structurally different polymers might improve or tailor some of the properties and its cost performance, making a blend that is able to meet specific end-use requirements (Kfoury et al., 2013). Furthermore, polymer blending can be performed in standard processing equipment such as twin-screw extruders. Generally, polymer blends possess unique physicomechanical properties, which are usually not present in their individual components. Polymer blends have a wide range of applications in fields such as automotive, household items, electrical, and electronics. Some of the commercially available biodegradable polymer blends are listed in Table 5.4. A blend of polylactic acid/lignin/fatty acid/wax shows mechanical properties similar to acrylonitrile butadiene styrene (ABS). It has good thermal insulation properties as well as excellent chemical resistance against polar substances (http://www.modernplastics.com/oct08/pdf/ecogehrpla.pdf). The properties of STARCLA™ (20 μm film) are comparable to polyethylene (http://www.showa-denko.com/wp-content/uploads/2014/Starcla_2014.pdf). BASF is producing biodegradable PBAT/PLA (55/45 wt.%) blend under the trade name Ecovio®. The mechanical performance of Ecovio® is similar to the properties of high-density polyethylene (Siegenthaler et al., 2012).

As discussed previously, high cost and inferior mechanical properties relative to several commodity plastics are the downside of most of the biodegradable polymers and, therefore, wide-scale adaptability of these polymers have been significantly hampered. In order to effectively overcome these drawbacks, melt blending is a fast, economical, and convenient approach compared to developing a novel polymeric material through synthetic polymerization technique (Koning et al., 1998). For example, slow crystallization behavior and poor impact resistance of PLA hinders its range of industrial applications. Researchers have found that crystallization and cold crystallization were enhanced by incorporating PBS into PLA systems (Shibata et al., 2006). Yokohara and Yamaguchi (2008) have conducted studies on the structure–properties correlation of PLA/PBS blend. They found that the cold-crystallization temperature of PLA was reduced in the blend compared to neat PLA. This result indicated that

Table 5.4 List of companies engaged in the production of some biodegradable polymer blends

Blends	Applications	Trade name	Manufacturer
PLA/PBAT	Blown films	Ecovio®	BASF, Germany
PLA/copolymer	Injection molding: housewares, tackle	Cereplast sustainable®	Cereplast Inc., USA
PCL-based blend	Films and injection-molded items	Mater-Bi®	Novamount, Italy
PLA/copolyester blend	Blown film and coextrusion	Bio-Flex®	FKuR, Germany
PLA-based blend	Extrusion and blown film	Starcla™	Showa Denko Europe
Biodegradable copolyester blend	Blown film and injection-molded parts	Terraloy™	Teknor Apex
PLA/PBS or PBSA	Food service ware	Ingeo™ AW 240D	NatureWorks, USA
PBAT/PHA	Fiber and nonwoven	BioTuf™ 970	Heritage Plastics Inc., USA
PLA/lignin/fatty acid/wax	Furniture, toys, medical, food industry, and replacement of PET	PLA-L	Modern Plastics, USA

Modified after Soroudi and Jakubowicz (2013), http://www.showa-denko.com/products/biodegradable-polymer-resin/, and http://www.teknorapex.com/division/bioplastics/product_lines.

PBS acted as a nucleating agent in the PLA matrix. Blends of biodegradable polymers are being used for short-term applications, such as absorbable medical implantations, compostable food packaging containers, and agricultural mulch films. Kim and co-workers (2014a,b) have reported the durability of talc-filled biodegradable PBS/PBAT (55/45 wt.%) blends under marine environments. Better elastic properties were observed in biodegradable materials compared to that of the recycled PE.

5.7.1 Miscible biodegradable polymer blends

At thermodynamic equilibrium, mixtures of polymers that exist in a single phase at the nanometer scale are referred to as miscible (soluble) blends (Manias and Utracki, 2014). Miscibility of polymers in the blend system can be detected through glass transition temperature (T_g) and phase morphology. Miscible polymer blends show single glass transition temperature. In most cases, the miscibility of two polymers depends on the molecular weight, chemical structure of the polymer, and crystallinity (Fekete et al., 2005). In the polymer blend system, some of the properties are determined by miscibility, interaction, and phase morphology. For example, the mechanical properties of the miscible blends are found to follow the rule of mixture (Robeson, 2007). A blend is considered partially miscible if there exists phase separation but each polymer-rich phase contains a sufficient amount of the other polymer to alter the properties of that phase (e.g., the glass transition temperature) (Robeson, 2007). Normally, polymers are miscible with each other up to a certain limit. The Flory–Huggins theory is most frequently used to calculate mutual solubility of polymer blends using an interaction parameter. The Flory–Huggins interaction parameter (χ) can be obtained by the equation (Szabó et al., 2004)

$$\chi_{12} = \frac{V_r}{RT}(\delta_1 - \delta_2)$$

where V_r is reference volume, T is absolute temperature, δ_1 and δ_2 are the solubility parameters of the components, and R is the gas constant. These solubility parameters can be calculated using group contribution methods (Small, 1953). If the interaction parameter (χ) value is less than 0.5, it can be considered as miscible blend (Pillin et al., 2006). The Flory–Huggins theory always yields negative interaction parameter values for miscible blends. For instance, a PHB/poly(epichlorohydrin-*co*-ethylene oxide), PEEO, blend has an interaction parameter value of −0.089 (Zhang et al., 2000) while a PLA/PBS blend has an interaction parameter value of −0.15 (Park and Im, 2002). Another study showed limited miscibility of PHB with 20% PBS blend (Qiu et al., 2003a). This miscibility was evidenced by a shift in glass transition temperature and depression of the PHB melting temperature. Bhatia et al. (2007) found that a PLA/PBS blend system has partial miscibility when the PBS composition was lower than 20 wt.%. This result has been confirmed by rheological properties. However, tensile strength and modulus of the PLA/PBS blends are lower compared to neat PLA. The brittleness of the PLA is reduced with the addition of PBS, thus making it a suitable material for packaging applications (Bhatia et al., 2007). Park and Im (2002) studied blends of two typical biopolyesters (PLA and PBS) by melt blending through an extruder. In this polymer blend system, a single glass transition temperature was observed, which was attributed to enhanced miscibility, between the polymer phases. Spherulites morphology of the blends was observed through optical polarizing microscopy and they concluded that crystallization-induced phase separation occurred when the PBS volume increased by more than 40 wt.% in the blend system. Moreover, PLA/poly(ethylene glycol) (Sheth et al., 1997; Sungsanit et al., 2012), PLA/poly(ethylene oxide) (Pillin et al., 2006; Nijenhuis et al., 1996), and PHB/poly(ethylene oxide) (Avella and Martuscelli, 1988) blends were able to form miscible blend system.

5.7.2 Immiscible biodegradable polymer blends

If mixtures of two polymers exhibit separate phase morphology (i.e., matrix-droplet and cocontinuous morphology), they are referred to as immiscible blends (Ravati and Favis, 2010), as shown in Figure 5.1a. In the binary blends, the matrix-droplet phase morphology (Figure 5.1b) changes to cocontinuous morphology (Figure 5.1c) with an increase in the volume fraction of the dispersed phase. Immiscible polymer blends exhibits different glass transition temperatures characteristic of their pure blend components and core-shell morphology. The core-shell morphology of the immiscible blends depends on the viscosity of the individual blend components, processing aspects, composition, and compatibility between the components. Inherently, some biodegradable polymers (PBS, PBSA, PBAT, certain PHAs, and PCL) have good toughness

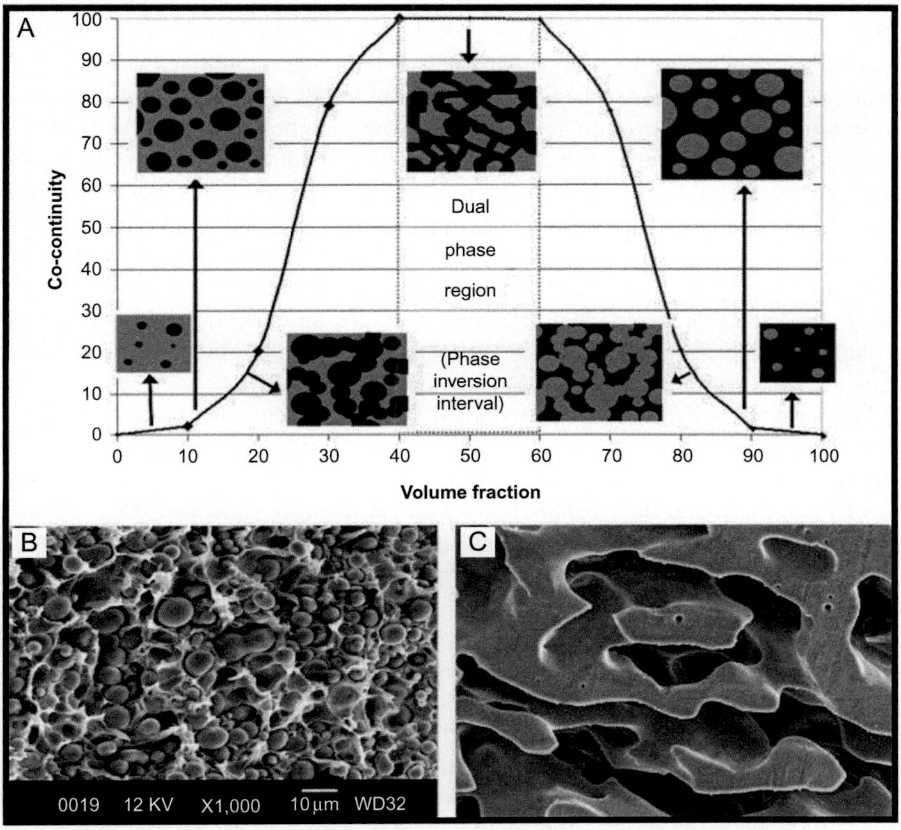

Figure 5.1 (a) Schematic representation of the evolution of morphology in a binary immiscible blend, (b) matrix/dispersed morphology, and (c) cocontinuous morphology. Adapted with kind permission from Ravati and Favis (2010), Copyright © 2015, Elsevier, License number 3591550612924.

and elongation. Therefore, they are considered as a promising candidate for modifying brittle biodegradable polymers while preserving the biodegradability. However, a large amount of literature shows that most of the polymer blends are immiscible due to poor dispersion of the inclusion phase in the continuous phase, weak interfacial adhesion, and instability between the components (Kfoury et al., 2013). For instance, biodegradable polymer blends of PHBV/PCL (Qiu et al., 2005), PHBV/PBAT (Javadi et al., 2010a), PHBV/PBS (Qiu et al., 2003b), PHBV/PLA (Nanda et al., 2011), PLA/PBS (Yokohara and Yamaguchi, 2008), PLA/PBAT (Signori et al., 2009; Jiang et al., 2005), PLLA/PCL (Vilay et al., 2010), PLA/PHB (http://www.metabolix.com/sites/default/files/ANTEC-PLA_PVC.pdf), PCL/PBS (John et al., 2002), and PBS/PBAT (Muthuraj et al., 2014) are all immiscible blends. In order to improve the interaction between the pure polymers in the blend system, compatibilization has proven to be an effective strategy for the immiscible polymer blend systems.

Jiang et al. (2005) studied the toughening of PLA through melt blending with PBAT. Dynamic mechanical analysis (DMA) results showed two independent T_g peaks corresponding to their individual components, suggesting that the PLA/PBAT blends were immiscible. Mechanical properties of the PLA blends showed the greatest improvement in elongation and toughness with incorporation of PBAT from 5 to 20 wt.%. These improvements were correlated to debonding induced by shear yield and weak interfacial adhesion between the components in the blends. Wu et al. (2012) studied the PBS/PLA blend morphology, rheological properties, and interfacial tension between the polymers. They observed that PBS/PLA blends were thermodynamically not miscible and possessed cocontinuous phase morphology in the 50/50 wt.% blend. DSC analysis reveled that PBS has no significant effect on melting point and crystallization behavior of PLA but, on the other hand, PLA had little effect on melting point and cold crystallization of PBS.

Furthermore, Zhou et al. (2013) studied PBS/PLA blends for biomedical applications. They also made similar observations that PBS/PLA blends were immiscible and their mechanical properties followed the rule of mixture. The hydrolysis behavior of PBS, PLA, and their blends was investigated under simulated body fluid for 16 months at 37 °C. The PBS/PLA blends lost their tensile properties earlier than the neat polymers due to faster hydrolysis reaction facilitated by the interface present between the blend components.

Throughout the hydrolysis time, the molecular weight reduction of the blends was faster than that of parent polymers. A blend of PHBV/PCL has been studied by Jenkins et al. (2007) and they suggested that the blend is not miscible. Melt-blended immiscible PHBV/PLA blend was studied by Nanda et al. (2011). The mechanical performances of the blends were dependent on processing parameters and composition of the constituents. The elongation at break observed in PHBV/PLA blend (Nanda et al., 2011) increased slightly compared to neat PLA and PHBV. Similar trends were observed in compostable PLA/PHBV blend by Ma et al. (2013b). It was demonstrated that matrix yielding and interfacial debonding were responsible for the increased toughness in PLA/PHBV blends. According to Qiu et al. (2003b), a PHBV/PBS blend is not miscible due to the independent glass transition temperatures of two phases.

Some of the immiscible blends are considered compatible blends when they show fine phase morphology and satisfactory performance in their mechanical properties (Koning et al., 1998). These types of blends show their characteristic T_g, but most often the T_g values are shifted toward other components in the blends, indicating improved compatibility. The performance of the compatible, immiscible polymer blends depends on many factors, including their interaction, component properties, composition, and compatibility between the polymers phases. Among all the parameters, interaction is a factor of greater importance because it determines the thickness of the interface between the components. John et al. (2002) have investigated the compatibility and miscibility of PBS, PBAT, and PCL blends. From DSC analysis, the blends of PCL/PBAT and PCL/PBS are seen to have two melting points, which correspond to their individual melting points. This result clearly indicates that these blends are not completely miscible. However, from DMA analysis, a single T_g was observed for the PBS/PBAT blend. This is because the T_g values of both the neat PBS and PBAT are very close to each other. Moreover, these blends showed an interesting behavior when tested for tensile strength. The tensile strength of PBS/PBAT (70/30 wt.%) and PCL/PBS (30/70 wt.%) blends was higher than that of neat PBS, PBAT, and PCL. This improvement was attributed to the interaction between the blended components as well as a change in crystallinity. Recently, Muthuraj et al. (2014) have reported a similar type of synergistic effect in the PBS/PBAT blends.

5.7.3 Compatibilization of polymer blends

Compatibilization is an interfacial phenomenon in the heterogeneous polymer blends. This interfacial activity can be modified by reactive and nonreactive processing strategies. Compatibilization is a process by which the blend properties are enhanced while increasing adhesion between the phases, reducing the interfacial tension and stabilizing morphology (Utracki, 2002). To achieve these goals, there are several strategies to the method of compatibilization. The degree of compatibility between the components in the blends can be enhanced through the addition of compatibilizers (Marten et al., 2005; Gu et al., 2008; Witt et al., 1996). The resulting mechanical properties of the immiscible, compatible blend will have a balance of their parent polymers properties or show a synergistic improvement (Manias and Utracki, 2014). Many commercially available polymer blends on the market are mostly compatibilized using a compatibilizer (Imre and Pukánszky, 2013). Polymers with polar groups can interact between the components in the blend through nonbonded interactions such as dipole–dipole interactions and hydrogen bonding (Imre and Pukánszky, 2013). Therefore, biopolymers have more mutual miscibility than polyolefins. Compatibilization of polymer blends was reviewed in detail by Koning et al. (1998) and Imre and Pukánszky (2013). According to these sources, the process of adding a premade compatibilizer (block copolymers) that has a good affinity toward the blend components is referred to as physical compatibilization. Introducing a compatibilizer that can promote chemical reaction and/or specific interaction between the blend components is referred to as reactive or in situ compatibilization. Varieties of reactive compatibilizers have been identified in earlier publications (Sirisinha and Somboon, 2012; Kumar et al., 2010; Zhang et al., 2009; Lin et al., 2012; Shin and Han, 2013; Liu et al., 2010; Oyama, 2009; Harada et al., 2007; Takayama and Todo, 2006). The following section discusses these compatibilization methods.

Generally, premade graft or block copolymers are mostly used to modify the heterogeneous polymer blends interface by a nonreactive compatibilization method. In order to make effective compatibilization of the blends by block copolymers, the copolymer must have a maximum solubility with components in the blend. This leads to good interfacial adhesion while lowering the interfacial tension between the components. The molecular weight and concentration of compatibilizer (block copolymer) should be slightly higher than that of critical entanglement and critical micelle concentration, respectively (Liu and Baker, 1992). In addition, compatibilized blends exhibit dispersed particle size at a submicron level, which prevents coalescence during subsequent processing (Koning et al., 1998). Nonreactive biopolymer blends with block copolymers have been widely studied in the literature (Na et al., 2002; Chavalitpanya and Phattanarudee, 2013; Kim et al., 2000). However, nonreactive compatibilization is less efficient than reactive compatibilization (Koning et al., 1998). Reactive extrusion generates reactive sites, which interact with the other polymer component and localize themselves at the interface between the components in the blends. Such localization at an interface enhances the mechanical properties and decreases the domain size of the inclusion phase. Reactive extrusion of the polymer blends can be performed by one-step (in situ) or two-step extrusion. All of the components are introduced simultaneously in the one-step reactive extrusion (Ma et al., 2014). In the two-step reactive extrusion, the first step involves functionalization of polymer with reactive agents followed by the second step in which the functionalized polymers are blended with other components through extrusion (Ryan, 2002).

5.7.4 Nonreactive compatibilization of biodegradable polymer blends

Historically, immiscible polymer blends can be compatibilized through the addition of a graft or block copolymers (Utracki, 2002). The block copolymer acts as a compatibilizer between the components in the immiscible polymer blend systems. As a consequence, the block copolymer can reduce the interfacial tension between the components (Ryan, 2002), as shown in Figure 5.2.

There are many commodity polymer blends compatibilized with a block copolymer. For example, styrene acrylonitrile, SAN, and styrene butadiene styrene, SBS blends, have been compatibilized by the addition of a diblock copolymer (Fayt and Teyssie, 1986). These compatibilized blends exhibit good mechanical properties

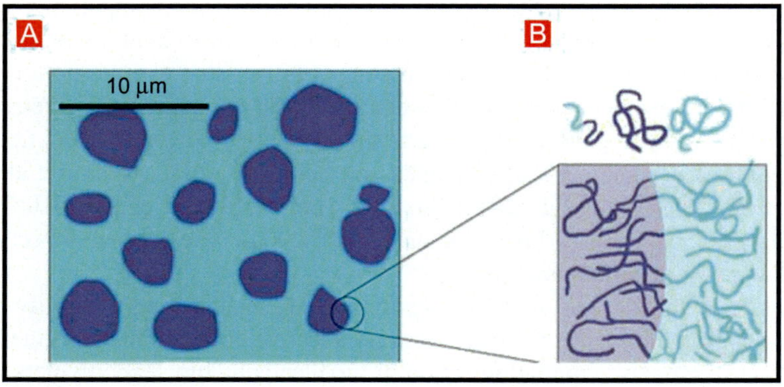

Figure 5.2 Adding a compatibilizing agent, such as a diblock copolymer, to a polymer blend can improve its stability, but is more likely to result in a dispersed morphology rather than a cocontinuous morphology. (a) A 2-D slice of a compatibilized blend with dispersed phase morphology, represented by a minority dark blue phase and a majority turquoise phase. (b) A molecular schematic showing how the diblock copolymers are segregated at the interface between the two phases.

compared to uncompatibilized blends. A similar type of compatibilization effect has been reported for polystyrene/low- and high-density PE blends via the addition of butadiene-styrene block copolymers (Fayt et al., 1982). In this blend system, the compatibility between the components has been found to be dependent on processing parameters. Recently, similar attempts have been made for bioplastic blends (Vilay et al., 2010; Na et al., 2002; Kim et al., 2000; Wu et al., 2010; Tsuji et al., 2003; Avella et al., 2000b; Ceccorulli et al., 2002; Wang and Hillmyer, 2001). LDPE-PLA block copolymers have been used as a compatibilizer in the PLA/LDPE blends in order to improve the properties of resulting blends (Wang and Hillmyer, 2001). These blends showed considerable increase in deformability with tensile strength lower than that of neat PLA. Another study also showed a remarkable increase in ductility with the addition of a block copolymer, PHB-block-poly(methyl methacrylate), PMMA, in the PHB/PMMA blend (Ceccorulli et al., 2002). PLLA/PCL (Vilay et al., 2010) and PBS/PCL (Can et al., 2014) blends were compatibilized with triblock copolymer, (polyethylene oxide–PP oxide–polyethylene oxide). These two studies concluded that the blends with compatibilizer achieved better compatibility and toughness compared to those of the uncompatibilized blends. The enhanced compatibility between the components was further confirmed by a T_g shift and decrease in PCL domain size in the matrix.

5.7.5 Reactive compatibilization of biodegradable polymer blends: some specific examples

Poor compatibility between the phases is a major drawback in multiphase immiscible polymer blends. Immiscible polymer blends can be compatibilized by introducing a compatibilizer. Extensive work has been conducted on improving the compatibility between the phases in heterogeneous biodegradable polymer blend systems.

5.7.5.1 Reactive compatibilization of PLA/PBAT blends

Recently, significant research has been directed toward producing compatible PLA/PBAT blend through reactive processing. For instance, Sirisinha and Somboon (2012) investigated the reactive PLA/PBAT blends by adding 0–0.5 wt.% of dicumyl peroxide (DCP) for blown-film processing. This study was completed using fixed PLA/PBAT ratio (70/30 wt.%) and varied the DCP loading from 0 to 0.5 wt.%. Rheological, MFI, and melt strength properties were measured to investigate the synergistic effect of DCP addition into the blend system. The addition of DCP into a PLA/PBAT blend was found to increase the elasticity and viscosity. In addition, good melt strength of the blend allowed for easy processing when blowing the plastic into a film.

According to the literature (Kumar et al., 2010), solubility parameter values of PLA and PBAT are 10.1 and 22.95 $(cal/cm^3)^{1/2}$, respectively. These solubility parameter differences lead to the formation of immiscible blends as well as a reduction of their mechanical performance. These drawbacks are addressed by using glycidyl methacrylate (GMA) as a reactive compatibilizer in melt-blended PLA/PBAT. Kumar and coworkers (2010) studied the effect of mechanical, thermal, as well as morphological properties of PLA/PBAT blends with GMA. They incorporated GMA (3 and 5 wt.%) into the optimum PLA/PBAT blend system in order to improve compatibility between the components. They found that the impact strength of the PLA/PBAT/GMA (70/25/5) blends increased 72% as compared to virgin PLA. The increment in properties was attributed to the formation of ethylene acrylic ester at the interface with the addition of GMA. Morphological examinations confirmed that compatibility between the PLA and PBAT phases was in fact improved in the presence of GMA. Thermomechanical analysis (DMA) further corroborated the improvement noticed in compatibility between the PLA and PBAT and heterogeneous phase morphology of the PLA/PBAT blends. Furthermore, with the addition of GMA within PLA/PBAT blends, reduction in crystallinity and inward shift in melting point were observed by DSC analysis, supporting the enhanced compatibility between the two polymers.

Zhang et al. (2009) investigated reactive blending of PLA/PBAT with a reactive processing agent, a random terpolymer of ethylene, acrylic ester, and glycidyl methacrylate (abbreviated as T-GMA in their work). They prepared reactive blends of PLA/PBAT with T-GMA at a varying concentration of 1–10 wt.% by a melt blending technique. They found that the tensile toughness and impact strength of the blends were improved with 1 wt.% of T-GMA incorporation, but no changes occurred with increasing T-GMA concentrations. The toughness and impact strength improvement were attributed to increased miscibility between PLA and PBAT phase, which was confirmed by SEM analysis. From the DSC analysis, they found two glass transition temperatures from the blends, indicating that the blends were not homogenous after adding T-GMA. During melt blending, the PLA/PBAT blend can be compatibilized by a transesterification product with the help of terbutyl titanate (TBT) (Lin et al., 2012). The mechanical and thermomechanical properties of this compatibilized blend were improved significantly compared to the noncompatibilized PLA/PBAT blend. Best elongation at break (298%), impact strength (9 kJ/m^2), and tensile strength (45 MPa) of PLA/PBAT blend were obtained with 0.5% TBT concentration. The compatibility between PLA and PBAT has been improved through transesterification reaction products, which was demonstrated by SEM micrographs.

5.7.5.2 Reactive compatibilization of PLA/PBS blends

PLA/PBS blends have been compatibilized using lysine triisocynate (LTI) as a reactive processing agent (Harada et al., 2007). Significantly higher level of improvement was observed in the impact strength (two- to fourfold increase compared to PLA), and elongation at break with a decrease in melt flow rate. This improvement could possibly be due to the grafting or cross-linking of LTI with PBS or PLA in the blends. However, this reactive compatibilization was detrimental to tensile strength of PLA/PBS blends. A similar trend was observed in the PLLA/PBS blend with a DCP reactive agent (Wang et al., 2009). These blends achieved better interfacial adhesion and fine dispersed phase morphology owing to the increased compatibility between the blended components. Consequently, the PLLA/PBS blend with 0.1 phr of DCP showed a 12-fold increment in the Izod impact strength compared to virgin PLLA. This improvement was attributed to debonding through shear yielding. However, both flexural strength and flexural modulus continuously decreased with increasing DCP concentration. Yet, DCP is not an appropriate choice when trying to improve elongation at break of PLLA/PBS blends. Unlike PLLA/PBS blends, PLA/PBAT blends with 0.1–0.2 phr DCP showed significant improvements in elongation at break as well as increased impact toughness (Ma et al., 2014). Chen et al. (2005) investigated the compatibility between the PLA and PBS blends with twice-functionalized nanoclay (TFC). Scanning electron microscopy (SEM) revealed that the PBS domain size gradually decreased when 2 wt.% TFC was incorporated into PLA/PBS blends, due to increased compatibility between the polymer phase in the blend system. Elongation at break was increased with the addition of TFC into PLA/PBS blends. Flory–Huggins theory suggested that TFC has a high compatibility with the PLLA phase when compared to the PBS phase. In addition, there was no improvement in the elongation at break when nanoclay without functionalization was incorporated, which indicates that no compatibility existed between the PLA/PBS blends. Ojijo et al. (2013) have improved the toughness of PLA/PBSA blends by in situ compatibilization with triphenyl phosphite (TPP). Comparing the impact strength and elongation at break for neat PLA and PLA/PBSA blend containing 10% PBSA and 2% TPP, the impact strength and elongation at break of compatibilized PLA/PBSA blend showed 166% and 516% higher value compared to neat PLA. These improvements were attributed to shear yielding of the matrix, which was initiated by the debonding between the blended components. In addition, there was no significant change in the tensile strength and thermal stability of the compatibilized blends. The authors also suggested that the barrier properties, thermal properties, and mechanical properties of this toughened PLA/PBSA blend system could be improved by the addition of clay into such a system.

5.7.5.3 Reactive compatibilization of PLA/PHB and PHBV/PBS blends

Reactive compatibilized PLA/20 wt.% PHB (Mirel™ M4300) copolymers showed 1.3 J tensile toughness and 200% elongation at break (http://www.metabolix.com/sites/default/files/ANTEC-PLA_PVC.pdf). These values are significantly higher than neat PLA and nonreactive blends of PLA/20 wt.% PHB. It is important to note that this amorphous PHB is thermodynamically immiscible with PLA. However, reactively compatibilized PLA/PHB blend showed that the notched Izod impact strength of immiscible PLA/20 wt.% PHB copolymer blend is 20 times greater than that of virgin PLA (from 0.4 to ~8.2 ft.lb/in.) (http://www.metabolix.com/sites/default/files/ANTEC-PLA_PVC.pdf). The same blend displayed a tensile modulus of ~2 GPa and tensile strength of ~50 MPa. Therefore, the stiffness–toughness performance of this blend is superior compared to poly(butylene terephthalate) and PP.

PBS is another biodegradable thermoplastic polyester with tensile strength comparable to PP and stiffness similar to LDPE with fair processability using processing equipment of conventional polyolefin thermoplastics (Mohanty et al., 2000). PBS has a glass transition temperature of around −30°C, which gives it a high toughness compared to the polymer at room temperature. Also, the high elongation of PBS (more than 200%), makes it an ideal candidate to blend with a brittle biopolymer (Qiu et al., 2003b). Therefore, PBS seems to be one of the biodegradable polymers that can be effective in improving the toughness of PHBV. Ma et al. (2012) investigated the toughening mechanism of PHB/PBS blend and PHBV/PBS blend by using an in situ compatibilizer DCP through melt blending. They found that the PBS domain size was considerably reduced by incorporating DCP into those two blends. The domain size reduction was directly related to the interfacial adhesion between the components in the blend systems. During melt blending, the addition of DCP into the blends could generate a grafting reaction between the components (PHBV, PHB, and PBS) in the blends such as PHBV-g-PBS and PHB-g-PBS. The PHBV-g-PBS and PHB-g-PBS acted as a compatibilizer in the blend system, which, in turn, improved the interfacial adhesion between the polymer phases. Consequently, the mechanical properties of the PHB/PBS and PHBV/PBS blends were considerably increased. For example, after incorporation of DCP (0.5 wt.%), the elongation at break of PHBV/PBS (80/20) blend was increased by 39-fold, in comparison to the corresponding uncompatibilized blend. In addition, a significant improvement was observed in unnotched Izod impact toughness of the PHB/PBS blends with 0.5 wt.% DCP. Similarly, the poor compatibility between the PHB and PLA has been improved by using DCP as reactive agent (Dong et al., 2013). The strength and impact toughness of the compatibilized blends were improved in comparison to uncompatibilized PHB/PLA blends. The improved compatibility was attributed to the formation of cross-link network and/or copolymer (PHB-g-PLA) at the interfaces. Consequently, the inclusion phase (PLA) domain size was observed to have reduced significantly.

5.7.5.4 Reactive compatibilization of PLA/PCL blends

Shin and Han (2013) showed melt blending of PLA/PCL resulted in immiscible blend without compatibilizer and that the compatibility was improved by adding GMA into the blend. The improved compatibility has a direct relationship with mechanical properties and fine morphology. In addition, they have observed that GMA acted as a reactive monomer between the PLA/PCL blend interfaces. Liu et al. (2010) have studied super-toughened ternary blends with PLA/ethyl-butyl acrylate and glycidyl methacrylate copolymer (EBA-GMA)/ethylene methacrylic acid zinc ionomer (EMAA-Zn). This reactive blend exhibited remarkable improvements in notched Izod impact strength (from 38 to 770 J/m) and elongation at break (from 5% to 229%). These improvements were attributed to the compatibilization that occurred between the components in the resulting blends. In another study, reactive blending of PLA and ethylene-glycidyl methacrylate (EGMA) showed a dramatic improvement in impact strength (70 kJ/m^2) and elongation at break (>200%) compared to virgin PLA (Oyama, 2009). Zhang et al. (2014a) have studied a super-toughened multiphase PLA blend with series of renewable poly(ether-b-amide) (PEBA) copolymer and ethylene methyl acrylate-glycidyl methacrylate (EMA-GMA) terpolymer. This reactive multiphase blend (PLA/EMA-GMA/PEBA) (70/20/10 wt.%) system showed a massive improvement in the impact toughness (~500 J/m) and elongation at break (~75%). The authors concluded that the improvements are attributed to the combination of strong interfacial adhesion between the phases and massive shear yielding phenomena. Both the miscibility and impact energy of the PLA/PCL blends were dramatically improved with the addition of LTI (Takayama and Todo, 2006). This was due to reduction in size of the PCL phase in the PLA matrix; this behavior reduced the localized stress concentration triggered by interfacial failure. A blend of PLA/PCL (85/15) was melt processed with and without LTI (1 wt.%) as a reactive additive (Takayama and Todo, 2006). The average fracture energy of PLA/PCL/LTI was measured to be 5 times higher than that of PLA/PCL blend. Similar to other studies, the addition of LTI was found to increase the miscibility of PLA with PCL. The improved compatibility between the PLA and PCL contributed to the reduction of PCL domain size, which facilitated increased energy absorption during the fracture process. Wang et al. (1998) have improved the toughness of PLA/PCL blends by in situ compatibilization with TPP. The percentage elongation of the reactively compatibilized PLA/PCL/TPP (80/20/2 or 20/80/2 by weight) blends improved considerably when compared to neat PLA and corresponding uncompatibilized blend. This result suggested that a synergism existed for certain compositions of PLA/PCL blends. Moreover, the compatibilized blends showed higher susceptibility to enzymatic degradation when compared to neat PLA, PCL, and PLA/PCL blend without reactive agent.

5.8 Natural fibers

5.8.1 Classification of natural fibers

Natural fibers are classified based on their origin as it can be obtained from either plant or animal sources. The major component of animal fibers (e.g., wool, feather, angora fiber, and silk fiber) is protein, whereas the major component of plant fibers is cellulose (Dicker et al., 2014). Plant fibers (cellulosic/lignocellulosic fibers) can be classified based on their origin: seed/fruit fibers (e.g., cotton, coir, etc.), leaf fibers (e.g., sisal, pineapple), bast fibers (e.g., jute, keneaf, flax, etc.), agriculture residues (e.g., corn straw, wheat straw, corn stover, etc.), and grass/reed fibers (switchgrass, miscanthus, bamboo, etc.). Plant fibers (excluding cotton) are mainly composed of water-soluble organic components, lignin, waxes, cellulose, hemicellulose, and pectin. Based on the type of the biomass, the chemical compositions of fibers can vary widely (Nagarajan et al., 2013a).

5.8.2 Natural fibers: nature and behavior

The performance of the plant fiber depends on its internal structure, architecture, chemical composition, crystallinity, and microfibril orientation angle (Ahmad et al., 2015; Table 5.5). Among the natural fiber constituents, cellulose shows a higher crystalline nature than other constituents do. In contrast, lignin is amorphous in nature with a highly complex aromatic structure. Because of its stiffness, lignin acts as a protective barrier for cellulose in the plant. Hemicellulose is a highly branched polymer with a degree of polymerization 10–1000 times lower than cellulose (Al-Maadeed and Labidi, 2014). Pectin is usually found in primary cell wall of most plant fibers. Pectin is a more hydrophilic component in plant fiber due to the presence of

Table 5.5 Properties of some natural and synthetic fibers

Fibers	Density (g/cm^3)	Specific tensile strength (MPa)	Specific tensile modulus (GPa)	Elongation at break (%)	Cellulose content (wt.%)	Cellulose crystallinity (%)	Microfibril angle (Degree)	Moisture (wt.%)	Price (US$/kg)	References
Flax	1.5	535–1000	18.4–53	1.2–3.2	64–71	50–90	5–10	8–12	2.1–4.2	Dicker et al. (2014), Ahmad et al. (2015), and Shah (2013)
Hemp	1.47	372–608	47.3	2–4	70–74	50–90	2–6	6.2–12	1.0–2.1	Mohanty et al. (2000), Dicker et al. (2014), Ahmad et al. (2015), and Shah (2013)
Jute	1.3–1.5	269–548	6.85–20.6	1.5–1.8	61–72	50–80	8	12.5–13.7	0.35–1.5	Mohanty et al. (2000), Dicker et al. (2014), Ahmad et al. (2015), and Shah (2013)
Kenaf	1.5–1.6	641	36.55	1.6	31–39	–	–	6.2–12	0.26–0.52	Dicker et al. (2014) and Ahmad et al. (2015)
Ramie	1.5–1.6	147–625	29.3–85	2–3.8	68.6–76.2	–	7.5	12–17	1.5–2.5	Mohanty et al. (2000), Dicker et al. (2014), and Ahmad et al. (2015)
Sisal	1.45	366–441	6.5–15.2	3–7	66–78	50–70	10–25	11	0.6–0.7	Dicker et al. (2014), Ahmad et al. (2015), and Shah (2013)
Bamboo	1.1	454	32.6	1.4	26–60	40–60	8–11	9.16	0.45–0.5	Dicker et al. (2014)

Coir	1.2	146	3.3–5	30	32–43	27–33	30–49	11.36	0.25–0.5	Ahmad et al. (2015) and Shah (2013)
Cotton	1.6	179–373	3.44–7.9	7–8	82.7–91	–	–	7.85–8.5	2.1–4.2	Ahmad et al. (2015) and Shah (2013)
Abaca	1.5	267	8	3–10	56–63		20–25	15	0.345	Koronis et al. (2013) and Ahmad et al. (2015)
Pineapple	1.4–1.6	118–446	4–27	1.6	70–82	44–60	14	11.8	0.4–0.55	Mohanty et al. (2000), Shah (2013), and Akil et al. (2011)
Banana	1.35	444	13.2	3.36	44–64	45–55	10–12	10.71	–	Ahmad et al. (2015) and Shah (2013)
Miscanthus	1.41	–	6.7	–	38	–	–	–	–	Nanda et al. (2013) and Bourmaud and Pimbert (2008)
Switchgrass	1.40	37.7	6.4	–	32	–	–	–	–	Nanda et al. (2013) and Nagarajan et al. (2013b)
E-glass	2.55	1333	28.63	3.4	–	–	–	–	2	Koronis et al. (2013), Ahmad et al. (2015), and Akil et al. (2011)

carboxylic acid group. Generally, native cellulose has a strength >2 GPa and a stiffness of 138 GPa (Shah, 2013). The stiffness of natural fibers depends on the microfibril angle. Therefore, plant fibers with high cellulose content (~60–80%) and low microfibril angles have a strong reinforcing effect in composites. Both pectin and hemicelluloses play important roles in the water absorption, swelling, elasticity, wet strength, and fiber bundle integration of the fibers (Bogoeva-Gaceva et al., 2007).

5.8.3 Advantages and challenges in using natural fibers

Unlike synthetic fibers, natural fibers are lightweight and inexpensive. They possess high toughness, reduced tool wear, good insulation properties, reduced human health hazard, enhanced energy recovery, recyclability, and biodegradability (Mohanty et al., 2000). The specific modulus and strength of the natural fibers are comparable to glass fibers (Zini and Scandola, 2011; Al-Maadeed and Labidi, 2014; Mohanty et al., 2002), as shown in Table 5.5. Furthermore, the natural abundance, renewability, CO_2 neutrality, and lower density of natural fibers have caused an increasing demand for these fibers in material applications. However, there are some major drawbacks with respect to plant fibers, including complex supply chain, geographical availability, moisture absorption, and inconsistency. The inconsistency of the plant fibers is related to many factors, such as harvesting time, species type, soil quality, fertilization, moisture content of the fibers, location of the plant growth, as well as climatic conditions during plant growth (Dicker et al., 2014). Moisture absorption of the natural fiber can vary greatly depending on the fiber types. This moisture absorption creates performance problems for products made out of such natural fiber composites (NFCs). Due to low thermal stability, natural fibers are recommended for process/use below 200 °C (Koronis et al., 2013; Dicker et al., 2014). These issues can be effectively addressed by appropriate material selection/screening techniques (Mohanty et al., 2002; Shah, 2014).

5.9 Biocomposites

The term composites originated from the Latin word, *compositus*. In general, composites are materials made from two or more distinct constituents having significantly different physical and/or chemical properties. At least two components are required in order to prepare composites material: matrix, or continuous phase, and reinforcement, or discontinuous phase. The continuous phase acts as a binder, while the discontinuous phase acts as reinforcement. Very often, both continuous and discontinuous phases are easily distinguishable in the resulting composite materials. Generally, the mechanical performances of the reinforcements are higher than those of the matrix phases. Composite materials properties can be tailored to suit specific application requirements. Composites are generally classified based on their matrix sources such as metal, ceramic, and polymer (thermoplastic and thermoset). Based on their reinforcements, the composites can be further categorized as fiber (short or long) composites and particulate (powder, flacks, and spherical) composites, fabric-based composites, and long-fiber-reinforced thermoplastic composites. Most common fibers, such as aramid, carbon, and glass, are used as a reinforcement in conventional polymer matrices, such as PP, epoxy resins, unsaturated polyesters, and polyurethane (Wambua et al., 2003), while E-glass fibers dominate in the fiber-reinforced plastic composites (FRCs) market (Shah, 2013). Biocomposites offer a combination of high mechanical properties, light weight, moldability, and design flexibility. Thermoplastic polymer matrices usually dominate the class of NFCs.

FRCs have diversified applications in automotive interior parts, packaging, construction, furniture, consumer goods, electrical, musical instruments, wind turbine rotor blade, and decking industries (Shah, 2013). A combination of flax and carbon fiber-reinforced composites are used in sporting equipment (snowboards, tennis rackets, and bicycle frames) because of its superior vibration damping properties and mechanical performance (Dicker et al., 2014). In 2002, fiber-reinforced composites' share of the global market was 771 million kg and this has continued to grow extensively (Bogoeva-Gaceva et al., 2007). The amount increased to 8.7 million tons in 2011 (Shah, 2013). In 2002, fiber composites had a maximum share of 31% in automotive applications, followed by construction applications (26%) (John and Thomas, 2008). In Europe, this trend increased significantly because of mounting environmental concerns. Increasing interest in biodegradable and/or biologically derived materials has fueled vivid research in biocomposites (Zini and Scandola, 2011). Either a composite produced from the combination of nonbiodegradable polymer with a natural fiber/filler or a bioplastic with synthetic fibers/fillers can be called a biocomposite.

5.9.1 Advantageous of NFCs

NFCs have numerous benefits compared to synthetic fibers such as abundant availability, light weight, ecofriendly, nonabrasive, inexpensive, good fatigue, and specific strength (Akil et al., 2011). Biocomposites made from plant-based fiber and renewable resource-based polymers can be termed ecocomposites or all green composites (Zini and Scandola, 2011). Any biocomposites that are compostable, recyclable, environmentally friendly, and commercially adequate can help in waste management (Mohanty et al., 2002). Biocomposites are in huge demand for automotive uses, as they reduce both manufacturing costs and fuel consumption while meeting the consumer push for greener products (Koronis et al., 2013). Because of global warming, environmental concerns, and an increased interest in greener products, biocomposites based on bioplastics have been developed.

5.9.2 Attributes of NFCs

Properties of NFCs are strongly influenced by many factors, including stress transfer between the fiber–matrix, fiber orientation, fiber–matrix interaction, fiber volume, fiber length, and aspect ratio. There are some drawbacks associated with NFCs, like incompatibility with some polymer systems, poor wetting, moisture sensitive, nondestructive behavior, and lower thermal stability that could affect the composites performance (Koronis et al., 2013; Wambua et al., 2003). Natural fibers are mainly hydrophilic in nature making them incompatible with the hydrophobic polymer matrix and their addition can lead to a reduction in the mechanical performance of the resulting products (Mohanty et al., 2000). The polarity difference between the thermoplastics matrix and the natural fibers may result in poor fiber–matrix adhesion in the resulting composites. A lack of interaction/compatibility between reinforcement and matrix reduces the overall performance of the resulting materials (Bogoeva-Gaceva et al., 2007). The compatibility between the reinforcement and matrix can be improved by chemical surface modification (sodium hydroxide, permanganate, peroxide, maleic anhydride, organosilanes, and isocynates) and physical surface modification (cold plasma treatment, corona treatment) of the fibers (Bogoeva-Gaceva et al., 2007).

Another strategy is to improve the interfacial adhesion through the addition of a "compatibilizing/coupling agent" into the composites during processing (Koronis et al., 2013). Generally, the "compatibilizer/coupling" agents are used to modify the interface between the matrix and reinforcement in the composite structures. The interfacial modification may be in the form of a physical and/or chemical interaction between the components in the composites. Enhanced compatibility between the phases of the composites helps in improving fiber–matrix adhesion, resulting in improved performance of the resulting composites. NFCs with compatibilizers are reviewed in detail in subsequent sections.

5.10 Biocomposites based on biodegradable blends as matrix material: Some specific examples

There have been many reviews on NFCs (Dicker et al., 2014; Ahmad et al., 2015; Shah, 2014; Al-Oqla and Sapuan, 2014; Shah, 2013; Bogoeva-Gaceva et al., 2007; Avella et al., 2009; Wahit et al., 2012; John and Thomas, 2008; Koronis et al., 2013; Zini and Scandola, 2011; Satyanarayana et al., 2009). Most of these publications reviewed single polymer matrix-based biocomposites. Reviews of biocomposites fabricated using biodegradable blends as matrix material are very limited. Recently, biocomposite fabrication using blends of polymer matrices is part of the growing trend because blend matrices could provide an optimum stiffness–toughness balance, or tailored properties for resulting biocomposites. Currently, only a few biodegradable preblends are commercially available in the market; for instance, PHBV/PBAT blend, available under the name of ENMAT™ from Tianan Biologic Materials Company. Ltd., China (Javadi et al., 2010b). BASF commercialized a biodegradable polymer blend under the trade name Ecovia®. It is a blend of PLA/PBAT (45/55 wt.%) with 45 wt.% biobased content (Harnnecker et al., 2012). In addition, FKuR is producing PLA-copolyester blends under the trade name Bio-Flex. All of these blends are successful matrix material for the incorporation of natural fibers. Zhang et al. (2012) fabricated a biodegradable and renewable ternary blend from PHBV, PLA, and PBS. The ternary blends (PLA/PHBV/PBS) showed a unique stiffness–toughness balanced mechanical properties compared to their neat components. The authors concluded that the ternary blend (PLA/PHBV/PBS) system is a very promising candidate for biocomposite applications. Similarly, Muthuraj et al. (2014) have also developed a binary blend of PBS and PBAT with good thermal, mechanical, and thermomechanical properties to be used as a matrix material for composite applications. These research efforts are indicative of the fact that the multiphase polymer blend provides a simple and an effective way to develop a new polymer matrix system for NFCs.

5.10.1 Biocomposites based on PHBV blends

As stated previously while reviewing the properties and performance of PHBV, it is a brittle, biodegradable polymer. By blending PHBV with tough, biodegradable polymers, their mechanical and thermal properties can be tailored without sacrificing the biodegradability (Nanda et al., 2011). Therefore, a number of researchers, as shown in Table 5.6, have extensively studied the toughened PHBV blend-based biocomposites. For example, Javadi et al. (2010b) investigated the preparation and performance evaluation of solid and microcellular preblend of PHBV/PBAT and its recycled wood fiber (RWF)-based composites with and without silane treatments. There were no significant changes in the SEM morphology of the PHBV/PBAT/silane-treated-RWF composites and PHBV/PBAT/untreated-RWF composites, suggesting that silane treatment was not worthy for improving thermal properties and interfacial bonding between the fiber and matrix. However, the addition of 10% RWF into PHBV/PBAT blends increased the crystallinity, storage modulus, specific tensile strength, and tensile modulus, and decreased strain at break and toughness compared to the neat PHBV/PBAT blend. Overall, the authors concluded that silane-treated-RWF-reinforced PHBV/PBAT composites did not induce any significant changes in morphology, mechanical, and thermal properties of PHBV/PBAT/RWF untreated-RWF composites.

Nagarajan et al. (2013b) studied the performance of biocomposites fabricated from a preblend of PHBV/PBAT (45/55%, ENMAT™) as matrix and switchgrass fiber (20–40 wt.%) as reinforcement. An obvious increase was observed in the modulus of the composites because of the reinforcing effect of the switchgrass fibers. However, the resulting composites showed poor interfacial adhesion between the fiber–matrix. Therefore, the interfacial adhesion was improved by adding poly(diphenylmethane diisocynate), PMDI, as a compatibilizer (0.5, 0.75, and 1 phr). SEM analysis indicated the occurrence of a better stress transformation between the fibers and matrix through improved interfacial adhesion, which lead to an increase in the mechanical performance of the resulting composites. The critical concentration of compatibilizer was found to be 0.75 phr. This critical concentration yielded 80%, 40%, and 56% improvements in notched Izod impact strength, flexural strength, and tensile strength, respectively. Furthermore, the composites with and without compatibilizer showed considerable improvements in storage modulus and heat resistance properties compared to the matrix. The melt flow of both compatibilized and uncompatibilized composites was sharply reduced in comparison to the PHBV/PBAT blend matrix. For example, with the addition of 30 wt.% switchgrass fibers into PHBV/PBAT matrix, MFI reduced to 7 g/10 min. It could be due to the restriction of polymer chain mobility in the presence of stiff fibers. The MFI of the composites with 30 wt.% switchgrass fibers and 0.75 phr compatibilizer is around 5 g/10 min. The authors suggest that this biocomposite can be injection-molded with the help of some processing additives.

Table 5.6 Recently developed biodegradable polymer blend matrix based biocomposites

Blend matrix	Fiber/filler		Compatibilizer/Coupling agent		Manufacturing process	References
	Type	Content	Name	Chemical structure		
PHBV/PBAT	Recycled wood fiber	10%	Gamma-methacryloxypropyltrimethoxysilane	$CH_2=C(CH_3)CO_2CH_2CH_2CH_2Si(OCH_3)_3$	Injection molding	Javadi et al. (2010b)
PHBV/PBAT	Switchgrass	20-40%	Poly diphenylmethane diisocyanate (PMDI)	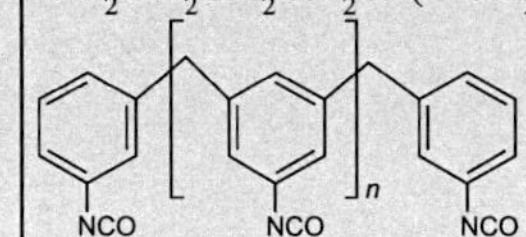	Injection molding	Nagarajan et al. (2013b)
PHBV/PBAT	Distiller's dried grains with solubles (DDGS)	20%	Poly diphenylmethane diisocyanate (PMDI)	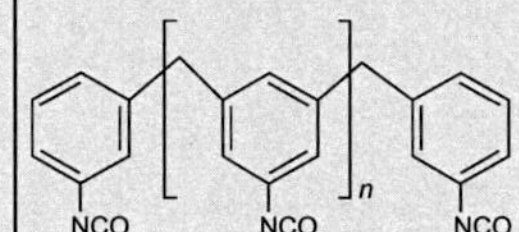	Injection molding	Zarrinbakhsh et al. (2013)
PHBV/PBAT	Miscanthus, switchgrass, wheat straw, corn stalk, and soy stalk	30%	NA	NA	Injection molding	Nagarajan et al. (2013a)
PHBV/PBAT	Jute, abaca and man-made cellulose	30%	NA	NA	Injection molding	Bledzki and Jaszkiewicz (2010)
PHBV/PBAT	Cellulose, Jute, abaca and flax	30%	NA	NA	Injection molding	Adam et al. (2013)
PHBV/PBAT	Kenaf	5 and 10%	NA	NA	Compression molding	Yang et al. (2014)
PHBV/PBS	Distiller's dried grains with solubles (DDGS)	30%	Poly diphenylmethane diisocyanate (PMDI)	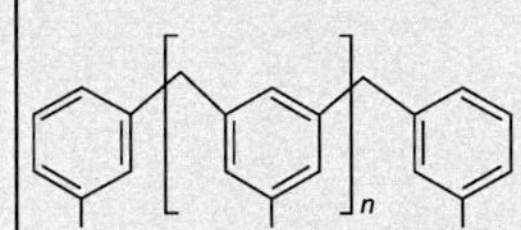	Injection molding	Zarrinbakhsh et al. (2011)

PHBV/PLA	Soy hull, switchgrass, and miscanthus	30%	NA	NA	Injection molding	Nanda et al. (2013)
PHBV/PLA	Soy fiber	30 and 50%	2,5-Bis(tert-butylperoxy)-2,5-dimethylhexane	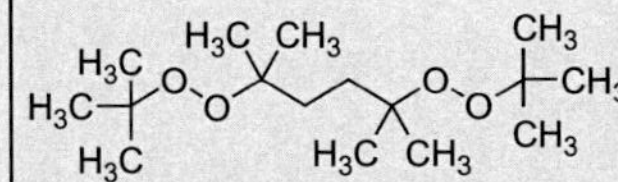	Injection molding	Nanda et al. (2012)
PLA/PBAT	Kenaf	10-50%	(3-aminopropyl) trimethoxysilane (APTMS)	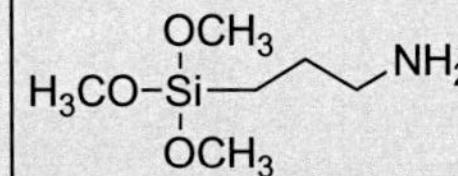	Compression molding	Sis et al. (2013)
PLA/PBAT	Curaua	5-20%	Maleic anhydride grafted polypropylene (MA-g-PP)	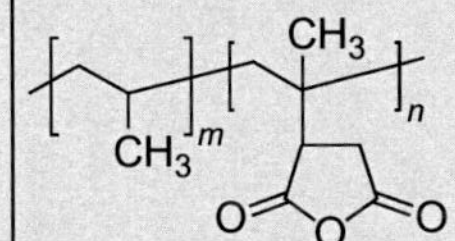	Compression molding	Harnnecker et al (2012)
PLA/PBAT	Softwood flour	30-50%	NA	NA	Injection molding	Sykacek et al. (2010)
PLA/PBAT	Wood flour	20-30%	NA	NA	Compression molding	Georgiopoulos et al. (2014)
PLA/PBAT	Ramie	10%	NA	NA	Compression molding	Yu and Li (2014)
PLA/PCL	Alkali treated palm fiber	10-25%	Dicumyl peroxide (DCP)	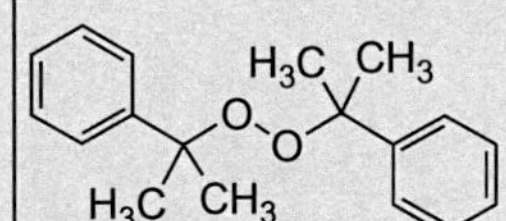	Injection molding	Ibrahim et al. (2014)
PLA/PCL	Silane treated Jute and Untreated Jute	50%	Trimethoxy (methyl) silane	$CH_3Si\ (OCH_3)_3$	Compression molding	Goriparthi et al. (2012)
PLLA/PBS	Flax	25.5%	NA	NA	Injection molding	Bourmaud et al. (2015)

Biocomposites were prepared from a PHBV/PBAT(45/55%) blend and distiller's dried grains with solubles (DDGS) using melt processing technique by Zarrinbakhsh et al. (2013). This study explored the influence of compatibilizer (0.5% and 1% PMDI) and lubricating agent (3% and 6% corn oil) on the performance of the resulting biocomposites. Use of PMDI as a compatibilizer was observed to increase mechanical properties significantly when compared to uncompatibilized composites. The composites with only 0.5% PMDI showed a maximum improvement in tensile strength (39%) and modulus (24%). There was no significant change of these mechanical properties while increasing PMDI content from 0.5% to 1%. Similar trends were observed in flexural properties. These improvements may be due to the good interfacial adhesion that occurred through cross-linking reactions between the components in the composites. This reaction was confirmed by MFI reduction, as well as by the increased melt viscosity of the composites. However, the addition of only corn oil reduced both the stiffness and strength of the composites. Interestingly, a combination of both corn oil and PMDI displayed a synergistic effect in impact strength and elongation at break. The authors attributed this improvement to be due to the lubrication effect of corn oil, which enhanced the chemical reaction between the components. Water absorption rate of all the biocomposites was higher than PHBV/PBAT blend matrix. However, the initial water absorption rate of the compatibilized composites was lower than that of the uncompatibilized composites. Reduction of free hydroxyl groups in the DDGS particles after chemically reacting with the matrix was believed to be the reason. The interfacial interaction between matrix and DDGS was justified by SEM analysis. In a few other studies, the biodegradability of DDGS/PBAT (Muniyasamy et al., 2013b), DDGS/PLA (Lu et al., 2014), and DDGS/PHA (Madbouly et al., 2014) composites has been investigated under different environmental conditions. These DDGS-based biocomposites exhibited superior biodegradation rate compared to their biodegradable polymer matrix.

In a different study, Nagarajan et al. (2013a) extensively studied the influence of adding 30 wt.% of various natural fibers (miscanthus, switchgrass, wheat straw, corn stalk, and soy stalk) into PHBV/PBAT (45/55 wt.%) blend by extrusion and injection molding technique. They have evaluated the water uptake behavior, mechanical, thermal, and thermomechanical properties. Among biocomposites with different types of fibers, miscanthus-based composites exhibited higher mechanical properties, heat deflection temperature, higher thermal stability, and comparatively low water absorption. The observed property differences between the composites were due to differences in fiber composition, fiber length distribution, and the interaction between the fibers–matrix. Moreover, SEM micrographs showed debonding and fiber pullout from the matrix, which was attributed to the inefficient stress transfer occurring between the components. Similar to this study, various types of agricultural residue-reinforced PLA composites have also been reviewed (Nyambo et al., 2014). The biodegradability rate of these natural fiber-reinforced composites can be expected to be much faster in comparison to their parent components (Pradhan et al., 2010).

The influences of man-made cellulose, jute, and abaca fiber on the mechanical properties of PHBV/PBAT (70/27.6% and 2.6% processing additive) blend were also investigated by Bledzki and Jaszkiewicz (2010). The composites samples were produced in an extrusion and injection molding process. The reinforcements had positive effects on tensile properties and notched Charpy impact strength. Among the abaca, jute fibers, and man-made cellulose, abaca fiber has a high tensile strength (980 MPa) and a high modulus (27–32 GPa) (Bledzki and Jaszkiewicz, 2010). However, abaca fiber-based composites underperformed compared to man-made cellulose and jute fiber composites. The composites with man-made cellulose showed a massive improvement (>500%) in notched Charpy impact strength and a 50% improvement in tensile strength over the PHBV/PBAT matrix. A SEM micrograph of PHBV/PBAT composites showed a poor adhesion between the fiber and matrix. This was attributed to the partial degradation of PHBV/PBAT blend during melt processing and post-molding shrinkage of the PHBV/PBAT blend matrix. Similar to a previous study, a massive increase in the tensile modulus was observed for the PHBV/PBAT/(70/27.6% and 2.6% processing additive) blend with jute fiber composites (Adam et al., 2013). The observed results were attributed to chemical composition, aspect ratio, and the geometry of the fibers. SEM analysis revealed that the compatibility between the fiber and the matrix was poor due to the shrinkage nature of PHBV in the matrix. Even still, PHBV/PBAT composite properties are comparable with PP biocomposites.

The effects of sodium hydroxide (NaOH) treated and pectinase retted kenaf fibers on PHBV/PBAT composite properties were studied using compression molding (Yang et al., 2014). This study investigated the DMA and crystallization behavior of composites with two different fiber lengths (5 and 10 mm). The dynamic mechanical properties of the composites showed that the storage modulus of the composites increased with the addition of fiber compared to the PHBV/PBAT matrix. The storage modulus of pectinase retted fiber-reinforced composites was higher compared to corresponding composites with NaOH retted fiber. The composites with 5 mm retted fibers showed a higher storage modulus than the 10 mm fiber and their hybrid (1:1) composites. In the presence of both pectinase and NaOH retted fibers, the glass transition temperature of PBAT increased and the damping factor of PHBV phase reduced. The fiber retting had an influence on crystallization, melting, and spherulite morphology of the PHBV/PBAT matrix, which was confirmed by polarizing optical microscopy (POM) analysis. POM was used to study the nucleation density and spherulite growth of PHBV/PBAT blends and their biocomposites. The nucleation density of the PHBV/PBAT blend increased with the addition of fibers and it lead to a change in the crystal morphology of the resulting composites. Consequently, the melting point of the composites was reduced as compared to the PHBV/PBAT matrix with an increase in the melting enthalpy. This suggested that the pectinase retted kenaf fibers have more surface activity and, thus, increased the crystallinity. The complex viscosity of the composites was higher than the matrix because of the restricted polymer chain mobility in the presence of rigid fibers. NaOH retted fiber composites showed less nucleation density and larger spherulites as compared to pectinase retted fiber composites. This was attributed to the enhanced roughness of the NaOH retted kenaf, which tend to form tiny spherulites.

Zarrinbakhsh et al. (2011) investigated green composites made from DDGS and an optimized blend of PHBV/PBS (70/30 wt.%) matrix. Mechanical, thermal, morphological, and thermomechanical properties of the injection-molded composites were studied. Interestingly, after water washing of DDGS, thermal stability was improved as the water-soluble molecules of lower thermal stability were removed effectively by water washing. The improved thermal stability allowed for higher processing temperatures without any thermal degradation of DDGS up to 180 °C. The water-washed DDGS composites showed an improved tensile and flexural strength compared to nonwashed DDGS composites. When polydiphenylmethane diisocynate (PMDI) was introduced as a compatibilizer into water-washed DDGS composite, there were significant improvements in flexural strength and tensile strength. The flexural and tensile modulus of the DDGS composites has an increasing trend in the following order: PHBV/PBS blend matrix < nonwashed DDGS composites < water-washed DDGS composites < water-washed/compatibilized DDGS composites. The improved tensile and flexural properties were attributed to the good interfacial adhesion between the fillers and matrix, which was confirmed through SEM. The impact strength of water-washed DDGS composites and water-washed/compatibilized DDGS composites was lower compared to nonwashed DDGS composites. It was hypothesized that the poor interfacial bonding between the phases needs more energy to break the fillers because the crack propagation occurred around the fillers rather than braking fillers.

Recently, Zhang et al. (2014b) studied preparation and performance evaluation of sustainable green composites from ternary PHBV/PBAT/epoxidized natural rubber (50/35/15 wt.%) blend matrix and miscanthus fibers. The selected ternary blend system has nonbreak behavior under notched Izod impact test. However, after the addition of 10 wt% miscanthus fiber into this ternary blend matrix, the impact strength of the resulting composites was observed around 273 J/m. A similar trend was observed in the elongation at break and tensile strength. This reduction was mainly due to weak interfacial interaction between the fiber–matrix as well as poor load transfer capability between the fiber–matrix. Interestingly, this drawback was overcome by adding 0.3 wt.% DCP as a reactive agent in the composites. In the presence of DCP, the polymers formed graft copolymers and partially cross-link network between the polymers, which improved the compatibility between the blend components. The authors also investigated the effect of 0.3 wt.% DCP while increasing fiber content up to 20 wt.% in the resulting composites. The impact energy of the composite with 20 wt.% fiber and 0.3 wt.% DCP still remained at 240.5 J/m. The flexural properties and storage modulus were found to increase with increasing fiber content up to 20 wt.% which is a fairly common observation in natural fiber-reinforced polymer composites.

5.10.2 Biocomposites based on PLA blends

PBAT, PBS, and PCL are the best candidates for increased toughness of PLA in their respective blends. Blending PLA with tough polymers (PBAT, PBS, and PCL) resulted in a higher toughness but lower stiffness in the blends in comparison to virgin PLA. The low stiffness can be addressed by adding fibers/filler to PLA blends. Therefore, many researchers have extensively studied toughened PLA-based biocomposites, as shown in Table 5.6. Commercially available preblends of PLA/PBAT, 45/55 wt.%, (BASF-Ecovio®) with chemically (alkali treatment) treated curaua fiber-reinforced biocomposites were studied by Harnnecker et al. (2012). There was a slight (4 °C) increase in T_g of the composites compared to the PLA/PBAT blend. This fact was attributed to the intermolecular (hydrogen bonds) interaction between the fibers and polymer matrix. Moreover, DSC thermograms showed an endothermic peak during the heating cycle for all the composites, whereas PLA/PBAT blend did not show any endothermic peak during the heating cycle. This was believed to be due to the nucleation effect of curaua fiber in the composites. Performance of these composite was studied with and without the addition of compatibilizer (2 wt.%); that is, maleic anhydride-grafted-polypropylene (MA-g-PP). Tensile, impact, and flexural strength of the compatibilized composites increased with increased fiber content from 5 to 20 wt.%. For example, the compatibilized composites with 20 wt.% curaua fiber showed a 56% increase in flexural strength and a 75% increase in the tensile strength.

The increased mechanical properties were attributed to the improvement of interfacial adhesion between the fibers and matrix, which was confirmed by cross-sectional fracture analysis. In addition, water absorption of the compatibilized composite was almost 100% lower compared to the uncompatibilized composite. Sis et al. (2013) examined the impact, flexural, and tensile properties of the PLA/PBAT/kenaf fiber composites with and without coupling agent ((3-aminopropyl) trimethoxysilane), APTMS, by melt blending. The tensile, flexural, and impact strength of PLA/PBAT/kenaf composites decreased with increasing fiber loading. This was because of inadequate fiber–matrix interaction and poor dispersion of fiber in the matrix. The composites were modified with different amounts (from 1 to 5 wt.%) of APTMS coupling agent. The optimum properties were obtained for PLA/PBAT/kenaf composites with the addition of 2 wt.% coupling agent. This composite showed 22%, 42%, and 63% improvement in Izod impact strength, tensile strength, and flexural strength, respectively. In addition, the tensile and flexural moduli of the composites with coupling agent were considerably higher than composites without coupling agent. The storage modulus of the composites with and without coupling agent exhibits similar trends, like the tensile and flexural modulus. This behavior was ascribed to good adhesion between the phases.

Unlike PLA/PBAT/kenaf and curaua fiber composites, PLA/PBAT/wood flour composites exhibited an increase in both tensile and flexural properties compared to the neat PLA/PBAT (45/55 wt.%) blend (Ecovio®) (Sykacek et al., 2010). With increasing fiber load from 30 to 50 wt.%, a 48.2% reduction was observed in the Charpy impact strength of the resulting composites compared to PLA/PBAT blend matrix. The loss in impact strength was attributed to low fiber aspect ratio of wood flour (Sykacek et al., 2010). The water resistance property of the composites was reduced because of the hygroscopic nature of the wood flour. However, the composites with an addition of 50 wt.% wood flour showed significant improvements in tensile strength (27.2%), modulus (174.3%), and flexural strength (20.6%) as compared to neat PLA/PBAT blend. These improvements were related to good fiber–matrix adhesion and better distribution of fibers in the matrix. This trend is in contradiction to the finding of Georgiopoulos et al. (2014). Another study observed similar results for PLA/PBAT (70/30 wt.%) biocomposites (Ludvik et al., 2007). More recently, a comparison study between PLA/30% ramie composites and PLA/PBAT (95/5, 90/10 and 85/15%) with 30 wt.% ramie composites has been investigated by Yu and Li (2014).

In general, strong interfacial bonding between fiber and the matrix decreases the water absorption of the NFCs. The Vicat softening temperature, flexural and tensile strength of the PLA/PBAT/ramie (85/15%) composites were lower than PLA-based composites, which was due to the soft elastomeric nature of the PBAT phase. Yet, the notched Izod impact strength of the PLA/PBAT/ramie composites was superior compared to PLA/ramie composites. This suggests that the PBAT phase induced stress concentration under impact load. Consequently, the PBAT phase absorbed more energy during failure of the PLA/PBAT composites. In order to maintain an energy dissipation source in the PLA/PBAT composite system, concentration of PBAT was considered an important factor. Furthermore, thermal stability of the PLA/PBAT-based composites was studied by thermogravimetric analysis. It was found that the thermal stability of PLA/PBAT-based ramie composites was higher than PLA-based composites because the PBAT phase enhanced the charring process for the composites.

Ibrahim et al. (2014) studied the performance of uncompatibilized and compatibilized PLA/PCL (10/90 wt.%)/alkali treated palm fiber composites with 10–25 wt.% fiber loading. With the addition of 0.01 phr DCP, the flexural modulus, impact strength, tensile strength, and flexural strength of the composites were superior when compared to uncompatibilized composites. These improvements were attributed to good fiber and matrix adhesion and good fiber distribution in the matrix. Fiber and matrix adhesion was confirmed by damping behavior as well as by SEM analysis. DCP increased the viscosity and cross-linking while reducing the brittleness of the composites. Consequently, the tensile modulus of compatibilized composites was lower than uncompatibilized composites in the entire fiber composition. This trend agreed with the trend noticed in storage modulus of the compatibilized composites. The storage modulus of the composites increased up to 15 wt.% fiber loading and then decreased with increasing fiber up to 25 wt.%. The increased storage modulus was due to the reinforcing effect of the fibers. The decreased storage modulus with increasing fiber loading was a result of free volumes that were created due to cross-linking. Furthermore, the thermal stability of the composites was higher than that of the matrix. Overall, the composites with 15 wt.% fiber load showed balanced mechanical, thermal, and thermomechanical properties. Recently, PLLA/PBS blends with 25.5% flax fiber composites showed well-balanced mechanical properties without any compatibilizer (Bourmaud et al., 2015).

The consequence of incorporating 30 wt.% miscanthus, switchgrass, and soy hull and their hybrids as a reinforcement in the PHBV/PLA (70/30 wt.%) blend matrix has been investigated by Nanda et al. (2013). The performance of the green composites was examined by means of their thermomechanical properties. Lack of interfacial adhesion between the components reduced significantly the tensile strength of all the composites when compared to PHBV/PLA matrix. Yet, the tensile and flexural moduli of all the composites were higher compared to PHBV/PLA blend. There was no significant difference observed in the notched Izod impact strength of all the composites compared to the PHBV/PLA matrix. Lower aspect ratio, and smaller cellulose and lignin content of soy hull in comparison to miscanthus and switchgrass, resulted in soy hull composites displaying inferior mechanical properties. All of the composites showed superior heat resistance properties compared to their matrix. Soy hull-based composites showed lower thermal stability compared to miscanthus- and switchgrass-based composites. This could probably be due to the less thermally stable compounds, such as pectin and proteins, present in the soy hull (Nanda et al., 2012). The melting temperature and melting enthalpy of the PLA and PHBV were not affected in the presence of all the fibers. Soy hull-based composites showed more water absorption compared to other composites. This was due to the smaller amount of lignin present in the soy hull (Nanda et al., 2012). This trend was changed in compatibilized PHBV/PLA/soy fiber composites.

5.11 NFCs market and their applications

Natural fiber-reinforced polymeric composites were produced early in 1908 (Zini and Scandola, 2011). Since then, NFCs have become attractive in both the automotive and building industries because of their low cost, low CO_2 emissions, and because they are lightweight. The specific mechanical properties of NFCs are superior to synthetic fiber (glass, carbon, and kevlar) composites (Al-Maadeed and Labidi, 2014). Currently, PP and polyurethane-based biocomposites are being used in automotive interior and exterior parts (John and Thomas, 2008; Koronis et al., 2013). In 2010, 430.7 million pounds of NFCs were produced and its worth was US$289.3 million. This is expected to reach US$531.3 million by 2016, which means an 11% growth rate over the next 5 years (Faruk et al., 2014). Automotive companies have set a goal and are working toward the possible replacement of some nonbiodegradable polymer products by using biocomposites and bioplastics. An average vehicle contains 113 kg of plastic parts and approximately 50% of automotive interior parts are made up of plastics (http://www.bioproductsatguelph.ca/newsevt/colloquium/assets/flashpaper/r_campbell.swf). Annually 10–11 million vehicles are discarded, and 25% of their parts cannot be recycled or reused because they are foams, rubber, and synthetic fibers (http://www.bioproductsatguelph.ca/newsevt/colloquium/assets/flashpaper/r_campbell.swf).

Biocomposites from PLA/kenaf and PBS/bamboo are used for making automotive spare tire covers and tailgate trim, respectively (Koronis et al., 2013; http://indiacompositesshow.com/bio-composites-in-automotive-applications/). There exists the possibility of producing durable car parts entirely from biodegradable materials. More specifically, a vehicle (Agri-Car) with 90% biodegradable materials has been jointly designed by the University of Akron and the Ohio State University (Zini and Scandola, 2011). In 2009, the University of Warwick developed a prototype World First Formula 3 racing car by using green composites (Zini and Scandola, 2011). In addition, the mechanical and heat resistance properties of PLA/kenaf biocomposites were almost equal to those of conventional polycarbonate/glass fiber composites. PLA-based biocomposites are used to produce prototype kayaks, wind turbine rotors, laptop computers, and cellular phones (Zini and Scandola, 2011; http://indiacompositesshow.com/bio-composites-in-automotive-applications/; Darder et al., 2007; http://www.mastersoflinen.com/eng/technique/19-secteurs-d-application). According to this resource (Ahmad et al., 2015), the NFC market is expected to grow by up to 20% and more than 50% in automotive and building applications, respectively. The global NFC market is expected to grow to US$531.3 million by 2016 (Faruk et al., 2014). This forecast is significantly higher than the 2010 world annual production (US$289.3 million) of the NFCs market. The European biocomposites market growth rate is predicted at 21% per year. In 2008, biocomposite production totalled 129,000 tons in Europe and this growth is expected to increase to 427,000 tons in 2014. Parallel to this, hemp and flax fibers are major players in the NFCs market share in Europe (Shah, 2014). In 2000, the North American natural fiber market was worth US$155 million and it is expected to reach US$1.38 billion by 2025 (Dicker et al., 2014).

5.12 Conclusions

Recently, utilization of biodegradable polymers has been accelerated due to the environmental concerns associated with disposal of nonbiodegradable plastics in landfills. Higher cost, thermal instability, low melt strength, and brittleness of biodegradable polymers currently limit their extended application. Great efforts have been taken to overcome these drawbacks by blending techniques as well as the incorporation of natural fibers. However, most of the polymer blends are thermodynamically immiscible due to a difference in solubility. Generally, the immiscible polymer blends lead to a weak interface between the blend partners. The poor interfacial adhesion between the blend components can be modified using various additive, compatibilizer, and compatibilization strategies. The compatibilization effect and its influence on the thermal, thermomechanical, morphological and mechanical properties of the blends have been reviewed in this chapter. With the increasing cost of fossil fuels, more research attention is being focused on developing renewable resource-based materials for various industrial applications. Natural fiber-reinforced biodegradable polymers provide a sustainable alternative to nonbiodegradable polymeric composites. It was found that biocomposites could be fabricated from a stiffness–toughness balanced blend matrix with various natural fibers. The difference in strength and modulus values between the natural fibers is attributed to the difference in individual fiber compositions. Recently, multiple binary blend matrix-based biocomposites have been studied and they all exhibit superior performance compared to single polymer-based biocomposites. Due to incompatibility between the matrix and fiber, several approaches have been explored to modify both matrix and fiber to enhance the compatibility between them. The mechanical properties of all the compatibilized composites are significantly higher in comparison to uncompatibilized one. However, there are only a limited number of applications for biodegradable polymers and their biocomposites. It is believed that significant research into existing processing technology will help to widen the application potential of biodegradable polymer-based blends and their biocomposites.

Acknowledgments

The authors are thankful to the Ontario Ministry of Agriculture, Food and Rural Affairs (OMAFRA)/University of Guelph–Bioeconomy for Industrial Uses Research Program; Ontario Research Fund, Research Excellence Program Round-4 (ORF-RE04) from the Ontario Ministry of Economic Development and Innovation (MEDI); the Natural Sciences and Engineering Research Council (NSERC) Discovery grant individual (to Mohanty); and the NSERC-AUTO21 NCE project to carry out this research. Rajendran Muthuraj is grateful to Vidhya Nagarajan for her editorial input into this chapter.

References

Adam, J., Korneliusz, B.A., Agnieszka, M., 2013. Dynamic mechanical thermal analysis of biocomposites based on PLA and PHBV—a comparative study to PP counterparts. J. Appl. Polym. Sci. 130 (5), 3175–3183.

Ahmad, F., Choi, H.S., Park, M.K., 2015. A review: natural fiber composites selection in view of mechanical, light weight, and economic properties. Macromol. Mater. Eng. 300 (1), 10–24.

Akil, H.M., Omar, M.F., Mazuki, A.A.M., Safiee, S., Ishak, Z.A.M., Abu Bakar, A., 2011. Kenaf fiber reinforced composites: a review. Mater. Des. 32 (8–9), 4107–4121.

Al-Maadeed, M.A., Labidi, S., 2014. Recycled polymers in natural fiber-reinforced pollymer composites. In: Hodzic, A., Shanks, R. (Eds.), Natural Fibre Composites: Materials, Processes and Properties. Woodhead Publishing Limited, Cambridge, UK, pp. 103–114 (Chapter 4).

Al-Oqla, F.M., Sapuan, S.M., 2014. Natural fiber reinforced polymer composites in industrial applications: feasibility of date palm fibers for sustainable automotive industry. J. Clean. Prod. 66, 347–354.

ASTM D6400-12, 2012. Standard Specification for Labeling of Plastics Designed to be Aerobically Composted in Municipal or Industrial Facilities. ASTM International, West Conshohocken, PA. www.astm.org.

Audic, J.-L., Chaufer, B., 2012. Properties of biodegradable poly(butylene adipate-*co*-terephtalate) and sodium caseinate blends. J. Appl. Polym. Sci. 125 (S2), E459–E467.

Auras, R., Harte, B., Selke, S., 2004. An overview of polylactides as packaging materials. Macromol. Biosci. 4 (9), 835–864.

Avella, M., Martuscelli, E., 1988. Poly-D-(−)(3-hydroxybutyrate)/poly(ethylene oxide) blends: phase diagram, thermal and crystallization behaviour. Polymer 29 (10), 1731–1737.

Avella, M., Martuscelli, E., Raimo, M., 2000a. Review properties of blends and composites based on poly(3-hydroxy)butyrate (PHB) and poly(3-hydroxybutyrate-hydroxyvalerate) (PHBV) copolymers. J. Mater. Sci. 35 (3), 523–545.

Avella, M., Errico, M.E., Laurienzo, P., Martuscelli, E., Raimo, M., Rimedio, R., 2000b. Preparation and characterisation of compatibilised polycaprolactone/starch composites. Polymer 41 (10), 3875–3881.

Avella, M., Buzarovska, A., Errico, M., Gentile, G., Grozdanov, A., 2009. Eco-challenges of bio-based polymer composites. Materials 2 (3), 911–925.

Babu, R., O'Connor, K., Seeram, R., 2013. Current progress on bio-based polymers and their future trends. Prog. Biomater. 2 (1), 8.

Bastioli, C., 1998. Properties and applications of mater-Bi starch-based materials. Polym. Degrad. Stab. 59 (1–3), 263–272.

Bhatia, A., Gupta, R.K., Bhattacharya, S.N., Choi, H., 2007. Compatibility of biodegradable poly(lactic acid) (PLA) and poly(butylene succinate) (PBS) blends for packaging application. Korea-Aust. Rheol. J. 19 (3), 125–131.

Bledzki, A.K., Jaszkiewicz, A., 2010. Mechanical performance of biocomposites based on PLA and PHBV reinforced with natural fibres – a comparative study to PP. Compos. Sci. Technol. 70 (12), 1687–1696.

Bogoeva-Gaceva, G., Avella, M., Malinconico, M., Buzarovska, A., Grozdanov, A., Gentile, G., Errico, M.E., 2007. Natural fiber eco-composites. Polym. Compos. 28 (1), 98–107.

Bourmaud, A., Pimbert, S., 2008. Investigations on mechanical properties of poly(propylene) and poly(lactic acid) reinforced by miscanthus fibers. Compos. A: Appl. Sci. Manuf. 39 (9), 1444–1454.

Bourmaud, A., Corre, Y.-M., Baley, C., 2015. Fully biodegradable composites: use of poly-(butylene-succinate) as a matrix and to plasticize L-poly-(lactide)-flax blends. Ind. Crop. Prod. 64, 251–257.

Breulmann, M., Künkel, A., Philipp, S., Reimer, V., Siegenthaler, K.O., Skupin, G., Yamamoto, M., 2000. Polymers, biodegradable. In: Ullmann's Encyclopedia of Industrial Chemistry. Wiley-VCH, Berlin, 29, 265–294.

Calabia, B., Ninomiya, F., Yagi, H., Oishi, A., Taguchi, K., Kunioka, M., Funabashi, M., 2013. Biodegradable poly(butylene succinate) composites reinforced by cotton fiber with silane coupling agent. Polymers 5 (1), 128–141.

Can, E., Bucak, S., Kınacı, E., Çalıkoğlu, A., Köse, T., 2014. Polybutylene succinate (PBS)-polycaprolactone (PCL) blends compatibilized with poly(ethylene oxide)-block-poly(propylene oxide)-block-poly(ethylene oxide)(PEO-PPO-PEO) copolymer for biomaterial applications. Polym.-Plast. Technol. Eng. 53 (11), 1178–1193.

Casarin, S.A., Malmonge, S.M., Kobayashi, M., Agnelli, J.A.M., 2011. Study on in-vitro degradation of bioabsorbable polymers poly(hydroxybutyrate-*co*-valerate)-(PHBV) and poly(caprolactone)-(PCL). J. Biomater. Nanobiotechnol. 2 (3), 207–215.

Ceccorulli, G., Scandola, M., Adamus, G., 2002. Compatibilizing effect of a graft copolymer on bacterial poly(3-hydroxybutyrate)/poly(methyl methacrylate) blends. J. Polym. Sci. B Polym. Phys. 40 (13), 1390–1399.

Chavalitpanya, K., Phattanarudee, S., 2013. Poly(lactic acid)/polycaprolactone blends compatibilized with block copolymer. Energy Procedia 34, 542–548.

Chen, G.-X., Kim, H.-S., Kim, E.-S., Yoon, J.-S., 2005. Compatibilization-like effect of reactive organoclay on the poly(L-lactide)/poly(butylene succinate) blends. Polymer 46 (25), 11829–11836.

Darder, M., Aranda, P., Ruiz-Hitzky, E., 2007. Bionanocomposites: a new concept of ecological, bioinspired, and functional hybrid materials. Adv. Mater. 19 (10), 1309–1319.

Del Gaudio, C., Fioravanzo, L., Folin, M., Marchi, F., Ercolani, E., Bianco, A., 2012. Electrospun tubular scaffolds: on the effectiveness of blending poly(ε-caprolactone) with poly(3-hydroxybutyrate-co-3-hydroxyvalerate). J. Biomed. Mater. Res. B Appl. Biomater. 100 (7), 1883–1898.

Deroiné, M., Le Duigou, A., Corre, Y.-M., Le Gac, P.-Y., Davies, P., César, G., Bruzaud, S., 2014a. Accelerated ageing and lifetime prediction of poly(3-hydroxybutyrate-*co*-3-hydroxyvalerate) in distilled water. Polym. Test. 39, 70–78.

Deroiné, M., Le Duigou, A., Corre, Y.-M., Le Gac, P.-Y., Davies, P., César, G., Bruzaud, S., 2014b. Seawater accelerated ageing of poly(3-hydroxybutyrate-*co*-3-hydroxyvalerate). Polym. Degrad. Stab. 105, 237–247.

Dicker, M.P.M., Duckworth, P.F., Baker, A.B., Francois, G., Hazzard, M.K., Weaver, P.M., 2014. Green composites: a review of material attributes and complementary applications. Compos. A: Appl. Sci. Manuf. 56, 280–289.

DiGregorio, B.E., 2009. Biobased performance bioplastic: Mirel. Chem. Biol. 16 (1), 1–2.

Dong, W., Ma, P., Wang, S., Chen, M., Cai, X., Zhang, Y., 2013. Effect of partial crosslinking on morphology and properties of the poly(β-hydroxybutyrate)/poly(D, L-lactic acid) blends. Polym. Degrad. Stab. 98 (9), 1549–1555.

Faruk, O., Bledzki, A.K., Fink, H.-P., Sain, M., 2014. Progress report on natural fiber reinforced composites. Macromol. Mater. Eng. 299 (1), 9–26.
Fayt, R., Teyssie, P., 1986. Molecular design of multicomponent polymer systems. 8. New acrylonitrile-butadiene-styrene-like resins from styrene-acrylonitrile-based mechanical blends. Macromolecules 19 (7), 2077–2080.
Fayt, R., Jérôme, R., Teyssié, P., 1982. Molecular design of multicomponent polymer systems. III. Comparative behavior of pure and tapered block copolymers in emulsification of blends of low-density polyethylene and polystyrene. J. Polym. Sci. Polym. Phys. Ed. 20 (12), 2209–2217.
Fekete, E., Földes, E., Pukánszky, B., 2005. Effect of molecular interactions on the miscibility and structure of polymer blends. Eur. Polym. J. 41 (4), 727–736.
Fujimaki, T., 1998. Processability and properties of aliphatic polyesters, 'BIONOLLE', synthesized by polycondensation reaction. Polym. Degrad. Stab. 59 (1–3), 209–214.
Garlotta, D., 2001. A literature review of poly(lactic acid). J. Polym. Environ. 9 (2), 63–84.
Georgiopoulos, P., Kontou, E., Niaounakis, M., 2014. Thermomechanical properties and rheological behavior of biodegradable composites. Polym. Compos. 35 (6), 1140–1149.
Goriparthi, B.K., Suman, K.N.S., Nalluri, M.R., 2012. Processing and characterization of jute fiber reinforced hybrid biocomposites based on polylactide/polycaprolactone blends. Polym. Compos. 33 (2), 237–244.
Grassie, N., Murray, E.J., Holmes, P.A., 1984. The thermal degradation of poly(-(D)-β-hydroxybutyric acid): part 3—the reaction mechanism. Polym. Degrad. Stab. 6 (3), 127–134.
Groot, W., Krieken, J.V., Sliekersl, O., Vos, A.D., 2011. Chapter 1: production and purification of lactic acid and lactide. In: Auras, R.A., Lim, L.-T., Selke, S.E., Tsuji, H. (Eds.), Poly(Lactic Acid): Synthesis, Structures, Properties, Processing, and Applications. John Wiley & Sons Inc., Hoboken, New Jersey, p. 3.
Gu, S.-Y., Zhang, K., Jie, R., Hui, Z., 2008. Melt rheology of polylactide/poly(butylene adipate-*co*-terephthalate) blends. Carbohydr. Polym. 74 (1), 79–85.
Harada, M., Ohya, T., Iida, K., Hayashi, H., Hirano, K., Fukuda, H., 2007. Increased impact strength of biodegradable poly(lactic acid)/poly(butylene succinate) blend composites by using isocyanate as a reactive processing agent. J. Appl. Polym. Sci. 106 (3), 1813–1820.
Harnnecker, F., dos Santos Rosa, D., Lenz, D., 2012. Biodegradable polyester-based blend reinforced with Curauá fiber: thermal, mechanical and biodegradation behaviour. J. Polym. Environ. 20 (1), 237–244.
Harris, A.M., Lee, E.C., 2010. Heat and humidity performance of injection molded PLA for durable applications. J. Appl. Polym. Sci. 115 (3), 1380–1389.
Harris, A.M., Lee, E.C., 2013. Durability of polylactide-based polymer blends for injection-molded applications. J. Appl. Polym. Sci. 128 (3), 2136–2144.
Henton, D.E., Gruber, P., Lunt, J., Randall, J., 2005. Chapter 16: polylactic acid technology. In: Mohanty, A.K., Misra, M., Drzal, L.T. (Eds.), Natural Fibers, Biopolymers, and Biocomposites. CRC Press LLC, Boca Raton, FL, pp. 527–577.
Hinüber, C., Häussler, L., Vogel, R., Brünig, H., Heinrich, G., Werner, C., 2011. Hollow fibers made from a poly(3-hydroxybutyrate)/poly(ε-caprolactone) blend. Express Polym. Lett. 5 (7), 643–652.
Huda, M.S., Drzal, L.T., Misra, M., Mohanty, A.K., 2006. Wood-fiber-reinforced poly(lactic acid) composites: evaluation of the physicomechanical and morphological properties. J. Appl. Polym. Sci. 102 (5), 4856–4869.
Ibrahim, A., Wahit, M., Yussuf, A., 2014. Effect of fiber reinforcement on mechanical and thermal properties of poly(ε-caprolactone)/poly(lactic acid) blend composites. Fibers Polym. 15 (3), 574–582.

Imre, B., Pukánszky, B., 2013. Compatibilization in bio-based and biodegradable polymer blends. Eur. Polym. J. 49 (6), 1215–1233.

Jacquel, N., Freyermouth, F., Fenouillot, F., Rousseau, A., Pascault, J.P., Fuertes, P., Saint-Loup, R., 2011. Synthesis and properties of poly(butylene succinate): efficiency of different transesterification catalysts. J. Polym. Sci. A Polym. Chem. 49 (24), 5301–5312.

Javadi, A., Kramschuster, A.J., Pilla, S., Lee, J., Gong, S., Turng, L.-S., 2010a. Processing and characterization of microcellular PHBV/PBAT blends. Polym. Eng. Sci. 50 (7), 1440–1448.

Javadi, A., Srithep, Y., Lee, J., Pilla, S., Clemons, C., Gong, S., Turng, L.-S., 2010b. Processing and characterization of solid and microcellular PHBV/PBAT blend and its RWF/nanoclay composites. Compos. A: Appl. Sci. Manuf. 41 (8), 982–990.

Jenkins, M.J., Cao, Y., Howell, L., Leeke, G.A., 2007. Miscibility in blends of poly(3-hydroxybutyrate-*co*-3-hydroxyvalerate) and poly(ε-caprolactone) induced by melt blending in the presence of supercritical CO_2. Polymer 48 (21), 6304–6310.

Jiang, L., Wolcott, M.P., Zhang, J., 2005. Study of biodegradable polylactide/poly(butylene adipate-*co*-terephthalate) blends. Biomacromolecules 7 (1), 199–207.

John, M.J., Thomas, S., 2008. Biofibres and biocomposites. Carbohydr. Polym. 71 (3), 343–364.

John, J., Mani, R., Bhattacharya, M., 2002. Evaluation of compatibility and properties of biodegradable polyester blends. J. Polym. Sci. A Polym. Chem. 40 (12), 2003–2014.

Kabasci, S., Bretz, I., 2011. Chapter 8. Succinic acid: synthesis of biobased polymers from renewable resources. In: Mittal, V. (Ed.), Renewable Polymers: Synthesis, Processing, and Technology. Scrivener Publishing LLC, Massachusetts, pp. 355–379.

Kanemura, C., Nakashima, S., Hotta, A., 2012. Mechanical properties and chemical structures of biodegradable poly(butylene-succinate) for material reprocessing. Polym. Degrad. Stab. 97 (6), 972–980.

Kfoury, G., Raquez, J.-M., Hassouna, F., Odent, J., Toniazzo, V., Ruch, D., Dubois, P., 2013. Recent advances in high performance poly(lactide): from "green" plasticization to super-tough materials via (reactive) compounding. Front. Chem. 1, 1–46.

Kim, H.-S., Kim, H.-J., 2008. Enhanced hydrolysis resistance of biodegradable polymers and bio-composites. Polym. Degrad. Stab. 93 (8), 1544–1553.

Kim, Y.J., Park, O.O., 1999. Miscibility and biodegradability of poly(butylene succinate)/poly(butylene terephthalate) blends. J. Polym. Environ. 7 (1), 53–66.

Kim, C.-H., Cho, K.Y., Choi, E.-J., Park, J.-K., 2000. Effect of P(lLA-*co*-εCL) on the compatibility and crystallization behavior of PCL/PLLA blends. J. Appl. Polym. Sci. 77 (1), 226–231.

Kim, S., Park, S.-w., Lee, K., 2014a. Fishing performance of environmentally friendly tubular pots made of biodegradable resin (PBS/PBAT) for catching the conger eel Conger myriaster. Fish. Sci. 80 (5), 887–895.

Kim, S., Park, S., Lee, K., 2014b. Fishing performance of an octopus minor net pot made of biodegradable twines. Turk. J. Fish. Aquat. Sci. 14, 21–30.

Koning, C., Van Duin, M., Pagnoulle, C., Jerome, R., 1998. Strategies for compatibilization of polymer blends. Prog. Polym. Sci. 23 (4), 707–757.

Koronis, G., Silva, A., Fontul, M., 2013. Green composites: a review of adequate materials for automotive applications. Compos. Part B 44 (1), 120–127.

Kumar, M., Mohanty, S., Nayak, S.K., Rahail Parvaiz, M., 2010. Effect of glycidyl methacrylate (GMA) on the thermal, mechanical and morphological property of biodegradable PLA/PBAT blend and its nanocomposites. Bioresour. Technol. 101 (21), 8406–8415.

Lim, L.T., Auras, R., Rubino, M., 2008. Processing technologies for poly(lactic acid). Prog. Polym. Sci. 33 (8), 820–852.

Lin, S., Guo, W., Chen, C., Ma, J., Wang, B., 2012. Mechanical properties and morphology of biodegradable poly(lactic acid)/poly(butylene adipate-*co*-terephthalate) blends compatibilized by transesterification. Mater. Des. 36, 604–608.

Liu, N.C., Baker, W.E., 1992. Reactive polymers for blend compatibilization. Adv. Polym. Technol. 11 (4), 249–262.

Liu, Q.-S., Zhu, M.-F., Wu, W.-H., Qin, Z.-Y., 2009. Reducing the formation of six-membered ring ester during thermal degradation of biodegradable PHBV to enhance its thermal stability. Polym. Degrad. Stab. 94 (1), 18–24.

Liu, H., Chen, F., Liu, B., Estep, G., Zhang, J., 2010. Super toughened poly(lactic acid) ternary blends by simultaneous dynamic vulcanization and interfacial compatibilization. Macromolecules 43 (14), 6058–6066.

Lopez, J.P., Girones, J., Mendez, J.A., Puig, J., Pelach, M.A., 2012. Recycling ability of biodegradable matrices and their cellulose-reinforced composites in a plastic recycling stream. J. Polym. Environ. 20 (1), 96–103.

Lu, H., Madbouly, S.A., Schrader, J.A., Srinivasan, G., McCabe, K.G., Grewell, D., Kessler, M.R., Graves, W.R., 2014. Biodegradation behavior of poly(lactic acid) (PLA)/distiller's dried grains with solubles (DDGS) composites. ACS Sustainable Chem. Eng. 2 (12), 2699–2706.

Ludvik, C.N., Glenn, G.M., Klamczynski, A.P., Wood, D.F., 2007. Cellulose fiber/bentonite clay/biodegradable thermoplastic composites. J. Polym. Environ. 15 (4), 251–257.

Lunt, J., 1998. Large-scale production, properties and commercial applications of polylactic acid polymers. Polym. Degrad. Stab. 59 (1–3), 145–152.

Ma, P., Hristova-Bogaerds, D.G., Lemstra, P.J., Zhang, Y., Wang, S., 2012. Toughening of PHBV/PBS and PHB/PBS blends via in situ compatibilization using dicumyl peroxide as a free-radical grafting initiator. Macromol. Mater. Eng. 297 (5), 402–410.

Ma, P., Ma, Z., Dong, W., Zhang, Y., Lemstra, P.J., 2013a. Structure/property relationships of partially crosslinked poly(butylene succinate). Macromol. Mater. Eng. 298 (8), 910–918.

Ma, P., Spoelstra, A.B., Schmit, P., Lemstra, P.J., 2013b. Toughening of poly(lactic acid) by poly(β-hydroxybutyrate-*co*-β-hydroxyvalerate) with high β-hydroxyvalerate content. Eur. Polym. J. 49 (6), 1523–1531.

Ma, P., Cai, X., Zhang, Y., Wang, S., Dong, W., Chen, M., Lemstra, P.J., 2014. In-situ compatibilization of poly(lactic acid) and poly(butylene adipate-*co*-terephthalate) blends by using dicumyl peroxide as a free-radical initiator. Polym. Degrad. Stab. 102, 145–151.

Madbouly, S.A., Schrader, J.A., Srinivasan, G., Liu, K., McCabe, K.G., Grewell, D., Graves, W.R., Kessler, M.R., 2014. Biodegradation behavior of bacterial-based polyhydroxyalkanoate (PHA) and DDGS composites. Green Chem. 16 (4), 1911–1920.

Madhavan Nampoothiri, K., Nair, N.R., John, R.P., 2010. An overview of the recent developments in polylactide (PLA) research. Bioresour. Technol. 101 (22), 8493–8501.

Manias, E., Utracki, L.A., 2014. Chapter 2: thermodynamics of polymer blends. In: Utracki, L.A., Wilkie, C.A. (Eds.), Polymer Blends Handbook. second ed. Springer, Dordrecht, pp. 171–289.

Marten, E., Müller, R.-J., Deckwer, W.-D., 2005. Studies on the enzymatic hydrolysis of polyesters. II. Aliphatic–aromatic copolyesters. Polym. Degrad. Stab. 88 (3), 371–381.

Mohanty, A.K., Misra, M., Hinrichsen, G., 2000. Biofibres, biodegradable polymers and biocomposites: an overview. Macromol. Mater. Eng. 276–277 (1), 1–24.

Mohanty, A.K., Misra, M., Drzal, L.T., 2002. Sustainable bio-composites from renewable resources: opportunities and challenges in the green materials world. J. Polym. Environ. 10 (1–2), 19–26.

Mohanty, A.K., Misra, M., Drzal, T.L., Selke, S.E., Harte, B.R., Hinrichsen, G., 2005. Natural fibers, and biocomposites: an introduction. In: Mohanty, A.K., Misra, M., Drzal, T.L.

(Eds.), Natural Fibers, Biopolymers, and Biocomposites. CRC Press, Taylor & Francis Group, Boca Raton, pp. 1–36 (Chapter 1).

Moraczewski, K., 2014. Characterization of multi-injected poly(ε-caprolactone). Polym. Test. 33, 116–120.

Muniyasamy, S., Anstey, A., Reddy, M.M., Misra, M., Mohanty, A., 2013a. Biodegradability and compostability of lignocellulosic based composite materials. J. Renew. Mater. 1 (4), 253–272.

Muniyasamy, S., Reddy, M.M., Misra, M., Mohanty, A., 2013b. Biodegradable green composites from bioethanol co-product and poly(butylene adipate-*co*-terephthalate). Ind. Crop. Prod. 43, 812–819.

Muthuraj, R., Misra, M., Mohanty, A., 2014. Biodegradable poly(butylene succinate) and poly(butylene adipate-*co*-terephthalate) blends: reactive extrusion and performance evaluation. J. Polym. Environ. 22 (3), 336–349.

Muthuraj, R., Misra, M., Mohanty, A.K., 2015. Hydrolytic degradation of biodegradable polyesters under simulated environmental conditions. J. Appl. Polym. Sci. 132 (27), 42189-42201.

Na, Y.-H., He, Y., Shuai, X., Kikkawa, Y., Doi, Y., Inoue, Y., 2002. Compatibilization effect of poly(ε-caprolactone)-b-poly(ethylene glycol) block copolymers and phase morphology analysis in immiscible poly(lactide)/poly(ε-caprolactone) blends. Biomacromolecules 3 (6), 1179–1186.

Nagarajan, V., Mohanty, A.K., Misra, M., 2013a. Sustainable green composites: value addition to agricultural residues and perennial grasses. ACS Sustainable Chem. Eng. 1 (3), 325–333.

Nagarajan, V., Misra, M., Mohanty, A.K., 2013b. New engineered biocomposites from poly(3-hydroxybutyrate-*co*-3-hydroxyvalerate) (PHBV)/poly(butylene adipate-*co*-terephthalate) (PBAT) blends and switchgrass: fabrication and performance evaluation. Ind. Crop. Prod. 42, 461–468.

Nanda, M.R., Misra, M., Mohanty, A.K., 2011. The effects of process engineering on the performance of PLA and PHBV blends. Macromol. Mater. Eng. 296 (8), 719–728.

Nanda, M.R., Misra, M., Mohanty, A.K., 2012. Mechanical performance of soy-hull-reinforced bioplastic green composites: a comparison with polypropylene composites. Macromol. Mater. Eng. 297 (2), 184–194.

Nanda, M.R., Misra, M., Mohanty, A.K., 2013. Performance evaluation of biofibers and their hybrids as reinforcements in bioplastic composites. Macromol. Mater. Eng. 298 (7), 779–788.

Nijenhuis, A.J., Colstee, E., Grijpma, D.W., Pennings, A.J., 1996. High molecular weight poly(L-lactide) and poly(ethylene oxide) blends: thermal characterization and physical properties. Polymer 37 (26), 5849–5857.

Nyambo, C., Nagarajan, V., Mohanty, A.K., Misra, M., 2014. Natural fiber composites from agricultural by-products: an overview. In: Netravali, A.N., Pastore, C.M. (Eds.), Sustainable Composites: Fibers, Resins and Applications. DEStech Publications, Inc., Lancaster, Pennsylvania, pp. 195–239 (Chapter 8).

Ojijo, V., Sinha Ray, S., Sadiku, R., 2013. Toughening of biodegradable polylactide/poly(butylene succinate-*co*-adipate) blends via in situ reactive compatibilization. ACS Appl. Mater. Interfaces 5 (10), 4266–4276.

Oyama, H.T., 2009. Super-tough poly(lactic acid) materials: reactive blending with ethylene copolymer. Polymer 50 (3), 747–751.

Park, J.W., Im, S.S., 2002. Phase behavior and morphology in blends of poly(L-lactic acid) and poly(butylene succinate). J. Appl. Polym. Sci. 86 (3), 647–655.

Parulekar, Y., Mohanty, A.K., 2006. Biodegradable toughened polymers from renewable resources: blends of polyhydroxybutyrate with epoxidized natural rubber and maleated polybutadiene. Green Chem. 8 (2), 206–213.

Pillin, I., Montrelay, N., Grohens, Y., 2006. Thermo-mechanical characterization of plasticized PLA: is the miscibility the only significant factor? Polymer 47 (13), 4676–4682.

Pillin, I., Montrelay, N., Bourmaud, A., Grohens, Y., 2008. Effect of thermo-mechanical cycles on the physico-chemical properties of poly(lactic acid). Polym. Degrad. Stab. 93 (2), 321–328.

Plant, T.R., 2020. The technology roadmap for plant/crop-based renewable resources (accessed January, 2015).

Pradhan, R., Misra, M., Erickson, L., Mohanty, A., 2010. Compostability and biodegradation study of PLA–wheat straw and PLA–soy straw based green composites in simulated composting bioreactor. Bioresour. Technol. 101 (21), 8489–8491.

Qiu, Z., Ikehara, T., Nishi, T., 2003a. Poly(hydroxybutyrate)/poly(butylene succinate) blends: miscibility and nonisothermal crystallization. Polymer 44 (8), 2503–2508.

Qiu, Z., Ikehara, T., Nishi, T., 2003b. Miscibility and crystallization behaviour of biodegradable blends of two aliphatic polyesters. Poly(3-hydroxybutyrate-*co*-hydroxyvalerate) and poly(butylene succinate) blends. Polymer 44 (24), 7519–7527.

Qiu, Z., Yang, W., Ikehara, T., Nishi, T., 2005. Miscibility and crystallization behavior of biodegradable blends of two aliphatic polyesters. Poly(3-hydroxybutyrate-*co*-hydroxyvalerate) and poly(ε-caprolactone). Polymer 46 (25), 11814–11819.

Ravati, S., Favis, B.D., 2010. Morphological states for a ternary polymer blend demonstrating complete wetting. Polymer 51 (20), 4547–4561.

Ren, J., 2010. Modification of PLA. Biodegradable Poly(Lactic Acid): Synthesis, Modification, Processing and Applications. Tsinghua University Press, Beijing (Springer, Heidelberg), pp. 38–141.

Risse, S., Tighzert, L., Berzin, F., Vergnes, B., 2014. Microstructure, rheological behavior, and properties of poly(lactic acid)/poly(butylene succinate)/organoclay nanocomposites. J. Appl. Polym. Sci. 131 (12), 40364–40371.

Robeson, L.M., 2007. Polymer Blends: A Comprehensive Review. Carl Hanser Verlag, Munich, pp. 1–7.

Rosa, D.d.S., Lenz, D.M., 2013. Chapter 16: biocomposites: influence of matrix nature and additives on the properties and biodegradation behavior. In: Chamy, R., Rosenkranz, F. (Eds.), Biodegradation-Engineering and Technology, InTech, Croatia, pp. 432–475. http://dx.doi.org/10.5772/56290.

Ryan, A.J., 2002. Polymer science: designer polymer blends. Nat. Mater. 1 (1), 8–10.

Sarazin, P., Roy, X., Favis, B.D., 2004. Controlled preparation and properties of porous poly(L-lactide) obtained from a co-continuous blend of two biodegradable polymers. Biomaterials 25 (28), 5965–5978.

Satyanarayana, K.G., Arizaga, G.G.C., Wypych, F., 2009. Biodegradable composites based on lignocellulosic fibers—an overview. Prog. Polym. Sci. 34 (9), 982–1021.

Shah, D., 2013. Developing plant fibre composites for structural applications by optimising composite parameters: a critical review. J. Mater. Sci. 48 (18), 6083–6107.

Shah, D.U., 2014. Natural fibre composites: comprehensive Ashby-type materials selection charts. Mater. Des. 62, 21–31.

Sheth, M., Kumar, R.A., Davé, V., Gross, R.A., McCarthy, S.P., 1997. Biodegradable polymer blends of poly(lactic acid) and poly(ethylene glycol). J. Appl. Polym. Sci. 66 (8), 1495–1505.

Shibata, M., Inoue, Y., Miyoshi, M., 2006. Mechanical properties, morphology, and crystallization behavior of blends of poly(L-lactide) with poly(butylene succinate-*co*-L-lactate) and poly(butylene succinate). Polymer 47 (10), 3557–3564.

Shin, B.Y., Han, D.H., 2013. Compatibilization of immiscible poly(lactic acid)/poly(ε-caprolactone) blend through electron-beam irradiation with the addition of a compatibilizing agent. Radiat. Phys. Chem. 83, 98–104.

Siegenthaler, K.O., Künkel, A., Skupin, G., Yamamoto, M., 2012. Ecoflex® and Ecovio®: biodegradable, performance-enabling plastics. In: Rieger, B., Künkel, A., Coates, G.W., Reichardt, R., Dinjus, E., Zevaco, Th.A (Eds.), Synthetic Biodegradable Polymers. Springer, Berlin, Heidelberg, pp. 91–136.

Signori, F., Coltelli, M.-B., Bronco, S., 2009. Thermal degradation of poly(lactic acid) (PLA) and poly(butylene adipate-*co*-terephthalate) (PBAT) and their blends upon melt processing. Polym. Degrad. Stab. 94 (1), 74–82.

Sinha Ray, S., Okamoto, K., Okamoto, M., 2003. Structure–property relationship in biodegradable poly(butylene succinate)/layered silicate nanocomposites. Macromolecules 36 (7), 2355–2367.

Sirisinha, K., Somboon, W., 2012. Melt characteristics, mechanical, and thermal properties of blown film from modified blends of poly(butylene adipate-*co*-terephthalate) and poly(lactide). J. Appl. Polym. Sci. 124 (6), 4986–4992.

Sis, A., Ibrahim, N., Yunus, W., 2013. Effect of (3-aminopropyl)trimethoxysilane on mechanical properties of PLA/PBAT blend reinforced kenaf fiber. Iran. Polym. J. 22 (2), 101–108.

Small, P.A., 1953. Some factors affecting the solubility of polymers. J. Appl. Chem. 3 (2), 71–80.

Soroudi, A., Jakubowicz, I., 2013. Recycling of bioplastics, their blends and biocomposites: a review. Eur. Polym. J. 49 (10), 2839–2858.

Stridsberg, K., Ryner, M., Albertsson, A.-C., 2002. Controlled ring-opening polymerization: polymers with designed macromolecular architecture. In: Degradable Aliphatic Polyesters. Springer Berlin, Heidelberg, pp. 41–65.

Sungsanit, K., Kao, N., Bhattacharya, S.N., 2012. Properties of linear poly(lactic acid)/polyethylene glycol blends. Polym. Eng. Sci. 52 (1), 108–116.

Sykacek, E., Schlager, W., Mundigler, N., 2010. Compatibility of softwood flour and commercial biopolymers in injection molding. Polym. Compos. 31 (3), 443–451.

Szabó, P., Epacher, E., Földes, E., Pukánszky, B., 2004. Miscibility, structure and properties of PP/PIB blends. Mater. Sci. Eng. A 383 (2), 307–315.

Takayama, T., Todo, M., 2006. Improvement of impact fracture properties of PLA/PCL polymer blend due to LTI addition. J. Mater. Sci. 41 (15), 4989–4992.

Terzopoulou, Z.N., Papageorgiou, G.Z., Papadopoulou, E., Athanassiadou, E., Alexopoulou, E., Bikiaris, D.N., 2014. Green composites prepared from aliphatic polyesters and bast fibers. Ind. Crop. Prod. 68, 60–79. http://dx.doi.org/10.1016/j.indcrop.2014.1008.1034.

Tserki, V., Matzinos, P., Pavlidou, E., Vachliotis, D., Panayiotou, C., 2006. Biodegradable aliphatic polyesters. Part I. Properties and biodegradation of poly(butylene succinate-*co*-butylene adipate). Polym. Degrad. Stab. 91 (2), 367–376.

Tsuji, H., Yamada, T., Suzuki, M., Itsuno, S., 2003. Part 7. Effects of poly(L-lactide-*co*-ε-caprolactone) on morphology, structure, crystallization, and physical properties of blends of poly(L-lactide) and poly(ε-caprolactone). Polym. Int. 52 (2), 269–275.

Utracki, L.A., 2002. Compatibilization of polymer blends. Can. J. Chem. Eng. 80 (6), 1008–1016.

Uyama, H., Kobayashi, S., 2002. Enzyme-catalyzed polymerization to functional polymers. J. Mol. Catal. B Enzym. 19–20, 117–127.

Van de Velde, K., Kiekens, P., 2002. Biopolymers: overview of several properties and consequences on their applications. Polym. Test. 21 (4), 433–442.

Vandewijngaarden, J., Murariu, M., Dubois, P., Carleer, R., Yperman, J., Adriaensens, P., Schreurs, S., Lepot, N., Peeters, R., Buntinx, M., 2014. Gas permeability properties of poly(3-hydroxybutyrate-*co*-3-hydroxyhexanoate). J. Polym. Environ. 22 (4), 501–507.

Verlinden, R.A.J., Hill, D.J., Kenward, M.A., Williams, C.D., Radecka, I., 2007. Bacterial synthesis of biodegradable polyhydroxyalkanoates. J. Appl. Microbiol. 102 (6), 1437–1449.

Vieira, M.G.A., da Silva, M.A., dos Santos, L.O., Beppu, M.M., 2011. Natural-based plasticizers and biopolymer films: a review. Eur. Polym. J. 47 (3), 254–263.

Vilay, V., Mariatti, M., Ahmad, Z., Pasomsouk, K., Todo, M., 2010. Improvement of microstructures and properties of biodegradable PLLA and PCL blends compatibilized with a triblock copolymer. Mater. Sci. Eng. A 527 (26), 6930–6937.

Vroman, I., Tighzert, L., 2009. Biodegradable polymers. Materials 2 (2), 307–344.

Wahit, M.U., Akos, N.I., Laftah, W.A., 2012. Influence of natural fibers on the mechanical properties and biodegradation of poly(lactic acid) and poly(ε-caprolactone) composites: a review. Polym. Compos. 33 (7), 1045–1053.

Wambua, P., Ivens, J., Verpoest, I., 2003. Natural fibres: can they replace glass in fibre reinforced plastics? Compos. Sci. Technol. 63 (9), 1259–1264.

Wang, Y., Hillmyer, M.A., 2001. Polyethylene-poly(L-lactide) diblock copolymers: synthesis and compatibilization of poly(L-lactide)/polyethylene blends. J. Polym. Sci. A Polym. Chem. 39 (16), 2755–2766.

Wang, L., Ma, W., Gross, R.A., McCarthy, S.P., 1998. Reactive compatibilization of biodegradable blends of poly(lactic acid) and poly(ε-caprolactone). Polym. Degrad. Stab. 59 (1–3), 161–168.

Wang, R., Wang, S., Zhang, Y., Wan, C., Ma, P., 2009. Toughening modification of PLLA/PBS blends via in situ compatibilization. Polym. Eng. Sci. 49 (1), 26–33.

Witt, U., Müller, R.-J., Deckwer, W.-D., 1996. Studies on sequence distribution of aliphatic/aromatic copolyesters by high-resolution ^{13}C nuclear magnetic resonance spectroscopy for evaluation of biodegradability. Macromol. Chem. Phys. 197 (4), 1525–1535.

Witt, U., Yamamoto, M., Seeliger, U., Müller, R.-J., Warzelhan, V., 1999. Biodegradable polymeric materials—not the origin but the chemical structure determines biodegradability. Angew. Chem. Int. Ed. 38 (10), 1438–1442.

Witt, U., Einig, T., Yamamoto, M., Kleeberg, I., Deckwer, W.D., Müller, R.J., 2001. Biodegradation of aliphatic–aromatic copolyesters: evaluation of the final biodegradability and ecotoxicological impact of degradation intermediates. Chemosphere 44 (2), 289–299.

Wu, Q., Wang, Y., Chen, G.-Q., 2009. Medical application of microbial biopolyesters polyhydroxyalkanoates. Artif. Cell. Blood Substit. Biotechnol. 37 (1), 1–12.

Wu, D., Zhang, Y., Yuan, L., Zhang, M., Zhou, W., 2010. Viscoelastic interfacial properties of compatibilized poly(ε-caprolactone)/polylactide blend. J. Polym. Sci. B Polym. Phys. 48 (7), 756–765.

Wu, D., Yuan, L., Laredo, E., Zhang, M., Zhou, W., 2012. Interfacial properties, viscoelasticity, and thermal behaviors of poly(butylene succinate)/polylactide blend. Ind. Eng. Chem. Res. 51 (5), 2290–2298.

Xu, J., Guo, B.-H., 2010. Microbial succinic acid, its polymer poly(butylene succinate), and applications. In: Chen, G.-Q. (Ed.), Plastics from Bacteria. Springer, Berlin, Heidelberg, pp. 347–388.

Yang, B., Nar, M., Visi, D.K., Allen, M., Ayre, B., Webber III, C.L., Lu, H., D'Souza, N.A., 2014. Effects of chemical versus enzymatic processing of kenaf fibers on poly(hydroxybutyrate-*co*-valerate)/poly(butylene adipate-*co*-terephthalate) composite properties. Compos. Part B 56, 926–933.

Yokohara, T., Yamaguchi, M., 2008. Structure and properties for biomass-based polyester blends of PLA and PBS. Eur. Polym. J. 44 (3), 677–685.

Yoo, E.S., Im, S.S., 1999. Melting behavior of poly(butylene succinate) during heating scan by DSC. J. Polym. Sci. B Polym. Phys. 37 (13), 1357–1366.

Yu, T., Li, Y., 2014. Influence of poly(butylenes adipate-*co*-terephthalate) on the properties of the biodegradable composites based on ramie/poly(lactic acid). Compos. A: Appl. Sci. Manuf. 58, 24–29.

Yu, L., Dean, K., Li, L., 2006. Polymer blends and composites from renewable resources. Prog. Polym. Sci. 31 (6), 576–602.

Zarrinbakhsh, N., Misra, M., Mohanty, A.K., 2011. Biodegradable green composites from distiller's dried grains with solubles (DDGS) and a polyhydroxy(butyrate-*co*-valerate) (PHBV)-based bioplastic. Macromol. Mater. Eng. 296 (11), 1035–1045.

Zarrinbakhsh, N., Mohanty, A., Misra, M., 2013. Improving the interfacial adhesion in a new renewable resource-based biocomposites from biofuel coproduct and biodegradable plastic. J. Mater. Sci. 48 (17), 6025–6038.

Zaverl, M., Seydibeyoğlu, M.Ö., Misra, M., Mohanty, A., 2012. Studies on recyclability of polyhydroxybutyrate-*co*-valerate bioplastic: multiple melt processing and performance evaluations. J. Appl. Polym. Sci. 125 (S2), E324–E331.

Zembouai, I., Bruzaud, S., Kaci, M., Benhamida, A., Corre, Y.-M., Grohens, Y., 2014. Mechanical recycling of poly(3-hydroxybutyrate-*co*-3-hydroxyvalerate)/polylactide based blends. J. Polym. Environ. 22 (4), 449–459.

Żenkiewicz, M., Richert, J., Rytlewski, P., Moraczewski, K., Stepczyńska, M., Karasiewicz, T., 2009. Characterisation of multi-extruded poly(lactic acid). Polym. Test. 28 (4), 412–418.

Zhang, M., Thomas, N.L., 2011. Blending polylactic acid with polyhydroxybutyrate: the effect on thermal, mechanical, and biodegradation properties. Adv. Polym. Technol. 30 (2), 67–79.

Zhang, L.L., Goh, S.H., Lee, S.Y., Hee, G.R., 2000. Miscibility, melting and crystallization behavior of two bacterial polyester/poly(epichlorohydrin-*co*-ethylene oxide) blend systems. Polymer 41 (4), 1429–1439.

Zhang, N., Wang, Q., Ren, J., Wang, L., 2009. Preparation and properties of biodegradable poly(lactic acid)/poly(butylene adipate-*co*-terephthalate) blend with glycidyl methacrylate as reactive processing agent. J. Mater. Sci. 44 (1), 250–256.

Zhang, K., Mohanty, A.K., Misra, M., 2012. Fully biodegradable and biorenewable ternary blends from polylactide, poly(3-hydroxybutyrate-*co*-hydroxyvalerate) and poly(butylene succinate) with balanced properties. ACS Appl. Mater. Interfaces 4 (6), 3091–3101.

Zhang, K., Nagarajan, V., Misra, M., Mohanty, A.K., 2014a. Supertoughened renewable PLA reactive multiphase blends system: phase morphology and performance. ACS Appl. Mater. Interfaces 6 (15), 12436–12448.

Zhang, K., Misra, M., Mohanty, A.K., 2014b. Toughened sustainable green composites from poly(3-hydroxybutyrate-*co*-3-hydroxyvalerate) based ternary blends and miscanthus biofiber. ACS Sustainable Chem. Eng. 2 (10), 2345–2354.

Zhou, J., Wang, X., Hua, K., Duan, C., Zhang, W., Ji, J., Yang, X., 2013. Enhanced mechanical properties and degradability of poly(butylene succinate) and poly(lactic acid) blends. Iran. Polym. J. 22 (4), 267–275.

Zini, E., Scandola, M., 2011. Green composites: an overview. Polym. Compos. 32 (12), 1905–1915.

Mechanical performance of microcellular injection molded biocomposites from green plastics: PLA and PHBV

6

H. Zhao, L.-S. Turng*†*
*Shandong University, Jinan, China; †University of Wisconsin–Madison, Madison, WI, USA

6.1 Introduction

Studies of biobased and biodegradable polymers have attracted increased attention due to the high demand to reduce dependency on petroleum-based polymers, decrease the accumulation of persistent plastic waste, and better control the emission of CO_2 in the environment. Environmentally friendly polymers such as poly(lactic acid) (PLA) and polyhydroxybutyrate-valerate (PHBV) are considered promising replacements for most polymeric products derived from fossil fuels, especially for commodity applications such as packaging, cushioning, construction, thermal and sound insulation, and plastic utensils. Due to their competitive material and processing costs, and comparable mechanical properties, significant research and development efforts have been made to develop a continuous or semicontinuous method of producing microcellular foams by extrusion and microcellular injection molding to enable widespread use of these specific polymers.

6.2 Biobased and biodegradable polymers PLA and PHBV

6.2.1 Poly(lactic acid)

PLA can be synthesized by either ring-opening polymerization of lactide or condensation polymerization of lactic acid monomers that are produced from renewable resources via a fermentation process. Lactic acid (LA) has two optically active configurations known as D-LA and L-LA. The L form is the most common in nature and, therefore, is the main constituent of commercially available PLA. Minor amounts of L-LA are typically used to control the crystallinity of PLA as the L-LA units disrupt the crystallization of L-LA chains (Gupta et al., 2007).

PLA has a high strength (50–70 MPa), high modulus (3 GPa), and high thermal plasticity, all of which are comparable to many petroleum-based plastics. Furthermore, it is biodegradable and biocompatible because it can be naturally recycled by biological processes. Due to these unique properties, as well as its affordable cost as compared to other biodegradable polymers, PLA represents an interesting alternative to nonrenewable polymers for applications with short life spans, particularly in

Biocomposites: Design and Mechanical Performance. http://dx.doi.org/10.1016/B978-1-78242-373-7.00015-9

biomedical applications such as sutures, bone screws, and scaffolding for tissue engineering (Burg et al., 2000; Chiellini and Solaro, 1996; Cui et al., 2014; Kramschuster and Turng, 2010). Over the past decade, PLA has attracted extensive industrial and academic interest as a potential substitute for petroleum-based polymers in both commodity and biomedical applications. This is not only due to its green and biodegradable features, but also because it releases no toxic components during manufacturing.

However, commercially available PLA resin cannot fully satisfy the requirements of industrial and medical applications due to its poor melt strength, brittleness, and low thermal resistance. The melt viscosity of PLA has low shear sensitivity and the melt itself has relatively poor strength. Its narrow processing window has also limited its widespread use in injection molding and foaming processes. To circumvent these shortcomings, extensive efforts have been made to improve its toughness and processability.

6.2.2 Polyhydroxybutyrate-valerate

Polyhydroxyalkanoates (PHAs) are a family of polyesters that are synthesized and intracellularly accumulated as a carbon and energy storage material by various microorganisms. PHAs are widely used for various biomedical applications, including drug delivery and tissue engineering scaffolds, due to their excellent biocompatibility and biodegradability. The first and most prevalent PHA is poly(β-hydroxybutyrate) (PHB). PHBV is a copolymer of PHB with randomly arranged 3-hydroxybutyrate (HB) groups and 3-hydroxyvalarate (HV) groups. PHB exhibits high stiffness and crystallinity. In order to increase its flexibility and processing capabilities, PHB is often copolymerized with PHV to form PHBV. Increasing PHV content in PHB also reduces the stiffness, melting point, and crystallinity of PHB.

In spite of improved mechanical and thermal properties, PHBV still exhibits some disadvantages including low strain-at-break, a narrow processing window, a slow crystallization rate, and a higher cost compared to petroleum-based synthetic polymers. In order to tailor its properties and decrease its total cost, several approaches have been proposed such as foaming blends or composites with biodegradable polymers, natural fillers, or inorganic fillers (Carli et al., 2011).

6.3 Principles, advantages, and challenges of microcellular injection molding

Microcellular injection molding is an emerging process capable of mass-producing parts with three-dimensional complex geometries and excellent dimensional stability. It is widely used for automotive components, electronic/electrical products, construction, outdoor products, and many other applications. Recently, it was reported that the world's largest injection molding machine equipped with a microcellular foaming unit began operation to produce large parts, such as automotive instrument panels. This high-tonnage molding machine has a clamping force of 5400 metric tons (5940 tons US).

The microcellular injection molding process uses supercritical fluid (SCF), usually nitrogen (N_2) or carbon dioxide (CO_2), as a plasticizer and physical blowing agent to produce

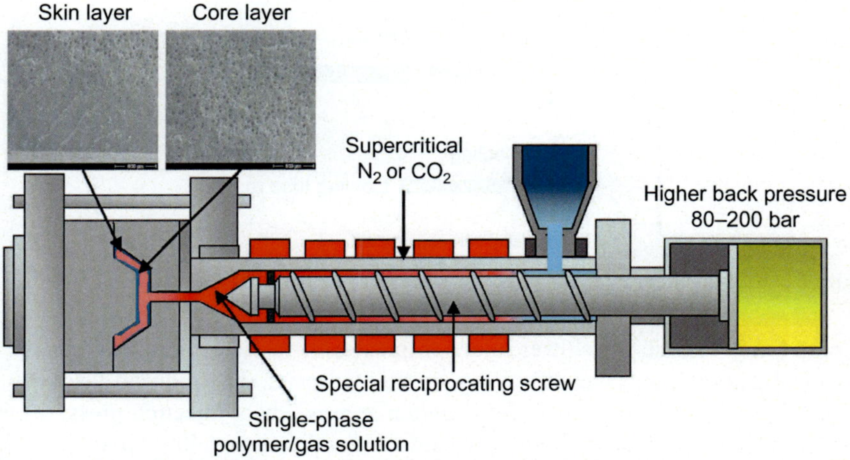

Figure 6.1 Microcellular injection molding machine and foaming mechanism.

lightweight foamed parts with reduced material, pressure, energy, and cycle times (see Figure 6.1). Compared with the conventional injection molding process, the microcellular injection molding process produces components with increased dimensional stability, less thermal degradation, and less use of material. The employment of SCF reduces the viscosity of the polymer melt due to the formation of a single-phase polymer/gas solution, thus enabling the polymer to be processed at lower temperatures and pressures.

There are three basic steps required for the production of microcellular plastics. First, a polymer/gas solution must be created. This involves the diffusion of high concentrations of gas (such as carbon dioxide or nitrogen) into the polymer melt to create a single-phase polymer/gas solution. Second, the nucleation of microvoids is required. This is accomplished by subjecting the polymer/gas solution to a thermodynamic instability by rapidly dropping the solution pressure and/or raising the solution temperature. The result is a drop in the equilibrium solubility of the gas in the polymer matrix, resulting in the nucleation of numerous microvoids. Finally, the uniform growth of microvoids must occur such that the density of the resulting cellular material is decreased (Baldwin et al., 1995). Cell growth is controlled by the gas diffusion rate and the stiffness of the polymer/gas solution. Cell growth occurs during the mold filling stage but mostly within the mold cavity. In general, cell growth is affected by the following factors: (a) time allowed for cells to grow, (b) state of supersaturation, (c) hydrostatic pressure applied to the polymer, (d) temperature of the system, and (e) viscoelastic properties of the single-phase polymer/gas solution.

Polymer foaming technology, invented by Suh et al., has been successfully commercialized by Trexel to injection mold microcellular foams after a series of modifications on certain components of a standard reciprocating-screw injection molding machine, such as the plasticizing unit, injection unit, hydraulic unit, clamping unit, and gas delivery unit. To produce microcellular foams, a new screw designed for better mixing and a new sealed barrel with gas injectors are used (Xu, 2011). In comparison with conventional injection molding, microcellular injection molding achieves

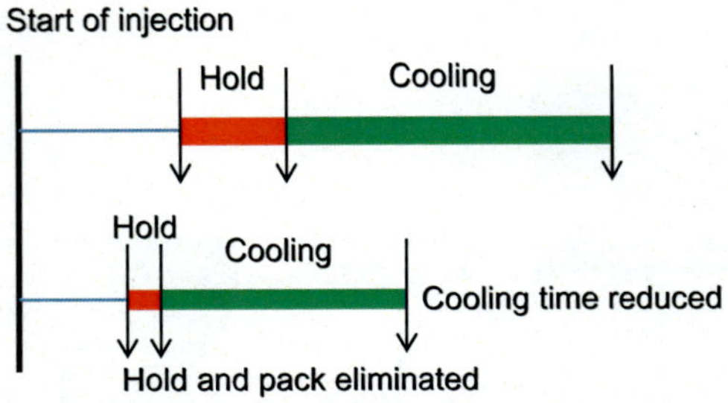

Figure 6.2 Microcellular injection molding process: cycle time reduction.

increased melt flowability, lower injection pressures, faster cycle times, greater dimensional stability, and weight savings in molded parts. The reduced viscosity after the addition of SCF allows faster injection speeds, lower injection pressures, and lower clamp tonnage requirements. The shot size for the microcellular foam process is usually smaller than that for the solid molding process, which yields a shorter recovery time. As shown in Figure 6.2, the pack and hold time is eliminated due to the internal gas pressure and the significantly smaller amount of material needed to be cooled (Chandra et al., 2005; Yuan and Turng, 2005). The uniform cell distribution and expansion allows for improved dimensional stability and diminished surface flaws.

Microcellular injection molding technology is being widely used in structural foaming because it offers greater control of cell nucleation and cell coalescence. Despite all of these advantages, the addition of SCF brings the complexity of flow behavior, material morphology, product surface quality, and modifying existing machinery, which has hampered the development and broad acceptance of the process. Moreover, the mass production of pure PLA or PHBV with uniform cell morphology is still quite challenging, mainly due to their low melt strengths.

6.4 Mechanical behavior of PLA- and PHBV-based blends and biocomposites

One of the most important properties of the polymer used in the microcellular injection molding process for industrial applications is sufficient mechanical strength, and no single-component polymer can meet all of the functional requirements. Therefore, the design and fabrication of multiphase polymer systems represent a viable strategy for developing multifunctional biodegradable polymer foams.

The mechanical performance of pure PLA and PHBV can be enhanced by modifying their physical properties through several approaches including copolymerization, blending, and plasticization (Bledzki and Faruk, 2005; Corre et al., 2011; Harris and Lee, 2008). Although copolymerization can be effective in improving mechanical properties, commercializing a new copolymer is often a long and costly process, thus making this a less viable option. The major drawback of plasticizers is that during long-term use, they have a tendency to migrate to the surface of the material, causing brittleness. Pilla et al. (2009a,b) investigated the effect of chain extenders (CE) on the foaming and mechanical behavior of linear PLA using supercritical nitrogen as the blowing agent.

They showed that the addition of 0.8 wt.% CE to PLA improved the foam morphology of the injection-molded PLA samples and promoted cell density. With the addition of CE, the tensile properties such as toughness, strain-at-break, and strength of the microcellular injection-molded bars increased but the modulus decreased. The microcellular samples had lower average values than the solid samples for toughness, strain-at-break, and tensile strength, but had higher average values for the moduli.

The blending of polymers is an effective, practical, and economic way of obtaining materials with desirable properties. Blends exhibit superior physical and mechanical properties over individual polymers. Blending PLA and PHBV with other polymers can substantially modify their mechanical and thermal properties, degradation rate, and permeability. Recently, extensive research and characterization work has been done on the microcellular injection molding of various PLA, PHBV, and composite blends. Applying the microcellular injection molding process showed that controlling the processing parameters and material formulations (chain branch extender or varying the molecular weight), adding fillers, and polymer blends could enhance the foaming ability and cell nucleation and growth (Li et al., 2011a; Lim et al., 2008; Martin and Averous, 2001; Modi et al., 2012). In particular, the addition of fillers such as nanoclay, which acts as a nucleating agent, leads to heterogeneous cell nucleation. Nanoclay platelets provide a large number of nucleation sites, thus leading to a higher cell density and a smaller cell size. These studies reported that a uniform foam morphology with an average cell size ranging from 3 to 40 μm was achieved (Matuana and Diaz, 2010; Nampoothiri et al., 2010).

6.4.1 PLA- and PHBV-based blends

PLA- and PHBV-based polymer blends have been extensively studied in order to reduce material costs, improve processability, tailor biodegradability, and engineer mechanical and thermal properties. Thus, biodegradable polymers such as poly(butylene adipate-co-terephthalate) (PBAT), PVOH, and TPU have been blended with PLA and PHBV. In addition, PLA and PHBV have been compounded with natural fibers such as recycled shopping bag fibers, coir fibers, flax fibers, and recycled wood fibers (RWFs) to reduce the total cost of PLA- and PHBV-based polymers while providing some improvements to their mechanical and physical properties. Table 6.1 shows the detailed blend information, the blowing agent types, cell size, cell density, and mechanical properties.

Peng et al. (2013) employed nitrogen and expandable thermoplastic microspheres (ETM) to produce foamed PLA parts. They showed that parts foamed with ETM exhibited better surface qualities than their supercritical nitrogen counterparts. The PLA bars foamed with nitrogen showed a better Young's modulus and tensile strength. With increasing ETM content, the strength and modulus of the PLA/ETM decreased sharply, while the stain-at-break improved. The bubble shell of the acrylic copolymer and the elastomeric EVA carrier of the ETM, the foamed spherical cell structure, and the overall effect of the ETM on the polymer matrix effectively modified their mechanical properties. They also examined PHBV/PBAT blends with a weight ratio of 45/55 using nitrogen and ETM as blowing agents (Peng et al.,2010). The results showed that microcellular bars foamed in nitrogen had a better strain-at-break.

Mi et al. (2013) blended PLA/TPU at different ratios with twin-screw extrusion and microcellular injection molding. They showed that their tensile and compressive

Table 6.1 Studies on the microstructure and mechanical properties of PLA- and PHBV-based blends

References	Blends/system	Blowing agents	Cell size (μm)	Cell density (cells/cm³)	Tests	Strength	Modulus	Strain-at-break (%)
Pilla et al. (2009a,b)	PLA	N_2	87 ± 13	1.38×10^6	Tensile	0.043 MPa/(Kg/m³)	1.57 MPa/(Kg/m³)	4
	PLA/chain extender	N_2	25 ± 7	3.1×10^7	Tensile	0.045 MPa/(Kg/m³)	1.58 MPa/(Kg/m³)	4.5
Peng et al. (2013)	PLA	N_2	18+	9.1×10^7	Tensile	38.9 MPa	1.82 GPa	4.2
		ETM	65–80	$1–3 \times 10^6$	Tensile	26.7–34.5 MPa	1.53–1.87 GPa	28–39
Kramschuster and Turng (2010)	PLA/PVOH	CO_2	25	NA	Tensile	0.5–0.8 MPa	5–10 GPa	8–15
Mi et al. (2013)	PLA/TPU	CO_2	100–250	$1.2–4 \times 10^5$	Tensile	5–30 MPa	10–1000 MPa	10–600+
					Compression	3–30 MPa	10–200 MPa	35–600+
Zhao et al. (2013a)	PLA/PHBV	N_2	25–150	$0.6–21 \times 10^6$	Tensile	30–35 MPa	1.25–1.5 GPa	2.5–12
Javadi et al. (2010a)	PHBV/PBAT	NA	239 ± 110	2.56×10^4	Tensile	9.1×10^{-3} MPa/(Kg/m³)	0.5–2.2 MPa/(Kg/m³)	51.8
Peng et al. (2010)	PHBV/PBAT	N_2	20–200	NA	Tensile	49 MPa	315 MPa	6.2
		ETM	80–200	NA	Tensile	6.48–10 MPa	345–441 MPa	2.6–4.1

stresses decreased with the addition of TPU. Solid TPU samples and blended samples did not break during the tensile tests even when the strain reached 600%. The strain-at-break increased with increasing TPU content for the microcellular injection-molded samples and the samples lost yield behavior when the TPU content increased to 75%. The tensile and compressive moduli decreased as the TPU content increased. The tensile and compressive moduli of the microcellular injection-molded samples ranged from 7 to 1007 MPa, and 11 to 200 MPa, respectively. The properties reported are the actual readings measured for the solid and microcellular specimens without taking into account the weight reduction of the microcellular specimens.

Javadi et al. (2010a) produced PHBV/PBAT blends with different weight ratios using the microcellular injection molding process. Adding PBAT into PHBV increased the cell size and decreased the cell density in the microcellular samples. With an increasing PBAT content, the toughness and strain-at-break increased while the modulus and strength decreased.

Kramschuster and Turng (2010) added salt particles to the blends and leached the foamed samples in water after the injection molding process to produce a porous interconnected foam structure as a scaffold for tissue engineering. Because the PVOH was used in granule shape, further mixing of the PLA with PVOH granules in a twin-screw extruder further reduced the size of the PVOH, while increasing its dispersion, in the PLA matrix. Consequently, the finely dispersed PVOH improved the uniformity of the blend's final foam morphology.

6.4.2 PLA/PHBV blends

Both PLA and PHBV are biodegradable polymers. Blending PLA with PHBV provides a practical way of improving or tailoring the structure and properties of the blends, especially their foam morphologies, without compromising their biodegradability. PHBV was also used to improve the performance and properties of PLA while helping to control the cell morphology in microcellular PLA/PHBV blends using a novel microcellular injection molding process. A number of studies have reported preparing PLA/PHBV blends by melt mixing or solvent casting processes. Nanda et al. (2011) studied the effects of processing parameters on the properties of PLA/PHBV blends and found that the elongation of the optimized blends significantly improved. They also found that the two polymers were immiscible in the blends, and the heat distortion properties of PLA can be significantly improved by the incorporation of PHBV into PLA. Zhang et al. studied the blends of PLA and PHB with different weight ratios and found that the biodegradability of PLA improved with increasing PHB content at room temperature. Richards et al. (2008) prepared PLA and PHBV foams by a batch foaming method and found that although PLA and PHBV were immiscible, the presence of small quantities of PHBV (25 wt.%) led to low-density foams with finer, more uniform cells.

Zhao et al. (2013a) showed that blending 30 wt.% PHBV with PLA increased the cell density of the injection-molded foam samples. The incorporation of PHBV decreased the tensile strength of solid PLA; the tensile strength decreased from 54.99 to 42.23 MPa by increasing the PHBV content from 0% to 45%, respectively. The reduction in tensile strength of the solid PLA/PHBV blends may be attributed to the relatively low strength of PHBV and the relatively weak polymer interfaces. The tensile strength of the microcellular specimens was found to be less than that of their solid counterparts; however, it was still more than 30 MPa. Furthermore, there were no significant differences in the tensile strengths of microcellular PLA, PHBV, and PLA/PHBV blends with different weight ratios of PHBV. The addition of PHBV did not cause any significant differences in the moduli either. Also, similar to the tensile strength, the moduli of the microcellular specimens were found to be less than that of their solid counterparts, which can be attributed to the presence of voids in the microcellular specimens. The microstructure information is shown in Figure 6.3. A few large bubbles were found in

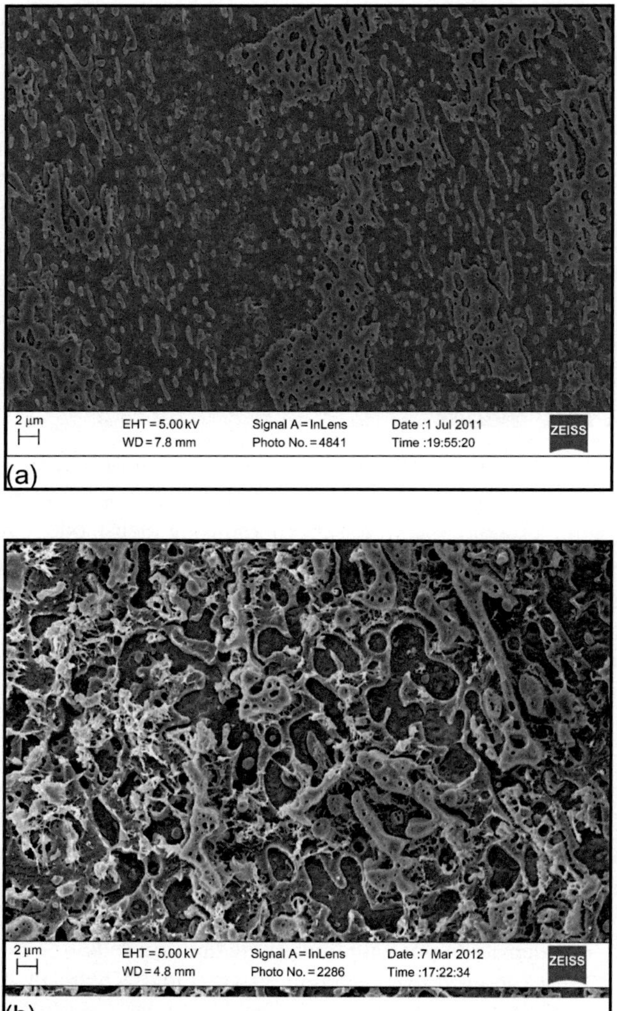

Figure 6.3 Representative SEM images of the fracture surfaces of solid and microcellular PLA/PHBV blends: (a) 70:30 (solid), (b) 55:45 (solid).

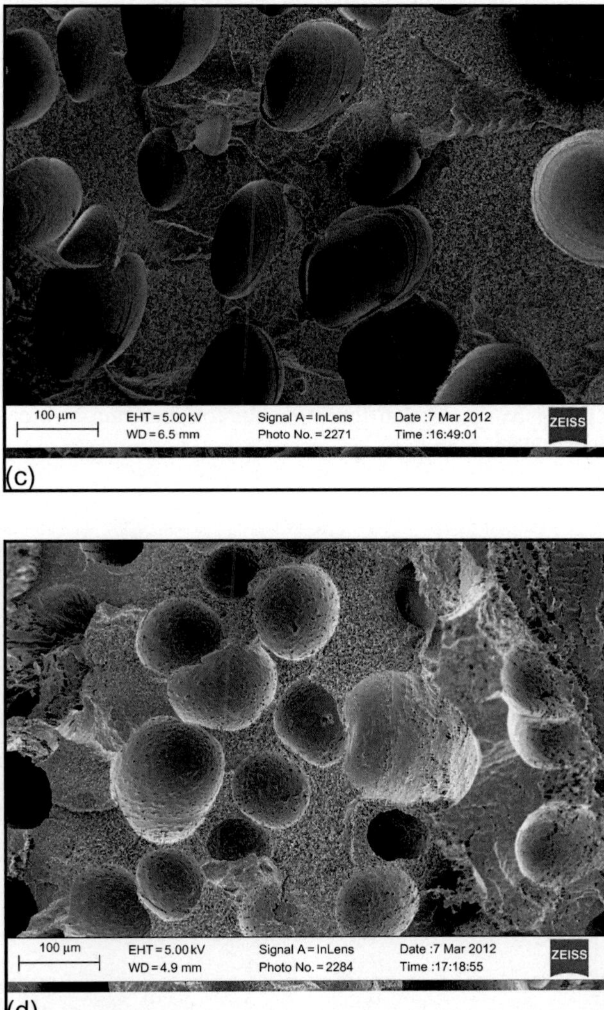

Figure 6.3 Continued. (c) 70:30 (microcellular), and (d) 55:45 (microcellular). Reprinted with permission from Zhao et al. (2013a). Copyright 2013, American Chemical Society.

the microcellular injection-molded specimens. These large bubbles may become points of stress concentrations, thereby potentially decreasing the mechanical properties of the specimens (Burg et al., 2000; Chiellini and Solaro, 1996).

The strain-at-break of neat PLA and PHBV were only about 3.5%. These materials had no necking in the tensile tests and exhibited a distinct yield point (maximum load) with subsequent failure by neck instability. However, the incorporation of PHBV in the PLA/PHBV blend enhanced the strain-at-break as compared to pure PLA and PHBV. The blends showed yielding and stable neck growth though cold drawing. The strain-at-break values of solid PLA/PHBV (85:15) and microcellular PLA/PHBV (70:30) were significantly higher than the other solid and microcellular specimens. This was expected because solid PLA/PHBV (85:15) had a better polymer interface and exhibits a single phase. However, microcellular PLA/PHBV (70:30) and PLA/PHBV (55:45) had larger strain-at-break values than their solid counterparts. For microcellular PLA/PHBV (85:15), the decrease in strain-at-break as compared to solid specimens may be attributed to the existence of large bubbles that may have acted as stress concentrators in the specimens. However, by increasing the amount of PHBV, a morphology with a much smaller cell size and higher cell density was obtained, especially for the microcellular PLA/PHBV (70:30) specimens. As shown in Figure 6.4, this modified morphology led to a bimodal cell morphology with cells as small as 1 μm. The large bubbles, which could be points of stress concentration, were eliminated. Thus, the combined effects of interfacial adhesion and foam morphology (cell size and cell density) yielded different results for the strain-at-break values of the microcellular PLA/PHBV blends.

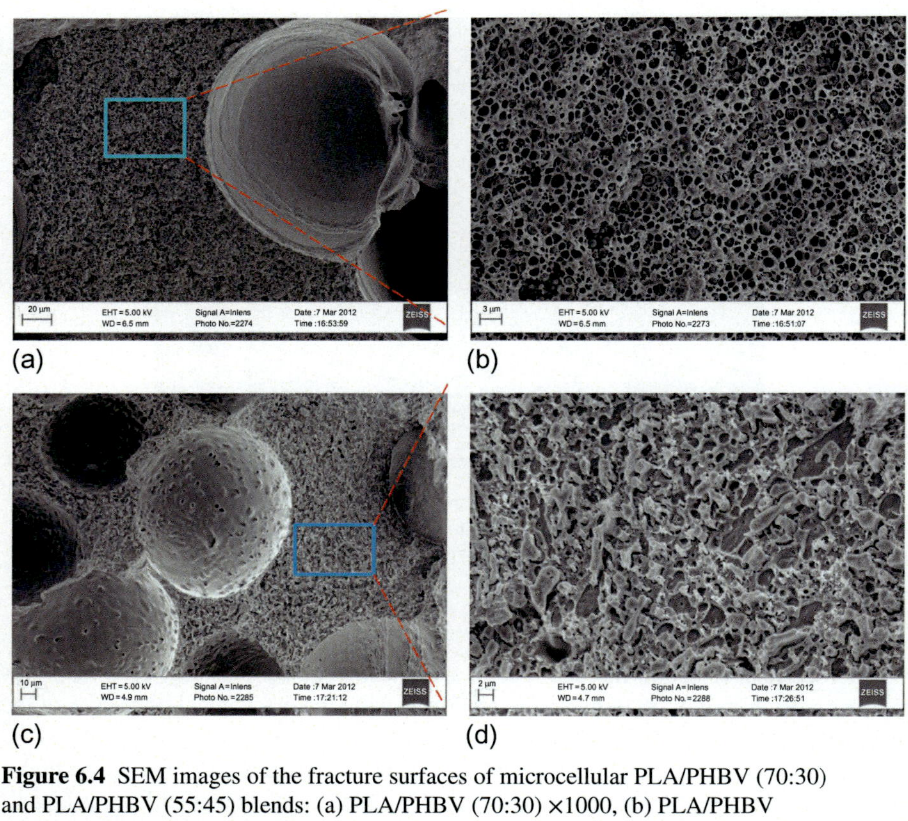

Figure 6.4 SEM images of the fracture surfaces of microcellular PLA/PHBV (70:30) and PLA/PHBV (55:45) blends: (a) PLA/PHBV (70:30) ×1000, (b) PLA/PHBV (70:30) ×5000, (c) PLA/PHBV (55:45) ×1000, and (d) PLA/PHBV (55:45) ×5000. Reprinted with permission from Zhao et al. (2013a). Copyright 2013, American Chemical Society.

6.4.3 PLA- and PHBV-based composites

In order to manufacture fully degradable biocomposite foams with enhanced mechanical properties, a few researchers have investigated the behavior of PLA and PHBV foams using several natural fibers/additives including flax fibers, coir fibers, and recycled paper shopping bag fibers (Bledzki and Jaszkiewicz, 2010). These studies have shown that natural fibers can be embedded in the PLA or PHBV matrix as an excellent reinforcement to improve mechanical properties. Table 6.2 shows different research groups' examination of the foaming of PLA/PHBV/fiber composites. Table 6.2 also lists information on the composites, the blowing agent types, cell size, cell density, and mechanical properties for each study.

Kramschuster et al. (2007a) investigated PLA's microcellular injection molding behavior with recycled paper shopping bag fibers. Fiber contents of 10 and 30 wt.% were used, with silane as the coupling agent. They showed that the addition of up to 30 wt.% fiber enhanced the cell density of the foamed samples as well as their storage moduli, specific moduli, and specific tensile strengths. Ameli et al. studied PLA's foam injection molding behavior with talc particles up to 5 wt.%. They showed that the addition of talc improved crystallinity and created a more uniform foam morphology with larger cell densities and smaller cell sizes, with a minimum average cell size of 38 μm, and high void fractions in the range of 50–65%. They also demonstrated that the toughening effect of talc, as well as the induced crystallinity in conjunction with the improved cell morphology, enhanced the mechanical properties of PLA–talc foamed samples.

Javadi et al. (2010b) produced microcellular PHBV components with treated/untreated coir fibers. They found that adding 10 wt.% untreated coir fibers decreased cell size and increased cell density. A further decrease in cell size and increase in cell density was observed with the incorporation of surface-modified coir fibers in the PHBV matrix. The storage moduli increased in composites with both untreated and treated coir fibers compared to neat PHBV samples. Adding both treated and untreated coir fibers increased the toughness and strain-at-break, and decreased the tensile strength slightly, for both solid and microcellular specimens. The coir fibers did not cause significant changes in the moduli of microcellular specimens.

Javadi et al. (2010c) also found that the addition of 10% RWF reduced the cell size and increased the cell density of microcellular components. Furthermore, the addition of 10% RWF generally increased the specific modulus and tensile strength, but decreased the toughness and strain-at-break in both solid and microcellular components. Moreover, unlike the neat PHBV/PBAT blend, microcellular PHBV/PBAT/RWF composites showed higher toughness and strain-at-break compared to their solid counterparts. In addition, higher toughness and strain-at-break was observed in the PHBV/PBAT/untreated RWF composite compared with the PHBV/PBAT/saline-treated RWF composite, particularly in microcellular components.

Pilla et al. (2009c) compounded PLA with 1%, 10%, and 20% flax fibers to investigate the effect of fiber and silane content on the cell morphology and mechanical properties of microcellular samples. Silane was also used as a coupling agent to create strong interface bonding between the PLA and the fibers. The foaming results showed

Table 6.2 **Studies on the microstructure and mechanical properties of PLA- and PHBV-based composites**

References	Composites	Blowing agents	Cell size (μm)	Cell density (cells/cm^3)	Tests	Strength	Modulus	Strain-at-break (%)
Kramschuster et al. (2007a)	PLA/recycled shopping bag fibers	N_2	3–50	NA	Tensile	0.5–0.8 GPa	5–10 GPa	8–15
Javadi et al. (2010b)	PHBV/coir fiber	NA	8–76	$1.37–1.55 \times 10^6$	Tensile	$2–2.5 \times 10^{-2}$ MPa/(Kg/m^3)	1.5–1.7 MPa/(Kg/m^3)	1.8–2.0
Pilla et al. (2009c)	PLA/flax fiber	N_2	2–8	$5–20 \times 10^7$	Tensile	0.035–0.045 MPa/(Kg/m^3)	1.75–2.25 MPa/(Kg/m^3)	3–4
Ameli et al. (2013)	PLA/talc	N_2	20	–	–	–	–	–
Javadi et al. (2010c)	PHBV/PBAT/RWF	NA	57–61	$2.5–2.87 \times 10^6$	Tensile	$9.3–9.8 \times 10^3$ MPa/(Kg/m^3)	$3.2–3.5 \times 10^{-1}$ MPa/(Kg/m^3)	20.9–30.7

that the average cell size decreased while the cell density increased with fiber content. An average cell size of 3 μm was achieved when 20 wt.% fibers were used. It was reported that neither the fiber content nor the silane treatment had much influence on the strain-at-break of the microcellular samples. The specific moduli of both solid and microcellular samples increased with fiber content. The specific strength of the solid and microcellular PLA–flax composites were only slightly lower than that of their solid and microcellular pure PLA counterparts. Overall, the toughness, strain-at-break, and specific strength of the microcellular samples were found to be lower than that of their solid counterparts. The storage moduli of the PLA–flax composites with 10% and 20% fiber content was higher than that of pure PLA.

6.4.4 PLA- and PHBV-based nanocomposites

Besides natural fibers, the effect of adding various inorganic fillers to PLA and PHBV matrices has also been investigated by various research groups. Nanotechnology involves materials that possess one or more of the dimensions in the nanometer scale. Generally, polymer nanocomposites are defined as a material consisting of two immiscible phases of which one is a polymer and the other is a filler (dispersed phase) with a nanoscale feature in at least one dimension. Polymer nanocomposites often show an excellent balance between strength and toughness compared to their individual components. The incorporation of nanoparticles into the polymer matrix provides the possibility of increasing and modulating some of the properties of a neat polymer, as exemplified by the thermal stability, mechanical and gas barrier performance, melt viscosity, biocompatibility, biodegradation rate, and so on (Bordes et al., 2009; Jack et al., 2009; Pluta et al., 2007; Ray et al., 2002).

The mechanical properties of porous polymer foams are often not sufficient or suitable for hard parts as the stiffness and compressive strength is lower than that of solid components. Nanofillers, such as clay or nanotubes, offer enormous advantages over traditional microscale fillers due to their higher aspect and surface area-to-volume ratios, improved adhesion between nanoparticles and polymers, and lowered amount of loading necessary to achieve equivalent properties. It has been shown that polymer nanocomposites with strong interactions between the polymer matrix and the nanofillers also exhibit an enhanced melt viscosity and elasticity by forming a rubber-like interconnected structure without significantly sacrificing impact properties (Bourbigot et al., 2008; Krishnamachari et al., 2009). Thus, the improved melt viscosity and elasticity of PLA, PHBV, and their nanocomposites help control cell growth in the microcellular injection molding process.

Although a high degree of uniform nanofiller dispersion was not achieved in either process, nanocomposites fabricated via the microcellular process showed better dispersion than those made via the conventional process. This finding is attributed to the use of SCF in the microcellular process, which helps to further disperse the fillers due to its rapid plasticization effect (Cui et al., 2012; Kharbas et al., 2003). Kramschuster et al. (2007b) found that the strain-at-break and the specific toughness of the microcellular PLA nanocomposites showed more improvement than the neat PLA foamed samples. They attributed this improvement to the homogenous foam morphology and to the smaller cell sizes caused by the nanoclay particles. With increasing nanofiller content, the average pore size decreased and the cell wall thickness remained almost the same at a low clay loading level. In contrast, at a high loading level, both the average pore size and the pore wall thickness increased. This phenomenon is attributed to the change of melt viscosity of nanocomposites and the heterogeneous nucleation behavior of the clay at low loading levels. Moreover, mass production of polymer nanocomposites and nanocomposite foams depend on reliable and affordable processing methods. In order to make these highly desirable materials commercial available, the challenges in preparing exfoliated nanocomposites with required mechanical properties in large quantities and at a low cost need to be overcome (Homminga et al., 2005). Table 6.3 outlines investigations of these nanocomposites with regard to blowing agent, cell size, cell density, and mechanical properties.

Hwang et al. (2009) found that the incorporation of MMT into the PLA matrix decreased the tensile strength. MMT made the nanocomposites more brittle than neat PLA under both solid and foamed molding, with the tensile strength of the foamed materials slightly less than that of the solid molded materials. As the nanoclay content increased from 0.5 to 2 wt.%, the stiffness of the nanocomposites decreased slightly compared to pure PLA. However, if the nanoclay content was larger than 3 wt.%, the stiffness was slightly larger than that of pure PLA. Recently, Ameli et al. (2013) reported that, during microcellular foam injection molding of PLA clay nanocomposites, not only the nanoparticles but also the induced crystallization promoted cell nucleation and a homogenous cell morphology. They showed that an average cell size of less than 50 μm was induced in the PLA nanocomposites when an expansion ratio of 2.8-fold was achieved. They also found that nanoclay enhanced the flexural and impact properties and the strain-at-break of microcellular injection-molded PLA with a high void fraction.

Li et al. (2011b) studied microcellular injection molding of PLA/PBAT/nanoclay composites. They showed that nanoclay platelets improved the interfacial adhesion between PLA and PBAT. Consequently, the cell density was significantly enhanced through the presence of nanoclay particles and by the improved interfacial adhesion of the two polymers.

Pilla et al. (2010) investigated hyperbranched polyesters and nanoclays on the properties of microcellular injection-molded PLA. They found that the addition of HBPs and/or HBPs with nanoclay decreased the average cell size and increased the cell density. The stress–stain plots of all of the microcellular PLA/HBP blends showed considerable strain softening and cold drawing, indicating a ductile fracture mode. Samples with HBPs showed higher values of strain-at-break and specific toughness. Pilla et al. (2007) also investigated the microcellular foaming behavior of PLA carbon nanotube nanocomposites. They explained how a small amount of carbon nanotubes noticeably promoted a foamed cell density due to its effect on cell nucleation. They demonstrated that the microcellular foaming process further dispersed the nanotubes within the PLA, which was caused by the dissolved plasticization effect.

Zhao et al. (2013b) showed that the addition of both 30% PHBV and nanoclay slightly decreased the tensile strength of solid PLA/PHBV blends. The slight reduction in the tensile strength of the PLA/PHBV blends and nanocomposites may be

Table 6.3 Studies on the microstructure and mechanical properties of PLA- and PHBV-based nanocomposites

References	Nanocomposites	Blowing agents	Cell size (μm)	Cell density (cells/cm^3)	Tests	Strength	Modulus	Strain-at-break (%)
Hwang et al. (2009)	PLA/MMT	CO_2	35–109	3×10^9–1×10^{11}	Tensile	1.0–3.4 MPa/(Kg/m^3)	NA	1.0–5.5
Javadi et al. (2010c)	PHBV/PBAT/nanoclay	NA	73 ± 48	1.89×10^6	Tensile	9.3×10^{-3} MPa/(Kg/m^3)	3.5×10^{-1} MPa/(Kg/m^3)	20.8 ± 1.9
Pilla et al. (2010)	PLA/HBPs/nanoclay	N_2	10–38	1.5–16×10^7	Tensile	0.025–0.045 MPa/(Kg/m^3)	1.3–1.6 MPa/(Kg/m^3)	3–20
Li et al. (2011b)	PLA/PBAT/nanoclay	N_2	25–45	2.5–5.5×10^7	Tensile	30–34 MPa	0.9–1.1 GPa	NA
Zhao et al. (2013b)	PLA/PHBV/nanoclay	N_2	25–30	1.5–2.9×10^7	Tensile	30–35 MPa	1.25–1.75 GPa	2.5–12.5

attributed to the relatively weak polymer–polymer and polymer–filler interfaces. The tensile strength of the microcellular specimens was found to be less than that of their solid counterparts. However, there were no significant differences in the strengths of the microcellular PLA/PHBV blends and nanocomposites. The increase of PHBV content did not cause a significant difference in the Young's modulus. However, with the addition of nanoclay to the PLA/PHBV (85:15) blend, the moduli of the solid and microcellular PLA/PHBV/clay nanocomposites increased by 20.15% and 18.98%, respectively. For the PLA/PHBV (70:30) blend, the moduli of the solid and microcellular specimens increased by 18.26% and 22.21%, respectively. Also, similar to the tensile strength, the Young's modulus of the microcellular specimens was found to be less than that of their solid counterparts, which could be attributed to the presence of large voids in the microcellular specimens. These large voids may become regions of stress concentration and, thus, could adversely affect the mechanical properties.

Due to the relatively weak polymer interface, the strain-at-break of the PLA/PHBV blends gradually decreased with the addition of PHBV and nanoclay, similar to the tensile strength results. Also, compared to solid samples, the strain-at-break of the microcellular specimens decreased as well. However, microcellular PLA/PHBV (70:30) had a much higher strain-at-break than the other specimens. This might have been due to the combination of material formulation and foam morphology. Judging by the appearance of the microcellular injection-molded tensile bars, the microcellular PLA/PHBV (70:30) specimens appeared much whiter and more ductile than the PLA/PHBV (85:15) specimens and had higher cell densities. During tensile testing, it was also found that some of the tensile bars broke around the neck of the tensile test bar, which was far away from the gate and may have had the greatest degree of foaming. Thus, the standard deviation for the elongation-to-break of the PLA/PHBV (70:30) specimens was higher than the other specimens tested for strain-at-break.

6.5 Conclusions and outlook for the future

In this chapter, methods of synthesizing PLA and PHBV have been presented along with a brief introduction to the microcellular injection molding process and the mechanical performance of their blends and composite foams. The novelty of microcellular injection molding opens a wide range of opportunities for the preparation of porous parts made from biobased and biodegradable polymers and their nanocomposites with enhanced and multifunctional properties, without the use of organic solvents. The porous polymer/nanocomposites prepared using this process have demonstrated improved mechanical strength, biocompatibility, and biodegradation rates. Further investigation into the structure and properties of biobased and biodegradable microcellular injection-molded PLA/PHBV may find many potential applications in the fields of structural components, packaging, and biomedical devices.

Despite the obvious advantages, however, the challenge remains to design a porous polymer with adequate mechanical properties, a microstructure with proper cell size and even cell distribution, and good surface quality. Surface defects such as swirl marks, silver streaks, surface blistering, post-blow, and large surface roughnesses limit the widespread application of microcellular injection molding. Several attempts such as using water as the blowing agent, coblowing agents (N_2 and CO_2) (Di Maio et al., 2005), gas-laden pellet extrusion (Lee et al., 2011), particle leaching (Cui et al., 2014), film insulation (Lee and Turng, 2010), and gas counterpressure have been employed to enhance the properties and surface quality of produced parts. However, further investigation and process optimization are still required to control the bubble nucleation and final microstructure and properties of microcellular foams.

Acknowledgments

The authors would like to acknowledge the financial support of the National Natural Science Foundation of China (No. 51403118), China Postdoctoral Science Foundation (2014 M561923), the Foundation for the Excellent Middle-Aged and Young Scientists of Shandong Province (BS2014ZZ010), the Fundamental Research Funds of Shandong University (No. 2014GN001 and 2014QY003-12), the Opening Project of State Key Laboratory of Molecular Engineering of Polymers (No. K2014-07), and the Key Laboratory of Polymer Processing Engineering, Ministry of Education (No. 2013006). The authors would also like to acknowledge the Wisconsin Institute for Discovery and the Polymer Engineering Center at UW–Madison for the use of their facilities and their technical support.

References

Ameli, A., Jahani, D., Nofar, M., Jung, P.U., Park, C.B., 2013. Processing and characterization of solid and foamed injection-molded polylactide with talc. J. Cell. Plast. 49 (4), 351–374.

Baldwin, D.F., Shimbo, M., Suh, N.P., 1995. The role of gas dissolution and induced crystallization during microcellular polymer processing – a study of poly(ethylene-terephthalate) and carbon-dioxide systems. J. Eng. Mater.-T. ASME 117 (1), 62–74.

Bledzki, A.K., Faruk, O., 2005. Effects of the chemical foaming agents, injection parameters, and melt-flow index on the microstructure and mechanical properties of microcellular injection-molded wood-fiber/polypropylene composites. J. Appl. Polym. Sci. 97 (3), 1090–1096.

Bledzki, A.K., Jaszkiewicz, A., 2010. Mechanical performance of biocomposites based on PLA and PHBV reinforced with natural fibres – a comparative study to PP. Compos. Sci. Technol. 70 (12), 1687–1696.

Bordes, P., Pollet, E., Averous, L., 2009. Nano-biocomposites: biodegradable polyester/nanoclay systems. Prog. Polym. Sci. 34 (2), 125–155.

Bourbigot, S., Fontaine, G., Bellayer, S., Delobel, R., 2008. Processing and nanodispersion: a quantitative approach for polylactide nanocomposite. Polym. Test. 27 (1), 2–10.

Burg, K.J., Porter, S., Kellam, J.F., 2000. Biomaterial developments for bone tissue engineering. Biomaterials 21 (23), 2347–2359.

Carli, L.N., Crespo, J.S., Mauler, R.S., 2011. PHBV nanocomposites based on organomodified montmorillonite and halloysite: the effect of clay type on the morphology and thermal and mechanical properties. Compos. Part A Appl. Sci. Manuf. 42 (11), 1601–1608.

Chandra, A., Gong, S.Q., Yuan, M.J., Turng, L.S., Gramann, P., Cordes, H., 2005. Microstructure and crystallography in microcellular injection-molded polyamide-6 nanocomposite and neat resin. Polym. Eng. Sci. 45 (1), 52–61.

Chiellini, E., Solaro, R., 1996. Biodegradable polymeric materials. Adv. Mater. 8 (4), 305–313.

Corre, Y.M., Maazouz, A., Duchet, J., Reignier, J., 2011. Batch foaming of chain extended PLA with supercritical CO_2: influence of the rheological properties and the process parameters on the cellular structure. J. Supercrit. Fluids 58 (1), 177–188.

Cui, Z., Zhao, H., Peng, Y., Kaland, M., Turng, L.-S., Shen, C., 2012. Morphology and properties of porous and interconnected poly(epsilon-caprolactone) matrices using solid and microcellular injection molding. J. Biobased Mater. Bioenergy 6 (3), 259–268.

Cui, Z., Zhao, H., Peng, Y., Han, J., Turng, L.-S., Shen, C., 2014. Fabrication and characterization of highly porous chitosan/poly(DL lactic-co-glycolic acid) nanocomposite scaffolds using electrospinning and freeze drying. J. Biobased Mater. Bioenergy 8 (3), 281–291.

Di Maio, E., Mensitieri, G., Iannace, S., Nicolais, L., Li, W., Flumerfelt, R.W., 2005. Structure optimization of polycaprolactone foams by using mixtures of CO_2 and N_2 as blowing agents. Polym. Eng. Sci. 45 (3), 432–441.

Gupta, B., Revagade, N., Hilborn, J., 2007. Poly(lactic acid) fiber: an overview. Prog. Polym. Sci. 32 (4), 455–482.

Harris, A.M., Lee, E.C., 2008. Improving mechanical performance of injection molded PLA by controlling crystallinity. J. Appl. Polym. Sci. 107 (4), 2246–2255.

Homminga, D., Goderis, B., Dolbnya, I., Reynaers, H., Groeninckx, G., 2005. Crystallization behavior of polymer/montmorillonite nanocomposites. Part I. Intercalated poly(ethylene oxide)/montmorillonite nanocomposites. Polymer 46 (25), 11359–11365.

Hwang, S.S., Hsu, P.P., Yeh, J.M., Chang, K.C., Lai, Y.Z., 2009. The mechanical/thermal properties of microcellular injection-molded poly-lactic-acid nanocomposites. Polym. Compos. 30 (11), 1625–1630.

Jack, K.S., Velayudhan, S., Luckman, P., Trau, M., Grondahl, L., Cooper-White, J., 2009. The fabrication and characterization of biodegradable HA/PHBV nanoparticle–polymer composite scaffolds. Acta Biomater. 5 (7), 2657–2667.

Javadi, A., Kramschuster, A.J., Pilla, S., Lee, J., Gong, S.Q., Turng, L.S., 2010a. Processing and characterization of microcellular PHBV/PBAT blends. Polym. Eng. Sci. 50 (7), 1440–1448.

Javadi, A., Srithep, Y., Pilla, S., Lee, J., Gong, S., Turng, L.S., 2010b. Processing and characterization of solid and microcellular PHBV/coir fiber composites. Mat. Sci. Eng. C-Mater. 30 (5), 749–757.

Javadi, A., Srithep, Y., Lee, J., Pilla, S., Clemons, C., Gong, S.Q., Turng, L.S., 2010c. Processing and characterization of solid and microcellular PHBV/PBAT blend and its RWF/nanoclay composites. Compos. Part A Appl. Sci. Manuf. 41 (8), 982–990.

Kharbas, H., Nelson, P., Yuan, M.J., Gong, S.Q., Turng, L.S., Spindler, R., 2003. Effects of nano-fillers and process conditions on the microstructure and mechanical properties of microcellular injection molded polyamide nanocomposites. Polym. Compos. 24 (6), 655–671.

Kramschuster, A., Turng, L.S., 2010. An injection molding process for manufacturing highly porous and interconnected biodegradable polymer matrices for use as tissue engineering scaffolds. J. Biomed. Mater. Res. B Appl. Biomater. 92B (2), 366–376.

Kramschuster, A., Pilla, S., Gong, S., Chandra, A., Turng, L.S., 2007a. Injection molded solid and microcellular polylactide compounded with recycled paper shopping bag fibers. Int. Polym. Process. 22 (5), 436–445.

Kramschuster, A., Gong, S., Turng, L.-S., Li, T., Li, T., 2007b. Injection-molded solid and microcellular polylactide and polylactide nanocomposites. J. Biobased Mater. Bioenergy 1 (1), 37–45.

Krishnamachari, P., Zhang, J., Lou, J.Z., Yan, J.Z., Uitenham, L., 2009. Biodegradable poly(lactic acid)/clay nanocomposites by melt intercalation: a study of morphological, thermal, and mechanical properties. Int. J. Polym. Anal. Charact. 14 (4), 336–350.

Lee, J., Turng, L.S., 2010. Improving surface quality of microcellular injection molded parts through mold surface temperature manipulation with thin film insulation. Polym. Eng. Sci. 50 (7), 1281–1289.

Lee, J.J., Turng, L.S., Dougherty, E., Gorton, P., 2011. A novel method for improving the surface quality of microcellular injection molded parts. Polymer 52 (6), 1436–1446.

Li, K., Peng, J., Turng, L.S., Huang, H.X., 2011a. Dynamic rheological behavior and morphology of polylactide/poly(butylenes adipate-co-terephthalate) blends with various composition ratios. Adv. Polym. Technol. 30 (2), 150–157.

Li, K., Cui, Z., Sun, X., Turng, L.-S., Huang, H., 2011b. Effects of nanoclay on the morphology and physical properties of solid and microcellular injection molded polyactide/poly(butylenes adipate-co-terephthalate) (PLA/PBAT) nanocomposites and blends. J. Biobased Mater. Bioenergy 5 (4), 442–451.

Lim, L.T., Auras, R., Rubino, M., 2008. Processing technologies for poly(lactic acid). Prog. Polym. Sci. 33 (8), 820–852.

Martin, O., Averous, L., 2001. Poly(lactic acid): plasticization and properties of biodegradable multiphase systems. Polymer 42 (14), 6209–6219.

Matuana, L.M., Diaz, C.A., 2010. Study of cell nucleation in microcellular poly(lactic acid) foamed with supercritical CO_2 through a continuous-extrusion process. Ind. Eng. Chem. Res. 49 (5), 2186–2193.

Mi, H.-Y., Salick, M.R., Jing, X., Jacques, B.R., Crone, W.C., Peng, X.-F., Turng, L.-S., 2013. Characterization of thermoplastic polyurethane/polylactic acid (TPU/PLA) tissue engineering scaffolds fabricated by microcellular injection molding. Mat. Sci. Eng. C-Mater. 33 (8), 4767–4776.

Modi, S., Koelling, K., Vodovotz, Y., 2012. Miscibility of poly(3-hydroxybutyrate-co-3-hydroxyvalerate) with high molecular weight poly(lactic acid) blends determined by thermal analysis. J. Appl. Polym. Sci. 124 (4), 3074–3081.

Nampoothiri, K.M., Nair, N.R., John, R.P., 2010. An overview of the recent developments in polylactide (PLA) research. Bioresour. Technol. 101 (22), 8493–8501.

Nanda, M.R., Misra, M., Mohanty, A.K., 2011. The effects of process engineering on the performance of PLA and PHBV blends. Macromol. Mater. Eng. 296 (8), 719–728.

Peng, J., Li, K., Cui, Z., Turng, L.-S., Peng, X.-F., 2010. Comparisons of microcellular PHBV/PBAT parts injection molded with supercritical nitrogen and expandable thermoplastic microspheres: surface roughness, tensile properties, and morphology. Cell. Polym. 29 (6), 327–341.

Peng, J., Srithep, Y., Wang, J., Yu, E., Turng, L.-S., Peng, X.-F., 2013. Comparisons of microcellular polylactic acid parts injection molded with supercritical nitrogen and expandable thermoplastic microspheres: surface roughness, tensile properties, and morphology. J. Cell. Plast. 49 (1), 33–45.

Pilla, S., Kramschuster, A., Gong, S., Chandra, A., Turng, L.S., 2007. Solid and microcellular polylactide-carbon nanotube nanocomposites. Int. Polym. Process. 22 (5), 418–428.

Pilla, S., Kim, S.G., Auer, G.K., Gong, S.Q., Park, C.B., 2009a. Microcellular extrusion-foaming of polylactide with chain-extender. Polym. Eng. Sci. 49 (8), 1653–1660.

Pilla, S., Kramschuster, A., Yang, L.Q., Lee, J., Gong, S.Q., Turng, L.S., 2009b. Microcellular injection-molding of polylactide with chain-extender. Mater. Sci. Eng. C Biomim. Supramol. Syst. 29 (4), 1258–1265.

Pilla, S., Kramschuster, A., Lee, J., Auer, G.K., Gong, S.Q., Turng, L.S., 2009c. Microcellular and solid polylactide–flax fiber composites. Compos. Interfaces 16 (7–9), 869–890.

Pilla, S., Kramschuster, A., Lee, J., Clemons, C., Gong, S.Q., Turng, L.S., 2010. Microcellular processing of polylactide-hyperbranched polyester–nanoclay composites. J. Mater. Sci. 45 (10), 2732–2746.

Pluta, M., Jeszka, J.K., Boiteux, G., 2007. Polylactide/montmorillonite nanocomposites: structure, dielectric, viscoelastic and thermal properties. Eur. Polym. J. 43 (7), 2819–2835.

Ray, S.S., Maiti, P., Okamoto, M., Yamada, K., Ueda, K., 2002. New polylactide/layered silicate nanocomposites. 1. Preparation, characterization, and properties. Macromolecules 35 (8), 3104–3110.

Richards, E., Rizvi, R., Chow, A., Naguib, H., 2008. Biodegradable composite foams of PLA and PHBV using subcritical CO(2). J. Polym. Environ. 16 (4), 258–266.

Xu, J., 2011. In: Microcellular Injection Molding, vol. 9. Wiley & Sons, Inc., Hoboken, NJ.

Yuan, M.J., Turng, L.S., 2005. Microstructure and mechanical properties of microcellular injection molded polyamide-6 nanocomposites. Polymer 46 (18), 7273–7292.

Zhao, H., Cui, Z., Sun, X., Turng, L.-S., Peng, X., 2013a. Morphology and properties of injection molded solid and microcellular polylactic acid/polyhydroxybutyrate-valerate (PLA/PHBV) blends. Ind. Eng. Chem. Res. 52 (7), 2569–2581.

Zhao, H., Cui, Z., Wang, X., Turng, L.-S., Peng, X., 2013b. Processing and characterization of solid and microcellular poly(lactic acid)/polyhydroxybutyrate-valerate (PLA/PHBV) blends and PLA/PHBV/clay nanocomposites. Compos. Part B 51, 79–91.

Mechanical performance of poly(propylene carbonate)-based blends and composites

7

A.B. Kousaalya, B.I. Biddappa, S. Rai, S. Pilla
Clemson University, Greenville, SC, USA

7.1 Introduction

Tremendous technological advancements in the last century have led to innovations that have dramatically improved the quality of human lives. However, these innovations coupled with other anthropogenic activities have contributed substantially to increase in GHG emissions resulting in global warming. Global warming has left an impacting and irreversible effect on earth's climate that is evidenced by irrational weather and rising sea levels. These unseasonal and unusual phenomena have resulted in an imbalanced ecosystem that challenges the very survival of the human habitat. Among several causes, the most contributing affect to global warming is CO_2, a GHG, the majority of which is the result of fossil fuel combustion. Thus, a number of efforts have been put in place to sequester CO_2 and, thereby, mitigate global warming (Herzog and Golomb, 2004; Lal, 2004; Schrag, 2007; Socolow, 2005; House et al., 2006).

CO_2 sequestration can be achieved by burying it underneath earth or ocean as well as by mineral carbonation (Lal, 2008). However, these steps provide only a temporary solution. Furthermore, they cause unintended consequences. For example, CO_2 injected in the ocean will affect the entire marine ecosystem, including fishes consumed by humans, thereby impacting the latter (Seibel and Walsh, 2001; Seibel, 2003). CO_2 storage inside earth's surface—propagated as CCS—can lead to earthquakes (Zoback and Gorelick, 2012; Mcclure, 2012; Bardin, 2012). CCS can also lead to mixing of CO_2 with the underground water system—a safe drinking water source (Little and Jackson, 2010). Also, the risk of CO_2 leaking into the earth's atmosphere on account of poor infrastructure or maintenance will always exist (Keeling et al., 2011). At best, such solutions may only postpone global warming and its after effects as opposed to permanently solving it. Hence, focus on efforts toward utilizing CO_2 to develop useful products has gained attention.

A number of researchers are investigating routes for direct utilization of CO_2 either through catalytic or noncatalytic means (Hacker, 2003; Otsuka et al., 2003; Shah et al., 2001, 2004; Azad et al., 2011). Prominent among these efforts is the utilization of CO_2 as a feedstock for polymer synthesis or, in other words, fixing CO_2 in polymers (Qin and Wang, 2010; Sugimoto and Inoue, 2004; Ree et al., 1999a; Lai et al., 2005). The first ever work reported on synthesis of PACs via copolymerization of CO_2 and epoxide was by Inoue et al. (1969a,b). Although their pioneering work has unprecedented

Biocomposites: Design and Mechanical Performance. http://dx.doi.org/10.1016/B978-1-78242-373-7.00016-0

effect on global warming, it was not until the recent past that these polymers realized commercial application (Empower Materials: Products, n.d.). Examples of some CO_2-based copolymers are polyethylene carbonate (PEC), polypropylene carbonate (PPC), polybutylene carbonate (PBC), and polycyclohexene carbonate (PCHC). A majority of these polymers are biodegradable. Thus, utilization of CO_2 in processing of polymers not only mitigates global warming but also advocates sustainability.

This chapter first presents a brief discussion on synthesis of CO_2-based polymers, with a special focus on PPC, the most vastly investigated CO_2-based polymer to date. The chapter then presents the mechanical performance, including static and dynamic mechanical properties, of various PPC-based blends and composites. Finally, the chapter concludes by highlighting the current and potential future applications of PPC-based blends and composites.

7.2 Synthesis of CO_2-based polymers

The general mechanism for synthesis of CO_2-based polymers is shown in Scheme 7.1. As shown, copolymerization of CO_2 with oxirane (epoxide group) occurs in the presence of an organometallic catalyst. It leads to the formation of either cyclic or acyclic

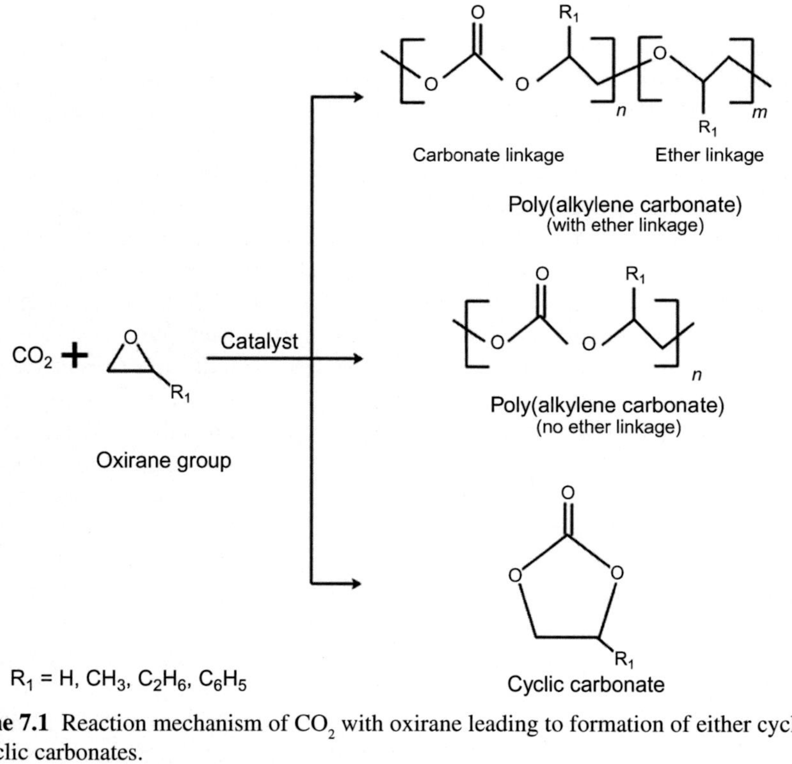

Scheme 7.1 Reaction mechanism of CO_2 with oxirane leading to formation of either cyclic or acyclic carbonates.
Redrawn from Taherimehr and Pescarmona (2014).

carbonates, based on the nature of the catalyst chosen and the reaction conditions (Taherimehr and Pescarmona, 2014; Pescarmona and Taherimehr, 2012).

During the copolymerization reaction, the crucial parameters governing the catalytic performance and, thereby, the reaction mechanism is the nature of the metal center and the strength of the metal-oxygen bond (Pescarmona and Taherimehr, 2012). Thus, the catalytic performance further dictates the predominance of a specific reaction path, yielding either acyclic or cyclic carbonate. The most common mechanism in the copolymerization reaction involves the activation of either epoxide and/or the CO_2 group in the presence of an organometallic catalyst. The activation of the epoxide group takes place by the coordination of epoxide to the metal, followed by a nucleophilic attack on the epoxide ring, resulting in ring-opening. CO_2 then reacts with the opened ring, which contains metal-oxygen bond, to form the carbonate group. The formed carbonate species can either propagate and react further with epoxide and CO_2 to form acyclic carbonate, or undergo ring closure via backbiting reaction to form a cyclic carbonate (Taherimehr and Pescarmona, 2014; Pescarmona and Taherimehr, 2012). As shown in Scheme 7.2, the formation of cyclic or acyclic carbonate is determined by the ability of the nucleophilic species in leaving the reaction as well as the strength of the metal-oxygen bond. When the nucleophilic species possess good leaving ability, along with the weak metal-oxygen bond that can be easily displaced by a nucleophile or a solvent molecule, backbiting reaction is favored, leading to an increase in selectivity towards the formation of cyclic carbonate. On the other hand, if the nucleophilic species possesses

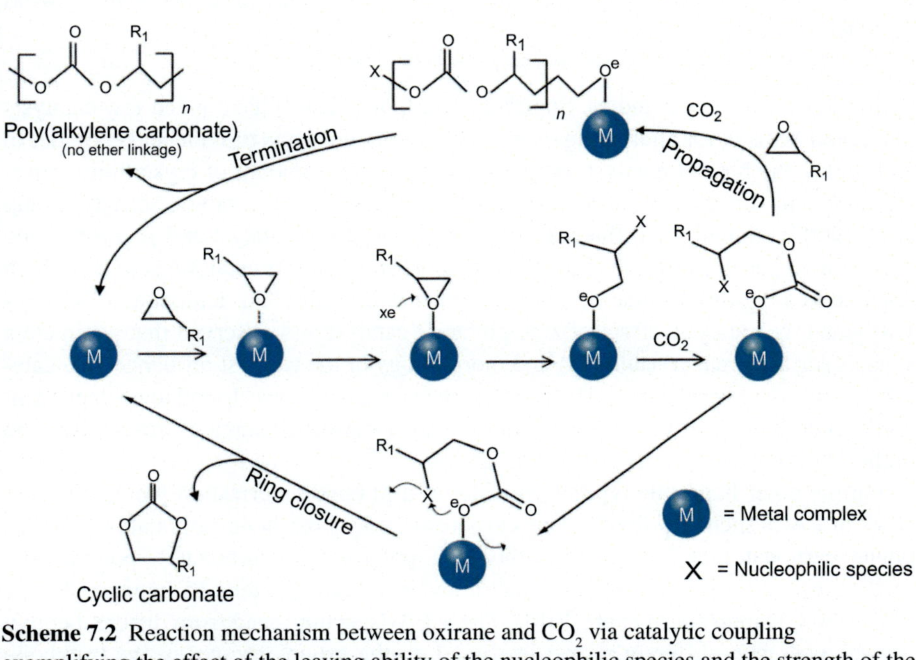

Scheme 7.2 Reaction mechanism between oxirane and CO_2 via catalytic coupling exemplifying the effect of the leaving ability of the nucleophilic species and the strength of the metal oxygen bond in the formation of cyclic or acyclic carbonates.
Redrawn from Taherimehr and Pescarmona (2014).

poor leaving ability along with intermediate strength of the metal-oxygen bond, further insertion of the carbonate species is preferred leading to chain growth, thus forming acyclic carbonate (Taherimehr and Pescarmona, 2014; Pescarmona and Taherimehr, 2012; Lu and Darensbourg, 2012; Lu et al., 2012). If the metal-oxygen bond is stronger, then the formation of either cyclic or acyclic carbonate cannot take place, thus leading to termination of the reaction without the formation of any product.

As the reaction of CO_2 with epoxide takes place via alternating copolymerization (Sugimoto and Inoue, 2004), ideally the reaction should proceed with alternate insertion of CO_2 and epoxide in the growing polymer chain. However, sometimes during the reaction, consecutive insertion of two epoxide groups can take place in the polymer chain, leading to the formation of ether linkage, which will significantly affect the properties (Gao et al., 2012a). This also leads to reduction in the CO_2 content in the polymer chain, compared to the theoretical value (Taherimehr and Pescarmona, 2014). Hence, various parameters such as the nature of epoxide, catalyst, reaction conditions (time and temperature), presence of impurities, and CO_2 pressure determines the type of copolymer, yield, and its physicochemical properties (Taherimehr and Pescarmona, 2014).

Among the listed parameters, the catalyst plays a crucial role in determining the formation of the cyclic or acyclic carbonate. Thus, a good catalyst should possess the following characteristics (Luinstra, 2008):

- The reaction should result in an alternating copolymer with minimum or no ether linkage and any by-product, thus avoiding the step of further purification.
- Control over molecular weight, because polymers with higher molecular weight possess better mechanical properties.
- The catalyst should be nontoxic and easy and safe to handle.

Based on numerous studies, Sugimoto and Inoue (2004) have listed few catalysts that could be used for alternating copolymerization of carbon dioxide and epoxide. In addition to the type of catalyst, properties such as size, amount of crystallinity, morphology, and surface area of the catalyst also influences the catalytic activity. Meng et al. (2002) studied the influence of particle size of the catalyst and found that the reduction in particle size (up to a certain extent) increased its catalytic activity, which resulted in a high molecular weight copolymer with high yield. Kim et al. (2005) investigated the catalytic effect of zinc glutarate catalyst and observed that parameters such as surface area, crystallinity, and morphology of the catalyst influenced the catalytic activity to a great extent. Particularly, the surface area was found to be dependent on the morphology of the catalyst, and the higher the surface area of the catalyst, the higher the catalytic activity (Kim et al., 2005).

On the other hand, the type of epoxide used in copolymerization yields different types of PACs such as poly(ethylene carbonate), poly(propylene carbonate), poly(butylene carbonate), poly(pentene carbonate), poly(hexene carbonate), poly(styrene carbonate), poly(cyclohexene carbonate), and poly(cyclopentene carbonate) (Thorat et al., 2003; Darensbourg et al., 2013). These PACs, synthesized from different epoxy groups, possesses different properties based on the groups present in the backbone and side chains. While PEC shows rubber-like properties at room temperature, PCHC is brittle. PPC possesses higher elongation at break and impact strength. Also, when compared to PPC, PCHC displays better mechanical properties, particularly tensile modulus and tensile strength (Taherimehr and Pescarmona, 2014).

Gao et al. (2012a) investigated the effect of ether linkage on mechanical properties and found that for a linear polycarbonate, as the amount of ether linkage increased, the material became more ductile with significant loss in tensile strength (Gao et al., 2012a). Rigidity of the polymer could be improved by the presence of stiff and bulky groups like cyclohexene, which possess high glass transition temperature and also remain brittle. Presence of bulky groups that are flexible leads to poor mechanical strength and low glass transition temperature, due to reduced chain interactions. These groups also improve the toughness of the polymer (Qin and Wang, 2010). Thus, linear carbonate is highly ductile due to the presence of ether linkages, while cyclic carbonate is highly brittle due to the presence of cyclic rings. Among all the investigated PACs derived from CO_2, PCHC possesses the highest T_g of ~105 °C (Thorat et al., 2003). Table 7.1 lists the mechanical properties of various polymers synthesized from CO_2. It can be observed that linear polymers possess poor mechanical strength compared to polymers containing cyclic bulky group.

From Table 7.1, one can understand that PPC possesses a balance of the mechanical properties—modulus, percentage of elongation at break, and impact strength—compared to other PACs. Also, it is used in many applications due to its unique properties and is relatively inexpensive. Hence, this chapter focuses on the synthesis of PPC, its blends and composites, and their mechanical properties.

Table 7.1 List of various polycarbonates synthesized from CO_2 and their mechanical properties

Polyalkylene carbonate	Tensile modulus (MPa)	Tensile strength (MPa)	Elongation at break (%)	Impact strength (J/cm)	References
Poly(ethylene carbonate)	3–8	–	>600	–	Thorat et al. (2003)
Poly(propylene carbonate)	2–2590	2–4	12–1100	–	Yang and Hu (2008) and Xu et al. (2006)
Poly(butylene carbonate)	2190	37	–	–	Qin and Wang (2010)
Poly(cyclohexene carbonate)	2460–2707	11.8–29.4	0.5–1.3	1.31	Thorat et al. (2003) and Hsu and Tan (2002)
Poly(styrene carbonate)	2400	54	–	–	Qin and Wang (2010)
Poly(4-vinyl-1-cyclohexene carbonate)	2110	36.7	1.3	1.02	Hsu and Tan (2002)

7.3 Poly(propylene carbonate)

Synthesis of PPC via copolymerization of CO_2 and PO is now well established. As explained, selection of the appropriate catalyst and the reaction condition plays a vital role, not only in producing a linear polymer with limited ether linkage, but also in determining the properties of the final polymer. Hence, major research dealt with the development of suitable catalyst for the production of PPC with improved properties (Ree et al., 1999a; Meng et al., 2002; Kim et al., 2005; Eberhardt et al., 2004; Darensbourg and Wilson, 2012). Various reviews are found in the literature that discuss different catalysts and their corresponding reaction mechanisms (Sugimoto and Inoue, 2004; Pescarmona and Taherimehr, 2012; Eberhardt et al., 2004). Among the available catalysts, heterogeneous catalysts are found to be more suitable in producing high molecular weight PPC than homogeneous catalysts. However, heterogeneous catalysts are difficult to synthesize. Further, among the various heterogeneous catalysts, zinc glutarate is proven to produce high molecular weight PPC with high yield due to its high catalytic activity (Qin and Wang, 2010; Meng et al., 2002), which is nearly 26 times higher than that of diethylzincglutaric acid system (Qin and Wang, 2010).

7.3.1 Synthesis of PPC

In order to synthesize PPC, the first step is the selection of the appropriate catalyst. Once the catalyst is selected, copolymerization can be carried out in an autoclave equipped with a mechanical stirrer. Various researchers have explained the synthesis of PPC from CO_2 (Ree et al., 1999a,b; Meng et al., 2002; Zhu et al., 2002, 2003), irrespective of the chosen catalyst. First, the catalyst is dried at 100 °C for 24 h and then introduced into the autoclave (connected to a vacuum line) as quickly as possible, where it is further dried (at 100 °C) under vacuum for 24 h. Before the start of the reaction, the autoclave is evacuated and purged with CO_2; this cycle is repeated until the autoclave is saturated with CO_2. Once the autoclave is ready, purified PO is introduced using a syringe. Subsequently, the autoclave is pressurized to 5.2 MPa with further CO_2. Copolymerization is then performed for 40 h at 60 °C under constant stirring. Once the reaction is complete, the pressure and temperature of the autoclave are brought to atmospheric conditions. Completion of this reaction results in a viscous mixture, which is further dried under vacuum. The dried product is then dissolved in methylene chloride. After dissolution, excess solvent and catalyst present in the dried viscous mixture are removed. Excess solvent is removed by using a rotary evaporator and catalyst by washing the solution with 200 ml of dilute HCl (5 vol.%) with distilled water. Finally, PPC precipitates out, which is free of catalyst and solvent. As-synthesized PPC is filtered and dried under vacuum at room temperature for 2 days. The resulting filtrate on distillation, to remove methanol and methylene chloride, yields a methanol-soluble product (Ree et al., 1999a; Meng et al., 2002; Zhu et al., 2002, 2003; Ree et al., 1999b). As-synthesized PPC is amorphous, with its glass transition temperature close to room temperature.

As listed in Table 7.1, the as-synthesized PPC possesses inferior mechanical properties, specifically tensile strength. Hence, a number of attempts have been made to improve the said properties either via blending with high-strength polymers, or reinforcing them with high-strength fibers to yield tunable property composites. These blends and composites can be processed by either solution blending or melt compounding. The process conditions for melt compounding of various PPC blends and composites are summarized in Table 7.2. The melt compoundings were carried out between 140 and 190 °C, while the majority were in the range of 140–160 °C for less than 10 min. These temperatures and times were selected to ensure that PPC doesn't degrade during melt compounding.

Table 7.2 Processing parameters employed in various PPC blends and composites

Material	Equipment	Temperature (°C)	Time (min)	Speed (rpm)	References
PPC/PLA	Single-screw extruder	110/115/120		20	Ma et al. (2006)
MA-PPC/PLA	Rheomix internal mixer	150	5	50	Yao et al. (2011a)
MA-PPC/PLA/PVAc	Torque rheometer	150	7	50	Gao et al. (2012b)
PPC/PHBV (glycidyl methacrylate and dicumyl peroxide as reactants)	Brabender mixer	175	4	50	Li et al. (2004a, 2005)
PPC/PHBV	Brabender mixer	175	5	50	Li et al. (2004a, 2005)
PPC/PLA/PHBV	Thermo kinetic mixer	175			Pilla and Turng (2013)
PPC/PBS	Brabender mixer	140	5	40	Zhang et al. (2007)
PPC/PBS	Rheomix twin rotary mixer	150	10	30	Pang et al. (2008)
PPC/PBS (TTI as reactive coupling agent)	Corotating twin-screw extruder	165–185		100	Chen et al. (2013)
PPC/EVOH	Twin rotatory mixer	170	5	50	Wang et al. (2007)
UHMW-PPC/PVA/EVOH/(PVA/EVOH)	Rheomix 600 internal mixer	185	7	50	Chen et al. (2010)
UHMW-PPC/PVA/EVOH	Rheomix 600 internal mixer	170	7	50	Chen et al. (2010)
PPC/HBP	Intensive batch mixer	140	5	60	Chen et al. (2011a)
PPC/corn starch	Rheomixer	120	10	50	Peng et al. (2005)
PPC/starch	Brabender Plasticorder batch mixer	150	6	30	Ge et al. (2004)

Continued

Table 7.2 Continued

Material	Equipment	Temperature (°C)	Time (min)	Speed (rpm)	References
PPC/starch acetate	Rheomix	170	8	50	Zeng et al. (2011)
PPC/S-g-PMA	Brabender Plasticorder batch mixer	150	7	60	Ge et al. (2005)
PPC/CAB	Polylab QC batch mixer	190	5	50	Xing et al. (2013)
PPC/PMMA/PVAc	Brabender type plastic mixer	155	5	100	Li and Shimizu (2009)
PPC/PMMA (use of EGMA, SAN-g-MA, and MA as compatibilizers)	Internal mixer	180	6	80	Yoo et al. (2013)
PPC/PS	Internal mixer	155	5	40	Huihong et al. (2010)
PPC/BEU, PPC/HDU, PPC/BPU	Batch intensive mixer	140	5	60	Chen et al. (2011b)
GF-reinforced PPC	Twin rotary mixer	150			Chen et al. (2009)
PPC-*Hildegardia populifolia*	Twin rotary mixer	180	5	30	Li et al. (2004b)
CSRF-reinforced PPC	Two-roll mixing mill	150	6		Chen et al. (2014)
PPC/WF	Batch mixer	150	6	60	Ge et al. (2006)
PPC/PWF	Rheomix batch kneader	140	5	30	Nörnberg et al. (2014)
PPC/OVMT	Twin rotary mixer	170		40	Xu et al. (2005)
PPC/OMMT	Brabender internal mixer	130	5	40	Xu et al. (2006)
PPC/OREC	Rheometer	120	10	50	Wan et al. (2006)
PPC/CNC	Rheomix internal mixer	135	6	50	Hu et al. (2013)
PPC/$CaCO_3$	Twin rotary mixer	180	5	30	Li et al. (2003)
PPC/EVOH/starch/$CaCO_3$	Rotor	155	9	30	Qiao et al. (2008)
PPC/PLA/MWCNT	Torque rheometer	150	10	50	Yang et al. (2013a)

7.3.2 PPC blends

Miscibility and compatibility are the two crucial parameters that dictate the final properties of any polymeric blend. Thus, blends are classified as fully miscible, partially miscible, and immiscible blends. Partially miscible and immiscible blends can either exhibit good compatibility among their constituent phases or remain incompatible. Compatibility of blends is identified by the macroscopically uniform physical properties they exhibit, irrespective of their miscibility. Compatibility within the blends can be enhanced by the addition of compatibilizer, which improves the interfacial adhesion between the two phases via intermolecular bonding and chain entanglements, thereby enhancing the mechanical properties of the blends. Conversely, immiscible and incompatible blends results in the deterioration of mechanical properties when compared with their individual counterparts, due to the presence of weaker interface, resulting in delamination. Thus, the formation of miscible or compatible blend is essential to achieve enhanced properties as compared to their individual counterparts (Sabzi and Boushehri, 2006). The processing conditions along with the ratio of polymers in the blend determines the miscibility and, thereby, the properties of the blend.

7.3.2.1 PPC-biopolymer blends

Owing to the biobased and biodegradable nature of PPC, various biopolymers like PLA (Ma et al., 2006; Yao et al., 2011a; Gao et al., 2012b), PHBV (Li et al., 2004a, 2005; Peng et al., 2003; Tao et al., 2009), PHB (Yang and Hu, 2008), PVA (Zhang et al., 2003), starch (Peng et al., 2005; Ge et al., 2004, 2005; Zeng et al., 2011), and EVOH (Wang et al., 2007) were blended with PPC to improve its properties without losing the bioresourced nature.

Zhang et al. (2002, 2003, 2006) prepared three types of PPC-based blends. In the first study, they blended PPC/PVPh via solution blending using THF as solvent. These blends exhibited miscibility due to the formation of hydrogen bonds between the two polymers, as one is a proton-donating while the other is a proton-accepting polymer (Zhang et al., 2002). In their second study, they blended PVA with BS-PPC via solution blending using DMSO as solvent (Zhang et al., 2003). The BS-PPC/PVA blends were prepared by first end capping PPC with BS and then reacting with PVA. They found that these blends were miscible in the PVA-rich regime. Immiscibility in the PPC-rich regime was mainly due to the presence of BS, which exhibited repulsive effect due to sulfonyl group and the spatial effect due to benzene ring. These effects together prevented the interaction between the carbonyl group of PPC and the hydroxyl group of PVA. Thus coagulation of PVA occurred, leading to immiscibility (Zhang et al., 2003). In their third study, Zhang et al. (2006) blended MA-PPC with ethyl cellulose via solution blending using DCM as solvent. Although variation in properties was observed in PPC-rich and ethyl cellulose-rich blends, no discussion was presented on the compatibility and miscibility of these blended polymers.

Yang and Hu (2008) prepared PPC/PHB blends of varying compositions by solution blending using chloroform as solvent. Enhanced miscibility up to 30 wt.% PHB content was observed. However, further addition of PHB resulted in loss of miscibility, due to the crystallization of PHB.

Ma et al. (2006) prepared PPC/PLA blends in the composition range of 85/15 to 30/70 via melt blending. Although this blend showed partial miscibility, it exhibited compatibility to some extent due to the similarity in chemical nature of PPC and PLA, thereby enhancing blend properties. Enhanced compatibility was observed only when PPC was the continuous phase, due to strong interfacial adhesion between the PPC and PLA phases. The extent of compatibility reduced with increase in PLA content. In order to enhance the miscibility of PPC/PLA blends, Yao et al. (2011a) incorporated MA into PPC/PLA blends wherein MA content was varied from 0.15 to 1.5 wt.%. It was observed that addition of MA content up to 0.3 wt.% resulted in enhanced compatibility and miscibility between PLA and PPC phases with optimum compatibility, evidenced by the change in size of the dispersed phases, at 0.3 wt.%. Further increment in the MA content (>0.3 wt.%) reduced the compatibility of PPC/PLA, which is in agreement with their earlier study (Yao et al., 2011b). This optimum MA ratio (0.3%) was employed to investigate the effect of blend constituent phases (PLA and PPC) on the mechanical properties, as discussed later.

Gao et al. (2012b) investigated the effect of PLA (varying %) in the presence of PVAc (10 wt.%) as compatibilizer in MA-PPC/PLA blends, by preparing blends with and without PVAc. Partial compatibility exhibited by MA-PPC/PLA blends (without compatibilizer), was enhanced with increase in PLA content, as evidenced by change in the morphology of PPC from sea-island to cocontinuous to droplet. Addition of 10 wt.% PVAc to MA-PPC/PLA, resulted in better uniform dispersion of PPC in MA-PPC/PVA/PVAc as compared to MA-PPC/PVA. PVAc was found mostly in the PPC/PLA interface with higher affinity toward PLA, resulting in the formation of bridges between PPC/PLA. These effects of bridging and uniform dispersion found in MA-PPC/PLA/PVAc, irrespective of the PLA percentage, results in enhanced compatibility as compared to MA-PPC/PLA blends.

Various researchers blended PPC with stiff and biodegradable polymer, PHBV (Luinstra, 2008; Li et al., 2004a, 2005; Peng et al., 2003; Tao et al., 2009). Both Peng et al. (2003) and Tao et al. (2009) prepared PPC/PHBV blends of varying compositions (100/0 to 0/100) by solution blending using chloroform as solvent. They found that these blends possess complete immiscibility due to difference in the structural order of the blended polymers, one being highly amorphous (PPC) and the other being highly crystalline (PHBV). This difference in the structural order hinders the reaction between the molecules to enhance the miscibility. Li et al. (2004a, 2005) prepared a (70:30) PPC/PHBV blend by adapting two different methodologies: reactive blending and mechanical blending. In general, reactive blending improves the miscibility in highly immiscible blends by generating strong intermolecular interactions resulting in better interfacial compatibility, thereby enhancing the mechanical and thermal properties (Li et al., 2005). Thus, Li et al. (2004a, 2005) carried out reactive blending with the addition of 1 wt.% GMA and 0.2 wt.% DCP, as reactive catalysts, during blending of PPC and PHBV, while mechanical blending was carried out by mere melting of PPC and PHBV. In the case of reactive blending, addition of GMA and DCP led to transesterification reaction resulting in grafting of PHBV onto PPC. This PHBV-g-PPC enhanced the miscibility in the blend as evidenced by the presence of uniformly dispersed, indistinguishable small droplets of PHBV. Furthermore, reduction in the percentage crystallinity of about 12.3% was observed in reactive blends. On the other hand, mechanically blended samples possessed distinct PHBV particles of spherulite morphology with ~1.4–2.8 μm in diameter.

Pilla and Turng (2013) blended both PLA and PHBV with PPC to obtain ternary blends via melt compounding. These ternary blends were also found to possess immiscibility. Despite their immiscible nature, these ternary blends were observed to be far superior to binary blends in terms of their toughness and strain-at-break, mainly due to the synergistic effect of the dispersed components (Pilla and Turng, 2013).

Blends of PPC with biodegradable PBS were investigated by Zhang et al. (2007), Pang et al. (2008), and Chen et al. (2013). Zhang et al. (2007) processed varying compositions of PPC/PBS blends in PPC-rich regime via melt compounding. The blend with 10 wt.% PBS exhibited partial miscibility, while further addition of PBS showed complete immiscibility due to the onset of crystallization, leading to phase separation. However, in a similar study carried out by Pang et al. (2008), enhancement in miscibility with increase in PBS content was observed. This enhancement in miscibility was mainly due to reduction in viscosity exhibited by these blends, as a result of higher processing temperatures and times. In another study, Chen et al. (2013) introduced reactive coupling agent, TTI, to improve the miscibility of PPC/PBS blends. The addition of 0.36 wt.% of TTI resulted in the reduction in crystallinity of PBS phase from 18.9% to 12.6%, thereby enhancing the compatibility of the blends.

Wang et al. (2007) produced a biodegradable blend of PPC/EVOH by melt compounding. These blends were immiscible when the EVOH content was lower. However, when the EVOH content increased beyond 40 wt.%, the miscibility enhanced and highly miscible blends were produced due to reaction between the constituent materials.

Chen et al. (2010) prepared ternary blends of UHMW-PPC, PVA, and EVOH via melt blending. First, PVA and EVOH were compounded at varying percentages. A good compatibility was observed from the DSC results for EVOH content above 30 wt.%. Hence, a 70/30 ratio of PVA/EVOH was chosen for further processing of ternary blends, UHMW-PPC/PVA/EVOH, wherein UHMW-PPC was varied from 100 to 50 wt.%. These ternary blends showed good compatibility across various phases due to the reaction between the carbonyl group of PPC and hydroxyl groups of PVA/EVOH.

PPC/HBP blends were melt compounded by Chen et al. (2011a). These blends were found to be miscible to some extent, due to the hydrogen bond formation via reaction between the O–H or N–H group of HBP, which donates a hydrogen bond to the carbonate group of PPC. The phase morphology of the blends was observed under SEM after selective etching of HBP phase using methanol as etchant. The blend containing 2.5 wt.% of HBP exhibited a uniform phase microstructure, indicating miscibility. However, cavities were observed in the blend containing 5 wt.% of HBP, whose size increased with increase in HBP indicating phase separation due to aggregation of HBP resulting in immiscibility. This was mainly due to the domination of the intra-associated interaction between the N–H and O–H bonds in HBP over the inter-associated interaction between PPC and HBP.

Peng et al. (2005), Ge et al. (2004), and Zeng et al. (2011) investigated PPC/starch blends using unmodified and modified starch derived from corn. Peng et al. (2005)

prepared PPC blends with the addition of CS, via melt mixing in the ratio of 100/0 to 60/40. Good interfacial adhesion was observed between PPC and starch, due the presence of large amounts of −OH groups in the starch, which can readily react with the carbonyl group of PPC resulting in hydrogen bond formation. Ge et al. (2004) processed PPC/starch blends from CS with varying weight ratios of 65/35 to 30/70 by melt compounding. However, with higher starch content, obvious voids and gaps between the hydrophobic PPC matrix and hydrophilic starch particles were observed resulting in agglomeration of starch. This incompatibility resulted in a brittle PPC/starch blend (Ge et al., 2004). In order to improve compatibility and interfacial adhesion, starch was chemically modified prior to blending with PPC by Zeng et al. (2011). Initially, unmodified starch was dried, and chemical modification was carried out by acetylation, resulting in transformation of the hydroxyl group of starch to an acetyl group, producing starch acetate. The acetyl group enhanced the hydrophobicity of starch, thereby enhancing the compatibility of blends due to strong interfacial adhesion. The degree of substitution (DS) of the acetyl group in starch had significant influence on the compatibility, as the extent of hydrophobicity is dependent on the amount of acetyl group present. When the DS was 0.51 or lower, a two-phase microstructure was observed, which transformed to a single-phase microstructure with increase in DS. Ge et al. (2005) synthesized PPC/S-g-PMA blends of varying composition via melt compounding. Prior to grafting of starch with PMA, presence of two-phase microstructure of PPC and starch indicated immiscibility in the blends. However, grafting of PMA to starch dramatically improved the interfacial adhesion, resulting in enhanced compatibility.

Xing et al. (2013) produced a blend of PPC/CAB by melt blending in the entire composition range (100/0 to 0/100). Even though these blends are thermodynamically immiscible, they exhibited partial miscibility due to hydrogen bond interactions.

7.3.2.2 PPC-synthetic polymer blends

In addition to biobased and biodegradable polymers, commercially available synthetic (nonbiodegradable) polymers like BPA (Fei et al., 2004), PMMA (Li and Shimizu, 2009; Yoo et al., 2013; Yang et al., 2013b), and PS (Huihong et al., 2010) have also been blended with PPC. Li and Shimizu (2009) prepared PPC/PMMA blends by melt compounding wherein they used PVAc to compatibilize the immiscible PPC and PMMA. The addition of PVAc enhanced the miscibility and the dispersion of PMMA, which was evidenced by the change in morphology of the dispersed phase from sea-island to cocontinuous structure. Yoo et al. (2013) investigated the effect of various compatibilizers, namely, EGMA, SAN-g-MA, and MAH, in enhancing the compatibility of PPC/PMMA blend. Varying amounts of compatibilizers were added to 70/30 PPC/PMMA. With the addition of the compatibilizers, the domain size of PMMA was found to decrease drastically, thus enhancing the compatibility at the interface. For the PPC/PMMA blend containing SAN-g-MAH, the droplet size was found to significantly reduce from 3.4 to 0.9 μm, while for blends containing EGMA and MAH, the reduction was only 1.8 and 2.3 μm, respectively. Thus, among the compatibilizers, SAN-g-MAH was found to be the most effective as it enhanced the miscibility. The enhanced miscibility was due to the combined effect of the polar–polar interaction between PPC and MAH along with the intermolecular forces between acrylonitrile of SAN and PMMA. Yang et al. (2013b) prepared PPC/PMMA blends by in situ synthesis of PMMA in PPC using supercritical CO_2 as solvent. They observed fine dispersion of PMMA in PPC. Even though the blends lacked miscibility, properties were enhanced due to the nanosize effect of dispersed PMMA, resulting in enhanced interaction in the interface. With this synthesis procedure, they showed that the mechanical properties of the blend could be significantly improved with the addition of very low fraction (~5 wt.%) of PMMA.

Huihong et al. (2010) blended PS with PPC via melt compounding and observed partial miscibility between PPC and PS. Huang et al. (1997) blended highly ductile PPC with highly brittle epoxy via melt blending to obtain balanced mechanical properties. Initially, PPC was melt mixed at 80 °C with uncured epoxy. Once uniform mixing occurred, curing agent was added to cure the epoxy. Partial miscibility within the constituent phases were observed.

Chen et al. (2011b) prepared blends of PPC with three types of urethanes: BEU, BPU, and HDU with ratios of 1–15 wt.%. Formation of intermolecular hydrogen bond, a key parameter for miscibility of constituent phases, was observed in all three urethane-PPC blends. Among these, the most effective hydrogen-bond interaction was observed for PPC/BPU blend. Furthermore, the highest hydrogen-bonding fraction (5.2%) was observed for the PPC/BPU blend with BPU content of 15 wt.%. The lower hydrogen-bond interaction for HDU and BEU blends is due to entrapment of O–H and N–H groups, participating in the intermolecular hydrogen bonding, in the crystalline phases of HDU and BEU.

Fei et al. (2004) prepared PPC/BPA blends (varying from 0 to 100 wt.%) by solution blending using chloroform as solvent. At low BPA content (i.e., <30 wt.%), these blends demonstrated miscibility due to the hydrogen-bond interaction between BPA and PPC molecules. However, as the BPA content was increased (i.e., >30 wt.%), BPA crystallized resulting in immiscibility.

Among the investigated binary and ternary blends involving PPC with biopolymers and synthetic polymers, the majority have exhibited partial miscibility or immiscibility due to the difference in the glass transition temperatures (T_g) of the involved polymers. PPC exhibits T_g very close to room temperature, while other polymers have higher T_g. This difference in glass transition temperatures also results in more inferior material properties than anticipated. In general, the partial miscibility or compatibility was found mainly due to the hydrogen-bond interactions between the two phases.

7.3.3 Poly(propylene carbonate) composites

In order to prepare PPC composites, various filler/reinforcements, such as GF (Chen et al., 2009), lignocellulose fiber (Li et al., 2004b; Chen et al., 2014), wood flour (Ge et al., 2006; Nörnberg et al., 2014), organovermiculite (Xu et al., 2005), montmorillonite (Xu et al., 2006), cellulose nanocrystal (Hu et al., 2013), calcium carbonate ($CaCO_3$) (Li et al., 2003), and ZnO (Seo et al., 2011), were investigated. Good interfacial adhesion between the matrix and filler/reinforcement plays a vital role in determining the properties of the final composite. To achieve good adhesion, several pretreatment methods such as silylation, grafting, and others, are employed to reinforcements (Nair and Thomas, 2003; Tang and Kardos, 1997; Rout et al., 2001). Also, there exists a critical level of filler/reinforcement loading with regard to improvement in properties. Addition of reinforcement beyond critical loading level tends to deteriorate the properties due to potential agglomeration, resulting in loss in tenacity as a result of poor interfacial adhesion (Yao et al., 2011b). The critical filler/reinforcement loading limit varies for different types of filler/reinforcement and matrix materials.

7.3.3.1 Macro/microcomposites

Chen et al. (2009) prepared GF-reinforced PPC via melt blending. At lower fiber loading (i.e., 10 wt.%), the GFs were uniformly dispersed within PPC. However, at high fiber loading level (i.e., >20 wt.%), agglomeration was observed, resulting in the formation of both fiber-rich and fiber-poor regions, thereby deteriorating the strength.

Li et al. (2004b) prepared PPC–HPF biocomposites. Two types of fibers were investigated, namely, untreated and alkali treated. They found that the untreated fibers possess a hemicellulose layer, which enhanced the interfacial bonding. Poor interfacial adhesion observed during the addition of treated fibers was due to the loss of the hemicellulose layer during alkali treatment. Chen et al. (2014) prepared biocomposites by the addition of plant fiber, CSRF, as reinforcement to PPC. The fibers were dispersed within the matrix, which exhibited intertwining on higher fiber loading. CSRF possessed hydroxyl groups from the cellulose, hemicellulose, and lignin layers present on the fibers. These hydroxyl groups readily reacted with carbonyl groups of PPC to form hydrogen bonds, thus enhancing the interfacial bond between the matrix and reinforcement.

Ge et al. (2006) prepared another cost-competitive, biodegradable composite by the addition of BWF at different weight ratios ranging from 10 to 40 wt.%, to PPC. Intermolecular hydrogen bonds were formed due to interactions between the carbonyl groups of PPC and the hydroxyl groups of BWF, enhancing the reinforcement effect of the composites. However, increase in BWF content led to agglomeration, thereby resulting in decreased interfacial adhesion. Nörnberg et al. (2014) reinforced PWF to PPC (without and with MA grafting) to prepare biocomposite. PPC was grafted with MA to achieve better wettability with PWF. Obvious voids were found in the composites, with PPC without MA grafting. The grafting of PPC with MA led to enhancement of properties due to the intimate contact of functionalized PPC with PWF.

Li et al. (2003) prepared PPC–$CaCO_3$ composites by mixing micro- and nanosized $CaCO_3$. Micron-sized $CaCO_3$ particles were found to disperse uniformly within PPC, thus resulting in good bonding between the filler and matrix phase. However, nanosized $CaCO_3$ were agglomerated resulting in inhomogeneous dispersion, leading to deterioration of properties (Li et al., 2003). Feng et al. (2014) prepared PPC–ESP (varying from 1 to 5 wt.%) biocomposite via solution blending using chloroform as solvent, where the ESP had particle size <40 μm. ESP is the natural form of $CaCO_3$ (95 wt.%) with the presence of some organic content (5 wt.%), making it a natural bioceramic composite. Better dispersion and interfacial adhesion between the filler and matrix was observed due to the presence of organic content, with the addition of up to 4 wt.% of ESP. However, further increase in ESP content resulted in agglomeration of the filler, thereby resulting in poor dispersion and interfacial adhesion.

Qiao et al. (2008) synthesized a unique biodegradable composite, PPC/EVOH/starch/$CaCO_3$, by the addition of both organic starch and inorganic $CaCO_3$. The PPC/EVOH ratio was kept constant at 60/40 and the amounts of starch and $CaCO_3$ were varied to understand their influence on the properties of the composite. $CaCO_3$ was found to be well dispersed within the matrix, while starch improved the interfacial adhesion due to formation of hydrogen bond. Due to high surface energy of $CaCO_3$, uniform dispersion with good adhesive interaction at the interface was observed. These combined effects enhanced the properties of the composite.

7.3.3.2 Nanocomposites

Xu et al. (2005, 2006) prepared two different PPC-based nanocomposites with the addition of OVMT and OMMT as reinforcements. In their first study, OVMT-reinforced PPC nanocomposites were prepared via melt blending (Xu et al., 2005). These nanocomposites possessed homogenously dispersed intercalated layers of OVMT, along with the presence of small clusters of OVMT layers in PPC. Thermal degradation of these composites were observed during melt compounding (Xu et al., 2005). In another study, OMMT-reinforced PPC nanocomposites were prepared (Xu et al., 2006). These nanocomposites possessed both intercalated and exfoliated morphology as shown in Figure 7.1. The intercalation was due to strong hydrogen bonding between the polar surfaces of OMMT and carbonyl group of PPC, while the exfoliated

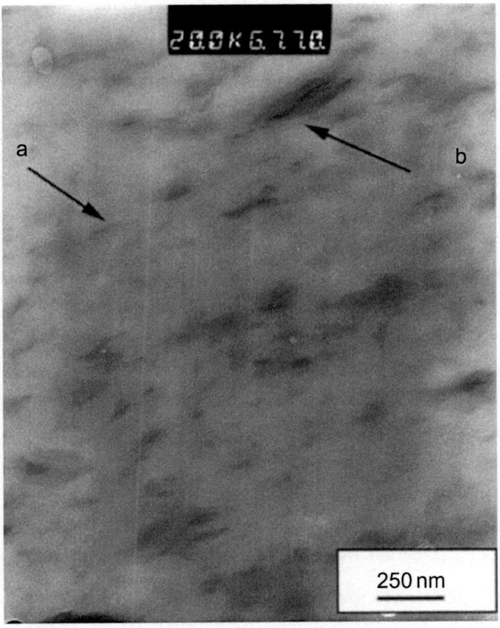

Figure 7.1 TEM micrograph indicating the presence of intercalated and exfoliated OMMT in PPC/2% OMMT nanocomposite.
Reprinted with permission from Xu et al. (2006).

morphology could be due to strong shear force during mixing that has disassociated and further dispersed OMMT within PPC. Shi and Gan (2007) prepared PPC/OMMT nanocomposites by the solution intercalation method, specifically to eliminate the thermal degradation observed by Xu et al. (2005). Initially, MMT was modified by surfactant CTAB producing OMMT. Varying amounts of OMMT (0–6 wt.%) were dispersed in benzene prior to the addition of PPC. The composites were prepared by adding PPC to OMMT/benzene solution and mechanically stirred for 3 h at 40 °C, followed by vacuum drying to remove excess solvent. The composites exhibited the intercalated-and-flocculated morphology. Optimal mechanical properties were achieved with the addition of 4 wt.% OMMT.

Wan et al. (2006) investigated PPC nanocomposites with the addition of OREC (varying from 2 to 10 wt.%) to PPC via direct melt intercalation. The OREC was observed to disperse finely in the PPC matrix, forming an intercalated and partially exfoliated morphology. However, improvement in mechanical properties was only observed until an addition of up to 4 wt.% of OREC; further addition led to deterioration in mechanical properties.

Du et al. (2006) prepared PPC/MgAl-LDH composites. Nanoscale dispersions were observed for MgAl-LDH content less than 10 wt.%, thereby exhibiting intercalation and exfoliation morphology.

Hu et al. (2013) fabricated PPC–CNC nanocomposites of varying compositions. Initially, a master batch of PPC/CNC composites containing 2 wt.% CNC was prepared via solution casting using THF as solvent. The master batch was then diluted with pure PPC to prepare different concentrations (0.1–1.5 wt.%) of PPC–CNC nanocomposites. In order to achieve compatibility between CNC, MA-PPC was used for the preparation of the nanocomposites. Uniform dispersion of CNC in the nanometric range was observed, with no agglomeration even at 1 wt.% CNC. The effect of varying amounts (0.1–1.5%) of CNCs on the mechanical properties was studied. It was also observed that addition of CNCs increased the molecular weight of the composites.

Gao et al. (2011, 2014) incorporated GO nanosheets into PPC via solution mixing. In their first study (Gao et al., 2011), THF was used as solvent to produce PPC/GO composites. First, GO (varying from 0.1 to 1 wt.%) dispersed in H_2O was added to PPC–THF solution with continuous stirring, resulting in coagulation of PPC/GO composites. GO was found to be well dispersed in PPC. Although the addition of GO enhanced the T_g of the composites from 29 to 39.5 °C, the properties were deteriorated at room temperature. In their second study (Gao et al., 2014) they had changed the solvent to DMSO and increased the GO content. The dispersibility of GO was found to be enhanced even at high filler loading of 20 wt.% and the properties were enhanced even at temperatures of 75 °C. Bian et al. (2012) modified GO using TDI to produce MGO and then prepared MGO/PPC nanocomposites, containing varying amounts of MGO (0.5–10 wt.%), by the solution intercalation method. When the MGO content was 3 wt.%, superior properties were observed, which deteriorated with further increase in MGO content. Barreto et al. (2013) prepared MWCNT and MA-PPC composites by solution blending with 0.5%, 3.0%, and 5.0% MWCNT. Two different dispersants, B10 and L10, were used to enhance the dispersibility of MWCNT. The composites were prepared in two stages: the first stage involved dispersion of the filler in the solvent, while the second stage involved incorporation of the polymer in the solvent. They found that B10 enhanced the dispersibiltiy of MWCNT in the composite as compared to L10.

Yang et al. (2013a) investigated MWCNTs-based nanocomposites using both PPC and PPC/PLA polymeric systems. For the PPC/PLA blend system, the constituent phases were varied and the optimal blend ratio was found to be 90/10. MWCNTs were then reinforced at varying compositions in both pure PPC and the optimal PPC/PLA blend. Uniform dispersion was observed for MWCNTs in both the polymeric systems. Particularly, in the PPC/PLA blend, MWCNTs formed bridge-like structures at the interface as can be seen from Figure 7.2. These nanobridges enhanced the load transfer between the two constituent phases, thereby resulting in improved mechanical strength while maintaining ductility. Furthermore, it was observed that the melting sequence (i.e., PPC/PLA/MWCNTs and PPC/MWCNT/PLA) during processing had a significant effect on the formation of nanobridges. This is due to the difference in the interfacial tension between the constituent phases. Among the investigated melting sequences, only PPC/PLA/MWCNT resulted in the formation of nanobridges, thereby enhancing the stated properties.

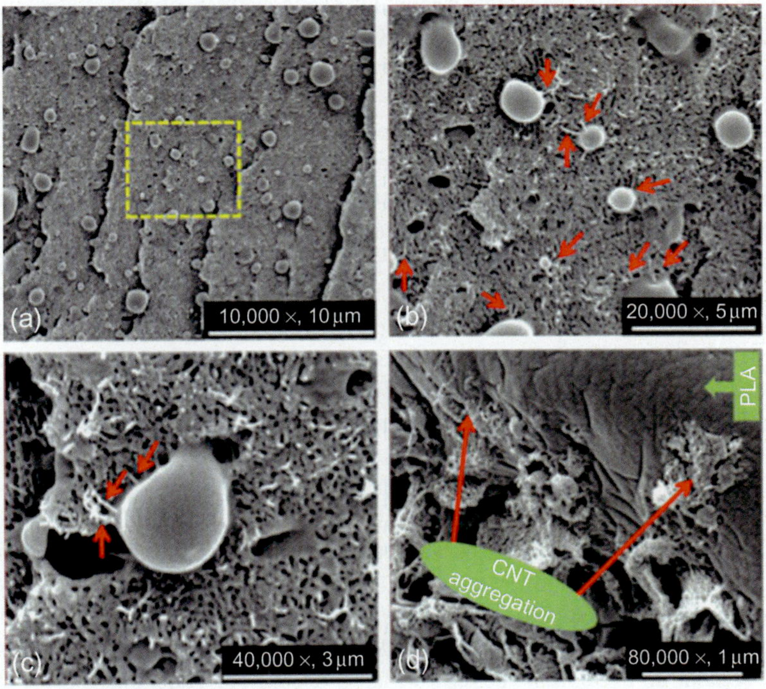

Figure 7.2 SEM of cryogenically fractured and etched 90 PPC/10 PLA/1.5 CNT. (a) Morphology of the etched surface, (b)–(d) enlarged SEM images exemplifying the presence of nanobridges.
Reprinted with permission from Yang et al. (2013a).

Seo et al. (2011) prepared composites with ZnO as fillers with five different compositions varying from 0 to 10 wt.% of ZnO via solution blending. Fine dispersion of ZnO was observed up to 5%, beyond which it agglomerated.

7.3.4 Mechanical properties

7.3.4.1 Static mechanical properties

Static mechanical properties like tensile strength, Young's modulus, elongation at break, and toughness were investigated for various PPC blends and composites. These properties were measured by varying the constituent of the secondary phase, either a polymer or a filler/reinforcement. Most of the blends and composites of PPC showed significant improvement in their strength and Young's modulus, while compromising on ductility. The properties of blends are not only dictated by that of the constituent phases and their proportions but also by the phase morphology within the blends (Ma et al., 2006). From the literature, it can be seen that PPC exhibits a broad range of properties that can be achieved by a mere change in the reaction conditions and the catalyst. As explained earlier, the catalytic activity can be enhanced not only by changing the type of catalyst, but also by changing the material properties of the catalyst; thereby resulting in variation of properties. Also, these properties are dependent on T_g, which is a result of processing conditions. For example, the Young's Modulus of PPC with T_g of 30 °C varies from 200 to 1000 MPa, while that of PPC with T_g of 40 °C varies from 700 to 1400 MPa (Luinstra, 2008). Pure PPC possesses high ductility (330–1200%) but low tensile strength (7–30 MPa) (Luinstra, 2008; Darensbourg and Wilson, 2012; Ge et al., 2005; Luinstra and Borchardt, 2012). In general, the tensile test of all the reported blends and composites were carried out at a strain rate of 10 mm/min, unless otherwise stated.

Yang and Hu (2008) observed that the addition of PHB enhanced the strength of PPC but reduced its ductility. In fact, the ductility decreased from 1090% to 2.7% with the addition of 40 wt.% PHB, due to the crystallization of PHB within the blend.

Various researchers have studied the variation in mechanical behavior of PPC with the addition of PLA (without and with compatibilizer). Ma et al. (2006) determined the effect of PLA (without compatibilizer) at a strain rate of 50 mm/min. They evaluated the mechanical properties (Young's modulus and yield strength) by both experimental investigation and theoretical calculation. Weight average method, which assumes perfect interfacial adhesion, was used to calculate the properties theoretically. Both Young's modulus and yield strength increased with increase in PLA content with a slight mismatch between the experimentally determined and theoretically predicted values. Until 60 wt.% PLA, the experimentally determined values were superior to the theoretical prediction. However, when PLA was increased beyond 60 wt.%, negative deviation from the theoretically calculated values was observed, due to the poor interfacial adhesion between the PPC and PLA phases. The toughness increased with increase in PLA content up to 50 wt.%, beyond which it decreased. Thus, phase inversion was observed in the composition range of 50–70 wt.% PLA in the PLA/PPC blends.

Yao et al. (2011a) investigated the influence of PLA in the presence of compatibilizer MA. Varying amounts of MA were added to PPC/PLA (30/70) to understand its influence on the mechanical properties of PPC/PLA blends. It was observed that with increase in MA content, the ductility increased while strength decreased, due to the plasticizing effect of MA. Further, at a constant ratio of 0.3 wt.% MA, the effect of PLA was investigated by adding varying ratios of PLA to PPC. Up to 30 wt.% of PLA, the strength of PPC/PLA blends decreased; beyond 30 wt.%, the strength enhanced.

Gao et al. (2012b) examined the effect of PLA in the presence of PVAc as compatibilizer on the mechanical properties of PPC/PLA blends. In PPC/PLA blend (without compatibilizer), increase in PLA content resulted in significant improvement in strength while compromising on ductility. The addition of 10 wt.% PVAc as compatibilizer to PPC/PLA dramatically increased the strength along with increase in percent of elongation at break. For example, addition of 10 wt.% PVAc to (70/30) PPC/PLA blends resulted in increase in tensile strength by 65% with respect to binary blend, while the increase in percent of elongation at break was 118% for PPC/PLA and 187% for PPC/PLA/PVAc. Hence, a balance of strength and ductility was obtained for the ternary blend of PPC/PLA/10 PVAc, as a result of effective stress transfer between PPC and PLA phases via the interface-localized PVAc, thereby preventing the crack initiation at the interface. Matrix shear yielding results in the formation of highly stretched and fibrillated ligaments. The finer phase morphology observed in the PPC/PLA/PVAc ternary blend is far more favorable in triggering large-scale shear yielding compared to the coarse phase morphology observed in the PPC/PLA binary blend (Jang et al., 1985). The remarkable improvement in phase dispersion and the bridge effect of the interface-localized PVAc leading to enhanced interfacial bonding are the main reasons for the significant improvement in the mechanical properties of PPC/PLA/PVAc blends. Even in PPC dominant compositions of the ternary blend, the tensile strength (28.5 MPa) and Young's modulus (1434 MPa) were comparable to the values of polypropylene, a widely used polymer (Arranz-Andrés et al., 2007).

Li et al. (2005) investigated the influence of processing route (reactive blending and mechanical blending) on the mechanical properties of PPC/PHBV blend (70/30). The reactive blended samples possessed superior properties compared to mechanically blended PPC/PHBV blend. The percent of elongation at break of reactive blended samples was 18 times higher compared to mechanically blended samples. The energy absorbed by reactive blended samples (4236 mJ) was about 10 times higher than the mechanically blended sample (74 mJ). Thus, enhanced properties of the reactive blended samples were due to the enhanced miscibility observed as a result of addition of reactive coupling agent as explained earlier, during reactive blending.

Zhang et al. (2007) studied the effect of addition of PBS on the mechanical properties of PPC at a strain rate of 50 mm/min. The impact strength of the blend increased from 31.7 (pure PPC) to 35.3 J/m with the addition of 10 wt.% PBS, which deteriorated with further addition of PBS up to 40 wt.%. Ductility of the blends deteriorated gradually until 20 wt.% PBS, while further addition of PBS resulted in exponential deterioration resulting in a brittle polymeric blend. Thus, optimal properties were achieved only for the blend of (90/10) PPC/PBS, which was highly miscible. Increase in PBS content resulted in onset of crystallization, thereby resulting in immiscibility. A similar study was carried out by Pang et al. (2008) in determining the mechanical properties of the PPC/PBS blends at a strain rate of 25 mm/min. They observed an increase in both yield strength and tensile strength with increase in PBS content accompanied by good ductility. Orientation-strengthening effect of PBS played a vital role in enhancing the yield strength of the blends in PPC-rich regime. Chen et al. (2013) studied the effect of addition of reactive coupling agent TTI (0–0.54 wt.%) to PPC/PBS blend (65/35) on the mechanical properties at a strain rate of 50 mm/min. Samples of dimension 100 × 15 × 15 were cut from films processed via calendaring process using a single-screw extruder with screw speed of 70 rpm. Both tensile strength and ductility of the blends were observed to increase with increase in TTI content up to 0.36%, above which it deteriorated. To understand the influence of various process parameters during calendaring, films were processed by varying the die temperature (160–210 °C), and calendaring roller speed (15–35 rpm). The mechanical properties were measured in both MD and TD. The properties increased with increase in die temperature and reached an optimal value at the die temperature of 200 °C. Any further increase in die temperature resulted in deterioration of properties due to onset of degradation. The properties were highly anisotropic and were superior in MD compared to TD, mainly due to the orientation of the crystalline phases toward MD. Similarly, increase in calendar roller speed enhanced the properties in MD while it deteriorated in TD.

The mechanical performance of PPC/EVOH blends was studied by Wang et al. (2007) at a strain rate of 50 mm/min. The strength was almost constant with increase in EVOH content up to 30 wt.% while reduction in ductility was observed, due to the rigid reinforcement behavior of EVOH in the immiscible PPC/EVOH blend. Increase in EVOH content in the range of 40–60 wt.%, enhanced the miscibility, thereby the tensile strength, without any change in ductility. The enhanced miscibility, due to the strong molecular interactions between PPC and EVOH, resulted in the combined effect of rigid EVOH contributing to strength and ductile PPC contributing to ductility. Thus, the best comprehensive properties were shown by PPC/EVOH blends with EVOH content ranging between 40 and 60 wt.%.

Chen et al. (2010) investigated the mechanical behavior of UHMW-PPC/PVA/EVOH at a strain rate of 50 mm/min. As explained earlier, the binary blend PVA/EVOH at the ratio of 70:30 was chosen for blending with PPC. Varying ratio of PVA/EVOH (70/30) was added to PPC and a dramatic increase was observed in tensile strength due to the reinforcement effect. However, a sharp decrease in the elongation at break from 350% (pure PPC) to ~5%, with addition of 10 wt.% PVA/EVOH to PPC was observed. Further addition of PVA/EVOH showed no visible change in the percent of elongation at break.

Chen et al. (2011a) investigated the performance of PPC/HBP blends. A similar trend was observed in both yield strength and elongation at break; the properties first increased with increase in HBP content to reach a peak value, and then decreased with further addition of HBP. However, the peak value was obtained at a different percentage of HBP for both the properties. The peak tensile strength was observed with the addition of 0.5 wt.% HBP, while the peak percent of elongation at break was observed at 2.5 wt.%. The strong hydrogen bond interactions enhanced the interfacial adhesion between PPC and HBP and led to the simultaneous occurrence of both the reinforcing and the toughening effects of PPC. However, for HBP content beyond 5 wt.%, both properties were observed to deteriorate to levels poorer than those of PPC, as a result of aggregation of HBP resulting in immiscible blends.

Tensile properties of PPC/starch composites were determined by Peng et al. (2005) and Ge et al. (2004). Both studies showed an increase in stiffness on addition of starch to PPC, due to the stiffening effect exhibited by starch granules. Peng et al. (2005) observed a decrease in tensile strength and increase in Young's moduli with increase in starch content. The experimentally determined and theoretically estimated tensile strength values were compared. The theoretical values were estimated using a geometric model developed by Nicolais and Narkis (1971) that assumes uniform dispersion of filler particles. The experimental values were found to be much higher than the theoretical predictions, indicating good interfacial adhesion between the two phases. Ge et al. (2004) observed an increase in Young's modulus with increase in starch. However, percent of elongation-at-break dramatically reduces from 641% (pure PPC) to 1.87% for 65/35 PPC/starch resulting in brittle behavior, due to the agglomeration of starch as explained earlier. Zeng et al. (2011) characterized the influence of addition of starch acetate in PPC on the mechanical behavior, at a strain rate of 50 mm/min. The tensile strength was observed to increase with increasing starch content, while the impact strength reduced. However, these properties were found to be highly dependent on the DS of the hydroxyl group by the acetyl group. When the DS of starch content was varied, the strength increased with increasing DS and reached a peak value for 0.51 DS. On further increase in DS, the strength deteriorated, yet was found to be above pure PPC, irrespective of the percentage of starch in the blend. This increase in strength up to DS of 0.51 was mainly due to the enhanced interfacial adhesion and the rigidity of starch acetate. However, when the DS was increased further, starch acetate dispersed well in PPC matrix forming a single-phase microstructure, which resulted in breakdown in the continuity of the ductile PPC phase thereby loss in toughness. However, the strength and ductility were much better than pure PPC due to their enhanced compatibility. Ge et al. (2005) investigated the tensile properties of PPC/S-g-PMA. Tensile strength increased with increasing content of S-g-PMA up to 35 wt.%, and saturated with further addition. With the addition of S-g-PMA to PPC, enhanced toughness due to hydrogen bonding was observed. The starch had resulted in stiffening effect and significantly enhanced the modulus upto the addition of 10 wt.%, while addition of S-g-PMA resulted in saturation.

Xing et al. (2013) carried out tensile testing of PPC/CAB at a strain rate of 20 mm/min. With increase in the CAB content in PPC/CAB blends, dramatic increase in both tensile strength and elastic moduli were observed due to the strengthening effect of CAB. The PPC/CAB (50/50) blend possessed tensile strength of 27.7 MPa, which was 21 times higher than that of pure PPC. Even though reduction in ductility was observed with increase in CAB, the blends possessed ductility to some extent due to good compatibility between PPC and CAB.

Li and Shimizu (2009) characterized the mechanical properties of PPC/PMMA (70/30) in the presence of PVAc as compatibilizer. The elastic modulus and tensile strength was observed to increase with increase in PVAc up to the addition of 5 phr. With further addition of PVAc, the modulus showed deterioration while the strength was found to be saturated. However, the elongation at break was very high (400%) for all the PPC/PMMA/PVAc blends due to the compatibility between PPC and PMMA. Yoo et al. (2013) characterized the effect of three types of compatibilizer (SAN-g-MAH, EGMA, MAH) on the mechanical performance of 70/30 PPC/PMMA blend. Binary blend (70/30 PPC/PMMA) (without any compatibilizer) exhibited tensile strength of 26.9 MPa. With increase in the compatibilizer ratio, the strength increased initially and reduced subsequently. However, the compatibilizer ratio at which the reduction in the strength was observed was different for different compatibilizers. In case of SAN-g-MAH and EGMA as compatibilizers, addition of 5 phr yielded maximum strength, while MAH as compatibilizer yielded maximum strength at 0.5 phr. The decrease in tensile strength (above 5 phr of SAN-g-MAH and 0.5 phr of MAH in PPC/PMMA), was mainly due to the plasticizing effect of the compatibilizer in the PPC/PMMA blend. However, with EGMA as compatibilizer, the strength decreased slightly and then saturated. SAN-g-MAH was found to provide higher strength compared to other compatibilizers due to the enhanced miscibility observed in PPC/PMMA/SAN-g-MAH, as compared to other compatibilizers. Yang et al. (2013b) studied the mechanical behavior of PPC/PMMA blends processed via super critical in situ solution synthesis at a strain rate of 20 mm/min. With the addition of 5 wt.% PMMA, a fourfold increase in tensile strength was observed; from 8.5 (pure PPC) to 34.2 MPa (5 wt.% PMMA). An increase in yield strength by 180% was achieved just by adding 3.5 wt.% PMMA, which is comparable to the PPC/PMMA 70/30 blends processed via other techniques. Thus, significant improvement in properties with the addition of less amount of PMMA was due to the fine nanosize dispersion of PMMA.

Huihong et al. (2010) characterized the mechanical properties of PPC/PS blends at a strain rate of 50 mm/min. A threefold increase in tensile strength was observed with the addition of 40 wt.% PS due to the partial miscibility exhibited between the phases. However, this was accompanied by a drastic reduction in the percent of elongation at break from 740% (pure PPC) to 82.9% for 60/40 PPC/PS blend.

Chen et al. (2011b) investigated the mechanical properties of urethane (BEU, BPU, HDU)-based PPC. Pure PPC exhibited a tensile strength of 43 MPa, Young's modulus of 1122 MPa, and a lower elongation at break of 13.5%. Significant toughening and strengthening effect was observed in the case of PPC/BEU blends, which was evident by 37% increase in strength with the addition of 1 wt.% of BEU, while the ductility was unaffected. With further increase in BPU, strength decreased along with increase in elongation at break. However, the strength was still superior to that of pure PPC until the addition of 10 wt.% BEU. For BPU content of 5 wt.% in the blend, the elongation at break was nearly twice that of pure PPC (a 100% increase), while its tensile strength was 50 MPa. This indicates the simultaneous achievement of the toughening and the reinforcing effects, providing further confirmation of the importance of hydrogen-bond interactions in miscibility and compatibility of the blends. But when the BPU content was increased beyond 10 wt.%, the strength was inferior to that of pure PPC. For PPC/HDU blends, the strength initially increased and then decreased. When the HDU content was varied over the range 5–10 wt.%, both distinct yielding and stable neck growth through cold drawing were observed. Although HDU had a lower Young's modulus compared to BPU and BEU, transition had occurred from brittle to ductile for the blend. PPC/HDU blend containing 10 wt.% of HDU exhibited an elongation at break of about 727%—nearly 53 times higher than that for pure PPC—while its tensile strength was 30 MPa, comparable to that of the LDPE. The molecular-level miscibility between HDU and PPC, combined with the decrease in T_g values for the blends, were considered the principal causes for the satisfactory plasticizing of the blend. Due to enhancement in mobility of the chain segment, the plasticizer was observed to behave like a lubricant that has been mixed with polymers, thereby reducing cohesion of the molecular chains. During tensile testing, the molecular chains were observed to gradually entangle and get oriented. This, in turn, dispersed the fracture energy more effectively and, thereby, delayed the final rupturing of the material. The toughening of PPC by HDU was not accompanied by any drastic reduction in the mechanical stress of the blend. However, an increase in HDU content to beyond 10 wt.% led to the domination of hydrogen bonding and crystallinity over the dilution effect of plasticizing. This led to a reduction in the elongation at break and an increase in tensile strength.

In the case of composites, not only interfacial adhesion, but other parameters such as the fiber content, fiber aspect ratio, nature of the fiber, and fiber agglomeration play a vital role in determining the mechanical properties of the composite. Chen et al. (2009) investigated the mechanical behavior of GF-reinforced PPC. They demonstrated that the use of GFs as reinforcing material led to an initial increase in tensile strength with increase in GF content, up to 20 wt.%. Tensile strength was observed to increase from 17.32 to 29.75 MPa, a significant increment (~72%). However, subsequent increase in GF content beyond the critical value led to a decrease in tensile strength, although it continued to remain higher compared to that of pure PPC. This was attributed to the counterbalance of two phenomena caused by the increase of GF content: increased surface fracture energy and the extension in sizes of voids or GF aggregation. These two effects were observed to play a critical role in determining the tensile strength of PPC/GF composites until the addition of 40 wt.% GF. The tensile modulus was observed to increase dramatically with an increase in GF content up to 40 wt.%. For PPC/40 wt.% GF, modulus was 4023 MPa, nearly 500% higher compared to pure PPC. This indicated a substantial increase in rigidity of the composites with increase in GF content. However, an addition of just 10 wt.% GF was observed to cause a dramatic reduction in the elongation at break. The hardness of PPC/GF composites increased with an increase in GF content, indicating the increase in both mechanical strength and stiffness (Chen et al., 2009).

Li et al. (2004b) investigated the tensile properties of PPC reinforced with HPF (both untreated and treated fibers). Both the composites exhibited similar properties. The tensile strength increased by 50% for 10 wt.% fiber loading. The Young's modulus values were initially predicted theoretically and then compared with the experimental data. It was observed that the theoretical calculations underpredicted the Young's modulus. The superior mechanical strength and modulus of the composites was due to the combined effect of higher aspect ratio of fibers and the good interfacial bonding between the matrix and reinforcement. The percent of elongation reduced drastically with the addition of fiber and remained constant with increase in fiber content.

Chen et al. (2014) measured the tensile strength of CSRF-reinforced PPC at a strain rate of 2 mm/min. They observed an increase in the tensile strength of PPC/CSRF composites at a decreasing rate with increase in CSRF content, reaching a maximum at 60 wt.% CSRF content due to the reinforcement effect. The Young's modulus and impact strength also showed an increase with increase in fiber content. The elongation at break was observed to reduce dramatically from ~486% for pure PPC to ~40% for 10 wt.% fiber addition and further to ~4% for 60 wt.% fiber addition.

Ge et al. (2006) studied the mechanical properties of PPC/BWF composites and observed a maximum increase of 16% in tensile strength with the addition of 10 wt.% of BWF. When the BWF content was increased to 20 wt.%, the strength deteriorated and was found to be saturated with further addition. This deterioration can be rightly attributed to the inhomogeneity in the distribution along with poor interfacial adhesion. Heavy agglomeration of wood particles was observed with increase in BWF due to the presence of lignin and OH group, which had less affinity toward the polymer. The modulus was found to increase with increase in BWF, resulting in the stiffening effect with significant loss in ductility (Ge et al., 2006).

Nörnberg et al. (2014) investigated the mechanical properties of PPC/PWF composites. The Young's modulus was found to increase with increase in wood flour content, while the tensile strength increased only until the addition of 30 wt.% of PWF and decreased with further addition. The elongation at break and impact strength decreased dramatically with increase in PWF content. The properties exhibited by PPC/PWF composites were found to be superior to the PPC/BWF composites investigated by Ge et al. (2006).

Li et al. (2003) investigated the variation in mechanical properties of PPC by addition of $CaCO_3$ along with varying the size and the content for the optimized particle size. 20 wt.% $CaCO_3$ with varying size (38 μm, 10 μm, 65 nm) was added to PPC to investigate the effect of filler size on the properties. They found that reducing the particle size of $CaCO_3$, reduced the strength and modulus while it increased the toughness and ductility due to the reduction in reinforcing effect. Thus, varying amounts of $CaCO_3$ (particle size 38 μm) was added to PPC to understand the variation in mechanical properties. With increasing $CaCO_3$ content up to 10 wt.%, the modulus and strength increased. With further increase in $CaCO_3$ content, the strength slightly reduced and then saturated. Feng et al. (2014) characterized the mechanical properties of PPC/eggshell composite film, at a strain rate of 100 mm/min. Tensile strength and Young's modulus increased while the strain at break and impact strength decreased with increase in filler content till 4 wt.%. At filler content of 4 wt.%, while the tensile strength of the film was 23.2 MPa (about an 81% increase over that of the matrix), the modulus was 1378 MPa (about a 67% increase over that of the matrix). Larger surface area of PPC resulted in its better dispersion and better filler–matrix interaction, which in turn led to improvement in strength and modulus. However, a decrease in tensile strength and modulus with an increase in filler content beyond 5 wt.% was observed, due to agglomeration of filler, which weakens the filler–matrix interface. Increase in filler content in the PPC matrix also led to reduction in ductility due to hardening and stiffening of the composite. Moreover, in all cases, the composite films showed higher strength and modulus compared to the pure PPC.

Qiao et al. (2008) investigated the mechanical properties PPC/EVOH/starch/$CaCO_3$. From the work of Wang et al. (2007), it is clear that 60 PPC/40 EVOH blend exhibits good mechanical properties. Thus, varying amounts of starch were added to 60 PPC/40 EVOH blend. A gradual increase in tensile strength with increasing starch content up to 30 phr was observed, due to chemical interactions between PPC, EVOH, and starch. However, further increase in starch content led to a decrease in tensile strength due to aggregation of starch. To understand the effect of $CaCO_3$, varying amounts of $CaCO_3$ were added to PPC/EVOH/starch blends, containing 60 PPC/40 EVOH and varying the amount of starch from 20 to 40 phr. Composites with 20 phr starch showed an increase in tensile strength with increase in $CaCO_3$ content. However, for composites with starch content at 30 phr, no change in tensile strength with increasing $CaCO_3$ content was observed. For composites with starch content at 40 phr, tensile strength was found to be optimal at $CaCO_3$ content of 5 phr. Increasing starch content also made the composites with fixed $CaCO_3$ content more brittle. Incorporation of $CaCO_3$ also reduced the toughness of the PPC/EVOH/starch/$CaCO_3$ composites. Due to the high surface energy of fine $CaCO_3$ fillers, enhanced interaction between the constituent materials was observed, which led to the formation of new quasi-cross-linking domains. These domains can easily transfer the stress due to their uniform dispersion, resulting in enhancement in strength with increase in $CaCO_3$ content. However, agglomeration of $CaCO_3$ occurs when the $CaCO_3$ content exceeds a critical level, resulting in deterioration of strength.

Xu et al. (2005) measured the tensile properties of PPC/OVMT composite at a strain rate of 5 mm/min. Up to the addition of 4 wt.% of OVMT, there was no reinforcement effect and the strength of the material remained almost constant. However, the strain at break had reduced dramatically. A stepwise decrease was observed in the yield strength and modulus with increasing OVMT content along with increase in ductility.

Xu et al. (2006) investigated the mechanical properties of PPC/OMMT composite at a strain rate of 5 mm/min. They had found that the optimal value of Young's modulus for the nanocomposite was obtained at the addition of 4 wt.% OMMT. At this loading level, Young's modulus was found to be 3.27 GPa, about 27% higher than that of pure PPC. Up to a loading level of 4 wt.% OMMT, Young's modulus showed an increase with increasing OMMT content, as an indication of the reinforcement effect caused by the nanoscale dispersion of OMMT layers within the PPC matrix. With further addition of OMMT, the Young's modulus deteriorated. Addition of small amounts of OMMT caused a large increase in the elongation at break. For instance, nanocomposites containing 2 wt.% of OMMT showed an elongation at break of 323%; about 363% higher than that of pure PPC. Further increase in OMMT content led to a sharp decrease in elongation at break of the nanocomposites. Shi and Gan (2007) also observed similar results in the PPC/OMMT composites irrespective of the difference in the processing method.

Wan et al. (2006) tested the mechanical properties of PPC/OREC nanocomposites with OREC content varying in the range 0–10 wt.%, at a strain rate of 50 mm/min. The tensile and impact strength of the composites showed increase with increase in OREC content up to 4 wt.%, due to enhanced dispersion of OREC in PPC. Further addition resulted in deterioration of both tensile and impact strength due to possible agglomeration of OREC. However, these properties were superior to pure PPC until the addition of 8 wt.% OREC. With further increase, the impact strength was found to be the same as that of pure PPC, while tensile strength was inferior to pure PPC.

Hu et al. (2013) characterized the mechanical behavior of PPC/CNC composites at a strain rate of 20 mm/min. They showed that a significant increase in strength and modulus was observed with increasing CNC. Incorporation of small amounts of CNCs content (0.1 wt.%) led to a nearly tenfold increase in tensile strength and a sevenfold increase in Young's modulus, compared to pure PPC due to reinforcing effect. Gao et al. (2011) demonstrated the mechanical behavior of PPC with the addition of GO at a strain rate of 100 mm/min. A significant improvement in tensile strength and Young's modulus through the addition of GO as a reinforcing agent was observed. An increase of GO content from 0 to just 1 wt.% led to a sharp increase in the tensile strength by 1000% (from 1.5 to 17 MPa) and increase in the Young's modulus by 1800% (from 50 to 960 MPa). Bian et al. (2012) examined the performance of PPC/MGO composites and observed an increase in both the tensile strength and the Young's modulus with the addition of MGO till 3 wt.%. They observed maximum tensile strength of 29.5 MPa, for addition of 3 wt.% MGO—200% higher than that of pure PPC and a higher Young's modulus of 1.93 GPa, much higher than the Young's modulus for pure PPC, 0.87 GPa. The improvement in mechanical properties could be attributed in part to two reasons. The first was the fine dispersion of disordered MGO platelets in the PPC matrix; the second was the enhanced interaction between the PPC and MGO resulting in grafting of MGO on to PPC. A further increase in MGO content beyond 3 wt.% was observed to cause deterioration in the mechanical properties, due to aggregation of some graphitic layers. This graphitic aggregate reduced the effective interfacial area, resulting in deterioration of properties. Nonetheless, all MGO/PPC composites were observed to exhibit improved mechanical properties compared to pure PPC (Bian et al., 2012).

Yang et al. (2013a) studied the mechanical behavior of PPC/PLA, PPC/MWCNT, and PPC/PLA/MWCNT. In the case of PPC/PLA blend, it was observed that addition of 10 wt.% PLA led to an increase in yield strength from 2 to 10 MPa without any change in the elongation at break. However, an increase in PLA content to 20 wt.% showed further increase in yield strength, but was accompanied with a dramatic reduction in elongation at break (from 690% to 370%). The elongation at break went below 100% with further increase in PLA content. The PPC/MWCNT composite showed that an increase in MWCNT content from 0 to 2 wt.% enhanced the yield strength modestly from 2 to 12 MPa, while the decrease in ductility was insignificant as compared to the decrease in ductility observed in PPC/PLA blends. Thus, 90 PPC/10 PLA/MWCNT composites with varying MWCNT were tested. They observed the change in properties of high PPC/PLA (90/10) with the addition of MWCNT. The yield strength showed an increase from 2 (pure PPC) to 23 MPa (90 PPC/10 PLA/1.5 MWCNT) without much compromise on the percent of elongation at break. As explained earlier, the microstructure of these composites were heavily influenced by the melting sequence and this change in morphology led to a significant change in properties. While the composite without CNT nanobridges (90 PPC/2 CNT/10 PLA) showed a small increase in yield strength from 10 to 17 MPa, the composite with CNT nanobridges (90 PPC/10 PLA/2 CNT) showed nearly 200% increase in its yield strength from 10 to 29 MPa. This significant difference in the yield strength of the two composite confirms that CNT nanobridges played a vital role in efficiently transferring the load between the two phases. Figure 7.3 shows a comparison of the tensile behavior of 90 PPC/10

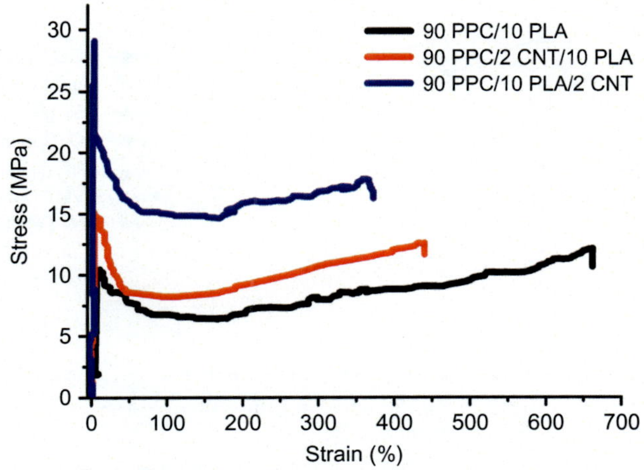

Figure 7.3 Stress–strain curves of 90 PPC/10 PLA, 90 PPC/10 PLA/2 CNT (without nanobridge), and 90 PPC/2 CNT/10 PLA (with nanobridge).
Reprinted with permission from Yang et al. (2013a).

PLA, 90 PPC/10 PLA/2 CNT (without nanobridge) and 90 PPC/2 CNT/10 PLA (with nanobridge). Thus, when compared to PP (with yield strength of 32 MPa), a widely used synthetic polymer, the yield strength of 90 PPC/10 PLA/2 CNT was found to be quite similar (yield strength of 29 MPa). However, a substantial difference was observed in their elongation at break; while for PP it was slightly above 100%, for 90 PPC/10 PLA/2 CNT, it was as high as 350% (Yang et al., 2013a).

7.3.4.2 *Summary of mechanical performance*

Although substantial improvements in mechanical properties were observed, the magnitude of increase was significantly dictated by the secondary constituent, which is a polymer or filler/reinforcement, and their interfacial adhesion. Certain secondary constituents, particularly carbon-based nanofillers, enhanced the properties at low filler content (1 wt.%), while certain secondary constituents (PHB) showed insignificant improvement even with the addition of 40 wt.%. Figures 7.4–7.9 show consolidated mechanical properties (modulus, elongation at break, and strength) exhibited by various PPC blends and composites. These graphs were prepared as a guideline for materials engineers to select appropriate filler (type and content) based on the required property for specific application(s). Figures 7.4 and 7.5 correspond to the Young's modulus of various secondary phase constituents. Among various secondary constituents, GF was found to significantly improve the Young's modulus of the composites. A maximum Young's modulus of 4 GPa was achieved with the addition of 40 wt.% GF. Figures 7.6 and 7.7 consolidate the change in ductility of the PPC blends and composites, while Figures 7.8 and 7.9 summarize the variation in tensile strength with varying filler content for different types of fillers used in PPC. The general trend shows that, with increase in filler content, the tensile strength and modulus enhanced while the elongation at break decreased. However, addition of certain secondary constituents

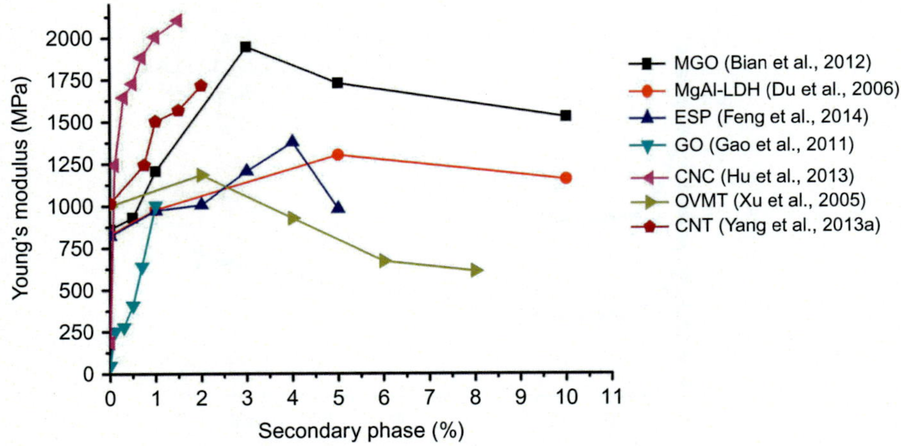

Figure 7.4 Variation of Young's modulus with different secondary constituents in PPC at lower loading.

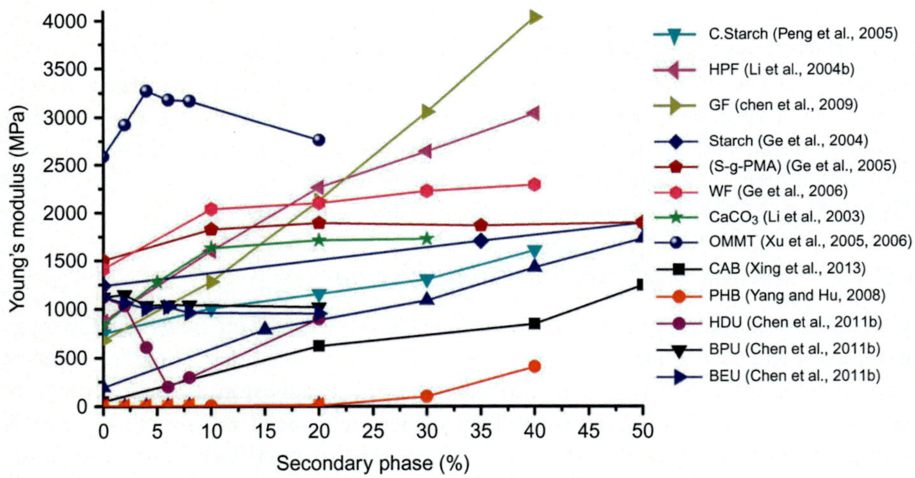

Figure 7.5 Variation of Young's modulus with different secondary constituents in PPC at higher loading.

such as urethane based polymers to PPC deviated from this trend. Increase in strength and toughness of any material is a greater challenge to be achieved. However, in PPC-based blend, by the addition of ternary phases that act as load transfer agents between the immiscible phases, it resulted in improved strength and toughness.

The huge variation in properties for pure PPC is due to the difference in the molecular weight of the synthesized PPC that, in turn, is the result of the processing parameters. Both the thermal and the mechanical performance of a material are strongly influenced by its molecular weight (Qin and Wang, 2010). Figure 7.10 shows the variation in Young's modulus with increasing molecular weight.

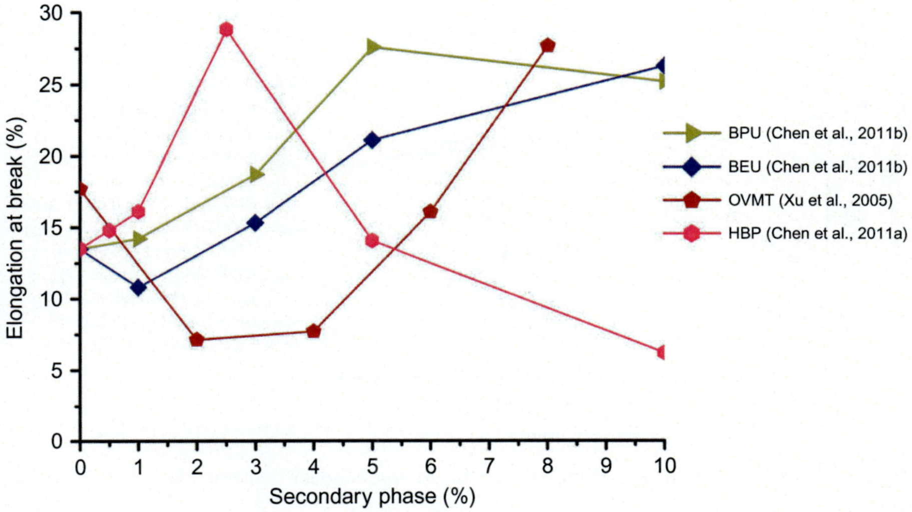

Figure 7.6 Variation in percent of elongation at break with different secondary constituents in PPC at lower loading.

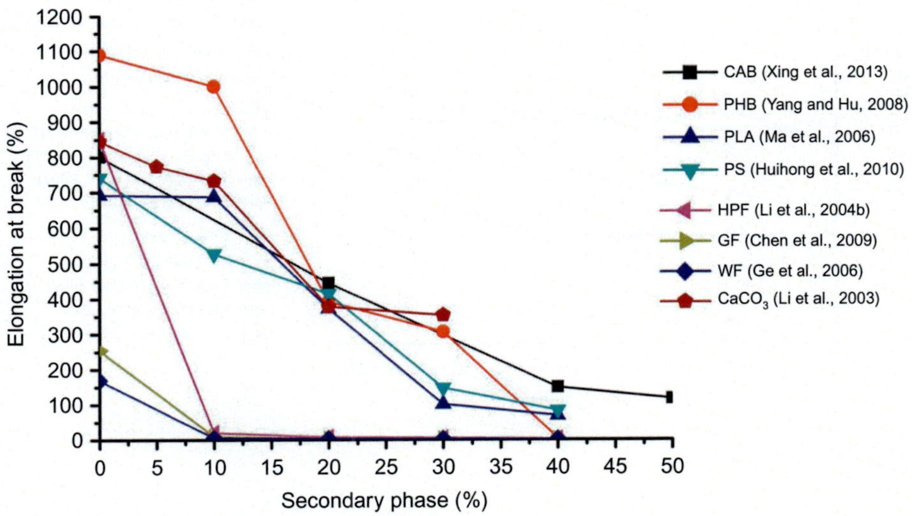

Figure 7.7 Variation in percent of elongation at break with different secondary constituents in PPC at higher loading.

7.3.4.3 Dynamic–mechanical properties

Dynamic–mechanical properties like storage modulus, loss modulus, and tan δ were determined for PPC blends and composites. While storage modulus demonstrates elastic behavior, loss modulus exemplifies the viscous behavior of the polymer. Similar to static mechanical properties, dynamic–mechanical properties of PPC blends and composites improved significantly with varying content of the secondary constituent.

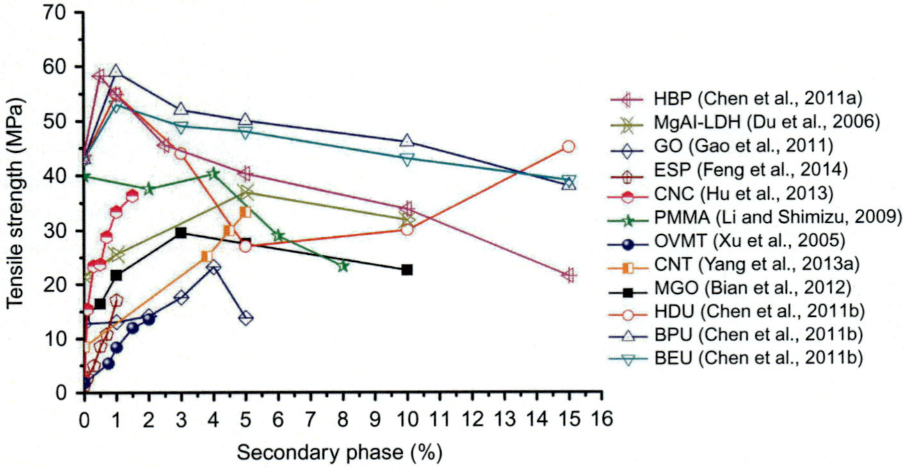

Figure 7.8 Variation in tensile strength with different secondary constituents in PPC at lower loading.

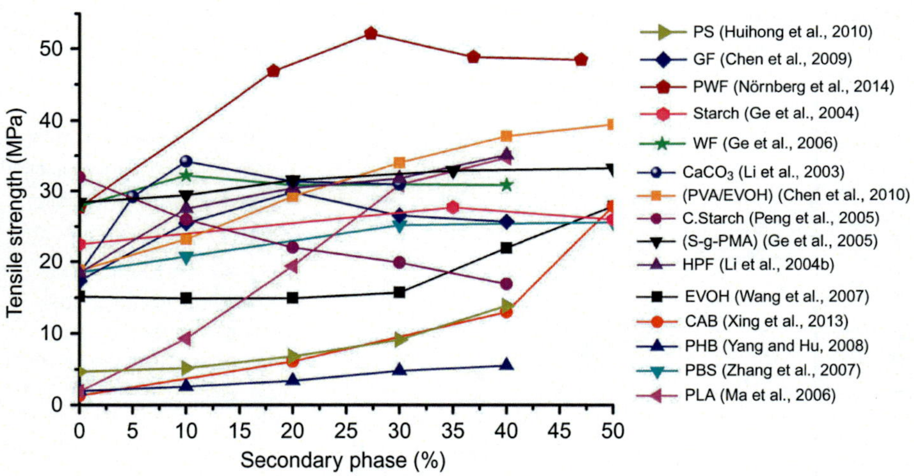

Figure 7.9 Variation in tensile strength with different secondary constituents in PPC at higher loading.

Dynamic–mechanical properties of most blends exhibit a similar trend as static mechanical properties. The storage modulus generally increases with increase in the percentage of secondary constituent (polymer as blend, fillers/reinforcement to make composite), while it decreases dramatically with increase in temperature, and a complete loss of properties is observed at the T_g, which is generally close to 40 °C. However, these secondary constituents can significantly alter the T_g of the blend/composite.

Zhang et al. (2007) found that with increase in PBS content, in PPC/PBS blend, the height of the loss factor peak decreased and become broader. This proves that the blend possessed improved strength with lower damping, along with compositional

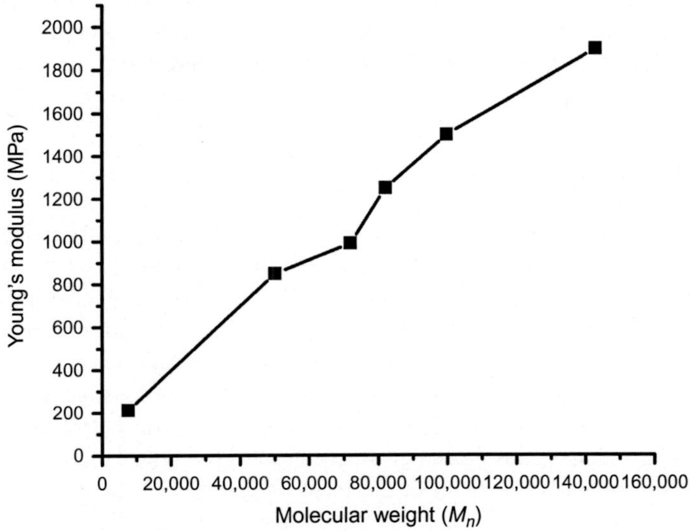

Figure 7.10 Variation in Young's modulus of PPC with different molecular weight (M_n).

heterogeneity with increasing PBS content. Increase in storage modulus due to the reinforcing effect of PBS was observed with increase in PBS content. A similar trend was observed by Pang et al. (2008) in PPC/PBS blends in both PPC-rich and PBS-rich regimes. Dynamic mechanical behavior of PPC/EVOH blends investigated by Wang et al. (2007), PPC/starch blends by Ge et al. (2004) and Peng et al. (2005), PPC/CAB by Xing et al. (2013), PPC/PMMA via in situ synthesis using supercritical CO_2 as solvent by Yang et al. (2013b), PPC/wood flour by Ge et al. (2006) and Nörnberg et al. (2014), PPC/OMMT by Xu et al. (2006), and PPC/MGO by Bian et al. (2012) had shown the similar trend of increase in storage modulus with increase in secondary constituent. PPC/PMMA/PVAc blends investigated by Li and Shimizu (2009) showed an increase in relaxation peak temperature with the addition of 5 phr PVAc. These ternary blends with the addition of PVAc were found to be thermally stable until 90 °C, indicating the enhanced interfacial adhesion, and observed a similar trend in the property enhancement. PPC/WF investigated by Nörnberg et al. (2014) observed that addition of fiber content of 20 wt.%, increased T_g to 41 °C, and further addition increased to a maximum value of 43.7 °C. Gao et al. (2011, 2014) observed the increases in T_g of PPC by 10.5 °C—from 29 to 39.5 °C with the addition of GO. Yang et al. (2013a) studied the dynamic mechanical behavior of PPC/PLA/MWCNT blends. They observed an increase of more than 7 °C in T_g upon the addition of 10 wt.% PLA.

7.4 Applications

Owing to its outstanding ductility, PPC offers a wide spectrum of opportunities in a variety of engineering applications. However, due to its low strength, decomposition temperature, and glass transition temperature, it is currently used as sacrificial binder.

With advancements such as the studies discussed in this chapter, it is foreseen that PPC and its blends and composites will leverage for a number of high-value applications such as automotive, civil, packaging, biomedical, and so on.

7.5 Conclusions

Over the past decade, utilization of CO_2 as a raw material in synthesizing polymers has gained wide attention as it provides a great solution not only to the major threat of global warming but also to white pollution. However, the high cost of these novel polymers coupled with their low strength and low glass transition temperature, are limiting their widespread application. Thus, improvement in these properties while reducing the cost will open new avenues for these polymers. A feasible and facile route to tailor the material properties of CO_2-based polymers is via producing blends and composites. This chapter has presented various routes for synthesizing the CO_2-based polymers with special focus on PPC, the most intriguing CO_2-based polymer with a wide range of tunable mechanical properties. Further, various PPC-based blends and composites, their processing and mechanical (static and dynamic) properties, were discussed. Finally, this chapter has presented a summary of static mechanical performance of PPC-based blends and composites with discussion on the design criteria to enhance the mechanical properties. This opens new directions for research in PPC-based materials for varied applications. It is foreseen that with continuous advancements, the day is not far off when the research into CO_2-based polymeric blends and composites will be commercialized, especially in high-end applications such as automotive, civil, packaging, biomedical, and others. A realization of this will not only benefit the environment but also mankind.

Acknowledgments

The authors would like to acknowledge the assistance of Mr. Jithin Joy and Sai Aditya Pradeep in plotting the graphs.

Abbreviations

BEU	1,6-bis(hydroxyethyl urethane) hexane
BPA	bisphenol-A
BPU	1,6-bis(hydroxyisopropyl urethane) hexane
BS-PPC	benzenesulfonyl-capped PPC
BWF	bamboo wood flour
CAB	cellulose acetate butyrate
$CaCO_3$	calcium carbonate
CCS	carbon capture and storage
CNC	cellulose nanocrystal

CNT	carbon nanotubes
CO_2	carbon dioxide
CS	corn starch
CSRF	caulis spatholobi residue fiber
DMA	dynamic mechanical analysis/analyzer
DMSO	dimethyl sulfoxide
EGMA	poly(ethylene-coglycidyl methacrylate)
ESP	egg shell powder
EVOH	ethylene-vinyl alcohol
GF	glass fiber
GHG	greenhouse gas
GMA	glycidyl methacrylate
GO	graphene oxide
HBP	hyperbranched poly(ester-amide)
HCl	hydrochloric acid
HDU	1,6-bis(methyl urethane) hexane
HPF	*Hildegardia populifolia* fiber
MAH	maleic anhydride
MD	machine direction
MgAl-LDH	MgAl-layered double hydroxide
MGO	graphitic oxide
MWCNT	multiwalled carbon nanotubes
OH	hydroxide
OMMT	organomodified montmorillonite
OREC	organophilic rectorite
OVMT	organovermiculite
PBC	poly(butylene) carbonate
PBS	poly(butylene succinate)
PCHC	poly(cyclohexene) carbonate
PEC	poly(ethylene) carbonate
PHB	poly(hydroxy butyrate)
PHBV	poly(β-hydroxybutyrate-co-β-hydroxyvalerate)
PLA	poly lactic acid
PMMA	polymethyl methacrylate
PO	propylene oxide
PPC	poly(propylene) carbonate
PS	polystyrene
PVA	poly vinyl alcohol
PVAc	poly vinyl acetate
PVPh	poly(p-vinylphenol)
SAN-g-MAH	poly(styrene-g-acrylonitrile)-maleic anhydride
S-g-PMA	starch-g-poly(methyl acrylate)
TD	transverse direction
T_g	glass transition temperature
THF	tetrahydrafuran
TTI	triphenylmethane triisocyanate
UHMW-PPC	ultra high molecular weight PPC
ZnO	zinc oxide

References

Arranz-Andrés, J., Peña, B., Benavente, R., Pérez, E., Cerrada, M.L., 2007. Influence of isotacticity and molecular weight on the properties of metallocenic isotactic polypropylene. Eur. Polym. J. 43, 2357–2370. http://dx.doi.org/10.1016/j.eurpolymj.2007.03.034.

Azad, A.-M., McDaniel, E., Al-batty, S., 2011. A novel paradigm in greenhouse gas mitigation. Environ. Prog. Sustain. Energy 30, 733–742. http://dx.doi.org/10.1002/ep.10515.

Bardin, J., 2012. Earthquakes likely from carbon dioxide storage, not fracking. Los Angeles Times.

Barreto, C., Altskär, A., Fredriksen, S., Hansen, E., Rychwalski, R.W., 2013. Multiwall carbon nanotube/PPC composites: preparation, structural analysis and thermal stability. Eur. Polym. J. 49, 2149–2161. http://dx.doi.org/10.1016/j.eurpolymj.2013.05.009.

Bian, J., Wei, X.W., Gong, S.J., Zhang, H., Guan, Z.P., 2012. Improving the thermal and mechanical properties of poly(propylene carbonate) by incorporating functionalized graphite oxide. J. Appl. Polym. Sci. 123, 2743–2752. http://dx.doi.org/10.1002/app.

Chen, W., Pang, M., Xiao, M., Wang, S., Wen, L., Meng, Y., 2009. Mechanical, thermal, and morphological properties of glass fiber-reinforced biodegradable poly(propylene carbonate) composites. J. Reinf. Plast. Compos. 29, 1545–1550. http://dx.doi.org/10.1177/0731684409336370.

Chen, X., Wang, S.J., Xiao, M., Han, D.M., Meng, Y., 2010. Miscibility, properties and morphology of biodegradable blends of UHMW-PPC/PVA/EVOH. J. Polym. Res. 18, 715–720. http://dx.doi.org/10.1007/s10965-010-9467-4.

Chen, L., Qin, Y., Wang, X., Li, Y., Zhao, X., Wang, F., 2011a. Toughening of poly(propylene carbonate) by hyperbranched poly(ester-amide) via hydrogen bonding interaction. Polym. Int. 60, 1697–1704. http://dx.doi.org/10.1002/pi.3132.

Chen, L., Qin, Y., Wang, X., Zhao, X., Wang, F., 2011b. Plasticizing while toughening and reinforcing poly(propylene carbonate) using low molecular weight urethane: role of hydrogen-bonding interaction. Polymer (Guildf.) 52, 4873–4880. http://dx.doi.org/10.1016/j.polymer.2011.08.025.

Chen, G.J., Wang, Y.Y., Wang, S.J., Xiao, M., Meng, Y.Z., 2013. Orientation microstructure and properties of poly(propylene carbonate)/poly(butylene succinate) blend films. J. Appl. Polym. Sci. 128, 390–399. http://dx.doi.org/10.1002/app.38150.

Chen, F., Liu, H., Zhao, Y., Feng, Y., Guo, R., Wu, Z., et al., 2014. Caulis spatholobi residue fiber reinforced biodegradable poly(propylene carbonate) composites: the effect of fiber content on mechanical and morphological properties. Polym. Compos. 35, 208–216. http://dx.doi.org/10.1002/pc.22652.

Darensbourg, D.J., Wilson, S.J., 2012. What's new with CO_2? Recent advances in its copolymerization with oxiranes. Green Chem. 14, 2665. http://dx.doi.org/10.1039/c2gc35928f.

Darensbourg, D.J., Wei, S.-H., Yeung, A.D., Ellis, W.C., 2013. An efficient method of depolymerization of poly(cyclopentene carbonate) to its comonomers: cyclopentene oxide and carbon dioxide. Macromolecules 46, 5850–5855. http://dx.doi.org/10.1021/ma401286x.

Du, L., Qu, B., Meng, Y., Zhu, Q., 2006. Structural characterization and thermal and mechanical properties of poly(propylene carbonate)/MgAl-LDH exfoliation nanocomposite via solution intercalation. Compos. Sci. Technol. 66, 913–918. http://dx.doi.org/10.1016/j.compscitech.2005.08.012.

Eberhardt, R., Allmendinger, M., Zintl, M., Troll, C., Luinstra, G.A., Rieger, B., 2004. New zinc dicarboxylate catalysts for the CO_2/propylene oxide copolymerization reaction: activity enhancement through Zn(II)-ethylsulfinate initiating groups. Macromol. Chem. Phys. 205, 42–47. http://dx.doi.org/10.1002/macp.200350081.

Empower Materials: Products, n.d. http://empowermaterials.com/products/index.htm (accessed October 07, 2014).

Fei, B., Chen, C., Peng, S., Zhao, X., Wang, X., Dong, L., 2004. FTIR study of poly(propylene carbonate)/bisphenol A blends. Polym. Int. 53, 2092–2098. http://dx.doi.org/10.1002/pi.1633.

Feng, Y., Ashok, B., Madhukar, K., Zhang, J.J., Reddy, K.O., Rajulu, A.V., 2014. Preparation and characterization of polypropylene carbonate bio-filler (eggshell powder) composite films. Int. J. Polym. Anal. Charact. 19, 637–647. http://dx.doi.org/10.1080/1023666X.2014.953747.

Gao, J., Chen, F., Wang, K., Deng, H., Zhang, Q., Bai, H., et al., 2011. A promising alternative to conventional polyethylene with poly(propylene carbonate) reinforced by graphene oxide nanosheets. J. Mater. Chem. 21, 17627. http://dx.doi.org/10.1039/c1jm14300j.

Gao, F., Zhou, Q., Dong, Y., Qin, Y., Wang, X., Xiaojiang, Z., et al., 2012a. Ether linkage in poly(1,2-propylene carbonate), a key structure factor to tune its performances. J. Polym. Res. 19, 9877. http://dx.doi.org/10.1007/s10965-012-9877-6.

Gao, J., Bai, H., Zhang, Q., Gao, Y., Chen, L., Fu, Q., 2012b. Effect of homopolymer poly(vinyl acetate) on compatibility and mechanical properties of poly(propylene carbonate)/poly(lactic acid) blends. Express Polym. Lett. 6, 860–870. http://dx.doi.org/10.3144/expresspolymlett.2012.92.

Gao, J., Bai, H., Zhou, X., Yang, G., Xu, C., Zhang, Q., et al., 2014. Observation of strong nano-effect via tuning distributed architecture of graphene oxide in poly(propylene carbonate). Nanotechnology 25, 025702. http://dx.doi.org/10.1088/0957-4484/25/2/025702.

Ge, X.C., Li, X.H., Zhu, Q., Li, L., Meng, Y.Z., 2004. Preparation and properties of biodegradable poly(propylene carbonate)/starch composites. Polym. Eng. Sci. 44, 2134–2140. http://dx.doi.org/10.1002/pen.20219.

Ge, X.C., Xu, Y., Meng, Y.Z., Li, R.K.Y., 2005. Thermal and mechanical properties of biodegradable composites of poly(propylene carbonate) and starch–poly(methyl acrylate) graft copolymer. Compos. Sci. Technol. 65, 2219–2225. http://dx.doi.org/10.1016/j.compscitech.2005.04.052.

Ge, X.C., Zhu, Q., Meng, Y.Z., 2006. Fabrication and characterization of biodegradable poly(propylene carbonate)/wood flour composites. J. Appl. Polym. Sci. 99, 782–787. http://dx.doi.org/10.1002/app.22557.

Hacker, V., 2003. A novel process for stationary hydrogen production: the reformer sponge iron cycle (RESC). J. Power Sources 118, 311–314. http://dx.doi.org/10.1016/S0378-7753(03)00076-4.

Herzog, H., Golomb, D., 2004. Carbon capture and storage from fossil fuel use. In: Cleveland, C.J. (Ed.), Encyclopedia of Energy. Elsevier, New York, pp. 277–287. http://dx.doi.org/10.1016/B0-12-176480-X/00422-8.

House, K.Z., Schrag, D.P., Harvey, C.F., Lackner, K.S., 2006. Permanent carbon dioxide storage in deep-sea sediments. Proc. Natl. Acad. Sci. USA 103, 12291–12295. http://dx.doi.org/10.1073/pnas.0605318103.

Hsu, T.-J., Tan, C.-S., 2002. Block copolymerization of carbon dioxide with cyclohexene oxide and 4-vinyl-1-cyclohexene-1,2-epoxide in based poly(propylene carbonate) by yttrium–metal coordination catalyst. Polymer (Guildf.) 43, 4535–4543. http://dx.doi.org/10.1016/S0032-3861(02)00264-1.

Hu, X., Xu, C., Gao, J., Yang, G., Geng, C., Chen, F., et al., 2013. Toward environment-friendly composites of poly(propylene carbonate) reinforced with cellulose nanocrystals. Compos. Sci. Technol. 78, 63–68. http://dx.doi.org/10.1016/j.compscitech.2013.02.002.

Huang, Y., Wang, J., Liao, B., Chen, M., Cong, G., 1997. Epoxy resins toughened by poly(propylene carbonate). J. Appl. Polym. Sci. 64, 2457–2465. http://dx.doi.org/10.1002/(SICI)1097-4628(19970620)64:12<2457::AID-APP20>3.0.CO;2-X.

Huihong, L., Lisha, P., Qiang, L., Nai, X., Lingbin, L., Sujuan, P., et al., 2010. Preparation and characterization of poly(propylene carbonate)/polystyrene composite films by melt-extrusion method. E-Polymers 10, 390–398.

Inoue, S., Koinuma, H., Tsuruta, T., 1969a. Copolymerization of carbon dioxide and epoxide. J. Polym. Sci. Part B Polym. Lett. 7, 287–292.

Inoue, S., Koinuma, H., Tsuruta, T., 1969b. Copolymerization of carbon dioxide and epoxide with organometallic compounds. Die Makromol. Chem. 130, 210–220. http://onlinelibrary.wiley.com/doi/10.1002/macp.1969.021300112/abstract.

Jang, B.Z., Uhlmann, D.R., Vander Sande, J.B., 1985. The rubber particle size dependence of crazing in polypropylene. Polym. Eng. Sci. 25, 643–651. http://dx.doi.org/10.1002/pen.760251011.

Keeling, R.F., Manning, A.C., Dubey, M.K., 2011. The atmospheric signature of carbon capture and storage. Philos. Trans. A Math. Phys. Eng. Sci. 369, 2113–2132. http://dx.doi.org/10.1098/rsta.2011.0016.

Kim, J.-S., Kim, H., Yoon, J., Heo, K., Ree, M., 2005. Synthesis of zinc glutarates with various morphologies using an amphiphilic template and their catalytic activities in the copolymerization of carbon dioxide and propylene oxide. J. Polym. Sci. Part A Polym. Chem. 43, 4079–4088. http://dx.doi.org/10.1002/pola.20905.

Lai, M.F., Li, J., Liu, J.J., 2005. Thermal and dynamic mechanical properties of poly(propylene carbonate). J. Therm. Anal. Calorim. 82, 293–298. http://dx.doi.org/10.1007/s10973-005-0892-2.

Lal, R., 2004. Carbon sequestration, terrestrial. In: Cleveland, C.J. (Ed.), Encyclopedia of Energy. Elsevier, New York, pp. 289–298. http://dx.doi.org/10.1016/B0-12-176480-X/00539-8.

Lal, R., 2008. Carbon sequestration. Philos. Trans. R. Soc. Lond. B Biol. Sci. 363, 815–830. http://dx.doi.org/10.1098/rstb.2007.2185.

Li, Y., Shimizu, H., 2009. Compatibilization by homopolymer: significant improvements in the modulus and tensile strength of PPC/PMMA blends by the addition of a small amount of PVAc. ACS Appl. Mater. Interfaces 1, 1650–1655. http://dx.doi.org/10.1021/am900314k.

Li, X.H., Tjong, S.C., Meng, Y.Z., Zhu, Q., 2003. Fabrication and properties of poly(propylene carbonate)/calcium carbonate composites. J. Polym. Sci. Part B Polym. Phys. 41, 1806–1813.

Li, J., Lai, M.F., Liu, J.J., 2004a. Effect of poly(propylene carbonate) on the crystallization and melting behavior of poly(hydroxybutyrate-co-hydroxy valerate). J. Appl. Polym. Sci. 92, 2514–2521.

Li, X.H., Meng, Y.Z., Wang, S.J., Rajulu, A.V., Tjong, S.C., 2004b. Completely biodegradable composites of poly(propylene carbonate) and short, lignocellulose fiber *Hildegardia populifolia*. J. Polym. Sci. Part B Polym. Phys. 42, 666–675.

Li, J., Lai, M.F., Liu, J.J., 2005. Control and development of crystallinity and morphology in poly(β-hydroxybutyrate-co-β-hydroxyvalerate)/poly(propylene carbonate) blends. J. Appl. Polym. Sci. 98, 1427–1436. http://dx.doi.org/10.1002/app.22117.

Little, M.G., Jackson, R.B., 2010. Potential impacts of leakage from deep CO_2 geosequestration on overlying freshwater aquifers. Environ. Sci. Technol. 44, 9225–9232. http://dx.doi.org/10.1021/es102235w.

Lu, X.-B., Darensbourg, D.J., 2012. Cobalt catalysts for the coupling of CO_2 and epoxides to provide polycarbonates and cyclic carbonates. Chem. Soc. Rev. 41, 1462–1484. http://dx.doi.org/10.1039/c1cs15142h.

Lu, X., Ren, W., Wu, G., 2012. CO_2 copolymers from epoxides: catalyst activity, product selectivity, and stereochemistry control. Acc. Chem. Res. 45, 1721–1735. http://dx.doi.org/10.1021/ar300035z.

Luinstra, G., 2008. Poly(propylene carbonate), old copolymers of propylene oxide and carbon dioxide with new interests: catalysis and material properties. Polym. Rev. 48, 192–219. http://dx.doi.org/10.1080/15583720701834240.

Luinstra, G.A., Borchardt, E., 2012. Material properties of poly(propylene carbonates). Adv. Polym. Sci. 245, 29–48. http://dx.doi.org/10.1007/12.

Ma, X., Yu, J., Wang, N., 2006. Compatibility characterization of poly(lactic acid)/poly(propylene carbonate) blends. J. Polym. Sci. Part B Polym. Phys. 44, 94–101. http://dx.doi.org/10.1002/polb.20669.

Mcclure, M., 2012. Carbon capture and storage likely to cause earthquakes, say Stanford researchers. http://news.stanford.edu/news/2012/june/carbon-capture-earthquakes-061912.html (accessed October 07, 2014).

Meng, Y.Z., Du, L.C., Tiong, S.C., Zhu, Q., Hay, A.S., 2002. Effects of the structure and morphology of zinc glutarate on the fixation of carbon dioxide into polymer. J. Polym. Sci. Part A Polym. Chem. 40, 3579–3591. http://dx.doi.org/10.1002/pola.10452.

Nair, K.C.M., Thomas, S., 2003. Effect of interface modification on the mechanical properties of polystyrene-sisal fiber composites. Polym. Compos. 24, 332–343. http://dx.doi.org/10.1002/pc.10033.

Nicolais, L., Narkis, M., 1971. Stress–strain behavior of styrene-acrylonitrile/glass bead composites in the glassy region. Polym. Eng. Sci. 11, 194–199. http://dx.doi.org/10.1002/pen.760110305.

Nörnberg, B., Borchardt, E., Luinstra, G.A., Fromm, J., 2014. Wood plastic composites from poly(propylene carbonate) and poplar wood flour – mechanical, thermal and morphological properties. Eur. Polym. J. 51, 167–176. http://dx.doi.org/10.1016/j.eurpolymj.2013.11.008.

Otsuka, K., Yamada, C., Kaburagi, T., Takenaka, S., 2003. Hydrogen storage and production by redox of iron oxide for polymer electrolyte fuel cell vehicles. Int. J. Hydrogen Energy 28, 335–342. http://dx.doi.org/10.1016/S0360-3199(02)00070-8.

Pang, M.Z., Qiao, J.J., Jiao, J., Wang, S.J., Xiao, M., Meng, Y.Z., 2008. Miscibility and properties of completely biodegradable blends of poly(propylene carbonate) and poly(butylene succinate). J. Appl. Polym. Sci. 107, 2854–2860. http://dx.doi.org/10.1002/app.27252.

Peng, S., An, Y., Chen, C., Fei, B., Zhuang, Y., Dong, L., 2003. Miscibility and crystallization behavior of poly(3-hydroxyvalerate-co-3-hydroxyvalerate)/poly(propylene carbonate) blends. J. Appl. Polym. Sci. 90, 4054–4060. http://dx.doi.org/10.1002/app.12970.

Peng, S., Wang, X., Dong, L., 2005. Special interaction between poly(propylene carbonate) and corn starch. Polym. Compos. 26, 37–41. http://dx.doi.org/10.1002/pc.20070.

Pescarmona, P.P., Taherimehr, M., 2012. Challenges in the catalytic synthesis of cyclic and polymeric carbonates from epoxides and CO_2. Catal. Sci. Technol. 2, 2169. http://dx.doi.org/10.1039/c2cy20365k.

Pilla, S., Turng, L.-S., 2013. Biobased and biodegradable ternary blends: investigation of properties via experimental and theoretical routes. In: SPE, ANTEC Tech. Conf.

Qiao, J., Du, F., Pang, M., Xiao, M., Wang, S., Meng, Y., 2008. Fabrication and properties of degradable PPC/EVOH/starch/$CaCO_3$ composites. J. Wuhan Univ. Technol. Sci. Ed. 23, 362–366. http://dx.doi.org/10.1007/s11595-007-3362-9.

Qin, Y., Wang, X., 2010. Carbon dioxide-based copolymers: environmental benefits of PPC, an industrially viable catalyst. Biotechnol. J. 5, 1164–1180. http://dx.doi.org/10.1002/biot.201000134.

Ree, M., Bae, J.Y., Jung, J.H., Shin, T.J., 1999a. A new copolymerization process leading to poly(propylene carbonate) with a highly enhanced yield from carbon dioxide and propylene oxide. J. Polym. Sci. Part A Polym. Chem. 37, 1863–1876.

Ree, M., Bae, J.Y., Jung, J.H., Shin, T.J., 1999b. A green copolymerization of carbon dioxide and propylene oxide. Korean Polym. J. 7, 333–348.

Rout, J., Tripathy, S.S., Misra, M., Mohanty, A.K., Nayak, S.K., 2001. The influence of fiber surface modification on the mechanical properties of coir-polyester composites. Polym. Compos. 22, 468–476. http://dx.doi.org/10.1002/pc.10552.

Sabzi, F., Boushehri, A., 2006. Compatibility of polymer blends. I. Copolymers with organic solvents. J. Appl. Polym. Sci. 101, 492–498. http://dx.doi.org/10.1002/app.23092.

Schrag, D.P., 2007. Preparing to capture carbon. Science 315, 812–813. http://dx.doi.org/10.1126/science.1137632.

Seibel, B.A., 2003. Biological impacts of deep-sea carbon dioxide injection inferred from indices of physiological performance. J. Exp. Biol. 206, 641–650. http://dx.doi.org/10.1242/jeb.00141.

Seibel, B.A., Walsh, P.J., 2001. Carbon cycle. Potential impacts of CO_2 injection on deep-sea biota. Science 294, 319–320. http://dx.doi.org/10.1126/science.1065301.

Seo, J., Jeon, G., Jang, Eu.S., Bahadar Khan, S., Han, H., 2011. Preparation and properties of poly(propylene carbonate) and nanosized ZnO composite films for packaging applications. J. Appl. Polym. Sci. 122, 1101–1108. http://dx.doi.org/10.1002/app.34248.

Shah, N., Panjala, D., Huffman, G.P., 2001. Hydrogen production by catalytic decomposition of methane. Energy Fuel 15, 1528–1534. http://dx.doi.org/10.1021/ef0101964.

Shah, N., Wang, Y., Panjala, D., Huffman, G.P., 2004. Production of hydrogen and carbon nanostructures by non-oxidative catalytic dehydrogenation of ethane and propane. Energy Fuel 18, 727–735. http://dx.doi.org/10.1021/ef034100c.

Shi, X., Gan, Z., 2007. Preparation and characterization of poly(propylene carbonate)/montmorillonite nanocomposites by solution intercalation. Eur. Polym. J. 43, 4852–4858. http://dx.doi.org/10.1016/j.eurpolymj.2007.09.024.

Socolow, R.H., 2005. Can we bury global warming? Sci. Am. 293, 49–55. http://dx.doi.org/10.1038/scientificamerican0705-49.

Sugimoto, H., Inoue, S., 2004. Copolymerization of carbon dioxide and epoxide. J. Polym. Sci. Part A Polym. Chem. 42, 5561–5573. http://dx.doi.org/10.1002/pola.20319.

Taherimehr, M., Pescarmona, P.P., 2014. Green polycarbonates prepared by the copolymerization of CO_2 with epoxides. J. Appl. Polym. Sci. 121, 41141. http://dx.doi.org/10.1002/app.41141.

Tang, L.-G., Kardos, J.L., 1997. A review of methods for improving the interfacial adhesion between carbon fiber and polymer matrix. Polym. Compos. 18, 100–113. http://dx.doi.org/10.1002/pc.10265.

Tao, J., Song, C., Cao, M., Hu, D., Liu, L., Liu, N., et al., 2009. Thermal properties and degradability of poly(propylene carbonate)/poly(β-hydroxybutyrate-co-β-hydroxyvalerate) (PPC/PHBV) blends. Polym. Degrad. Stab. 94, 575–583. http://dx.doi.org/10.1016/j.polymdegradstab.2009.01.017.

Thorat, S.D., Phillips, P.J., Semenov, V., Gakh, A., 2003. Physical properties of aliphatic polycarbonates made from CO_2 and epoxides. J. Appl. Polym. Sci. 89, 1163–1176. http://dx.doi.org/10.1002/app.12355.

Wan, C., Yu, J., Shi, X., Huang, L., 2006. Preparation of poly(propylene carbonate)/organophilic rectorite nanocomposites via direct melt intercalation. Trans. Nonferrous Met. Soc. China 16, s508–s511. http://dx.doi.org/10.1016/S1003-6326(06)60245-8.

Wang, X.L., Du, F.G., Jiao, J., Meng, Y.Z., Li, R.K.Y., 2007. Preparation and properties of biodegradable polymeric blends from poly(propylene carbonate) and poly(ethylene-co-vinyl alcohol). J. Biomed. Mater. Res. B Appl. Biomater. 83, 373–379. http://dx.doi.org/10.1002/jbm.b.30806.

Xing, C., Wang, H., Hu, Q., Xu, F., Cao, X., You, J., et al., 2013. Mechanical and thermal properties of eco-friendly poly(propylene carbonate)/cellulose acetate butyrate blends. Carbohydr. Polym. 92, 1921–1927. http://dx.doi.org/10.1016/j.carbpol.2012.11.058.

Xu, J., Li, R.K.Y., Xu, Y., Li, L., Meng, Y.Z., 2005. Preparation of poly(propylene carbonate)/organo-vermiculite nanocomposites via direct melt intercalation. Eur. Polym. J. 41, 881–888. http://dx.doi.org/10.1016/j.eurpolymj.2004.10.033.

Xu, J., Li, R.K.Y., Meng, Y.Z., Mai, Y.W., 2006. Biodegradable poly(propylene carbonate)/montmorillonite nanocomposites prepared by direct melt intercalation. Mater. Res. Bull. 41, 244–252. http://dx.doi.org/10.1016/j.materresbull.2005.08.019.

Yang, D.Z., Hu, P., 2008. Miscibility, crystallization, and mechanical properties of poly (3-hydroxybutyrate) and poly(propylene carbonate) biodegradable blends. J. Appl. Polym. Sci. 109, 1635–1642. http://dx.doi.org/10.1002/app.28002.

Yang, G., Geng, C., Su, J., Yao, W., Zhang, Q., Fu, Q., 2013a. Property reinforcement of poly(propylene carbonate) by simultaneous incorporation of poly(lactic acid) and multi-walled carbon nanotubes. Compos. Sci. Technol. 87, 196–203. http://dx.doi.org/10.1016/j.compscitech.2013.08.010.

Yang, G., Hu, X., Su, J., Geng, C., Yao, W., Zhang, Q., et al., 2013b. Significant reinforcement of poly(propylene carbonate): nanostructured polymer composites of poly(propylene carbonate)/poly(methyl methacrylate) via a supercritical carbon dioxide route. J. Supercrit. Fluids 82, 200–205. http://dx.doi.org/10.1016/j.supflu.2013.08.003.

Yao, M., Deng, H., Mai, F., Wang, K., Zhang, Q., Chen, F., et al., 2011a. Modification of poly(lactic acid)/poly(propylene carbonate) blends through melt compounding with maleic anhydride. Express Polym. Lett. 5, 937–949. http://dx.doi.org/10.3144/expresspolymlett.2011.92.

Yao, M., Mai, F., Deng, H., Ning, N., Wang, K., Fu, Q., 2011b. Improved thermal stability and mechanical properties of poly(propylene carbonate) by reactive blending with maleic anhydride. J. Appl. Polym. Sci. 120, 3565–3573. http://dx.doi.org/10.1002/app.33565.

Yoo, S.J., Lee, S.H., Jeon, M., Lee, H.S., Kim, W.N., 2013. Effects of compatibilizers on the mechanical, morphological, and thermal properties of poly(propylene carbonate)/poly(methyl methacrylate) blends. Macromol. Res. 21, 1182–1187. http://dx.doi.org/10.1007/s13233-013-1159-6.

Zeng, S., Wang, S., Xiao, M., Han, D., Meng, Y., 2011. Preparation and properties of biodegradable blend containing poly(propylene carbonate) and starch acetate with different degrees of substitution. Carbohydr. Polym. 86, 1260–1265. http://dx.doi.org/10.1016/j.carbpol.2011.06.023.

Zhang, Z., Mo, Z., Zhang, H., Zhang, Y., Na, T., An, Y., et al., 2002. Miscibility and hydrogen-bonding interactions in blends of carbon dioxide/epoxy propane copolymer with poly(p-vinylphenol). J. Polym. Sci. Part B Polym. Phys. 40, 1957–1964. http://dx.doi.org/10.1002/polb.10252.

Zhang, Z., Mo, Z., Zhang, H., Wang, X., Zhao, X., 2003. Crystallization and melting behaviors of PPC-BS/PVA blends. Macromol. Chem. Phys. 204, 1557–1566. http://dx.doi.org/10.1002/macp.200350012.

Zhang, Z., Zhang, H.H., Zhang, Q., Zhou, Q., Mo, Z., Zhao, X., et al., 2006. Thermotropic liquid crystallinity, thermal decomposition behavior, and aggregated structure of poly(propylene carbonate)/ethyl cellulose blends. J. Appl. Polym. Sci. 100, 584–592. http://dx.doi.org/10.1002/app.23338.

Zhang, H., Sun, X., Chen, Q., Ren, M., Zhang, Z., Zhang, H., et al., 2007. Miscibility, crystallization and mechanical properties of PPC/PBS blends. Chin. J. Polym. Sci. 25, 589–597.

Zhu, Q., Meng, Y., Tjong, S., Zhao, X., Chen, Y., 2002. Thermally stable and high molecular weight poly(propylene carbonate)s from carbon dioxide and propylene oxide. Polym. Int. 51, 1079–1085. http://dx.doi.org/10.1002/pi.847.

Zhu, Q., Meng, Y., Tjong, S., Zhang, Y., Wan, W., 2003. Catalytic synthesis and characterization of an alternating copolymer from carbon dioxide and propylene oxide using zinc pimelate. Polym. Int. 52, 799–804. http://dx.doi.org/10.1002/pi.1157.

Zoback, M.D., Gorelick, S.M., 2012. Earthquake triggering and large-scale geologic storage of carbon dioxide. Proc. Natl. Acad. Sci. USA 109, 10164–10168. http://dx.doi.org/10.1073/pnas.1202473109.

Processing, performance, and applications of plant and animal protein-based blends and their biocomposites

8

T.H. Mekonnen, M. Misra, A.K. Mohanty
University of Guelph, Guelph, ON, Canada

8.1 Introduction to protein-based biomaterials

Biomass feedstock is a viable alternative to finite fossil fuel resources to provide many of the same benefits for energy and material applications (Wei et al., 2012). Nature-derived biomaterials consist of naturally occurring macromolecules such as proteins, cellulose, starch, pectin, lignin, chitin, and various other polysaccharides. Among these proteins, starch, cellulose, and lignin are extensively researched as a renewable feedstock for biomaterial development. From a biochemical point of view, proteins contribute to biological functions within a cell, along with other biological macromolecules. From the chemistry perspective, proteins are random copolymers of amino acids linked by peptide bonds that range in length from a few to thousands of amino acid units. The capability of proteins to form a cohesive and continuous matrix draws attention as a biomaterial. Moreover, the large number of functional groups in proteins associated with the amino acid residues and end group of protein chains offers an excellent range of chemical modification pathways.

Emerging technology needs in medicine, security, energy, and sustainability have encouraged scientists and engineers to look to proteins as a source of building blocks and design inspiration for the next generation of polymeric materials (Olsen, 2013). Also, societal pressure to develop a bioeconomy as a result of recent concerns over environmental safety and interest in the utilization of renewable resources for energy and material spurs a significant motive for the utilization of protein as biomaterial. The biofuel industry that consumes lipids (e.g., soy oil, cotton oil) for biodiesel, and starch (e.g., corn and wheat starch) for bioethanol, resulted in less expensive protein residue that has a potential for biomaterial development. The utilization of such residues could also avoid the controversial competition of food versus industrial utilization of agricultural commodities.

While protein-based materials constitute amazing natural accomplishments in the form of strong fibers, body armors (mollusk shell, and bones), defence and aggression (claws, teeth, tusks, horns, and antlers), support (bones, mollusk shell, and hooves) and mastication (teeth), underwater adhesives, and so on, a growing community of researchers are also designing materials at the interface of proteins and polymers

Biocomposites: Design and Mechanical Performance. http://dx.doi.org/10.1016/B978-1-78242-373-7.00017-2

(Olsen, 2013). Historically, proteins were used for centuries to prepare adhesives until petroleum-based adhesives started to dominate the market in the early twentieth century (Liu and Li, 2007). For instance, traditionally, bone glues have been used in carpentry and blood has been used as an additive in plywood–glue mixes (Pizzi, 2006). In the last decade, investigation has been harnessed to utilize protein-based materials for similar application and several more.

The use of proteins to produce biomaterials including biodegradable plastics, edible films and sheets (Swain et al., 2004; Gennadios et al., 1996), microencapsulating agents and active compounds in pharmaceutical applications (Dangaran et al., 2009; Elzoghby et al., 2011), adhesives (Mekonnen et al., 2014b), polymer blend and composite materials (Mohanty et al., 2002; Mekonnen et al., 2013a), wound dressing and bionanocomposites (Tian, 2012) have been extensively reported in the literature. Despite several research efforts, protein-based materials remain challenged by low strength and high sensitivity to moisture, with the notable exception of keratin (Zhang et al., 2012). This chapter provides an overview of protein-based composite materials: various resources and characteristics, processing, performance, and challenges.

8.2 Plant and animal proteins: structure, properties, and sources

Within the myriad of biopolymers present in nature, proteins are considered to be one of the most sophisticated groups in terms of chemistry. At the level of covalent bonds, proteins are poly(amino acid) polymers synthesized from a set of 20 naturally occurring monomers, with an almost limitless number of sequential arrangements. As such, proteins are different from the other common natural polymers–polysaccharides, which are usually a polymer of few monomers. As an illustration, common polysaccharides like starch and cellulose are a polymer of single monomer unit glucose, and chitin is a polymer of *N*-acetylglucosamine. In comparison to synthetic polymers, proteins demonstrate far more complexity due to their sequence-specific chemistry. The specific monomer sequences in proteins enable them to fold into complex hierarchical structures, which give rise to their function and differentiate them from synthetic polymers (Olsen, 2013). Figure 8.1 shows the difference between protein and synthetic polymer in terms of chain sequence and conformation.

The structure of proteins is commonly described in terms of four hierarchical levels of organization: primary, secondary, tertiary, and quaternary structures. The primary structure is the linear sequence of amino acids, starting with N-terminal and ending with C-terminal amino acids. This primary structure is held together by covalent or peptide bonds, which are made during the process of protein biosynthesis or translation. The secondary structure describes the highly regular local conformation of amino acids in the protein chain—stabilized by hydrogen bonding. The two particularly common structural motives of the secondary structure are α-helix and β-sheet (Buxbaum, 2007). The NH and CO groups of proteins form hydrogen bonds between themselves in a regular pattern, and this creates the particular conformation of the chain to form α-helix. In β-sheet, successive separate chains are bonded in such a fashion to form

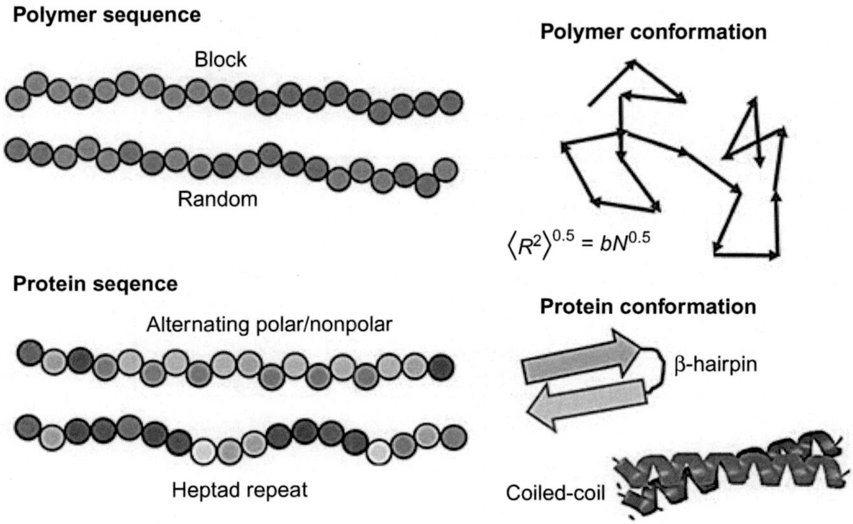

Figure 8.1 Difference between polymer and protein structure.
Adapted from Olsen (2013), Wiley, copyright © 2013.

pleated sheets. Tertiary structure describes the global conformation of protein or the arrangement of secondary structures in space. The quaternary structure on the other hand is a larger assembly of several polypeptide chains that come together to form a single functional protein. The quaternary structure is stabilized by the same noncovalent interactions and disulfide bonds as the tertiary structure.

Many natural structural proteins such as silks, elastins, collagens, keratins, and resilins have been shown to have distinguished mechanical, chemical, electrical, electromagnetic, and optical properties (Hu et al., 2012);, thus, they are highly sought after for protein-based biomaterial development. These structural proteins are characterized by long-range ordered molecular secondary structures (e.g., beta-pleated sheets, coiled sheets, or triple helices) that arise due to highly repetitive primary amino acid sequences within the proteins. These features promote self-assembly and formation of structural hierarchy that resulted in their functional roles as nature's material in an animal's body. The interactions that stabilize these proteins include hydrophobic–hydrophilic interactions, charge–charge (electrostatic) interactions, or hydrogen bond interactions, as shown in Figure 8.2 (Hu et al., 2012).

Also, many naturally occurring proteins are blend materials based on the combination of a variety of proteinaceous components. For example, silk fibers are composed of fibroins and sericins (Hardy and Scheibel, 2010); wheat gluten is a combination of gliadin, gluten, and glutenin; collagen (types I ~ XXVIII) is intertwined together (Hu et al., 2012); keratin proteins are composed of two protein chains with different compositions and molecular weights (types I and II) (Rouse and Van Dyke, 2010); soy protein is a mixture of proteins with different molecular properties with 7S and 11S making the majority composition (Cao et al., 2007); and corn zein is also composed of several structurally different types of prolamin proteins (Anderson and Lamsal, 2011). Additionally,

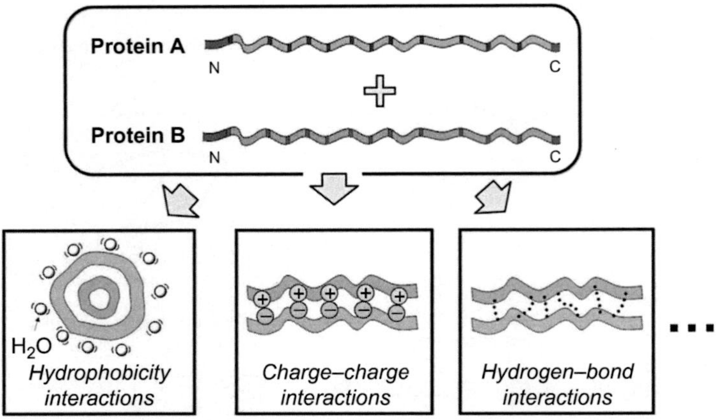

Figure 8.2 Types of interactions between chains of protein A and protein B that can generate stabilized structures.
Adapted from Hu et al. (2012), Elsevier, copyright © 2013.

it is important to note that many structural biological materials such as teeth, bones, eggshells, tusks, hoofs, and insect exoskeletons are also composite materials composed of a biopolymer matrix (structural proteins such as collagen, keratin, cellulose, chitin, etc.) and a reinforcing mineral phase (calcium carbonate, calcium phosphate, hydroxyapatite, or silica) (McKittrick et al., 2010; Meyers et al., 2008; Yang et al., 2013). Table 8.1 shows a review of the composition of some of the structural biological materials. Besides, there is structural hierarchy that displays organization at all levels, from the nanoscale to the macroscale, in these biological materials. The biopolymer imparts toughness and resilience, while the mineral increases hardness and stiffness.

For practical reasons, different natural proteins utilized for biocomposites application are distinguished as plant proteins and animal proteins based on their resource. Though it is possible to recover protein from single cells, there is no research that utilized single-cell protein for biomaterial application to the best knowledge of the authors. As a result of the biofuel activity in the last decade, there is a huge increment in residual plant protein production that has applications merely as low-value animal feed. Conversely, the emergence of bovine spongiform encephalopathy (mad cow disease) and government protocols in many countries, as a result, to prevent the disease-causing agents (prions) entry into the food/feed chain, generated a surplus of animal protein-rich biomass (such as blood meal, meat and bone meal, or specified risk materials) (Mekonnen et al., 2014a,c). Such biomass could be utilized for the production of concentrated animal proteins that have applications for biomaterial development including bioplastics and biocomposites (Mekonnen et al., 2014a,c). The physicochemical properties of proteins, such as charge, solubility, and chemical reactivity, are dependent on the chemical nature of the side-chain group. Amino acid side chains may be conveniently classified as aliphatic or aromatic; polar or nonpolar; acidic, basic, or polar uncharged (Reddy and Yang, 2011). Generally, the physicochemical and functional properties of proteins vary depending on the amino acid composition and sequence, and the composition and sequence themselves vary with the resources (Table 8.2).

Table 8.1 Principal components of common structural biological composites

Biological composite	Mineral			Organic				References
	Calcium carbonate	Calcium phosphate	Hydroxyapatite	Keratin	Collagen	Chitin	Other	
Bones	–	–	Yes	–	Yes	–	–	Meyers et al. (2008)
Horns	–	Yes	–	Yes	–	–	–	McKittrick et al. (2010)
Hoofs	–	Yes	–	Yes	–	–	–	McKittrick et al. (2010)
Teeth	–	–	Yes	–	–	–	Yes	McKittrick et al. (2010)
Shells	Yes	–	–	–	–	–	Yes	Meyers et al. (2008)
Tusks	–	–	Yes	–	–	–	Yes	Meyers et al. (2008)
Insect exoskeleton	–	–	–	–	–	Yes	Yes	Gunderson and Schiavone (1989)
Chicken eggshell	Yes	–	–	–	Yes	–	Yes	Hassan et al. (2013)

Table 8.2 **Selected properties of proteins derived from plant and animal resources**

	Plant proteins			Animal proteins		
Property	**Soy protein**	**Wheat gluten**	**Zein**	**Keratin**	**Collagen**	**MBM extract**
Average molecular weight (kDa)	26–120	25–130	17–44	20–64	300	60–250
Amino acids						
Polar	25.6	62	46.6	42	37.3	52
Nonpolar	74.4	38	53.4	58	62.7	48
Isoelectric pH	4.7	6	6.2	4.3	4.7	4.5
Solubility						
Water	Yes	No	No	No	Yes	Yes
Ethanol	Yes	No	Yes	No	Yes	No
Organic solvents	Yes	No	Yes	No	Yes	No
Moisture resistance	Low	Low	Average	High	Low	Low
Biomaterials						
Films	Yes	Yes	Yes	Yes	Yes	Yes
Plastic sheet	Yes	Yes	Yes	Yes	Yes	Yes
Composites	Yes	Yes	Yes	Yes	Yes	Yes
Processibility						
Thermoplastic	Yes	Yes	Yes	Yes	Yes	Yes
Thermoset	Yes	Yes	Yes	Yes	Yes	Yes
References	Reddy et al. (2010) and Reddy and Yang (2011)	Rombouts et al. (2009) and Reddy and Yang (2011)	Shukla and Cheryan (2001), Kim (2008), and Karthikeyan et al. (2012)	Yin et al. (2013) and Ullah and Wu (2013)	Sommer and Kunz (2012) and Reddy and Yang (2011)	Garcia and Phillips (2009) and Mekonnen et al. (2013d)

8.2.1 Plant proteins

Proteins exist in plants to provide structure and biological activity (Hernandez-Izquierdo and Krochta, 2008). Plant proteins are highly sought after as a feedstock for biomaterials, mainly because they are widely available, biodegradable, and can be made into fibers, films, hydrogels and micro- and nanoparticles for biomaterial applications. Plant proteins are generated as coproducts when cereal grains are processed for food or fuel applications. Zein and corn gluten, soy proteins and wheat proteins (gluten, gliadin, and glutenin) are among the major plant proteins highly studied for biomaterial applications. In addition, canola, cotton seed, peanuts, sorghum, millets, barley, and other cereal grains contain proteins, albeit to a lesser extent. Compared to structural animal proteins such as collagen and silk, plant proteins have lower molecular weights and have higher net charges (Reddy and Yang, 2011) as shown in Table 8.2. Also, most plant proteins have higher moisture sensitivity and lower mechanical strength than structural animal proteins.

Nonetheless, for applications that do not require exposure to high moisture or mechanical strength, plant proteins can be utilized owing to their abundance, low cost, and less potential to be immunogenic compared with animal proteins such as bovine collagen (Reddy and Yang, 2011). Plant proteins that are processed into a fiber for tissue engineering and other biomedical applications were also shown to have better moisture resistance and mechanical strength than films as a biomaterial (Reddy and Yang, 2007, 2011; Reddy et al., 2008). Electospinning of soy protein in combination with poly(ethylene oxide) were shown to form quality fibers and nanofiber membranes with superhydrophilicity and other interesting properties (Xu et al., 2012).

8.2.2 Animal proteins

Animal protein biomasses such as blood meal (Bier et al., 2013; Verbeek et al., 2011; Verbeek and van den Berg, 2011), meat and bone meal (MBM) (Park et al., 2000; Lukubira and Ogale, 2013), MBM from specified risk materials (El-Thaher et al., 2013b; Mekonnen et al., 2013b), collagen (Chen and Shana, 2010; Langmaier et al., 2008), chicken feather keratin (Ullah et al., 2011; Ullah and Wu, 2013; Jin et al., 2011b), and casein (Dangaran et al., 2009; Sohail et al., 2006) are reported in the literature for biomaterial applications including bioplastics and biocomposites.

Natural animal proteins such as collagen, fibrin, fibronectin, elastin, silk, and keratin are also widely used in tissue engineering, because of their good compatibility, biodegradability, and minimal immunological reactions compared with other synthetic materials (Gomes et al., 2012). Recombinant DNA technology approaches for the production of structural proteins with tuneable properties that can be used for tissue engineering and biomedical applications have also attracted a significant research effort more recently (Gomes et al., 2012; George and Ravindran, 2010). Successful electrospinning of animal proteins (gelatine, collagen, alpha-elastin, and tropoelastin) for generating tissue-engineered scaffolds were described. Also, the scaffolds made as a result were shown to support attachment and growth of human embryonic cells (Li et al., 2005). Selected properties of proteins derived from plant and animal resources for biomaterial application are shown in Table 8.2.

8.2.3 Plasticized proteins

Proteins are interesting biomaterials that confer a wide range of functional and film-forming properties as a function of various extrinsic or intrinsic conditions such as plasticizer type and concentration during processing. Plasticization of protein is usually conducted to improve processability, reduce brittleness, and modify the properties of the final structure (Hernandez-Izquierdo and Krochta, 2008). Plasticizers added to protein resins usually consist of low molecular weight, low volatility substances, which mainly compete for hydrogen bonding and electrostatic interactions with other protein chains to produce swelling (Sothornvit and Krochta, 2001; Mekonnen et al., 2013c). Common plasticizers of proteins that are extensively reported in the literature for protein-based material application include water, glycerol, ethylene glycerol, propylene glycerol, palmitic acid, diethyl tartarate, dibutyl tartarate, 1,2-butanediol, 1,3-butanediol, polyethylene glycol (PEG), sorbitol, and mantitoletc (Mekonnen et al., 2013c; Tummala et al., 2006; Cao et al., 2009).

The composition, size, and shape of plasticizers influences the mechanical, physical, thermal, moisture permeability, and aging behavior of proteinaceous plastics (Cao et al., 2009; Orliac et al., 2003). Cao et al. (2009) compared the plasticizing effect of PEG having different molecular sizes (300–20,000 g/mol) on gelatine films. The result showed that PEG plasticizers with lower molecular weight gave better plasticizing effect: higher elongation, lower water vapor permeability, and better visual effect. In contrast, higher molecular weight PEG resulted in an increase in the tensile strength and elastic modulus and a decrease in the elongation of gelatine films.

Polar functional groups such as =OH along the plasticizer chains are believed to develop protein–plasticizer hydrogen bonds replacing the protein–protein interactions in biopolymer films (Yang and Paulson, 2000). Thus, hydrogen bonding ability of plasticizers, such as PEGs, are expected to be affected by factors such as the number of hydroxyl groups per mole, molecular size, solubility, and polarity that will explain the observed variation. Recent studies by Ullah coworkers (Ullah et al., 2011; Ullah and Wu, 2013) also demonstrated that the variation in hydrogen-bonding interactions between plasticizers (glycerol, diethyl tartarate, propylene glycol, and diethyl tartarate) and keratin protein from poultry feather quills and poultry feather fiber could be responsible for the variation in plasticization efficacy of the plasticizers and the performance of the biomaterial as a result.

8.3 Protein biocomposites

This section discusses protein-based biocomposites, including protein–polymer blend, particle-reinforced protein biocomposites, and fiber-reinforced protein biocomposites. Native proteins are often quite stiff and brittle due to extensive intermolecular interactions between protein chains through hydrogen bonding, electrostatic forces, hydrophobic bonding, and disulphide cross-linking. As such, modifications are imperative to utilize proteins as a polymeric biomaterial. The most commonly utilized techniques for polymer modifications besides denaturation and plasticization are blending, grafting, cross-linking, and composite formation, which are all multicomponent systems.

Thermoplastic processing of proteins usually involves denaturation of the proteins to break down interactions that may impede processibility using reducing agents such as sodium sulphite and blending plasticizers during processing such as extrusion or injection molding (Verbeek and van den Berg, 2010; Hernandez-Izquierdo and Krochta, 2008). Conversely, thermosetting protein plastics processing takes place through chemically cross-linking that involves the formation of covalent bond bridges between proteins chains by using cross-linking agent. The cross-linkers chiefly target the reaction between themselves and protein functional groups—such as primary amines, carboxyl, hydroxyl, and sulfhydryl of amino acid residues—to provide mechanical strength and moisture resistance. Usually proteins processed as such are used as a matrix for reinforced composite application, or as a coblend with other polymers for biomaterial applications.

8.3.1 Protein–polymer blend biocomposites

Protein-based composite biomaterials have been actively pursued as they can encompass a range of physical properties to accommodate a broader spectrum of functional requirements. Blending is often an efficient and economic means to address many of the problems protein-based biomaterials face, including limited moisture resistance, fast degradation, and unsatisfactory performance, particularly in wet environment. Blending of polymers could offer the following advantages: providing materials with a full set of desired properties at low cost; extending the engineering resins' performance by capitalizing on the maximum possible performance of the blend; improving specific performances such as impact strength or solvent resistance; improving processibility, product uniformity, and scrap reduction, and so on (Yu et al., 2006). Overall, effective blends can provide advantage over the individual polymers in terms of processibility, performance, and cost.

Several techniques are available for developing protein–polymer blends. Melt process blending and aqueous blending are among these prominent blending techniques. While melt process blending of protein involves mixing of protein with another polymer to an adequate dispersion by using thermal processing, solvent blending encompasses solubilization of the protein and the polymer to be blended in a common solvent and mixing the solubilized dispersions together before further processing such as casting. Blends can be a mixture of two or more polymer chains that are not covalently bonded together but have interaction as a result of ion–ion interaction, hydrogen bonding, chemical interaction, other types of interactions, or a combination (John et al., 2002).

The blending of proteins with biodegradable thermoplastic polymers is particularly interesting because the resulting blends usually retain their biodegradability. Some of the studied potential polymers that could retain protein's biodegradability and improve performance include poly(butylene adipate-co-therephthalate) (PBAT) (Chen and Zhang, 2010), polyhydroxy(butyrate-co-valerate) (PHBV) (Zarrinbakhsh et al., 2013), poly(butylene succinate) (PBS) (Reddy et al., 2012b), polycaprolactone (PCL) (Zhong and Sun, 2001), poly(lactic acid) (Calabria et al., 2012; Liu et al., 2011; Zhang et al., 2006; Zhu et al., 2012), or combinations. The blending of less expensive natural macromolecules like protein with biodegradable and biobased polymers could also contribute to improve their cost competitiveness in the market. As an illustration, PLA synthesized by ring-opening polymerization of lactides obtained from fermentation of starch is still more expensive than its counterpart petroleum-based commodity plastics. An effective blending of PLA with less expensive proteins can make the blend cheaper than the pure PLA, while also improving the performance of both. The slow biodegradation rate of PLA and the too fast biodegradation of protein can also be compromised to an intermediate degradation.

Numerous studies have shown that the properties and performance of a polymer blend are largely determined by its interfacial interaction and phase morphology. In most cases, a compatibilizer or coupling agent is utilized to improve the interfacial adhesion of blends, which result in fine phase morphology and improved mechanical performance. Common compatibilizers for protein-polyesters include maleic anhydride, toluene diisocynate, glyceydyl methacrylate, methylene diphenyl diisocynate, poly(2-ethyl-2-oxazoline), and others. For instance, Chen and Zhang (2010) examined the phase morphology of soy protein concentrate and PBAT melt blended specimens with different compositions using atomic force microscopy (AFM) as shown in Figure 8.3, where the bright yellow is the SPC phase, and the dark brown area is the PBAT phase. In this figure, a and b reveals the percolated SPC phase structure in the blends having 50% and 30% SPC, respectively, and c shows the dispersed SPC phase in the matrix for the blend with 15% SPC. Also, the same study showed good phase interaction between the PBAT phase and the soy concentrate, which could be attributed to the compatibilization of the phases using 3 wt.% maleic anhydride.

It was recently reported that the addition of keratin to synthetic elastomers resulted in materials with good thermal and mechanical properties, decreased flammability, and enhanced resistance to thermo-oxidative aging (Janowska et al., 2013; Prochon et al., 2012). Wide ranges of protein–polymer blend mechanical performances are shown in Table 8.3. It can be seen from this table that the performance of the blend is very dependent on the characteristics of the polymer that is blended with the protein. For example, blending of protein with PLA resulted in a blend material that has high tensile strength and modulus, but limited elongation and impact strength, mainly

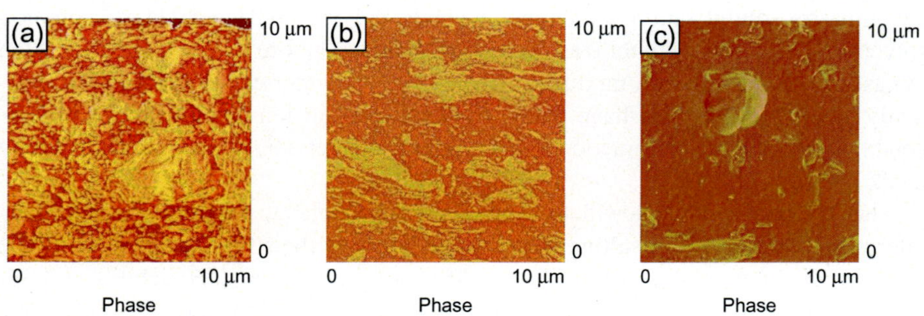

Figure 8.3 High-resolution AFM micrographs of PBAT/SPC blends containing 3 wt.% PBAT-g-MA on the basis of the total PBAT/SPC weight. (a) PBAT/SPC 50/50 (w/w). (b) PBAT/SPC 70/30 (w/w). (c) PBAT/SPC 85/15 (w/w).
Adapted from Chen and Zhang (2010), Elsevier, copyright © 2010.

Table 8.3 **Mechanical properties of protein blends**

Protein–polymer blend	TS (MPa)	E (MPa)	ε (%)	Impact strength (J/m)	References
SPI–PCL (50/50) (2 phr MDI)	23.5	2.4	12.5	–	Zhong and Sun (2001)
SPC–PLA (30/70) (2 phr MA)	52.2	4210	1.85	5.8	Zhu et al. (2012)
Soy protein[a]–PBAT (50/50) (3 phr maleic anhydride compatibilizer)	24.5	937	11.6	–	Chen and Zhang (2010)
Corn gluten[b]–PBAT	17	52	545	–	Reddy et al. (2014)
Plasticized soymeal[c] – PCL-PBS (30:35:35)	22		290	150	Reddy et al. (2012a)
Modified soymeal[d] – PBAT (40:60)	8.5	43	304	197.2	Zhou et al. (2013)

[a]Plasticized soy protein concentrate (SPC) – SPC (100 phr), sodium sulphite (0.5 phr), glycerol (10 phr), water (15 phr).
[b]Corn gluten was plasticized with 30 wt.% gluten and denatured with 12.5 phr urea.
[c]Plasticized soymeal:soymeal:glycerol:urea (57.5:30:12.5).
[d]Modified soymeal –70 wt.% soy meal, 30 wt.% glycerol, 10 phr water, 7.5 phr guanidine chlorine.

because of the PLA. Similarly, blending of proteins with PBAT resulted in a very ductile blend with elongation up to 197.5%, because of the characteristics of the PBAT (Zhou et al., 2013).

Conversely, protein–polymer conjugates are hybrid materials comprising natural polypeptides or proteins and synthetic polymer segments, which can benefit from the advantages of both components and avoid the disadvantages of each separate component. The advantage to be achieved includes the precise chemical structure and diverse functionalities of peptides/proteins and the stability and processibility of synthetic polymers to generate hybrid materials with potential for several applications by which the single component was not able to achieve (Dube et al., 2011; Shu et al., 2013; Jung et al., 2013). The disadvantages that could be avoided by forming conjugates includes the physical limitations of the peptide/protein (such as sensitivity to temperature, pH, organic solvents, and fast degradation) and possible limitations of the polymer (little reactivity, too slow degradation, etc.). Protein–protein blend composites can also be fabricated as films, gels, particles, and fibers for a diverse field of applications, as well as for electrical and optical devices (Hu et al., 2012; Cao et al., 2007). As such, protein–polymer conjugates received tremendous attention in the last few years.

8.3.2 Particle and fiber-reinforced protein biocomposites

Composites are materials formed with two or more components that are combined as a structural unit with one component as the continuous matrix and the other as the reinforcement or filler. Such materials often result in lightweight structures having high stiffness and tailored properties for specific applications, thereby reducing weight and saving energy (Mohanty et al., 2005c). Normally, the matrix is the material that holds the reinforcement together and has lower strength than that of the reinforcing material. In polymeric composites, the polymer in the form of either thermoplastic or thermoset acts as a matrix, and particles or fibers incorporated to improve its performance or reduce cost act as the reinforcement and filler phase, respectively. Biocomposites are composite materials where either the matrix, reinforcement, or both are partially or completely made of biobased resources.

While a wide range of proteins, including casein (Chen, 1995), corn zein (Kim et al., 2004), wheat (Micard et al., 2000), egg albumin (Gennadios et al., 1996), soy protein (Reddy et al., 2010), feather meal (Sharma et al., 2008), blood meal (Verbeek and Koppel, 2012), cotton seed protein (Yue et al., 2012), and MBM extract (Mekonnen et al., 2013b), have been investigated and were shown to produce bioplastics with fairly good properties, the modulus and strength are not high enough for load-bearing applications. Incorporation of reinforcing agents, in the form of particles or fibers could solve the limitations with regard to stiffness, rigidity, and load bearing as a result.

8.3.2.1 Particle-reinforced protein biocomposites

Reinforcement of polymers with particles can produce several desirable effects; for instance, an increase in stiffness, a reduction in the coefficient of thermal expansion, an improvement in creep resistance, and fracture toughness (Ahmed and Jones, 1990). The mechanical properties of reinforced composites results from a complex interplay between the properties of the resin, the filler, and the interfacial region. Also, it is evident that the mechanical properties of particulate-reinforced composites are affected by a number of parameters: the size, shape, aspect ratio, and distribution of the reinforcing particles (Fu et al., 2008). In the case of nonspherical particles, the degree of orientation with respect to the applied stress is also important (Ahmed and Jones, 1990). Reinforcing of various types of proteins with particles sized from millimeter to nanometer scale fibrils, whiskers, platelets, particulates, and so on have been known for a long time. For instance, microsized fillers (talc, alumina, calcium carbonate, etc.) were introduced to reinforce polymeric materials at reduced cost. Also, other particulates such as soybean husk, rice hull, wood flour, water bamboo husks, and others (Figure 8.4) can reinforce protein plastics. However, the high moisture absorption and the insufficient strength of the resulting materials limited their extensive applications.

More recently, protein polymeric matrices reinforced with nanosize particles (<100 nm) have attracted considerable attention. This novel class of composites emerges as so-called protein-nanocomposites. Successful reinforcement of various types of protein bioplastics and films with nanoparticles such as nanoclay, chitin nanowhiskers, cellulose nanocrystals, and carbon nanotubes (Chen and Zhang, 2006; Zheng et al., 2007; Lu et al., 2004; Li et al., 2014) over a range of loading percentage can be achieved through either compounding or casting methods. In general, an improvement in tensile strength and modulus can be achieved with an increase in loading of fillers until an optimum or climax loading point, where there is no more increment or even reduction of the tensile properties is observed. The increment in tensile

Figure 8.4 Macro photographs and bright field optical micrographs of fibers that illustrate the distribution of fiber diameter and length (inset micrographs are fiber silhouettes demonstrating the typical range in fiber diameter).
Adapted from Robertson et al. (2013), Wiley, copyright © 2013.

strength is associated with the high stiffness and strength of these fillers. Moreover, nanofillers have the advantage of having very minimal internal defect owing to their small diameter resulting in a high strength.

Composites reinforced with particles above the optimum loading point usually exhibit reduction in mechanical performance. This phenomenon is associated with poor dispersion of the particles in the matrix, thereby reducing effective stress transfer between the protein matrix and fillers. Figure 8.5 shows SEM images of the fractured surface of (a) pure soy plastic sheet and soy nanocomposite reinforced with (b) 5 wt.%, (c) 15 wt.%, and (d) 30 wt.% chitin nanowhiskers. It is evident from this

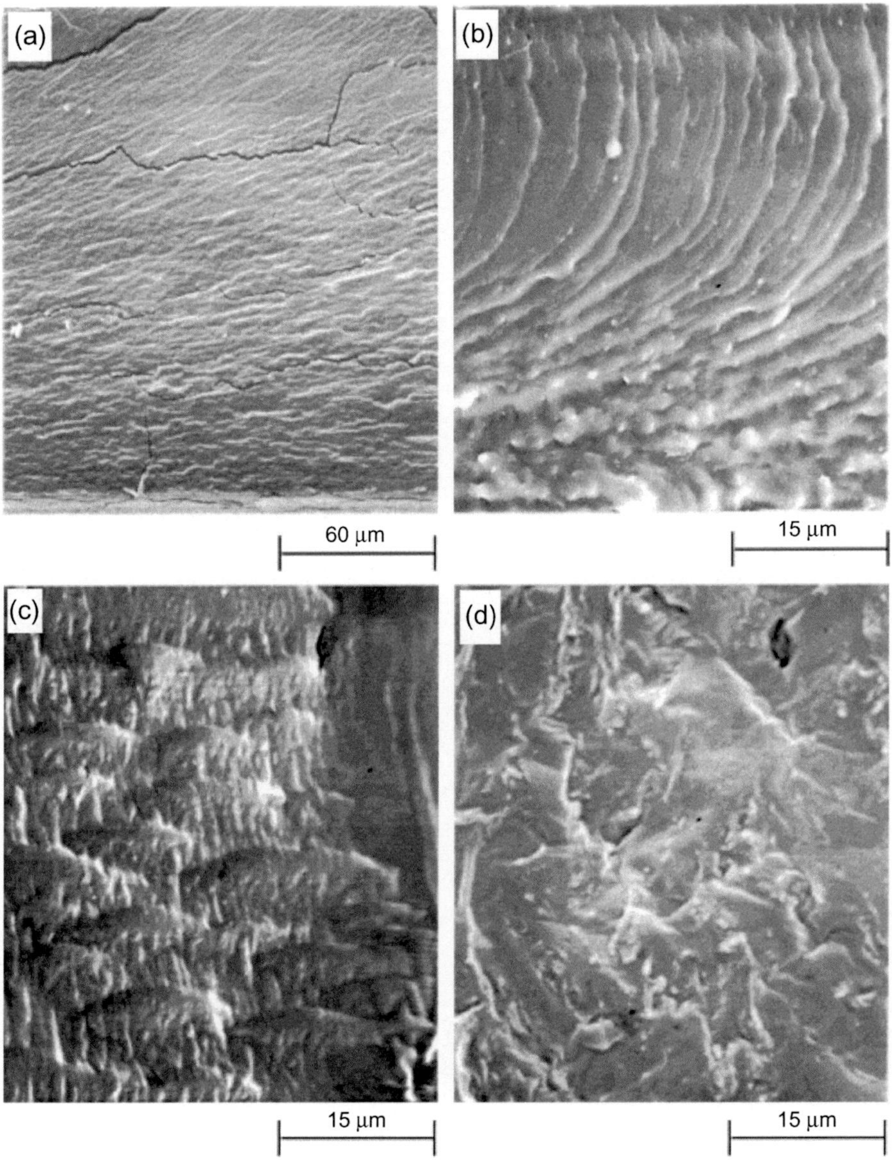

Figure 8.5 SEM images of glycerol plasticized soy protein isolate matrix: (a) no reinforcement; (b) reinforced with 5 wt.% chitin whiskers; (c) reinforced with 15 wt.% chitin whiskers; and (d) reinforced with 30 wt.% chitin whiskers.
Adapted from Lu et al. (2004), ACS, Copyright © 2004.

figure that the pure sheet exhibits a relatively uniform surface. A relatively uniform distribution of the chitin whisker in the soy protein matrix can also be observed when the chitin content is lower than 15 wt.%. However, as the chitin loading increases, the resulting nanocomposites show agglomerates of whiskers.

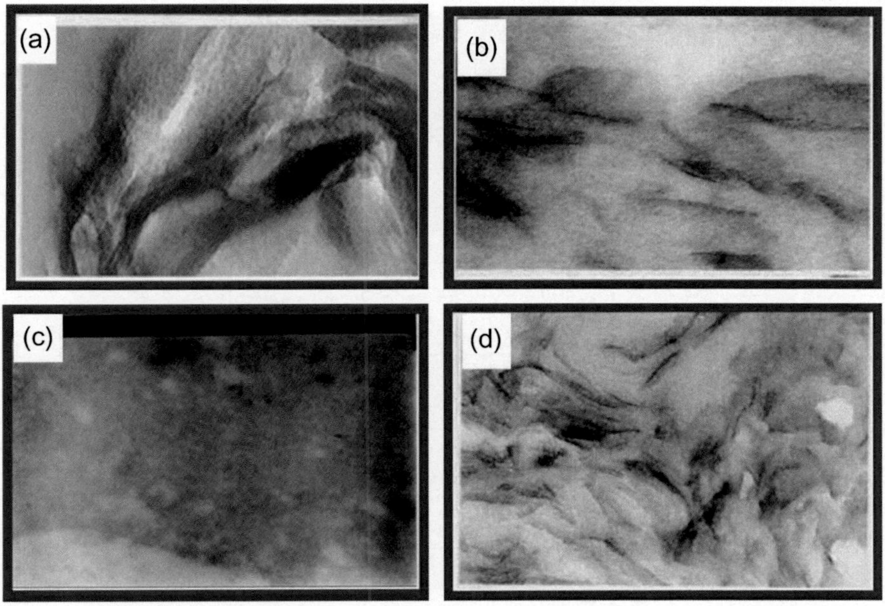

Figure 8.6 Transmission electron microscopy of (a) clay (pure montmorillonite) and nanocomposite films with (b) 0, (c) 5, and (d) 10 g of montomorilline per 100 g of soy-protein isolate at 200,000 magnification.
Adapted from Echeverria et al. (2014), Elsevier, Copyright © 2014.

A transmission electron microscopic (TEM) study of clay (montmorillonite) dispersion in soy protein isolate was also shown to have an excellent dispersion throughout the matrix up to a loading of 15 g per 100 g as shown in Figure 8.6. The TEM study also showed that tactoid ordering of the MMT molecules have been markedly displaced by the soy proteins, despite preservation of its ordering. Thus, the high degree of intercalation and consequent exfoliation of the two components observed could be the reason for a mechanical property improvement, resistance to water, and heat degradation observations of soy nanocomposites (Echeverria et al., 2014).

In general, the success of reinforcing particulates or nanofillers in improving the performance of protein-based materials heavily depend on the interactions that exist between the proteins and the fillers. For this, the thorough dispersion of the particles throughout the protein matrix and higher degree of intercalation by the protein components to the point of exfoliation is important. Also, the influence of nanoparticles on the complex protein–protein interaction, and protein–plasticizer interaction can influence the efficacy of improving the mechanical, thermal, or water resistance performance of the protein-based nanocomposite (Echeverria et al., 2014). Change in tensile strength, modulus, and elongation of the protein bioplastic with nanofiller reinforcement (such as montmorillonite, chitin nanowhisker, cellulose nanocrystals) is reviewed and presented in Table 8.4.

The use of cellulose whiskers has also been reported to reinforce proteins while preserving their biodegradability. Wang et al. (2006) have illustrated that the

Table 8.4 Comparison of mechanical properties of particle-reinforced proteins

Protein	Reinforcing agent	Tensile strength (MPa)	Modulus (MPa)	ε (%)	References
Soy protein isolate	Montmorillonite (20 wt.%)	14.5	580	4.5	Chen and Zhang (2006)
	Montmorillonite (16 wt.%)	15.4	530	7	Chen and Zhang (2006)
	Montmorillonite (4 wt.%)	10.5	265	63	Chen and Zhang (2006)
	Pure soy protein isolate	8.77	180	87	Chen and Zhang (2006)
Soy protein isolate	Chitin nanowhiskers (20 wt.%)	8.4	158	35	Lu et al. (2004)
	Chitin nanowhiskers (15 wt.%)	5.7	44	101	Lu et al. (2004)
	Chitin nanowhiskers (5 wt.%)	3.8	33	132	Lu et al. (2004)
	Pure soy protein isolate	3.3	26.4	205	Lu et al. (2004)
Collagen film	Cellulose nanocrystals (CNC) (10 wt.%)	11.5	760	22	Li et al. (2014)
	Cellulose nanocrystals (CNC) (5 wt.%)	24.2	905	32.5	Li et al. (2014)
	Pure collagen	10	700	6	Li et al. (2014)

reinforcement of soy protein with a suspension of cellulose nanowhiskers having an average length of 1.2 μm and diameter about 90 nm. Results of the same study (Wang et al., 2006) showed that the reinforcement of soy protein isolate by the whiskers have enhanced the tensile strength and modulus by 202% and 40%, respectively, with a whisker content of 20 wt.%. Moreover, an improvement in water resistance of the composite was observed that might be attributed to cross-linking networks caused by intermolecular hydrogen bonds between the cellulose whiskers and the SPI matrix.

8.3.2.2 Fiber-reinforced protein biocomposites

Natural fibers such as hemp, flax, jute, sisal, kenaf, banana fiber, miscanthus, henequen, bamboo, pineapple leaf fiber, grass, corn stalk, rice straw, and wheat straw have been used as reinforcing fiber to produce protein biocomposites. Also, various grades of fiberglass for reinforcing proteins are reported in the literature (Mekonnen et al., 2013a). The natural fibers impart several advantages to the protein composite including low cost, low density, acceptable mechanical properties, sustainability, and biodegradability (Mohanty et al., 2005b); more advantages than fiberglass or other synthetic fiber reinforcements. The reduction in fiber density and, thus, composite density is favorable in certain engineering applications. For instance, biocomposites for automotive parts where vehicle weight and fuel economy are often a concern, lower density is very desirable. Moreover, glass fibers are abrasive to tooling and can cause irritation and discomfort to personnel during manufacturing. Thus, natural fibers such as hemp and flax fibers can potentially compete with E-glass fibers, which serve as a reference because of their great importance in composite technology (Cheung et al., 2009; Fowler et al., 2006).

The mechanical performance of fiber-reinforced protein composites is an intrinsic property of the constituents; that is, the reinforcing fiber used, the formulation and nature of the matrix, and the interface between the matrix and matrix. Natural fiber-reinforced proteins exhibit significant improvement in tensile, flexural, and impact properties of protein-based plastics. For instance, in the comparison table (Table 8.5), ramie fiber at 0.45 weight fraction loading has increased the tensile modulus and flexural strength of 30% glycerol plasticized soy protein plastics by 260% and 285%, respectively (Lodha and Netravali, 2005). In another study, the addition of 0.35% weight fraction banana fiber to 50% glycerol plasticized soy protein plastic has shown a 606% increase in tensile modulus, while it's tensile strength reduced slightly (Kumar et al., 2008). Hemp fiber reinforcement of protein plastic is also very popular, mainly because of its good mechanical properties and its commercial availability. Mohanty et al. (2005a) has showed that the incorporation of only 15% hemp in soy protein–poly(ester amide) blend has increased the tensile strength, tensile modulus, and flexural strength by 148%, 831%, and 295%, respectively.

Mekonnen et al. (2013a) showed the wide disparity of mechanical performance of animal protein-based composites with a variation in matrix composition and fiber types. However, in most studies the modulus of all protein composites has increased with the use of either natural fiber or synthetic glass fibers, irrespective of the matrix type. Some of these studies showed a tensile strength reduction of protein and other composites with the incorporation of natural fibers (Mekonnen et al., 2013a; Sawpan et al., 2012; Karnani et al., 1997).

Table 8.5 Mechanical properties of plant and animal protein composites with different reinforcing fibers

Protein source	Tensile strength (MPa)	Tensile modulus (MPa)	Flexural strength (MPa)	References
SPI + 30% glycerol resin	–	124.7 ± 8.1	6.1 ± 0.5	Lodha and Netravali (2005)
SPI + 30% glycerol + Ramie fiber (0.45 weight fraction of fiber)	–	3420 ± 4.4	180.2 ± 33.7	Lodha and Netravali (2002)
SPI + 50% glycerol	1.78 ± 0.1	39.5 ± 3.0	–	Kumar et al. (2008)
SPI + 50% glycerol + Banana fiber (0.35 weight fraction of fiber)	1.55 ± 0.10	279 ± 8.1	–	Gupta and Kumar (2007)
Soy flour + 30% glycerol + poly(ester amide) (plasticized soy: poly(ester amide)—2:1 ratio)	8.0 ± 0.1	2.2 ± 0.0	8.1 ± 0.0	Mohanty et al. (2005a)
Soy flour + 30% glycerol + poly(ester amide) (plasticized soy: poly(ester amide)—2:1 ratio) + 15% hemp	18.0 ± 0.1	14.9 ± 0.1	26.5 ± 0.1	Mohanty et al. (2005a)
Soy flour + 30% glycerol + poly(ester amide) (plasticized soy: poly(ester amide)—2:1 ratio) + 30% hemp	19.9 ± 0.1	20.5 ± 0.2	32.0 ± 0.2	Mohanty et al. (2005a)
Hydrolyzed animal protein (20%) + Epoxy resin (80%)	73.8 ± 4.9	1300 ± 110	73.1 ± 5.1	Mekonnen et al. (2013a)
Hydrolyzed animal protein (20%) + Epoxy resin (80%) + 20% (v/v) chopped strand fiberglass mat	102.4 ± 24.4	8800 ± 270	188.9 ± 12.7	Mekonnen et al. (2013a)
Hydrolyzed animal protein (20%) + Epoxy resin (80%) + 20% (v/v) hemp mat	14.1 ± 3.65	3930 ± 70	32.5 ± 3.5	Mekonnen et al. (2013a)
Hydrolyzed animal protein (30%) + Epoxy resin (70%)	58.6 ± 6.8	1500 ± 60	66.3 ± 4.7	Mekonnen et al. (2013a)
Hydrolyzed animal protein (30%) + Epoxy resin (70%) + 20% (v/v) chopped strand fiberglass mat	97.3 ± 12.6	9000 ± 59	180.3 ± 17.9	Mekonnen et al. (2013a)
Hydrolyzed animal protein (30%) + Epoxy resin (70%) + 20% (v/v) Hemp mat	11.94 ± 2.5	4050 ± 60	34.5 ± 10.4	Mekonnen et al. (2013a)

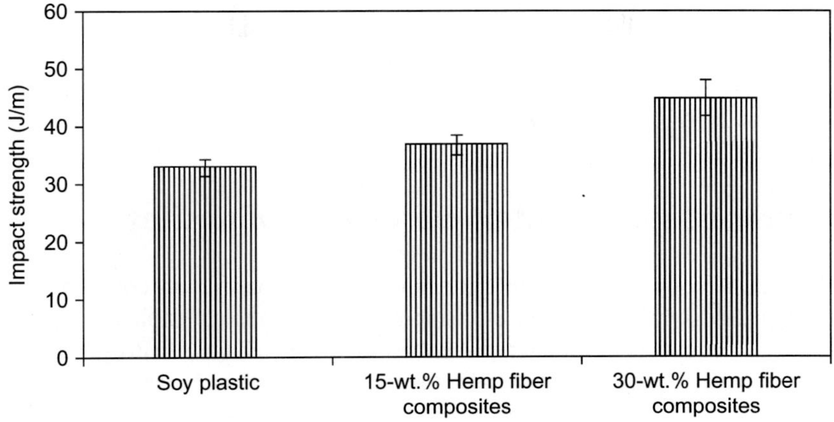

Figure 8.7 Impact strength of hemp fiber-reinforced soy-based green composites. Adapted from Mohanty et al. (2005a), Springer, Copyright © 2005.

The effect of fiber reinforcement on the impact strength and elongation at break of composites is also important. Impact strength of composite materials, in general, is one of the most challenging parameters to maintain while reinforcing polymeric matrices. While fibers or other reinforcement efforts improve tensile modulus and strength, it usually reduces the ductility/elongation of polymers that, in effect, hampers plastic deformation by restricting polymer chain movements and, as such, reduces the impact strength. In another case, fibers could increase impact strength of composites due to the fact that fibers reduce crack propagation rate by forcing crack around the fiber rather than the matrix and bridging the crack through the pull out (Crosby and Drye, 1987). For example, the improvement in notched Izod impact strength of soy protein plastics is presented in Figure 8.7. A 13% and 40% increase in impact strength with 15% and 30% hemp fiber loading, respectively, was observed in this study (Mohanty et al., 2005a).

On the top of mechanical performance, thermal stability of protein-based biocomposites is crucial not only to the design and development of end user applications, but also to select proper processing temperature such as curing temperature in the case of thermosets, and extrusion and injection molding temperature in thermoplastic processing of these biocomposites. Heat deflection temperature (HDT) is one of the common measurements of the ability of polymeric materials to withstand deformation at elevated temperatures under the application of load. With the incorporation of rigid or stiff reinforcement, the HDT of proteins biocomposites is expected to increase. About 23 and 35 °C increase in HDT was reported for a soy protein plastic reinforced with 15 and 30 wt.% hemp reinforcement, respectively (Mohanty et al., 2005a).

Thermogravimetric study of protein matrices and natural fiber-reinforced protein composites showed that the thermal stability or degradation temperatures vary based on various factors. Dastidar and Netravali (2012) showed that cross-linking of proteins can improve thermal stability. Mekonnen et al. (2013a) reported that amid the thermal resistance of protein composite is not of great concern for end user application, the degradation point can change based on the reinforcing agent used. For example, while the use of fiberglass improved the 5% degradation point of animal protein-based composite by 15%, reinforcing the same matrix by hemp fiber reduced it by 13.5%.

8.4 Processing of protein-based biocomposites

Protein-based bioplastics have inferior mechanical properties in comparison with commonly used synthetic polymers, principally due to their complex composition, extensive intermolecular interactions between protein chains, hydrophilic nature, impurities, and high environmental sensitivity. Thus, most protein biomaterial processing techniques target these issues via various modification approaches and processability optimization, including protein denaturation, plasticization, grafting, cross-linking, and others. Also, most protein processing involves thermal/heat treatments that disintegrate, denature, dissociate, unravel, and align them for further modifications.

Thermoplastics in general are materials that soften when heated and can then be formed while soft and cooled to form a rigid shape. In thermoplastics, the softening and forming processes are reversible. Conversely, thermosetting materials are materials that are formed while warm and set in the formed shape. Reheating and cooling do not allow thermosetting materials go soft again. Proteins, as matrices of biocomposites, can be processed through both thermoplastic and thermosetting processing techniques. Process selection for protein-based composites should also consider the size, shape, and other characteristics of the reinforcement/filler, beside the proteinaceous matrix. Moreover, the techniques used for the processing and manufacture of biocomposites are based largely on existing techniques for processing of synthetic composites. Some of these common techniques include hand layup, filament winding, pultrusion, extrusion, injection molding, compression molding, resin transfer molding, and sheet molding compounding (Fowler et al., 2006).

8.4.1 *Thermoplastic processing*

Thermoplastic processing of protein matrices for protein composites is accomplished by thermal processing of proteins in the presence of denaturants and plasticizers. The fundamental of thermoplastic processing is achieving a flowing protein, where protein molecule chains slide past each other and solidify to form a shape when the temperature is reduced. Denaturation of proteins is typically necessary to break the strong intermolecular hydrogen bonds, stacks of α-helices and β-sheets, and other interactions, and achieve a near-linear structure. Also, denaturation can unfold native globular proteins and, as a result, expose hidden functional groups for further modification. Denaturation of proteins can be induced by chemical or thermomechanical means. Chemicals such as concentrated aqueous solutions of inorganic or organic salts (urea, sodium dodecyl sulfate, sodium sulphite, guanidine hydrochloride, etc.) (Bhuyan, 2010; Li et al., 2010), fluorinated solvents, ionic liquids, or strong acids are known protein denaturants (Hardy and Scheibel, 2010). Plasticizers are also added to the protein matrices during thermoplastic processing to improve processibility, reduce brittleness, and modify the properties of the final structure (Mekonnen et al., 2013c).

In general, two types of thermoplastic processing are known for protein-based materials: wet processing and dry processing (Cuq et al., 1997; Verbeek and van den Berg, 2010). The wet process, or casting, involves dispersing and solubilizing the protein in large quantities of solvent followed by solvent removal by drying. The dry process involves mixing the proteins and appropriate additives, under low moisture conditions, followed by thermomechanical shaping using compression molding, extrusion, or injection molding. Commercial viability of protein-based plastics and composites is hinging on utilizing such common synthetic processing techniques (Verbeek and van den Berg, 2010).

During extrusion compounding of protein and reinforcing filler, the reinforcements are usually in the form of particulate or small fibers of dimension not more than a few centimeters. This is to achieve effective mixing of the protein matrix and reinforcing agents without causing jamming of the extruder barrels and to have good control over feeding of the extrusion process. Once the constituents have been thoroughly mixed, the compound can be either extruded directly to a final product, or palletized as a precursor to further extrusion or other processing such as injection molding or compression molding (Fowler et al., 2006). One of the major limitations of extrusion compounding is that only relatively short fibers can be used. If longer fibers are to be included, alternative methods may need to be employed. On the other hand, injection molding is the main process used by manufactures for making thermoplastic composite products, where complex shapes are needed in a cyclic, high-volume production. The advantages of using this process are the excellent dimensional tolerance and short cycle times, coupled with few postprocessing operations.

8.4.2 Thermoset processing

In thermoset protein composites, fibers or other reinforcing fillers are impregnated with thermoset protein resins and then exposed to high temperatures for curing. For protein-based thermosetting biocomposite development, the protein matrix resin is first prepared by incorporation of cross-linking agents to unfolded proteins. Chemical or other cross-linking agents such as enzymes or UV irradiation can be utilized (Vaz et al., 2003) as cross-linking agents of proteins. Common chemical cross-linking agents of proteins include aldehydes (formaldehyde, glutaraldehyde, glyoxal) (Wang et al., 2007; El-Thaher et al., 2013a), maleic anhydride (Jang et al., 2011), epoxy resin (Mekonnen et al., 2013a), carbodiimide, hydroxysuccinimide (Kim et al., 2004; Tropini et al., 2000, 2004), and others. Precured protein thermoset resin and reinforcing fillers can then be processed or manufactured through hand layup process, press molding, filament winding, or pultrusion. One unique advantage of thermoset processing of biocomposites over thermoplastics is that it provides the opportunity to use longer and continuous fibers or fiber mats. On the contrary, the relatively low thermal resistance of natural fibers and proteins could limit the curing temperature requirements during thermoset processing.

8.5 Modification of proteins for biocomposites development

8.5.1 *Protein matrix modification*

Modification of protein matrix for better performance could be in the form of denaturating and destructurizing native proteins, plasticization for ease of processibility, blending with other polymers, or grafting it with either monomers or polymers for better interaction with the reinforcing agents. Also, the addition of an appropriate compatibilizer into protein-based composites could provide an efficient dispersion of proteins as a matrix, thus obtaining higher mechanical performances (Chen and Zhang, 2010). The stability of a native protein is a function of surrounding environmental variables such as pH, temperature, ionic strength, and solvent composition as they dislocate various intramolecular bonds responsible for stability and integration of the protein (Dubey and Jagannadham, 2003). Systematic change of any of these parameters could lead to denaturation or destructurization of proteins.

Protein denaturation is the unfolding of the protein from a structure native state into an unstructured state with no or little fixed residual structure, close to a random coil (De Graaf, 2000). Protein denaturation induced by chemical modification could disrupt its secondary, tertiary, or quaternary structures, exposing the amino acid side chains, thereby introducing new interactions by means of hydrogen or ionic bonding (Yue et al., 2012). Urea, in combination with some water, is commonly used as a chemical denaturant for the preparation of protein-based bioplastics as it can destabilize the globular protein by forming strong hydrogen bonds with water molecules that surround the protein. This can disrupt the protein–hydrogen bonds, resulting in flexible peptide chains. In this case, the denatured protein would facilitate the formation of entanglements and cross-linked structures during the chemical cross-linking process, resulting in plastics with high tensile strength and water absorption resistance. The denaturation temperature of the protein strongly depends on the water content up to a water content of about 10–20%. Generally, protein denaturation temperature was shown to decrease from 200 to 120 °C down to 80 °C between 0% and 20% w/w water (Luescher et al., 1974).

Plasticizers of proteins are generally added to the protein matrix during thermoplastic processing such as extrusion or injection molding to improve processibility, reduce crystallinity and brittleness, impart flexibility, lower glass transition temperature and melting temperature, and modify the properties of the final structure (Hernandez-Izquierdo and Krochta, 2008; Sothornvit and Krochta, 2005). While most synthetic plastics have a glass transition temperature below 100 °C, proteins with less than 5% water show glass transition temperatures close to or above their decomposition temperatures (Tolstoguzov, 1993). Thus, the addition of plasticizers that allows processing at lower temperature and avoids thermal degradation is necessary. Common plasticizers for proteins for biomaterial application includes diethyl tartrate, glycerol, ethylene glycol, propylene glycol, oleic acid, sorbitol, mannitol (Mekonnen et al., 2013c). While blending of proteins with other polymers is a physical mixing of the two to obtain the requisite properties, grafting and cross-linking are among the major irreversible methods of chemically modifying protein properties.

8.5.2 Compatibilization and interface modification

Polymer blends could be miscible or immiscible. Miscible blends form solutions and there is no phase among the ingredient polymers. On the contrary, immiscible blends are characterized by two or more phases that are separated by interfaces (White and Bumm, 2011). Most pairs of polymers including blends of protein with polymers are immiscible with each other. Numerous studies have shown that the properties of a polymer blend are largely determined by its phase morphology. Additives such as compatibilizers are often added in small quantity to improve the compatibility of such immiscible polymer blends and, thus, improve the morphology and resulting performance of the blend. Similarly, it is often challenging to disperse particulate fillers or wet fibers effectively in a polymeric matrix of a composite. Thus, the use of compatibilizers or other interface modifiers is crucial to obtain optimum performance and cost characteristics.

Compatibilizers usually are macromolecular species exhibiting interfacial activities in heterogeneous polymer blends. Commonly, the chains of a compatibilizer have a block and graft copolymers structure or chemical reactive species with one constitutive block miscible with one blend component and a second block miscible with the other blend component (Brown, 1992; Koning et al., 1998). These block structures can be premade and added to the immiscible polymer blend, or they can also be generated in situ during the blending process. The in situ compatibilization can be achieved by reactive extrusion, where mutual reactivity of both blend components may occur. Reactive extrusion for such application is very attractive because it can act as a solvent-free continuous reactor that is able to process highly viscous materials.

Chemical modification of proteins with monomers using reactive extrusion technology is a well-studied approach. For instance, Ly et al. (1998) grafted vinyl monomers including methyl methacrylate, ethyl acrylate, butyl acrylate, and hexyl acrylate to soy protein using a free radical mechanism to form vinyl polymer. Also, Jin et al. (2011a, b) have recently demonstrated the graft polymerization of native chicken feather proteins with methyl acrylate for thermoplastic applications. As a result of this graft polymerization, the poor thermoplasticity of native chicken feathers has been improved, and further ductility with plasticization was observed. Other compatibilizers for protein–polymer blend and their effect are shown in Table 8.6.

Zarrinbakhsh et al. (2011) has shown the compatibilization between PHBV–PBS matrix and protein containing DDGS (about 35.76%) using pMDI. Figure 8.8 (a and b) shows the comparison between these blends without compatibilization, where a clear gap between the DDGS and the PHBV–PBS blend is visible, and Figure 8.8 (c and d) that shows a pMDI compatibilized DDGS–PHBV–PBS blend where a smooth fracture surface with almost no pullout of DDGS particles was visible. A possibility of

Table 8.6 Common compatibilizing agents for protein–polymer blends and protein composites

Blend or composite	Compatibilizing agent	Note	References
Soy protein–PLA	Maleic anhydride	Significant increase in tensile strength and good interfacial adhesion	Zhu et al. (2012)
Soy protein–PBAT	Dicumyl peroxide initiator—Maleic anhydride	Fine domain sizes and good dispersion of SPC and, hence, improved tensile and dynamic–mechanical properties	Chen and Zhang (2010)
DDGS–PHBV–PBS	Methylene diisocynate (pMDI)	Improved adhesion between DDGS and PHBV–PBS matrix	Zarrinbakhsh et al. (2011)
Soy protein–PCL	Methylene diisocynate	Mechanical properties of the blend were improved	Zhong and Sun (2001)
Blood meal–LLDPE	Maleic anhydride	Better compatibility as evidenced by a more homogeneous distribution of plasticized blood meal	Marsilla and Verbeek (2013)
Zein–PCL	Hexamethylene diisocynate	Hinders crystallization of PCL and allows better interaction with zein	Wu et al. (2003)

Abbreviations: PLA, poly(lactic acid); PBAT, poly(butylene adipate-co-terephthalate), PCL, poly(caprolactone); PBS, poly(butylene succinate); LLDPE, linear low-density polyethylene.

covalent bond between the DDGS proteins and (PHBV–PBS) system was suggested as a result of the pMDI compatibilization by the same study. The proposed mechanism was also substantiated by improved tensile modulus, flexural strength, and flexural modulus observation, in addition to the scanning electron microscopy morphology study shown in Figure 8.8.

Treatment or surface modification of reinforcing fillers such as natural fibers to make them more compatible with protein matrices is the other well-studied avenue to improve the interface and compatibility of protein composites. For example, Kumar et al. (2008) treated banana fiber with alkali solution, and the treated fibers resulted in an improvement in tensile strength and modulus of the protein composite. The tensile strength and modulus of the protein composite made out of the treated fibers exhibited 82% and 963% increase in tensile strength and modulus, respectively, compared to soy protein film without fibers. The the improvement of the mechanical properties of treated fibers and the soy composites they make was attributed to the decrease in the lignin content on the fiber surfaces and, as such, increase in surface roughness (Kumar et al., 2008) that could have led to better interaction in the form of mechanical interlocking or hydrogen bonding between the banana fibers and the soy protein functional groups.

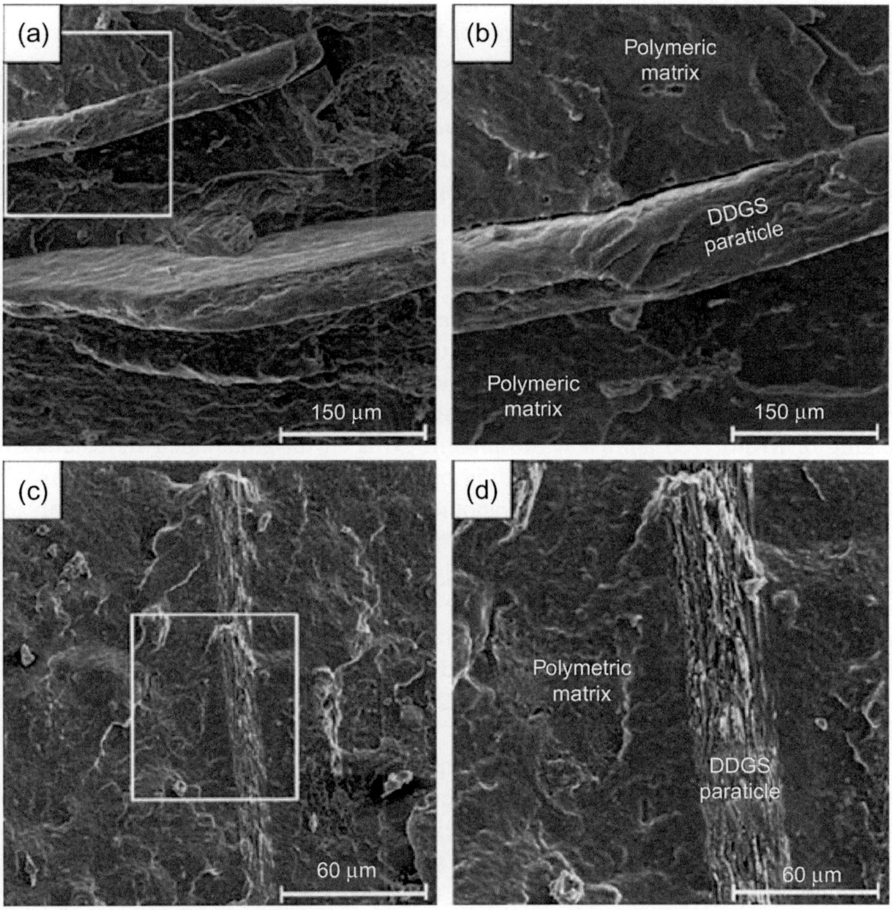

Figure 8.8 SEM micrographs of the interface between (a and b) DDGS–PHBV–PBS in uncompatibilized composite; (c and d) pMDI compatibilized composite.
Adapted from Zarrinbakhsh et al. (2011), Wiley Copyright © 2011.

8.6 Challenges and application

8.6.1 Water absorption performance of protein composites

While protein-based materials exhibit good barrier properties against oxygen, lipids, and aroma, water sensitivity in the form of poor barrier properties toward moisture and high moisture absorption constitutes major limitations to the industrial deployment of protein-based films, plastics, and composites. For example, Zheng et al. (2003) reported that soy protein sheets submerged in water for 20 h absorbed moisture up to 180% of their initial weight. The presence of hydrophilic protein functional groups, including $-NH_2$, −NH, −OH, −SH, and their tendency to associate with water is the molecular driver for this phenomenon. It is worthwhile mentioning here that protein-based biocomposites/materials have a high tendency to be washed or solubilized in extreme humid environments or when immersed in water. This in fact dramatically affects the mechanical properties of protein biocomposites. The diffusivity of water into protein composites (water absorption) as a result of its hydrophilicity of proteins functional groups triggers degradation of reinforcing agent–matrix interface and interfacial cracks. As a result, water-saturated protein biocomposites display inferior mechanical properties such as lower values of tensile modulus, flexural modulus, tensile strength, and flexural strength. For the case of natural fiber-reinforced protein biocomposites or other matrices with low fiber, water molecules could act as a plasticizing agent leading to a slightly higher value of tensile and flexural strain of the composite (Espert et al., 2004; Mekonnen et al., 2013a).

A possible solution to reduce the high moisture absorption and its deleterious effects could be the formation of covalent bond bridges between proteins hydrophilic moieties to form a thermoset using a cross-linking agent. For example, the cross-linking of hydrolyzed animal waste protein extract with glutaraldehyde resulted in only 5–10% weight loss of the resulting biobased plastic under 24h of water soaking (El-Thaher et al., 2013a). In an another study, the cross-linking of hydrolyzed waste animal proteins with epoxy functional groups of epoxy resins exhibited very low water absorption and solubility not only in water, but also in weak acid and weak base media (Mekonnen et al., 2013b). The reason for this improvement could be the formation of covalent bridge between amine (primary and secondary), hydroxyl, carboxyl, and sulfhydryl groups of hydrolyzed protein with epoxide functional groups of epoxy resin as shown in Figure 8.9 (Petrie, 2006; Ratna, 2003; El-Thaher et al., 2013b; Mekonnen et al., 2013b).

Zhong and Sun (2001) showed that blending of hydrophobic polymers to proteins could increase the water resistance of protein-based matrices. Indeed, the blending of 40wt.% poly(caprolactone) with soy protein, by using a 2wt.% MDI compatibilizer reduced the water absorption of soy plastic soaked for 26h from 74% to 36%. Increase of the poly(caprolactone) composition in the blend to 50wt.% further reduced the water absorption to 12%. Blocking some of the hydrophilic functional groups with hydrophobic functional groups could enhance the hydrophobicity of proteins for composite application. For example, esterification of protein amine groups with low molecular weight alcohols is known to reduce protein's water affinity (Fraenkeiconrat and Olcott, 1945). The use of nanoparticles as reinforcement in protein composites was also shown to lessen water absorption by restricting the chain mobility of protein molecules. Low chain mobility freedom in effect can reduce swelling, thereby rejecting further permeation of more water into the protein composites (Zheng et al., 2007).

8.6.2 Applications of protein-based composites

Protein-based composite biomaterials have been actively pursued as they can encompass a range of mechanical, thermal, and physical properties to accommodate a broader spectrum of functional requirements. By optimizing molecular interfaces between structural proteins, useful composite materials can be fabricated as films, gels, particles, and fibers, as well as for electrical and optical devices (Hu et al., 2012).

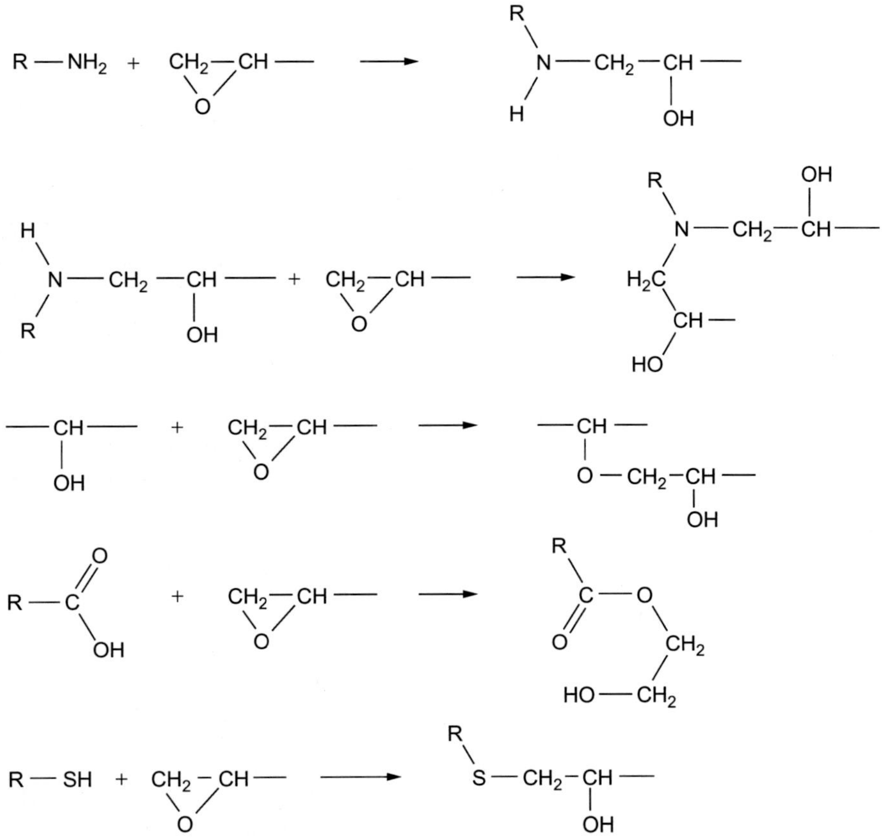

Figure 8.9 Cross-linking reactions between functional groups of proteins ($-NH_2$, –NH, –OH, –COOH and –SH) and epoxide functional group of epoxy resins. Adapted from Mekonnen et al. (2013b), Wiley, Copyright © 2013.

A nanocomposite of soy protein/boron nitride has showed chemical resistance, with enhanced thermal stability and substantial reduction in oxygen permeability that enable the material to have applications as covering and protective materials (Dash and Swain, 2013). Also, owing to improvement in mechanical performance with reinforcement, degradation behavior, bioactive and biocompatible character, as well as the possibility of a broad range of physical and chemical modifications to tailor their properties show that proteins have biomedical applications. Development of protein-based active food packaging composite films (Lee et al., 2008), barley protein–clay antimicrobial composite film (Shin et al., 2013), and rice bran-reinforced gelatine edible composite film (Shin et al., 2011) have been reported in the literature. Furthermore, the use of biodegradable protein composites for indoor construction, furniture adhesives, degradable construction practice, development of mulching trays, flower or planting pots, greenhouse films, and so on have also been suggested in the literature (Mekonnen et al., 2013a; Yue et al., 2012).

8.7 Summary

Protein-based composite materials have been demonstrated to have several applications, including biodegradable packaging films, plastic bags, adhesives, and construction. The wide availability of agricultural proteins, and generation of cheap protein by products such as corn gluten, soybean meal, and canola meal from biofuel industries, provided an excellent opportunity to develop cost-competitive protein-based composite materials. Nonetheless, the rapid technological development of protein biocomposites has yet to be translated into significant market impact, primarily due to performance limitations in terms of mechanical properties and moisture sensitivity. Besides the usual purpose of improving mechanical performance, reducing cost, improving processibility and moisture sensitivity, research needs to progress in areas such as the search for protein reinforcing agents that impart additional functions of flame retardancy, optical quality, electric conductivity or insulation, thermoxidative stability, chemical, and temperature. Also, reactive fillers that provide chemical integrity, moisture and chemical fume impermeability, and maintain biodegradability and biocompatibility to the protein matrices need to be further investigated.

Acknowledgments

The authors gratefully recognize the financial support provided bythe Ontario Ministry of Agriculture, Food, and Rural Affairs' (OMAFRA) New Directions and Alternative Renewable Fuels "Plus" Research Programs; to the OMAFRA-University of Guelph Bioeconomy-Industrial Uses Program; to the Ontario Research Fund (ORF) Research Excellence (RE) Round-4 from the Ontario Ministry of Economic Development and Innovation (MEDI); the Natural Sciences and Engineering Research Council of Canada (NSERC) Discovery grant individuals to Mohanty; the NSERC, Collaborative Research and Development (CRD) Grant program; the Grain Farmers of Ontario (GFO), Canada fund; AUTO21-NCE (Canada's automotive R&D program); the Hannam Soybean Utilization Fund (HSUF); the Canadian Foundation for Innovation's Leaders Opportunity Fund (CFI-LOF); and the Canadian Agricultural Adaptation Program (CAAP) from Agriculture and Agri Food Canada (AAFC) & Federal Economic Development Agency for Southern Ontario (FedDev Ontario).

References

Ahmed, S., Jones, F., 1990. A review of particulate reinforcement theories for polymer composites. J. Mater. Sci. 25, 4933–4942.

Anderson, T., Lamsal, B., 2011. Zein extraction from corn, corn products, and coproducts and modifications for various applications: a review. Cereal Chem. 88, 159–173.

Bhuyan, A., 2010. On the mechanism of SDS-induced protein denaturation. Biopolymers 93, 186–199.

Bier, J., Verbeek, C., Lay, M., 2013. Identifying transition temperatures in bloodmeal-based thermoplastics using material pocket DMTA. J. Therm. Anal. Calorim. 112, 1303–1315.

Brown, S., 1992. Chemical strategies for compatibilization of immiscible polymer blends – polyphenylene ether-nylon copolymer formation through nucleophilic displacement on aryloxytriazine-capped polyphenylene ethers. Abstr. Pap. Am. Chem. Soc. 204, 126.

Buxbaum, E., 2007. Fundamentals of Protein Structure and Function. Springer, New York, USA.

Calabria, L., Vieceli, N., Bianchi, O., De Oliveira, R., Do Nascimento, I., Schmidt, V., 2012. Soy protein isolate/poly(lactic acid) injection-molded biodegradable blends for slow release of fertilizers. Ind. Crop. Prod. 36, 41–46.

Cao, N., Fu, Y., He, J., 2007. Preparation and physical properties of soy protein isolate and gelatin composite films. Food Hydrocoll. 21, 1153–1162.

Cao, N., Yang, X., Fu, Y., 2009. Effects of various plasticizers on mechanical and water vapor barrier properties of gelatin films. Food Hydrocoll. 23, 729–735.

Chen, H., 1995. Functional properties and applications of edible films made of milk proteins. J. Dairy Sci. 78, 2563–2583.

Chen, H.I., Shana, Z.H., 2010. Stabilization of collagen by cross-linking with oxazolidine E-resorcinol. Int. J. Biol. Macromol. 46, 535–539.

Chen, P., Zhang, L., 2006. Interaction and properties of highly exfoliated soy protein/montmorillonite nanocomposites. Biomacromolecules 7, 1700–1706.

Chen, F., Zhang, J., 2010. In-situ poly(butylene adipate-co-terephthalate)/soy protein concentrate composites: effects of compatibilization and composition on properties. Polymer 51, 1812–1819.

Cheung, H., Ho, M., Lau, K., Cardona, F., Hui, D., 2009. Natural fibre-reinforced composites for bioengineering and environmental engineering applications. Compos. Part B 40, 655–663.

Crosby, J., Drye, T., 1987. Fracture studies of discontinuous fiber reinforced thermoplastic composites. J. Reinf. Plast. Compos. 6, 162–177.

Cuq, B., Gontard, N., Guilbert, S., 1997. Thermoplastic properties of fish myofibrillar proteins: application to biopackaging fabrication. Polymer 38, 4071–4078.

Dangaran, K., Tomasula, P., Qi, P., Embuscado, M., Huber, K., 2009. Structure and function of protein-based edible films and coatings. In: Edible Films and Coatings for Food Applications. Springer, New York, USA. pp. 25–56.

Dash, S., Swain, S., 2013. Effect of nanoboron nitride on the physical and chemical properties of soy protein. Compos. Sci. Technol. 84, 39–43.

Dastidar, T.G., Netravali, A.N., 2012. 'Green' crosslinking of native starches with malonic acid and their properties. Carbohydr. Polym. 90, 1620–1628.

De Graaf, L., 2000. Denaturation of proteins from a non-food perspective. J. Biotechnol. 79, 299–306.

Dube, N., Presley, A.D., Shu, J.Y., Xu, T., 2011. Amphiphilic peptide-polymer conjugates with side-conjugation. Macromol. Rapid Commun. 32, 344–353.

Dubey, V., Jagannadham, M., 2003. Differences in the unfolding of procerain induced by pH, guanidine hydrochloride, urea, and temperature. Biochemistry 42, 12287–12297.

Echeverria, I., Eisenberg, P., Mauri, A.N., 2014. Nanocomposites films based on soy proteins and montmorillonite processed by casting. J. Membr. Sci. 449, 15–26.

El-Thaher, N., Mekonnen, T., Mussone, P., Bressler, D., Choi, P., 2013a. Effect of electrolytes, water, and temperature on cross-linking of glutaraldehyde and hydrolyzed specified risk materials. Ind. Eng. Chem. Res. 52, 4987–4993.

El-Thaher, N., Mekonnen, T., Mussone, P., Bressler, D., Choi, P., 2013b. Non-isothermal DSC study of epoxy resins cured with hydrolyzed specified risk material. Ind. Eng. Chem. Res. 52, 8189–8199.

Elzoghby, A., El-Fotoh, W., Elgindy, N., 2011. Casein-based formulations as promising controlled release drug delivery systems. J. Control. Release 153, 206–216.

Espert, A., Vilaplana, F., Karlsson, S., 2004. Comparison of water absorption in natural cellulosic fibres from wood and one-year crops in polypropylene composites and its influence on their mechanical properties. Compos. A: Appl. Sci. Manuf. 35, 1267–1276.

Fowler, P., Hughes, J., Elias, R., 2006. Biocomposites: technology, environmental credentials and market forces. J. Sci. Food Agric. 86, 1781–1789.

Fraenkeiconrat, H., Olcott, H.S., 1945. Esterification of proteins with alcohols of low molecular weight. J. Biol. Chem. 161, 259–268.

Fu, S., Feng, X., Lauke, B., Mai, Y., 2008. Effects of particle size, particle/matrix interface adhesion and particle loading on mechanical properties of particulate–polymer composites. Compos. Part B 39, 933–961.

Garcia, R.A., Phillips, J.G., 2009. Physical distribution and characteristics of meat and bone meal protein. J. Sci. Food Agric. 89, 329–336.

Gennadios, A., Weller, C.L., Hanna, M.A., Froning, G.W., 1996. Mechanical and barrier properties of egg albumen films. J. Food Sci. 61, 585–589.

George, A., Ravindran, S., 2010. Protein templates in hard tissue engineering. Nano Today 5, 254–266.

Gomes, S., Leonor, I., Mano, J., Reis, R., Kaplan, D., 2012. Natural and genetically engineered proteins for tissue engineering. Prog. Polym. Sci. 37, 1–17.

Gunderson, S., Schiavone, R., 1989. The insect exoskeleton – a natural structural composite. JOM J. Miner. Met. Mater. Soc. 41, 60–62.

Gupta, A., Kumar, V., 2007. New emerging trends in synthetic biodegradable polymers – polylactide: a critique. Eur. Polym. J. 43, 4053–4074.

Hardy, J., Scheibel, T., 2010. Composite materials based on silk proteins. Prog. Polym. Sci. 35, 1093–1115.

Hassan, T., Rangari, V., Rana, R., Jeelani, S., 2013. Sonochemical effect on size reduction of $CaCO_3$ nanoparticles derived from waste eggshells. Ultrason. Sonochem. 20, 1308–1315.

Hernandez-Izquierdo, V., Krochta, J., 2008. Thermoplastic processing of proteins for film formation – a review. J. Food Sci. 73, R30–R39.

Hu, X., Cebe, P., Weiss, A., Omenetto, F., Kaplan, D., 2012. Protein-based composite materials. Mater. Today 15, 208–215.

Jang, Y., Huang, J., Li, K., 2011. A new formaldehyde-free wood adhesive from renewable materials. Int. J. Adhes. Adhes. 31, 754–759.

Janowska, G., Kucharska-Jastrzabek, A., Prochon, M., Przepiorkowska, A., 2013. Thermal properties and combustibility of elastomer-protein composites. Part II. Composites NBR-keratin. J. Therm. Anal. Calorim. 113, 933–938.

Jin, E., Reddy, N., Zhu, Z., Yang, Y., 2011a. Graft polymerization of native chicken feathers for thermoplastic applications. Abstr. Pap. Am. Chem. Soc. 241, 1.

Jin, E., Reddy, N., Zhu, Z., Yang, Y., 2011b. Graft polymerization of native chicken feathers for thermoplastic applications. J. Agric. Food Chem. 59, 1729–1738.

John, J., Mani, R., Bhattacharya, M., 2002. Evaluation of compatibility and properties of biodegradable polyester blends. J. Polym. Sci. A Polym. Chem. 40, 2003–2014.

Jung, B., Theato, P., Schlaad, H., 2013. Chemical strategies for the synthesis of protein–polymer conjugates. Adv. Polym. Sci. 253, 37–70.

Karnani, R., Krishnan, M., Narayan, R., 1997. Biofiber-reinforced polypropylene composites. Polym. Eng. Sci. 37, 476–483.

Karthikeyan, K., Lakra, R., Rajaram, R., Korrapati, P., 2012. Development and characterization of zein-based micro carrier system for sustained delivery of aceclofenac sodium. AAPS PharmSciTech 13, 143–149.

Kim, S., 2008. Processing and properties of gluten/zein composite. Bioresour. Technol. 99, 2032–2036.

Kim, S., Sessa, D.J., Lawton, J.W., 2004. Characterization of zein modified with a mild cross-linking agent. Ind. Crop. Prod. 20, 291–300.

Koning, C., Van Duin, M., Pagnoulle, C., Jerome, R., 1998. Strategies for compatibilization of polymer blends. Prog. Polym. Sci. 23, 707–757.

Kumar, R., Choudhary, V., Mishra, S., Varma, I., 2008. Banana fiber-reinforced biodegradable soy protein composites. Front. Chem. Chin. 3, 243–250.

Langmaier, F., Mokrejs, P., Kolomamik, K., Mladek, M., 2008. Plasticizing collagen hydrolysate with glycerol and low-molecular weight poly(ethylene glycols). Thermochim. Acta 469, 52–58.

Lee, J., Son, S., Hong, S., 2008. Characterization of protein-coated polypropylene films as a novel composite structure for active food packaging application. J. Food Eng. 86, 484–493.

Li, M., Mondrinos, M., Gandhi, M., Ko, F., Weiss, A., Lelkes, P., 2005. Electrospun protein fibers as matrices for tissue engineering. Biomaterials 26, 5999–6008.

Li, F., Li, X., Wang, W., 2010. Soy flour adhesive modified with urea, citric acid and boric acid. Pigm. Resin Technol. 39, 223–227.

Li, W., Guo, R., Lan, Y., Zhang, Y., Xue, W., Zhang, Y., 2014. Preparation and properties of cellulose nanocrystals reinforced collagen composite films. J. Biomed. Mater. Res. A 102, 1131–1139.

Liu, Y., Li, K., 2007. Development and characterization of adhesives from soy protein for bonding wood. Int. J. Adhes. Adhes. 27, 59–67.

Liu, B., Jiang, L., Zhang, J., 2011. Study of effects of processing aids on properties of poly(lactic acid)/soy protein blends. J. Polym. Environ. 19, 239–247.

Lodha, P., Netravali, A., 2002. Characterization of interfacial and mechanical properties of "green" composites with soy protein isolate and ramie fiber. J. Mater. Sci. 37, 3657–3665.

Lodha, P., Netravali, A., 2005. Characterization of stearic acid modified soy protein isolate resin and ramie fiber reinforced "green" composites. Compos. Sci. Technol. 65, 1211–1225.

Lu, Y., Weng, L., Zhang, L., 2004. Morphology and properties of soy protein isolate thermoplastics reinforced with chitin whiskers. Biomacromolecules 5, 1046–1051.

Luescher, M., Ruegg, M., Schindler, P., 1974. Effect of hydration upon thermal-stability of tropocollagen and its dependence on presence of neutral salts. Biopolymers 13, 2489–2503.

Lukubira, S., Ogale, A.A., 2013. Thermal processing and properties of bioplastic sheets derived from meat and bone meal. J. Appl. Polym. Sci. 130, 256–263.

Ly, Y., Johnson, L., Jane, J., 1998. Soy protein as biopolymer. In: Kaplan, D. (Ed.), Biopolymers from Renewable Resources. Springer, Berlin, Germany.

Marsilla, K., Verbeek, C., 2013. Properties of bloodmeal/linear low-density polyethylene blends compatibilized with maleic anhydride grafted polyethylene. J. Appl. Polym. Sci. 130, 1890–1897.

Mckittrick, J., Chen, P., Tombolato, L., Novitskaya, E., Trim, M., Hirata, G., Olevsky, E., Horstemeyer, M., Meyers, M., 2010. Energy absorbent natural materials and bioinspired design strategies: a review. Mater. Sci. Eng., C 30, 331–342.

Mekonnen, T., Mussone, P., Alemaskin, K., Sopkow, L., Wolodko, J., Choi, P., Bressler, D., 2013a. Biocomposites from hydrolyzed waste proteinaceous biomass: mechanical, thermal and moisture absorption performances. J. Mater. Chem. A 1, 13186–13196.

Mekonnen, T., Mussone, P., El-Daher, N., Choi, P., Bressler, D., 2013b. Thermosetting proteinacious plastics from hydrolyzed specified risk material. Macromol. Mater. Eng. 298, 1294–1303.

Mekonnen, T.H., Mussone, P.G., Khalil, H., Bressler, D.C., 2013c. Progress in bio-based plastics and plasticizing modifications: feature article. J. Mater. Chem. A 1, 13379–13398.

Mekonnen, T.H., Mussone, P.G., Stashko, N., Choi, P.Y., Bressler, D.C., 2013d. Recovery and characterization of proteinacious material recovered from thermal and alkaline hydrolyzed specified risk materials. Process Biochem. 48, 885–892.

Mekonnen, T., Mussone, P., Bressler, D., 2014a. Valorization of rendering industry wastes and co-products for industrial chemicals, materials and energy: review. Crit. Rev. Biotechnol. 1–12. http://dx.doi.org/10.3109/07388551.2014.928812.

Mekonnen, T., Mussone, P., Choi, P., Bressler, D., 2014b. Adhesives from waste protein biomass for oriented strand board composites: development and performance. Macromol. Mater. Eng. http://dx.doi.org/10.1002/mame.201300402.

Mekonnen, T., Mussone, P., El-Thaher, N., Choi, P., Bressler, D., 2014c. Subcritical hydrolysis and characterization of waste proteinaceous biomass for value added applications. J. Chem. Technol. Biotechnol. 90, 476–483. http://dx.doi.org/10.1002/jctb.4333.

Meyers, M., Chen, P., Lin, A., Seki, Y., 2008. Biological materials: structure and mechanical properties. Prog. Mater. Sci. 53, 1–206.

Micard, V., Belamri, R., Morel, M.H., Guilbert, S., 2000. Properties of chemically and physically treated wheat gluten films. J. Agric. Food Chem. 48, 2948–2953.

Mohanty, A., Misra, M., Drzal, L., 2002. Sustainable bio-composites from renewable resources: opportunities and challenges in the green materials world. Abstr. Pap. Am. Chem. Soc. 223, D70.

Mohanty, A., Tummala, P., Liu, W., Misra, M., Mulukutla, P., Drzal, L., 2005a. Injection molded biocomposites from soy protein based bioplastic and short industrial hemp fiber. J. Polym. Environ. 13, 279–285.

Mohanty, A.K., Liu, W., Tummala, P., Drzal, L.T., Misra, M., Narayan, R., 2005b. Soy Protein-Based Plastics, Blends, and Composites. Taylor and Francis group, Boca Raton, FL, USA.

Mohanty, A.K., Misra, M., Drazal, L.T., Selke, S.E., Harte, B.R., Hinrichsen, G., 2005c. Natural Fibers, Biopolymers, and Biocomposites: An Introduction. Taylor and Francis Group, Boca Raton, FL, USA.

Olsen, B., 2013. Engineering materials from proteins. AIChE J. 59, 3558–3568.

Orliac, O., Rouilly, A., Silvestre, F., Rigal, L., 2003. Effects of various plasticizers on the mechanical properties, water resistance and aging of thermo-moulded films made from sunflower proteins. Ind. Crop. Prod. 18, 91–100.

Park, S.K., Bae, D.H., Hettiarachchy, N.S., 2000. Protein concentrate and adhesives from meat and bone meal. J. Am. Oil Chem. Soc. 77, 1223–1227.

Petrie, E.M., 2006. Epoxy Adhesive Formulations. The McGraw-Hill companies, Inc., New York, USA.

Pizzi, A., 2006. Recent developments in eco-efficient bio-based adhesives for wood bonding: opportunities and issues. J. Adhes. Sci. Technol. 20, 829–846.

Prochon, M., Janowska, G., Przepiorkowska, A., Kucharska-Jastrzabek, A., 2012. Thermal properties and combustibility of elastomer-protein composites Part I. Composites SBR-keratin. J. Therm. Anal. Calorim. 109, 1563–1570.

Ratna, D., 2003. Modification of epoxy resins for improvement of adhesion: a critical review. J. Adhes. Sci. Technol. 17, 1655–1668.

Reddy, N., Yang, Y.Q., 2007. Novel protein fibers from wheat gluten. Biomacromolecules 8, 638–643.

Reddy, N., Yang, Y., 2011. Potential of plant proteins for medical applications. Trends Biotechnol. 29, 490–498.

Reddy, N., Tan, Y., Li, Y., Yang, Y., 2008. Effect of glutaraldehyde crosslinking conditions on the strength and water stability of wheat gluten fibers. Macromol. Mater. Eng. 293, 614–620.

Reddy, M., Mohanty, A.K., Misra, M., 2010. Thermoplastics from soy protein: a review on processing, blends and composites. J. Biobased Mater. Bioenergy 4, 298–316.

Reddy, M., Mohanty, A., Misra, M., 2012a. Biodegradable blends from plasticized soy meal, polycaprolactone, and poly(butylene succinate). Macromol. Mater. Eng. 297, 455–463.

Reddy, M.M., Mohanty, A.K., Misra, M., 2012b. Optimization of tensile properties thermoplastic blends from soy and biodegradable polyesters: Taguchi design of experiments approach. J. Mater. Sci. 47, 2591–2599.

Reddy, M., Misra, M., Mohanty, A., 2014. Biodegradable blends from corn gluten meal and poly(butylene adipate-co-terephthalate) (PBAT): studies on the influence of plasticization and destructurization on rheology, tensile properties and interfacial interactions. J. Polym. Environ. 22, 167–175.

Robertson, N.L., Nychka, J.A., Alemaskin, K., Wolodko, J.D., 2013. Mechanical performance and moisture absorption of various natural fiber reinforced thermoplastic composites. J. Appl. Polym. Sci. 130 (2), 969–980.

Rombouts, I., Lamberts, L., Celus, I., Lagrain, B., Brijs, K., Delcour, J., 2009. Wheat gluten amino acid composition analysis by high-performance anion-exchange chromatography with integrated pulsed amperometric detection. J. Chromatogr. A 1216, 5557–5562.

Rouse, J., Van Dyke, M., 2010. A review of keratin-based biomaterials for biomedical applications. Materials 3, 999–1014.

Sawpan, M.A., Pickering, K.L., Fernyhough, A., 2012. Analysis of mechanical properties of hemp fibre reinforced unsaturated polyester composites. J. Compos. Mater. 47, 1513–1525.

Sharma, S., Hodges, J.N., Luzinov, I., 2008. Biodegradable plastics from animal protein coproducts: feathermeal. J. Appl. Polym. Sci. 110, 459–467.

Shin, Y., Jang, S., Song, K., 2011. Preparation and mechanical properties of rice bran protein composite films containing gelatin or red algae. Food Sci. Biotechnol. 20, 703–707.

Shin, Y., Song, H., Jo, W., Lee, M., Song, K., 2013. Physical properties of a barley protein/nano-clay composite film containing grapefruit seed extract and antimicrobial benefits for packaging of *Agaricus bisporus*. Int. J. Food Sci. Technol. 48, 1736–1743.

Shu, J., Panganiban, B., Xu, T., 2013. Peptide-polymer conjugates: from fundamental science to application. Annu. Rev. Phys. Chem. 64, 631–651.

Shukla, R., Cheryan, M., 2001. Zein: the industrial protein from corn. Ind. Crop. Prod. 13, 171–192.

Sohail, S.S., Wang, B.W., Biswas, M.A.S., Oh, J.H., 2006. Physical, morphological, and barrier properties of edible casein films with wax applications. J. Food Sci. 71, C255–C259.

Sommer, I., Kunz, P., 2012. Collagen-based packaging films for nonfood applications and their comparison with commercial polyolefin (polypropylene block copolymer) films. J. Plast. Film Sheeting 28, 30–62.

Sothornvit, R., Krochta, J., 2001. Plasticizer effect on mechanical properties of beta-lactoglobulin films. J. Food Eng. 50, 149–155.

Sothornvit, R., Krochta, J.M., 2005. Plasticizers in edible films and coatings. In: Innovations in Food Packaging. Elsevier Science, London, United Kingdom, pp. 403–433.

Swain, S., Biswal, S., Nanda, P., Nayak, P., 2004. Biodegradable soy-based plastics: opportunities and challenges. J. Polym. Environ. 12, 35–42.

Tian, H., 2012. Processing and properties of soy protein/silica nanocomposites fabricated in situ synthesis. J. Compos. Mater. 46, 427–435.

Tolstoguzov, V., 1993. Thermoplastic extrusion – the mechanism of the formation of extrudate structure and properties. J. Am. Oil Chem. Soc. 70, 417–424.

Tropini, V., Lens, J.P., Mulder, W.J., Silvestre, F., 2000. Cross-linking of wheat gluten using a water-soluble carbodiimide. Cereal Chem. 77, 333–338.

Tropini, V., Lens, J.P., Mulder, W., Silvestre, F., 2004. Wheat gluten films cross-linked with 1-ethyl-3-(3-dimethylaminopropyl) carbodiimide and *N*-hydroxysuccinimide. Ind. Crop. Prod. 20, 281–289.

Tummala, P., Liu, W.J., Drzal, L.T., Mohanty, A.K., Misra, M., 2006. Influence of plasticizers on thermal and mechanical properties and morphology of soy-based bioplastics. Ind. Eng. Chem. Res. 45, 7491–7496.

Ullah, A., Wu, J., 2013. Feather fiber-based thermoplastics: effects of different plasticizers on material properties. Macromol. Mater. Eng. 298, 153–162.

Ullah, A., Vasanthan, T., Bressler, D., Elias, A.L., Wu, J.P., 2011. Bioplastics from feather quill. Biomacromolecules 12, 3826–3832.

Vaz, C.M., De Graaf, L.A., Reis, R.L., Cunha, A.M., 2003. In vitro degradation behaviour of biodegradable soy plastics: effects of crosslinking with glyoxal and thermal treatment. Polym. Degrad. Stab. 81, 65–74.

Verbeek, C., Koppel, N., 2012. Moisture sorption and plasticization of bloodmeal-based thermoplastics. J. Mater. Sci. 47, 1187–1195.

Verbeek, C., Van Den Berg, L., 2010. Extrusion processing and properties of protein-based thermoplastics. Macromol. Mater. Eng. 295, 10–21.

Verbeek, C., Van Den Berg, L., 2011. Development of proteinous bioplastics using bloodmeal. J. Polym. Environ. 19, 1–10.

Verbeek, C., Hicks, T., Langdon, A., 2011. Degradation as a result of UV radiation of bloodmeal-based thermoplastics. Polym. Degrad. Stab. 96, 515–522.

Wang, Y., Cao, X., Zhang, L., 2006. Effects of cellulose whiskers on properties of soy protein thermoplastics. Macromol. Biosci. 6, 524–531.

Wang, Y., Mo, X., Sun, X., Wang, D., 2007. Soy protein adhesion enhanced by glutaraldehyde crosslink. J. Appl. Polym. Sci. 104, 130–136.

Wei, T., Lei, L., Kang, H., Qiao, B., Wang, Z., Zhang, L., Coates, P., Hua, K., Kulig, J., 2012. Tough bio-based elastomer nanocomposites with high performance for engineering applications. Adv. Eng. Mater. 14, 112–118.

White, J.L., Bumm, S.H., 2011. Polymer blend compounding and processing. In: Isayev, A. (Ed.), Encyclopedia of Polymer Blends. Wiley-VCH, Weinheim, Germany, pp. 1–26.

Wu, Q.X., Yoshino, T., Sakabe, H., Zhang, H.K., Isobe, S., 2003. Chemical modification of zein by bifunctional polycaprolactone (PCL). Polymer 44, 3909–3919.

Xu, X., Jiang, L., Zhou, Z., Wu, X., Wang, Y., 2012. Preparation and properties of electrospun soy protein isolate/polyethylene oxide nanofiber membranes. ACS Appl. Mater. Interfaces 4, 4331–4337.

Yang, L., Paulson, A.T., 2000. Mechanical and water vapour barrier properties of edible gellan films. Food Res. Int. 33, 563–570.

Yang, W., Gludovatz, B., Zimmermann, E.A., Bale, H.A., Ritchie, R.O., Meyers, M.A., 2013. Structure and fracture resistance of alligator gar (*Atractosteus spatula*) armored fish scales. Acta Biomater. 9, 5876–5889.

Yin, X., Li, F., He, Y., Wang, Y., Wang, R., 2013. Study on effective extraction of chicken feather keratins and their films for controlling drug release. Biomater. Sci. 1, 528–536.

Yu, L., Dean, K., Li, L., 2006. Polymer blends and composites from renewable resources. Prog. Polym. Sci. 31, 576–602.

Yue, H., Cui, Y., Shuttleworth, P., Clark, J., 2012. Preparation and characterisation of bioplastics made from cottonseed protein. Green Chem. 14, 2009–2016.

Zarrinbakhsh, N., Misra, M., Mohanty, A., 2011. Biodegradable green composites from distiller's dried grains with solubles (DDGS) and a polyhydroxy(butyrate-co-valerate) (PHBV)-based bioplastic. Macromol. Mater. Eng. 296, 1035–1045.

Zarrinbakhsh, N., Mohanty, A., Misra, M., 2013. Improving the interfacial adhesion in a new renewable resource-based biocomposites from biofuel coproduct and biodegradable plastic. J. Mater. Sci. 48, 6025–6038.

Zhang, J.W., Jiang, L., Zhu, L.Y., 2006. Morphology and properties of soy protein and polylactide blends. Biomacromolecules 7, 1551–1561.

Zhang, M.-N., Huang, G.-R., Jiang, J.-X., 2012. Effects of chemical modification and molecular weight distribution on ironbinding ability of phytate-removal soybean protein isolate hydrolysate. Adv. J. Food Sci. Technol. 4, 78–83.

Zheng, H., Tan, Z., Zhan, Y., Huang, J., 2003. Morphology and properties of soy protein plastics modified with chitin. J. Appl. Polym. Sci. 90, 3676–3682.

Zheng, H., Ai, F., Wei, M., Huang, J., Chang, P., 2007. Thermoplastic soy protein nanocomposites reinforced by carbon nanotubes. Macromol. Mater. Eng. 292, 780–788.

Zhong, Z., Sun, X., 2001. Properties of soy protein isolate/polycaprolactone blends compatibilized by methylene diphenyl diisocyanate. Polymer 42, 6961–6969.

Zhou, X., Mohanty, A., Misra, M., 2013. A new biodegradable injection moulded bioplastic from modified soy meal and poly (butylene adipate-co-terephthalate): effect of plasticizer and denaturant. J. Polym. Environ. 21, 615–622.

Zhu, R., Liu, H., Zhang, J., 2012. Compatibilizing effects of maleated poly(lactic acid) (PLA) on properties of PLA/soy protein composites. Ind. Eng. Chem. Res. 51, 7786–7792.

Mechanical performance of polyethylene (PE)-based biocomposites

9

A.R. Kakroodi, Y. Kazemi, A. Cloutier, D. Rodrigue
Université Laval, Quebec City, QC, Canada

9.1 General introduction to natural fibers and their composites

9.1.1 *Natural fibers and synthetic fibers*

Polymers, including polyethylene (PE), usually have relatively low mechanical properties (modulus and strength) compared to other commodity and engineering materials. One of the most convenient methods to enhance their properties is through inclusion of dispersed phases (natural or synthetic) with high stiffness and strength (Valente et al., 2011; Qiu et al., 2014). The dispersed phase is usually in the form of long or short fibers or even very small particles (Yemele et al., 2010; Shinoj et al., 2011).

Artificial reinforcements (glass, talc, calcium carbonate, etc.) usually have higher mechanical properties compared to natural reinforcements (plant or vegetable fibers). However, over the last decades, natural fibers have attracted a growing attention as reinforcement in polymer composites (Beshay et al., 1985). Several investigations were devoted to reinforce thermosets or thermoplastics using natural fibers such as wood fiber, hemp, flax, jute, rice husk, and so on (Pickering et al., 2003; Panthapulakkal et al., 2005; Cheung et al., 2009; Oksman et al., 2009). Natural fibers are less expensive and are considered ecofriendly materials (Barone and Schmidt, 2005; Majid et al., 2010). Although mainly thermosetting resins were used initially as matrices for natural fiber-reinforced composites, today thermoplastic resins such as PE and polypropylene (PP) have received a great deal of attention due to their simple processability and recyclability (Saheb and Jog, 1999). The main purpose of introducing natural fibers in thermoplastics is to reduce raw material costs and for environmental issues. Other noticeable advantages of natural fibers include light weight, high stiffness/weight ratio, biodegradability, and a wide range of local availability, which make them interesting choices in different applications, especially in the automotive and building industries (George et al., 1998; Marcovich and Villar, 2003; Kim et al., 2006; Holbery and Houston, 2006).

Joshi et al. (2004) reported that the energy consumption for production of natural fibers is considerably lower in comparison with glass fiber. It was shown that production of natural fibers, such as flax or China reed fiber, consumes significantly less energy (9.55 and 3.64 MJ/kg, respectively) than glass fiber (54.7 MJ/kg). The authors

Biocomposites: Design and Mechanical Performance. http://dx.doi.org/10.1016/B978-1-78242-373-7.00003-2

also reviewed and compared the life cycle studies of selected natural fiber and glass fiber composites. The study covered the environmental aspects of both composites through their life cycle considering their production, use, and end of life management options (i.e., recycling, incineration, and disposal). They concluded that natural fibers are environmentally superior compared to glass fiber because (i) production of natural fibers leads to lower environmental impact compared to glass fiber; (ii) natural fibers can be used at high concentrations (due to their low price and density), which results in a lower amount of polymers (petroleum-based) in the final products; (iii) lower density of natural fibers results in better fuel efficiency for automotive applications; and (iv) although it's not recommended as a suitable technique, end of life incineration of natural fibers leads to energy recovery.

It is notable that some natural fibers (such as pineapple and banana leaf fibers) are not even considered as raw materials. Such materials are agricultural waste that can be purchased at very low prices. Such advantages, combined with low density and low abrasiveness, have recently driven a considerable number of researchers to develop engineered products containing naturally made fillers (Bledzki and Gassan, 1999; Pasquini et al., 2008).

Agricultural plant-based lignocellulosic fibers are usually grouped into three types depending on their source: (i) seed hair (cotton), (ii) bast fibers (jute and flax fiber), and (iii) leaf fibers (sisal and abaca). Mechanical properties of natural fibers vary with chemical composition, which depends on their type, age, and climatic conditions. Components of natural fibers include cellulose, hemicellulose, lignin, waxes, water-soluble components, and pectin. Each natural fiber is a bundle of cellulose fibers that are covered with hemicellulose and lignin as matrix (Li et al., 2007; Faruk et al., 2012).

Thermosets (Merlini et al., 2011; Jawaid et al., 2013), thermoplastics (Cao et al., 2012), and even elastomers (Kumar et al., 1995; Jacob et al., 2006) are commonly used as matrices for natural fiber composites. Among these, thermoplastics have attracted more attention due to their design flexibility and simple processing methods. Thermosets and elastomers, on the other hand, have cross-linked structures that do not allow them to be reprocessed by conventional methods. Among thermoplastics, however, only a handful of resins can be used as matrices for natural fibers. Because natural fibers are prone to thermal degradation at high temperature, thermoplastics with high processing temperatures (higher than 200 °C) cannot be used generally. As a result, PE (with low melting point) is one of the most commonly used matrices for natural fiber composites (Facca et al., 2006).

9.1.2 Compatibility of PE and natural fibers

Presence of high concentrations of hydroxyl groups on different components of natural fibers, especially cellulose, results in their hydrophilic behavior (Zhang et al., 2005). Low surface adhesion with hydrophobic polyolefin matrices (such as PE), which is a direct result of the hydrophilic behavior of natural fibers, is considered as the most important issue in the fabrication of natural fiber composites. The main disadvantage of low surface interaction is reduced mechanical properties due to limited stress transfer at the interface and also low homogeneity (fiber dispersion) in the composite (Balasuriya et al., 2001; George et al., 2001). A lack of compatibility between the fiber and the matrix leads to low surface interaction between each phase in the composite. As a result, the matrix is not able to efficiently transfer the applied loads to the reinforcing phase and failure mostly occurs at the interface (Jarukumjorn and Suppakarn, 2009; Shinoj et al., 2010).

Kakroodi et al. (2012) studied the morphological and mechanical properties of PE/flax fiber composites. Through SEM observations, they reported that inclusion of flax fibers in PE resulted in composites with poor surface adhesion between the matrix and fibers. Presence of gaps and voids between the reinforcements and matrix and also unbroken flax particles with clean surfaces (fiber pull-out) were evidence for low compatibility between the phases. They also reported that although inclusion of 30% flax fibers resulted in increased PE tensile modulus (from 292 to 347 MPa), tensile strength decreased significantly (from 22.0 to 13.5 MPa). Such reduction in tensile strength is frequently ascribed to stress concentrations caused by low compatibility between the hydrophobic matrix and hydrophilic natural fibers.

Chen and Porter (1994) fabricated natural fiber composites based on PE and kenaf fiber and studied the effects of kenaf fibers on crystallization kinetics of PE using differential scanning calorimetry (DSC). Kenaf fibers were expected to act as nucleating sites and increase the crystallinity of the PE matrix. However, they reported that this was not the case. The peak temperatures were around 107 °C for composites with different fiber contents. They concluded that because compatibility between PE and kenaf was low (due to difference in their polarities), nucleation of PE molecules on kenaf fibers was not significant. This can be ascribed to the presence of the voids at the fiber–matrix interface; that is, there is actually no contact between them (or at least a very limited one).

High water absorption in natural fiber composites is also a negative effect of hydrophilic fibers (Naik and Mishra, 2007; Kakroodi et al., 2013). Water absorption in natural fiber composites may occur due to two major mechanisms: (i) the presence of natural fiber in the compound, which increases the hydrophilic behavior of the composite; and (ii) the presence of filler particles in the matrix, disturbing the structural integrity in the material. This can produce voids at the interface and may increase the ability of water molecules to penetrate the composite through capillary transport (Espert et al., 2004). Low compatibility between natural fibers and matrix is expected to increase water uptake through both mechanisms. Moisture absorption in natural fiber composites may lead to several disadvantages including (i) low dimensional stability due to swelling, (ii) reduced mechanical properties due to the creation of residual stress in the composite, and (iii) increase in the rate of biological degradation of natural fibers (Wang et al., 2006).

Shinoj et al. (2010) studied the water absorption patterns in linear low-density polyethylene (LLDPE)/palm oil fiber composites. They reported that water absorption in alkali-treated composites showed Fickian behavior. An increase in fiber content was shown to increase both water uptake and time required for saturation. In composites containing fibers of 425–840 μm, increase in fiber content from 10% to 50%, increased ultimate water uptake from around 3% to over 20%. They also reported that reduction in palm oil fibers particle size resulted in decreased water absorption of the composites. Ultimate water uptake for composites with 50% fibers (75–177 μm) was smaller than 20%. This can be ascribed to increased composite homogeneity with reduction in particle sizes.

9.1.3 Compatibilization of PE/natural fiber composites

Two methods have been commonly used in order to modify the surface interactions between cellulosic fibers and polyolefins: (i) surface modification of the fiber and (ii) modification of the matrix phase through active groups grafting or the addition of a compatibilizing agent (Bledzki and Gassan, 1999).

9.1.3.1 Surface modification of natural fibers

The surface of natural fibers can be modified via different (chemical or physical) methods as described next.

Chemical surface treatment of natural fibers

Chemical surface treatment of natural fibers decreases their hydrophilic behavior through reduction in the number of hydroxyl groups on their surfaces (chemical agents are used to react with hydroxyl groups) (Biagiotti et al., 2004).

(i) Silane treatment is one of the most effective chemical modification techniques to increase surface adhesion between lignocellulosic fibers and hydrophobic polymers (Bisanda and Ansell, 1991; Cantero et al., 2003; Bengtsson et al., 2005). Silanes are chemical compounds (Si_nH_{2n+2}) that are also commonly used to compatibilize glass fiber with polymers. In the presence of moisture, silanols are produced through hydrolyzation of alkoxy groups. Silanol groups then react with hydroxyl groups on the surface of natural fibers and create chemical bonds with the fibers. As a result, hydrophilic behavior of the fiber decreases leading to enhanced fiber/matrix interaction (Xie et al., 2010). Abdelmouleh et al. (2007) studied the reinforcement of low-density polyethylene (LDPE) and natural rubber using different types of cellulosic fibers. Cellulose fibers were added to the matrices before and after chemical treatments using three silane coupling agents; namely, γ-methacryloxypropyltrimethoxy (MPS), γ-mercaptoproyltrimethoxy (MRPS), and hexadecyltrimethoxy-silane (HDS). Increased interaction between the matrix and cellulosic fibers after inclusion of MPS and MRPS was confirmed by mechanical characterizations. For instance, tensile strength of LDPE/cellulose fiber (50/50 wt.%) was improved by about 30% and 17% after compatibilizations using MPS and MRPS, respectively. HDS, however, caused only a slight improvement in the properties. Morphological observations revealed that surface treatment with silanes increases compatibility between LDPE and cellulosic fibers.

(ii) Maleic anhydride can also be used for surface covering of natural fibers in order to compatibilize them with polymers. In this approach, natural fibers are preimpregnated in a solution of maleic anhydride or maleic anhydride grafted polymers. Maleic anhydride groups react with hydroxyl groups on the surface of cellulosic fibers (Figure 9.1) which results in decreased hydrophilic behavior (Felix and Gatenholm, 1991; Zhang et al, 2002a; Pracella et al., 2010). Modified surface of lignocellulosic fibers allows better interaction with thermoplastic matrices through decreased hydrophilic behavior and also physical entanglement of polymer chains with the matrix molecules (e.g., maleic anhydride grafted polymers) (Bledzki and Gassan, 1999; Fung et al., 2003).

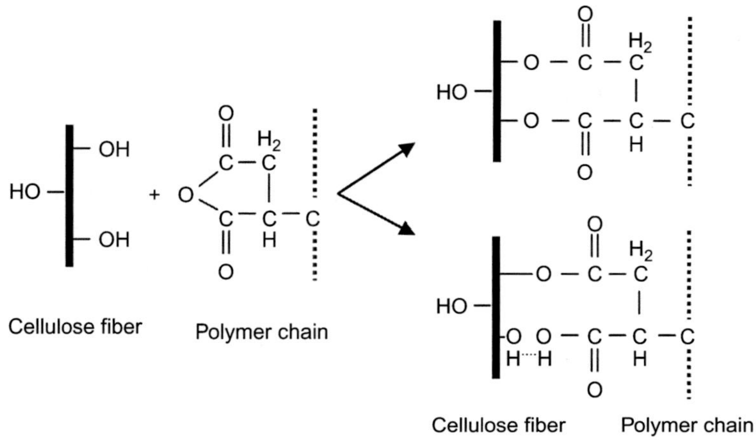

Figure 9.1 Chemical reaction between maleated polyethylene and cellulosic fibers (Felix and Gatenholm, 1991).

Physical surface treatment of natural fibers

Alkalization (or mercerization) is a common method for physical treatment of natural fiber surfaces. In this method, lignocellulosic fibers are immersed in aqueous NaOH solution (Seki et al., 2012; Lu and Oza, 2013a). Lignin, wax, and oils will be removed from the surfaces of natural fibers due to immersion in NaOH solution and a clean and porous cellulosic surface will be obtained (Arrakhiz et al., 2013). This treatment leads to increased fiber mechanical properties due to higher mechanical properties of cellulose compared to lignin. Furthermore, specific surface area of natural fibers is also increased, which leads to better interaction (mechanical interlocking) with the matrix. The main drawback of this approach is the increased hydrophilic behavior of natural fibers due to an increased number of exposed cellulose fibers on the surface. As a result, this method is usually proposed as a first step for other chemical modifications such as the addition of silanes or maleic anhydride (Bledzki and Gassan, 1999).

Surface covering of natural fibers using matrix (in the absence of a chemical reaction) is also a physical approach that leads to better wetting of fibers. Herrera-Franco and Aguilar-Vega (1997) produced composites based on LDPE and henequen fibers before and after preimpregnation of fibers in an LDPE–xylene solution. They reported that modification of fibers resulted in improvement in tensile strength of composites and also level of dispersion of fibers. In the case of composites containing 30% (by volume) henequen fiber, tensile strength increased from around 14 to 16.5 MPa due to fiber treatment.

9.1.3.2 Matrix modification

Modification of the matrix is also a common method to enhance surface interaction between lignocellulosic fibers and polyolefins. This type of compatibilization is usually achieved via two different approaches: (i) inclusion of a maleated polymer (compatibilizer) to the matrix (Oksman et al., 1998) and (ii) chemical modification of the matrix by grafting active groups (especially maleic anhydride) (Gao et al., 2012). In both methods, the active group (maleic anhydride) reacts with the hydroxyl groups on the natural fiber surface (reducing its hydrophilic behavior), while the polymeric chains entangle with the matrix molecules (Bledzki and Gassan, 1999; Keener et al., 2004). Efficiency of maleic anhydride-grafted polymers (as compatibilizer) depends on their acid number and molecular weight. Macromolecules with longer chains (higher molecular weight) have better ability to entangle with the matrix. The acid number, on the other hand, represents the concentration of maleic anhydride groups in maleated polymers (Lu et al., 2005). Table 9.1 presents a brief overview of the main works regarding compatibilization of natural fibers with PE matrix.

9.2 Hybridization of PE biocomposites

PE/natural fiber composites show good mechanical properties on a weight basis due to lower density of reinforcements, while other advantages include low cost, renewability, environmental friendly source of fiber, and lower equipment wear. In spite of these advantages, natural fiber composites usually suffer from some defects such as low-dimensional stability, high water sorption, and low mechanical properties in comparison with composites containing synthetic fibers (Kakroodi et al., 2013). Lower mechanical properties of such composites are the results of (i) different natures of the components (hydrophobic matrix and hydrophilic fiber) and (ii) lower mechanical properties of natural fibers in comparison with synthetic fibers such as glass.

9.2.1 Hybrid PE/natural fiber composites containing synthetic fibers

Fabrication of hybrid composites is a common technique to deal with the aforementioned disadvantages of natural fibers. Hybrid composites are materials with two or more types of reinforcing phases or fillers (Panthapulakkal and Sain, 2007). In PE/natural fiber composites, hybridization is usually performed using common fillers such as glass, calcium carbonate, talc, and so forth. The resulting products are thus a combination of the advantages of the different reinforcements used (reduced cost as well as improved mechanical and thermal properties) (Kretsis, 1987). Characteristics of the produced hybrid composites will be controlled by (i) the properties of the matrix; (ii) properties, sizes, and content of each reinforcement; and (iii) adhesion between the matrix and reinforcements.

Kalaprasad et al. (1996) studied hybrid composites based on LDPE and two different reinforcements of natural (sisal) and synthetic (glass fiber) origins. They reported that the inclusion of longitudinally oriented glass fibers (20%) resulted in an increase in composite's tensile strength from 7 to over 22 MPa. Elongation at break of PE, however, decreased from 63% to around 5%. Substitution of glass fibers using sisal caused a reduction in composites strength. Tensile strength of LDPE-based composites decreased from around 23 to around 18 MPa after substitution of 70% of the glass

Table 9.1 Compatibilization of PE/natural fiber composites

Fiber type	Compatibilization technique	Processing methods	Results	References
Henequen	Alkalization, silane treatment, preimpregnation using dilute PE solution	BRABENDER intensive mixer	Alkalization and preimpregnation increased mechanical interlocking between phases. Silane treatment increased interfacial load transfer	Valadez-Gonzalez et al. (1999)
Sisal	Washing the fibers followed by treatment with 3% stearic acid	Compression molding	Pretreatment of fibers improved interfacial shear strength (τ). Morphological observations showed less fiber pull-out after treatment	Torres and Cubillas (2005)
Wood and bagasse	Maleated coupling agent, carboxylated coupling agent, titanium derived mixture (TDM)	Kneading in a Haake rotor mixer	All compatibilizers improved compatibility between bagasse and matrix. TDM acted as lubricant in wood filled composite	Lei et al. (2007)
Curaua	Coupling agents (maleated PE, Poly(ethylene-*co*-vinyl acetate))	Extrusion	Maleated PE decreased thermal stability of composites more significantly than EVA	Araujo et al. (2008)
Coir	Silane, alkalization, etherification	Heated two-roll mill	Fiber treatment improved mechanical properties. Alkalization increased deformability of the composites, while silane made them more rigid	Arrakhiz et al. (2012)

Continued

Table 9.1 Continued

Fiber type	Compatibilization technique	Processing methods	Results	References
Wood fiber	Silane and isocyanate coupling agents	Extrusion	Increase in tensile properties after silane and isocyanate treatment	Raj et al. (1989)
Hardwood pulp	Silane treatment, benzoyl peroxide (BPO) treatment	Brabender plasticorder	Significant increase in strength after treatment with BPO. Small improvement in strength after silane treatment	Bataille et al. (1990)
Cellulose fiber	Maleation of PE matrix	Internal mixer	Even slight maleation of PE increased adhesion significantly. Inclusion of fibers to maleated PE increased strength significantly	Zhang et al. (2002b)
Wood flour	Maleated polypropylene (PP)	Extrusion	Coupling agent improved both stability and mechanical properties. It also increased dimensional stability of composites	Adhikary et al. (2007)
Pineapple leaf fiber	Silane, isocyanate and peroxide treatment	Solution mixing	Improved fiber–matrix interaction (after fiber treatment) produced increase in thermal stability	George et al. (1996)
Wood flour	Coupling agents (MAPE, MAPP, acrylic acid grafted PE, maleated SEBS)	Extrusion	All compatibilizers increased tensile strength and modulus. Maleated LLDPE gave maximum tensile and impact strength	Wang et al. (2003)

fibers with sisal. Based on the additive rule of hybrid mixtures and using the mechanical properties of individual composites, they showed that all the mechanical properties of hybrid composites (except elongation at break) have a positive hybrid effect. Other examples of PE-based hybrid composites containing a natural and a synthetic reinforcement are summarized in Table 9.2.

9.2.2 Hybrid PE/natural fiber composites containing natural fibers

Actually, very limited work has been conducted to produce hybrid composites based on two different kinds of natural fibers. Processing hybrid composites containing both short and long natural fibers of the same kind (autohybridization), on the other hand, provides materials with a wide variety of properties with no need to use different fibers and processing techniques. Gomez et al. (2005) investigated the effect of fiber length on the mechanical properties of agave fiber-reinforced composites. They observed that increasing fiber length led to higher Young's modulus, but lowered elongation at break. They also reported an optimum in impact strength for a length of 0.81 mm. In other words, incorporation of long and short reinforcing fibers can improve some properties over others. It is thus expected to optimize mechanical properties by producing hybrid composites from both long and short particles of the same origin.

9.2.3 Hybrid PE/natural fiber composites containing virgin or recycled rubber

Inclusion of a reinforcing phase in thermoplastics usually leads to significant reduction in impact strength, due to the negative effects of stress concentration (Oksman and Clemons, 1998). For example, PE-based composites are usually more brittle than PE itself. To increase the impact properties of such composites, elastomer phases including copolymers of PE or PP such as ethylene propylene diene monomer (EPDM) and styrene ethylene butylene styrene tri-block copolymer (SEBS) can be incorporated into the composites through melt mixing (Oksman et al., 1998; Clemons, 2010). Rubber particles with uniform distribution and small size (smaller than 1 μm in diameter) are expected to effectively increase the impact strength of the composites. Virgin elastomers can act in natural fiber composites in three ways: (i) they can act as a third phase in the blend, (ii) they can encapsulate the particles and act as an interfacial agent, and (iii) they can have a mixed behavior between (i) and (ii) (Oksman and Clemons, 1998).

Clemons (2010) studied the effect of elastomer addition into PE/PP/wood flour (WF) composites. He used EPDM and maleated EPDM (MA-EPDM) as impact modifier and reported that incorporation of 10% MA-EPDM led to a significant increase (63%) in impact strength of PE/PP (75/25) with 30% WF, while the same amount of EPDM proved to be less effective (46% increase). Higher impact modification of MA-EPDM is probably due to its compatibility with WF, leading to better ability for fiber encapsulation. Lower tensile modulus of the composites was also reported (32% reduction for MA-EPDM compared to 26% for EPDM).

Table 9.2 Hybrid composites containing natural/synthetic fibers

Fiber type	Compatibilization technique	Processing method	Results	References
Sisal and short glass fiber	Alkalization of sisal	Solution mixing	Inclusion of small amount of glass fiber in LDPE/sisal composites increased strength in longitudinally oriented composites	Kalaprasad et al. (1997)
Sisal and short glass fiber	None	Solution mixing	Sisal reduced thermal conductivity, while hybridization with glass increased conductivity	Kalaprasad et al. (2000)
Henequen and sand	Silane coupling agent, preimpregnation of fibers in HDPE–xylene solution	Extrusion	At low processing temperatures, tensile moduli were controlled by sand content, while at high temperatures, henequen content controlled the properties	Herrera-Franco et al. (1996)

Waste rubber in the form of ground tire rubber (GTR) can also be considered as a very interesting candidate for impact modification of reinforced thermoplastics. Such an approach is beneficial from both environmental and economical aspects. But two main drawbacks are low compatibility between GTR and polyolefins and large size of GTR particles (Grigoryeva et al., 2005). If the compatibility between GTR and matrix is low, (i) effective load transfer from the matrix to the rubber phase cannot be expected, and (ii) creation of gaps and voids at the rubber–matrix interface causes even higher stress concentration leading to premature failure (Kumar et al., 2002). In the past, no research work was successful in producing PE/waste rubber compounds with high compatibility between rubber and thermoplastic phases (Karger-Kocsis et al., 2013). But recently, Kakroodi and Rodrigue (2013a) proposed a new and successful approach to solve this problem. They used maleated PE as the sole matrix (not as a compatibilizer) for GTR compounds. They reported excellent mechanical properties (especially elongation at break). Using this technique, compounds with very high GTR concentrations (as high as 90% by weight) were produced with tensile elongation at break higher than 200% (a good thermoplastic elastomer is expected to have an elongation at break higher than 100%). Kakroodi et al. (2013) also produced hybrid natural fiber composites based on MAPE, hemp fibers, and GTR. Strong interfacial adhesion between MAPE and GTR led to substantial improvement in impact behavior of MAPE/hemp composites. Impact strength of composite containing 50% hemp fiber was improved from 134 to 201 J/m after adding only 10% GTR. Due to strong adhesion between the phases, they also reported that the composites showed significantly lower moisture uptake (lower than 7% for composite containing 60% hemp) in comparison with similar natural fiber composites.

9.3 Stability of PE biocomposites

Concerns regarding environmental issues caused by discarded polymeric products also apply to PE-based biocomposites (mostly due to the presence of PE). Such compounds must also be recycled at the end of their service life. Thermomechanical recycling is a common method regarding polymer recycling using a lower amount of energy (Beg and Pickering, 2008). In this method, the polymers are first granulated and then reprocessed to produce new products. The quality of the recycled products depends strongly on the stability of the materials against two factors: (a) degradation during their service life and (b) degradation during recycling (reprocessing) (Kakroodi and Rodrigue, 2013b).

During service life, environmental stressors such as heat, radiation, oxygen, humidity, or combinations of these factors cause undesirable changes in the properties of polymers and composites such as loss of ductility, change of color or dimensions, and so on (Al-Salem, 2009; White et al., 2011). These changes are usually the result of detrimental chemical modifications such as (Kakroodi and Rodrigue, 2013b):

- Chain scission and cross-linking of macromolecules, which can lead to decrease or increase in molecular weight, respectively.
- Change in molecular structure of polymer macromolecules. One possibility is formation of unsaturated C=C bonds due to C–H bonds breakage in two adjacent carbon atoms. Such

double bonds are highly prone to reaction with macroradicals (which are products of chain scission) resulting in cross-linking. Oxidation of macromolecules can also occur due to aging.

- Formation of small molecules such as H_2, CO, CO_2, and monomers. Once trapped in the polymer network, such molecules (except for H_2) cause crazes and even cracks.

Introduction of fillers in thermoplastics changes their properties including their resistance against degradation (Azuma et al., 2009). Presence of natural fibers can provide protection against degradation of polymer matrices caused by radiation. In the case of GTR-filled compounds, the vulcanized structure of the rubber phase and also the presence of a variety of stabilizers and carbon black are expected to increase stability (Kakroodi and Rodrigue, 2013b).

Reprocessing of polymers can lead to their degradation due to thermal and mechanical stresses (during extrusion and molding processes) (Al-Malaika and Amir, 1989). For natural fiber composites, fiber breakage during extrusion or injection-molding processes is a common result of reprocessing.

Absorption of moisture also causes degradation (especially in composites containing natural fibers) in different ways: (i) swelling of fibers due to absorption of moisture leads to internal stresses and also loss of dimensional stability; (ii) presence of water increases the rate of natural fibers biological degradation; (iii) small molecules of water act as plasticizers and reduce mechanical properties; and (iv) for composites with outdoor applications, freezing and thawing of water inside the composite's structure leads to the creation and growth of cracks (Panthapulakkal et al., 2006; Wang et al., 2006).

Joseph et al. (1995) studied the effects of degradation processes (thermal degradation in water and air) on mechanical properties and dimensional stability of LDPE/sisal composites before and after fiber treatment using cardanol derivative of toluene diisocyanate (CTDIC). They reported that exposure of composites containing 20% sisal fiber to boiling water resulted in significant reduction in tensile strength (25% reduction) and modulus (28% reduction). Composites containing treated fibers showed higher tensile properties and less noticeable loss of tensile strength (10% reduction) and modulus (20% reduction) after exposure to boiling water.

Stark and Matuana (2006) studied the stability of HDPE/WF (50/50 wt.%) composites against photodegradation via exposure to xenon-arc radiation. Color analysis of samples during the exposure period showed that lightness increased. Inclusion of stabilizers, ultraviolet absorber (UVA), and zinc ferrite pigment was reported to decrease the intensity of lightening caused by radiation. Measurement of flexural properties also showed a decrease in strength (over 10 MPa) and modulus (over 1.2 GPa) after simultaneous exposure to radiation and water spray for 3000 h. Composites with stabilizers showed improved retention in mechanical properties. In composites containing 1% UVA and 2% pigment, losses in tensile strength and modulus were 1 MPa and 0.6 GPa, respectively. Exposure of composites to radiation (without water spray), on the other hand, was shown to have the opposite effect on mechanical properties. It was shown that composite flexural modulus and strength increased.

Panthapulakkal et al. (2006) studied the effects of water absorption, freezing, and thawing, as well as photo-aging on HDPE/rice husk composites. They reported that composite compatibilization resulted in water absorption reduction as well as improved mechanical properties retention. After 1600 h, water absorption of composites with and without compatibilizer (polymer-based on ethylene and glycidylmethacrylate) were 8% and 14%, respectively. They performed the following freezing and thawing cycles: freezing at −21 °C for 24 h, followed by thawing at room temperature for 2 h, followed by immersion in water at room temperature for 22 h. After performing 12 cycles, they reported a decrease in flexural strength (over 35% for uncompatibilized composite) and modulus (over 65% for uncompatibilized composite). Again, loss in mechanical properties of compatibilized samples was less significant. In compatibilized composites, losses in flexural strength and modulus were around 20% and 30%, respectively.

9.4 Biocomposites based on recycled PE

During recent decades, production of plastic-based materials has attracted increasing attention due to their relatively low price, low density, easy processing conditions, durability, and good mechanical characteristics. Plastics are used in a wide range of applications such as coating, wiring, packaging, as well as the automotive and construction industries (Aguado et al., 2007). As a result, recycling has emerged as a global concern and an inevitable approach to deal with environmental impacts caused by waste plastics. Among all waste plastic materials, PE is considered as particularly interesting due to ease of separation from the waste stream in the form of blends containing both PE and PP (known as light fraction) (Strapasson et al., 2005).

Researchers have also investigated the possibility of such blends to serve as the matrix phase for natural fiber composites (Favaro et al., 2010; Lu and Oza, 2013b). Kazemi et al. (2013a) recently performed comprehensive work regarding the production of wood–plastic composites from the light fraction of municipal plastic waste. Ethylene octene copolymer (EOC) and maleated polyolefins (maleated PE and maleated PP) were used as compatibilizers to improve adhesion between PE and PP phases and also compatibilizing the PE/PP blend with wood particles, respectively. Morphological observations showed EOC effectiveness through reduction in domain sizes. Maleated polyolefins were also shown to improve compatibility between wood particles and the blend matrix, as well as all the mechanical properties of the resulting composites. For instance, tensile strength of composites containing 30% WF was improved by almost 29%.

Kazemi et al. (2013b) further performed design analysis of three-layered structural composites based on plastic waste and WF. The studied parameters included wood content, thickness of individual composite layers, as well as stacking sequence and configuration (symmetric and asymmetric structures). Through morphological observations performed by SEM and optical microscopy, they reported that the layers had adhesion. Mechanical characterizations showed that for symmetric structures, skin layers had more significant influence on flexural modulus. Interesting results were reported after measurement of flexural properties of asymmetric structures. It was shown that changing the direction of applied load had no effect on flexural modulus, but flexural deformability and strength were higher when the load was applied to the skin face having the highest wood content. For asymmetric structures with different layer thicknesses, the effect of wood content in each layer was shown to be more significant than layer thicknesses.

Dintcheva and La Mantia (1999) performed a work dealing with separation and recycling of the true light fraction of plastic waste streams (PE/PP blend) and studied the effects of WF addition. They reported that WF led to increased tensile modulus, while tensile strength, elongation at break, and impact strength decreased. For instance, tensile modulus increased from around 600 to around 900 MPa when wood content increased from 20% to 40%, while tensile strength decreased from 12.5 to 11.5 MPa. In a similar work, Dintcheva et al. (2001) also studied the effects of different filler types (namely wood fiber, glass fiber, and calcium carbonate) and processing equipment (discontinuous mixer, single- and twin-screw extruder) on the mechanical properties of light fraction-based composites. They reported that inclusion of 20% of all types of fillers led to similar effects on the mechanical properties of the composites. They also reported that although mechanical properties of light fraction (without filler) was not affected by the processing method used, wood fiber-based composites have higher mechanical performance after injection molding in comparison with compression molding. For instance, in composite with 20% wood, tensile strength increased from 11.5 MPa for compression-molded compounds to 20 MPa for injection-molded samples. This might be due to increased fiber alignment caused by higher levels of shear stress during injection molding.

9.5 Challenges and opportunities

As presented here, PE-based biocomposites have the potential to provide a unique scope to produce novel products meeting future demands from both an economical and environmental point of view. These products have the potential to substantially change the face of solid waste treatment, which is considered as a great challenge in today's plastics industry. Based on the above discussions, however, the successful incorporation of natural fibers strongly depends on the effective resolution of these challenges, the most important being low compatibility between the fibers and PE.

9.6 Conclusion

This chapter provided information about biocomposites containing PE as matrix and different natural fibers as reinforcements. The significance of natural fiber use was discussed with a focus on their ecofriendly and economical nature. Poor compatibility between polyolefins (particularly PE/PP) and natural fibers was introduced as the most important drawback in their composites. Most common modification techniques regarding improvement of bonding between natural fibers and PE were presented along with notable research work from the literature. It was shown that increasing compatibility between natural fibers and PE (using suitable compatibilization techniques) noticeably improves mechanical, thermal, morphological, and water absorption properties. Hybridization of natural fiber composites using (i) synthetic reinforcements, (ii) natural fibers with different sizes, and (iii) virgin or recycled rubber was discussed as promising techniques that provide tailored composites covering a wide range of characteristics. Stability of composites against environmental stressors (radiation, heat, ozone, water absorption, etc.) was discussed as an important factor on their service life and also recyclability. Finally, recycled PE (or PE/PP blend) was presented as a promising candidate for the production of natural fiber composites with even smaller environmental impact. Quality of adhesion between PE and PP was discussed as an influential factor regarding quality of composites containing recycled PE/PP blends.

References

Abdelmouleh, M., Boufi, S., Belgacem, M.N., Dufresne, A., 2007. Short natural-fiber reinforced polyethylene and natural rubber composites: effect of silane coupling agents and fibers loading. Compos. Sci. Technol. 67, 1627–1639.

Adhikary, K.B., Pang, S., Staiger, M.P., 2007. Dimensional stability and mechanical behaviour of wood-plastic composites based on recycled and virgin high-density polyethylene (HDPE). Compos. Part B 39, 807–815.

Aguado, J., Serrano, D.P., San Miguel, G., Castro, M.C., Madrid, S., 2007. Feedstock recycling of polyethylene in a two-step thermo-catalytic reaction system. J. Anal. Appl. Pyrolysis 79, 415–423.

Al-Malaika, S., Amir, E.J., 1989. Thermoplastic elastomers: part III—ageing and mechanical properties of natural rubber-reclaimed rubber/polypropylene systems and their role as solid phase dispersants in polypropylene/polyethylene blends. Polym. Degrad. Stab. 26, 31–41.

Al-Salem, S.M., 2009. Influence of natural and accelerated weathering on various formulations of linear low density polyethylene (LLDPE) films. Mater. Des. 30, 1729–1736.

Araujo, J.R., Waldman, W.R., De Paoli, M.A., 2008. Thermal properties of high density polyethylene composites with natural fibers: coupling agent. Polym. Degrad. Stab. 93, 1770–1775.

Arrakhiz, F.Z., El Achaby, M., Kakou, A.C., Vaudreuil, S., Benmoussa, K., Bouhfid, R., Fassi-Fehri, O., Qaiss, A., 2012. Mechanical properties of high density polyethylene reinforced with chemically modified coir fibers: impact of chemical treatments. Mater. Des. 37, 379–383.

Arrakhiz, F.Z., El Achaby, M., Malha, M., Bensalah, M.O., Fassi-Fehri, O., Bouhfid, R., Benmoussa, K., Qaiss, A., 2013. Mechanical and thermal properties of natural fibres reinforced polymer composites: doum/low density polyethylene. Mater. Des. 43, 200–205.

Azuma, Y., Takeda, H., Watanabe, S., Nakatani, H., 2009. Outdoor and accelerated weathering tests for polypropylene and polypropylene/talc composites: a comparative study of their weathering behavior. Polym. Degrad. Stab. 94, 2267–2274.

Balasuriya, P.W., Ye, L., Mai, Y.W., 2001. Mechanical properties of wood flake–polyethylene composites. Part I: Effects of processing methods and matrix melt flow behaviour. Compos. Part A 32, 619–629.

Barone, J.R., Schmidt, W.F., 2005. Polyethylene reinforced with keratin fibers obtained from chicken feathers. Compos. Sci. Technol. 65, 173–181.

Bataille, P., Allard, P., Cousin, P., Sapieha, S., 1990. Interfacial phenomena in cellulose/polyethylene composites. Polym. Compos. 11, 301–304.

Beg, M.D.H., Pickering, K.L., 2008. Reprocessing of wood fibre reinforced polypropylene composites. Part I: effects on physical and mechanical properties. Compos. Part A 39, 1091–1100.

Bengtsson, M., Gatenholm, P., Oksman, K., 2005. The effect of crosslinking on the properties of polyethylene/wood flour composites. Compos. Sci. Technol. 65, 1468–1479.

Beshay, A.D., Kokta, B.V., Daneault, C., 1985. Use of wood fibers in thermoplastic composites II: polyethylene. Polym. Compos. 6, 261–271.

Biagiotti, J., Puglia, D., Torre, L., Kenny, J.M., Arbelaiz, A., Cantero, G., Marieta, C., Llano-Ponte, R., Mondragon, I., 2004. A systematic investigation on the influence of the chemical treatment of natural fibers on the properties of their polymer matrix composites. Polym. Compos. 25, 470–479.

Bisanda, E.T.N., Ansell, M.P., 1991. The effect of silane treatment on mechanical and physical properties of sisal-epoxy composites. Compos. Sci. Technol. 41, 165–178.

Bledzki, A.K., Gassan, J., 1999. Composites reinforced with cellulose based fibers. Prog. Polym. Sci. 24, 221–274.

Cantero, G., Arbelaiz, A., Llano-Ponte, R., Mondragon, I., 2003. Effects of fiber treatment on wettability and mechanical behaviour of flax/polypropylene composites. Compos. Sci. Technol. 63, 1247–1254.

Cao, X.V., Ismail, H., Rashid, A.A., Takeichi, T., Vo-Huu, T., 2012. Maleated natural rubber as a coupling agent for recycled high density polyethylene/natural rubber/kenaf powder biocomposites. Polym.-Plast. Technol. Eng. 51, 904–910.

Chen, H.L., Porter, R.S., 1994. Composite of polyethylene and kenaf, a natural cellulose fiber. J. Appl. Polym. Sci. 54, 1781–1783.

Cheung, H.Y., Ho, M.P., Lau, K.T., Cardona, F., Hui, D., 2009. Natural fibre-reinforced composites for bioengineering and environmental engineering applications. Compos. Part B 40, 655–663.

Clemons, C., 2010. Elastomer modified polypropylene–polyethylene blends as matrices for wood flour–plastic composites. Compos. Part A 41, 1559–1569.

Dintcheva, N.T., La Mantia, F.P., 1999. Recycling of the "light fraction" from municipal post-consumer plastics: effect of adding wood fibers. Polym. Adv. Technol. 10, 607–614.

Dintcheva, N.T., La Mantia, F.P., Trotta, F., Luda, M.P., Camino, G., Paci, M., Di Maio, L., Acierno, D., 2001. Effects of filler type and processing apparatus on the properties of the recycled "light fraction" from municipal post-consumer plastics. Polym. Adv. Technol. 12, 552–560.

Espert, A., Vilaplana, F., Karlsson, S., 2004. Comparison of water absorption in natural cellulosic fibres from wood and one-year crops in polypropylene composites and its influence on their mechanical properties. Compos. Part A 35, 1267–1276.

Facca, A.G., Kortschot, M.T., Yan, N., 2006. Predicting the elastic modulus of natural fiber reinforced thermoplastics. Compos. Part A 37, 1660–1671.

Faruk, O., Bledzki, A.K., Fink, H.P., Sain, M., 2012. Biocomposites reinforced with natural fibers: 2000–2010. Prog. Polym. Sci. 37, 1552–1596.

Favaro, S.L., Lopes, M.S., Neto, A.G.V.C., Santana, R.R., Radovanovic, E., 2010. Chemical, morphological, and mechanical analysis of rice husk/post-consumer polyethylene composites. Compos. Part A 41, 154–160.

Felix, J.M., Gatenholm, P., 1991. The nature of adhesion in composites of modified cellulose fiber and polypropylene. J. Appl. Polym. Sci. 42, 609–620.

Fung, K.L., Xing, X.S., Li, R.K.Y., Tjong, S.C., Mai, Y.W., 2003. An investigation on the processing of sisal fiber reinforced polypropylene composites. Compos. Sci. Technol. 63, 1255–1258.

Gao, H., Xie, Y., Ou, R., Wang, Q., 2012. Grafting effects of polypropylene/polyethylene blends with maleic anhydride on the properties of the resulting wood–plastic composites. Compos. Part A 43, 150–157.

George, J., Bhagawan, S.S., Thomas, S., 1996. Thermogravimetric and dynamic mechanical thermal analysis of pineapple fibre reinforced polyethylene composites. J. Therm. Anal. Calorim. 47, 1121–1140.

George, J., Bhagawan, S.S., Thomas, S., 1998. Improved interactions in chemically modified pineapple leaf fibre reinforced polyethylene composites. Compos. Interfaces 5, 201–223.

George, J., Sreekala, M.S., Thomas, S., 2001. A review on interface modification and characterization of natural fiber reinforced plastic composites. Polym. Eng. Sci. 41, 1471–1485.

Gomez, D.M., Galindo, J.R., Gonzalez-Nunez, R., Rodrigue, D., 2005. Polymer composites based on agave fibers. In: SPE ANTEC Proceedings, Boston. pp. 1345–1350.

Grigoryeva, O.P., Fainleib, A.M., Tolstov, A.L., Starostenko, O.M., Lievana, E., Karger-Kocsis, J., 2005. Thermoplastic elastomers based on high-density polyethylene, ethylene-propylene-diene monomer rubber, and ground tire rubber. J. Appl. Polym. Sci. 95, 659–671.

Herrera-Franco, P., Aguilar-Vega, M.J., 1997. Effect of fiber treatment on the mechanical properties of LDPE–henequen cellulosic fiber composites. J. Appl. Polym. Sci. 65, 197–207.

Herrera-Franco, P., Valadez-Gonzalez, A., Cervantes-Uc, M., 1996. Development and characterization of a HDPE–sand–natural fiber composite. Compos. Part B 28, 331–343.

Holbery, J., Houston, D., 2006. Natural-fiber-reinforced polymer composites in automotive applications. JOM 58, 80–86.

Jacob, M., Francis, B., Thomas, S., Varughese, K.T., 2006. Dynamical mechanical analysis of sisal/palm oil hybrid fibre-reinforced natural rubber composites. Polym. Compos. 27, 671–680.

Jarukumjorn, K., Suppakarn, N., 2009. Effect of glass fiber hybridization on properties of sisal fiber-polypropylene composites. Compos. Part B 40, 623–627.

Jawaid, M., Abdul Khalil, H.P.S., Hassan, A., Dungani, R., Hadiyane, A., 2013. Effect of jute fibre loading on tensile and dynamic mechanical properties of oil palm epoxy composites. Compos. Part B 45, 619–624.

Joseph, K., Thomas, S., Pavithran, C., 1995. Effect of ageing on the physical and mechanical properties of sisal-fiber-reinforced polyethylene composites. Compos. Sci. Technol. 53, 99–110.

Joshi, S.V., Drzal, L.T., Mohanty, A.K., Arora, S., 2004. Are natural fiber composites environmentally superior to glass fiber reinforced composites? Compos. Part A 35, 371–376.

Kakroodi, A.R., Rodrigue, D., 2013a. Highly filled thermoplastic elastomers from ground tire rubber, maleated polyethylene and high density polyethylene. Plast. Rubber Compos. 42, 115–122.

Kakroodi, A.R., Rodrigue, D., 2013b. Degradation behaviour of maleated polyethylene/ground tire rubber thermoplastic elastomers with and without stabilizers. Polym Degrad. Stab. 98, 2184–2192.

Kakroodi, A.R., Bainier, J., Rodrigue, D., 2012. Mechanical and morphological properties of flax fiber reinforced high density polyethylene/recycled rubber composites. Int. Polym. Process. 27, 196–204.

Kakroodi, A.R., Kazemi, Y., Rodrigue, D., 2013. Mechanical, rheological, morphological and water absorption properties of maleated polyethylene/hemp composites: effect of ground tire rubber addition. Compos. Part B 51, 337–344.

Kalaprasad, G., Thomas, S., Pavithran, C., Neelakantan, N.R., Balakrishnan, S., 1996. Hybrid effect in the mechanical properties of short sisal/glass hybrid fiber reinforced low density polyethylene composites. J. Reinf. Plast. Compos. 15, 48–73.

Kalaprasad, G., Joseph, K., Thomas, S., 1997. Influence of short glass fiber addition on the mechanical properties of sisal reinforced low density polyethylene composites. J. Compos. Mater. 31, 509–527.

Kalaprasad, G., Pardeep, P., Mathew, G., Pavithran, C., Thomas, S., 2000. Thermal conductivity and thermal diffusivity analysis of low density polyethylene composites reinforced with sisal, glass and intimately mixed sisal/glass fibres. Compos. Sci. Technol. 60, 2967–2977.

Karger-Kocsis, J., Mészáros, L., Bárány, T., 2013. Ground tyre rubber (GTR) in thermoplastics, thermosets and rubbers. J. Mater. Sci. 48, 1–38.

Kazemi, Y., Cloutier, A., Rodrigue, D., 2013a. Mechanical and morphological properties of wood plastic composites based on municipal plastic waste. Polym. Compos. 34, 487–493.

Kazemi, Y., Cloutier, A., Rodrigue, D., 2013b. Design analysis of three-layered structural composites based on post-consumer recycled plastics and wood residues. Compos. Part A 53, 1–9.

Keener, T.J., Stuart, R.K., Brown, T.K., 2004. Maleated coupling agents for natural fiber composites. Compos. Part A 35, 357–362.

Kim, J.P., Yoon, T.H., Mun, S.P., Rhee, J.M., Lee, J.S., 2006. Wood–polyethylene composites using ethylene-vinyl alcohol as adhesion promoter. Bioresour. Technol. 97, 494–499.

Kretsis, G., 1987. A review of the tensile, compressive, flexural and shear properties of hybrid fibre-reinforced plastics. Composites 18, 13–23.

Kumar, R.P., Amma, M.L.G., Thomas, S., 1995. Short sisal fiber reinforced styrene-butadiene rubber composites. J. Appl. Polym. Sci. 58, 597–612.

Kumar, C.R., Fuhrmann, I., Karger-Kocsis, J., 2002. LDPE-based thermoplastic elastomers containing ground tire rubber with and without dynamic curing. Polym. Degrad. Stab. 76, 137–144.

Lei, Y., Wu, Q., Yao, F., Xu, Y., 2007. Preparation and properties of recycled HDPE/natural fiber composites. Compos. Part A 38, 1664–1674.

Li, X., Tabil, L.G., Panigrahi, S., 2007. Chemical treatments of natural fiber for use in natural fiber-reinforced composites: a review. J. Polym. Environ. 15, 25–33.

Lu, N., Oza, S., 2013a. Thermal stability and thermo-mechanical properties of hemp-high density polyethylene composites: effect of two different chemical modifications. Compos. Part B 44, 484–490.

Lu, N., Oza, S., 2013b. A comparative study of the mechanical properties of hemp fiber with virgin and recycled high density polyethylene matrix. Compos. Part B 45, 1651–1656.

Lu, J.Z., Wu, Q., Negulescu, I.I., 2005. Wood-fiber/high-density-polyethylene composites: coupling agent performance. J. Appl. Polym. Sci. 96, 93–102.

Majid, R.A., Ismail, H., Taib, R.M., 2010. Effects of polyethylene-g-maleic anhydride on properties of low density polyethylene/thermoplastic sago starch reinforced kenaf fibre composites. Iran. Polym. J. 19, 501–510.

Marcovich, N.E., Villar, M.A., 2003. Thermal and mechanical characterization of linear low-density polyethylene/wood flour composites. J. Appl. Polym. Sci. 90, 2775–2784.

Merlini, C., Soldi, V., Barra, G.M.O., 2011. Influence of fiber surface treatment and length on physic-chemical properties of short random banana-fiber reinforced castor oil polyurethane composites. Polym. Test 30, 833–840.

Naik, J.B., Mishra, S., 2007. Esterification effect of maleic anhydride on swelling properties of natural fiber/high density polyethylene composites. J. Appl. Polym. Sci. 106, 2571–2574.

Oksman, K., Clemons, C., 1998. Mechanical properties and morphology of impact modified polypropylene-wood flour composites. J. Appl. Polym. Sci. 67, 1503–1513.

Oksman, K., Lindberg, H., Holmgren, A., 1998. The nature and location of SEBS-MA compatibilizer in polyethylene–wood flour composites. J. Appl. Polym. Sci. 69, 201–209.

Oksman, K., Mathew, A.P., Långström, R., Nyström, B., Joseph, K., 2009. The influence of fibre microstructure on fibre breakage and mechanical properties of natural fibre reinforced polypropylene. Compos. Sci. Technol. 69, 1847–1853.

Panthapulakkal, S., Sain, M., 2007. Studies on the water absorption properties of short hemp-glass fiber hybrid polypropylene composites. J. Compos. Mater. 41, 1871–1883.

Panthapulakkal, S., Law, S., Sain, M., 2005. Enhancement of processability of rice husk filled high-density polyethylene composite profiles. J. Thermoplast. Compos. Mater. 18, 445–458.

Panthapulakkal, S., Law, S., Sain, M., 2006. Effect of water absorption, freezing and thawing, and photo-ageing on flexural properties of extruded HDPE/rice husk composites. J. Appl. Polym. Sci. 100, 3619–3625.

Pasquini, D., Teixeira, E.M., Curvelo, A.A.S., Belgacem, M.N., Dufresne, A., 2008. Surface esterification of cellulose fibres: processing and characterization of low-density polyethylene/cellulose fibres composites. Compos. Sci. Technol. 68, 193–201.

Pickering, K.L., Abdalla, A., Ji, C., McDonald, A.G., Franich, R.A., 2003. The effect of silane coupling agents on radiate pine fibre for use in thermoplastic matrix composites. Compos. Part A 34, 915–926.

Pracella, M., Minhaz-Ul Haque, M., Alvarez, V., 2010. Functionalization, compatibilization and properties of polyolefin composites with natural fibers. Polymers 2, 554–574.

Qiu, F., Wang, M., Hao, Y., Guo, S., 2014. The effect of talc orientation and transcrystallization on mechanical properties and thermal stability of the polypropylene/talc composites. Compos. Part A 58, 7–15.

Raj, R.J., Kokta, B.V., Maldas, D., Daneault, C., 1989. Use of wood fibers in thermoplastics. VII. The effect of coupling agents in polyethylene–wood fiber composites. J. Appl. Polym. Sci. 37, 1089–1103.

Saheb, D.N., Jog, J.P., 1999. Natural fiber polymer composites: a review. Adv. Polym. Technol. 18, 351–363.

Seki, Y., Sarikanat, M., Ezan, M.A., 2012. Effect of siloxane treatment of jute fabric on the mechanical and thermal properties of jute/HDPE. J. Reinf. Plast. Compos. 31, 1009–1016.

Shinoj, S., Panigrahi, S., Visvanathan, R., 2010. Water absorption pattern and dimensional stability of oil palm fibre-linear low density polyethylene composites. J. Appl. Polym. Sci. 117, 1064–1075.

Shinoj, S., Visvanathan, R., Panigrahi, S., Varadharaju, N., 2011. Dynamic mechanical properties of oil pal fibre (OPF)-linear low density polyethylene (LDPE) biocomposites and study of fibre–matrix interactions. Biosyst. Eng. 109, 99–107.

Stark, N.M., Matuana, L.M., 2006. Influence of photostabilizers on wood flour–HDPE composites exposed to xenon-arc radiation with and without water spray. Polym. Degrad. Stab. 91, 3048–3056.

Strapasson, R., Amico, S.C., Pereira, M.F.R., Sydensticker, T.H.D., 2005. Tensile and impact behavior of polypropylene/low density polyethylene blends. Polym. Test 24, 468–473.

Torres, F.G., Cubillas, M.L., 2005. Study of the interfacial properties of natural fiber reinforced polyethylene. Polym. Test 24, 694–698.

Valadez-Gonzalez, A., Cervantes-Uc, J.M., Olayo, R., Herrera-Franco, P.J., 1999. Effect of fiber surface treatment on the fiber–matrix bond strength of natural fiber reinforced composites. Compos. Part B 30, 309–320.

Valente, M., Sarasini, F., Marra, F., Tirillo, J., Pulci, G., 2011. Hybrid recycled glass fiber/wood flour thermoplastic composites: manufacturing and mechanical characterization. Compos. Part A 42, 649–657.

Wang, Y., Yeh, F.C., Lai, S.M., Chan, H.C., Shen, H.F., 2003. Effectiveness of functionalized polyolefins as compatibilizers for polyethylene/wood flour composites. Polym. Eng. Sci. 43, 933–945.

Wang, W., Sain, M., Cooper, P.A., 2006. Study of moisture absorption in natural fiber plastic composites. Compos. Sci. Technol. 66, 379–386.

White, C.C., Tan, K.T., Hunston, D.L., Nguyen, T., Benatti, D.J., Stanley, D., Chin, J.W., 2011. Laboratory accelerated and natural weathering of styrene-ethylene-butylene-styrene (SEBS) block copolymer. Polym. Degrad. Stab. 96, 1104–1110.

Xie, Y., Hill, C.A.S., Xiao, Z., Militz, H., Mai, C., 2010. Silane coupling agents used for natural fibre/polymer composites: a review. Compos. Part A 41, 806–819.

Yemele, M.C.N., Koubaa, A., Cloutier, A., Soulounganga, P., 2010. Effect of bark fibre content and size on the mechanical properties of bark/HDPE composites. Compos. Part A 41, 131–137.

Zhang, F., Endo, T., Qiu, W., Yang, L., Hirotsu, T., 2002a. Preparation and mechanical properties of composite of fibrous cellulose and maleated polyethylene. J. Appl. Polym. Sci. 84, 1971–1980.

Zhang, F., Qiu, W., Yang, L., Endo, T., Hirotsu, T., 2002b. Mechanochemical preparation and properties of a cellulose–polyethylene composite. J. Mater. Chem. 12, 24–26.

Zhang, M.Q., Rong, M.Z., Lu, X., 2005. Fully biodegradable natural fiber composites from renewable resources: all-plant fiber composites. Compos. Sci. Technol. 65, 2514–2525.

Performance of biomass filled polyolefin composites

10

C. Vasile, R.N. Darie-Niţă, E. Părpăriţă
"P. Poni" Institute of Macromolecular Chemistry, Iasi, Romania

10.1 Introduction

Polyolefins are the largest group of thermoplastics by volume of production and consumption and are often referred to as commodity thermoplastics. It is already known that plastics are the biggest problem in terms of environment protection, due to their high amount of waste generated and nonbiodegradability. A simple solution to transform them into degradable/disintegrable materials under the action of environmental factors is available via blending and composite technologies.

Degradation is a general term used to denote that a product will break down into smaller parts, leading to a significant change in the structure of a material and to a loss of properties (molecular weight or mechanical strength) and/or fragmentation. Degradation is highly dependent on the surrounding environment and the exposure time (Lucas et al., 2008).

Although it is important to avoid confusion and misunderstandings, it is difficult to find a universally acceptable definition of biodegradability. Different definitions have been adopted and standardized depending on particular criteria. The European standard EN 13432:2000 "Packaging. Requirements for packaging recoverable through composting and biodegradation. Test scheme and evaluation criteria for the final acceptance of packaging" defines *biodegradation* as "chemical breakdown of materials into CO_2, water and minerals; pursuant to the standard, at least 90% of the materials have to be broken down by biological action within 6 months" and *disintegration* as "the physical decomposition of a product into tiny pieces; after 12 weeks at least 90% of the product should be able to pass through a 2 × 2 mm mesh." The biodegradation of plastics under composting conditions could be measured by test method ISO 14855:2005 "Determination of the ultimate aerobic biodegradability of plastic materials under controlled composting conditions – Method by analysis of evolved carbon dioxide."

Biodegradation is a more specific term that explains the process of how a product degrades under the action of microorganisms or enzymes. Biodegradation is considered nothing more than catabolism (Vasile, 2009). In other words, biodegradation is defined as the transformation of a substance into new compounds through biochemical reactions or the actions of microorganisms such as bacteria (U.S. Geological Survey Science in the Decade, 2007–2017, 2007). When a product biodegrades it is digested by bacteria, fungi, or algae and it is broken down into biogas and biomass or humus. The biodegradable by-products are then naturally recycled back into the cycle of earth (Ross, 2012).

Biocomposites: Design and Mechanical Performance. http://dx.doi.org/10.1016/B978-1-78242-373-7.00012-3

A possible solution to minimize the polymer waste is to use *biodegradable polymers*, which can be degraded through the action of naturally occurring microorganisms. For example, in agriculture by using biodegradable mulch, films would not need to be collected after use and could simply be ploughed into the soil, where they would biodegrade.

The renewability of different biomass-based materials depends on the availability and life cycle of the raw materials. Biomass-based materials can be degraded by the enzymatic action of bacteria, fungi algae, and other organisms, into CO_2, CH_4, H_2O, and/or various inorganic compounds, so they have limited lifetime in the environment.

Polyolefin thermoplastics like polypropylene (PP), polyethylene (PE), or polystyrene (PS) were considered matrices for structural applications. In view of environmentally positive impact, part of these synthetic matrices have been replaced by different types of biopolymers such as starch, cellulose, chitosan, lignin or wood fibres, poly(lactic acid), polyhydroxyalkanoates or others. The incorporation of biomass into polyolefin matrix does not ensure the complete degradation of the host matrix, but induces fragmentation and disintegration.

Neat polyolefins, due to carbon–carbon backbone, are usually resistant to environmental degradation. By introducing digestible components or pro-oxidants, the degradation of the polymer chain is promoted, being disintegrated into microbe-digestible low molecular weight fragments.

Biodegradation of the polyolefins occurs by a combination of abiotic and biotic oxidation, controlled by the rate of initiation of the abiotic peroxidation. Some vicinal hydroperoxides are generated in abiotic peroxidation of the polyolefins, especially in the case of polypropylene, due to the susceptibility of the tertiary carbon atom to hydrogen abstraction via a hydrogen-bonded intermediate. Hydrogen-bonded vicinal hydroperoxides break down to small biodegradable molecules as alcohols, ketones, and carboxylic acids, as well as longer chain oxygen-modified breakdown products, which oxobiodegrade more slowly. These are biodegradable species, similar to products formed by hydrolysis of aliphatic polyesters and they are rapidly bioassimilated to give cell biomass (Wiles and Scott, 2006).

Polypropylene could not be classified as a biodegradable polymer, but PP can be used in obtaining ecocomposite materials. Tech-Wood International Netherlands developed Tech-Wood® ecocomposite containing 70% pinewood fibres and 30% compatibilized PP, suitable for construction elements (Netravali and Chabba, 2003).

Polyolefin/starch blends rapidly loose strength, disintegrate, and visually disappear when exposed to microorganisms; nevertheless, the polyolefin fraction persists in the environment.

Under composting or litter environments, biodegradable polyolefins undergo the following degradation steps (Vasile, 2009): (i) digestion of the starch component by microorganisms that weaken the polymer matrix; (ii) oxidative or photo-oxidative breakdown of the polymer chains; and (iii) microbial mineralization of the polymer fragments. Microorganisms further convert low molecular weight fragments into water, carbon dioxide and biomass.

Chandra and Rustgi (1997) presented the biodegradability of maleated linear low-density polyethylene (MA-g-LLDPE) and starch blends in two biotic environments: soil environment over a period of 6 months and mixed fungi inoculum (*Aspergillus niger*, *Penicillium funiculosum*, *Chaetomium globosum*, *Gliocladium virens* and *Pullularia pullulans*) for 28 days, concluding that blends exposed to a soil environment degraded more than in fungi alone. The percentage of crystallinity decreased as the starch content increased and biodegradation resulted in an increase of crystallinity in MA-g-LLDPE/starch blends.

Muthukumar et al. (2014) proved the biodegradation of high-density polyethylene (HDPE) and starch blends exposed to seawater for 6 months by altering its physical and chemical properties. The formation of C–O stretching band and a decrease in ester and ketone carbonyl bands were observed in FT-IR spectra. The polymer surface became hydrophilic and 17% weight loss was observed. In vitro biodegradation of the polymer exposed to sunlight for 150 days incubated with the isolated pure strain (*Exiguobacterium*) and combination of two strains (*Exiguobacterium* and *B. subtilis*) for 75 days, and showed a gravimetric weight loss of 4.7% and 12.1%, respectively, indicating synergy between the two organisms.

In the last few years there have been several attempts to find selective microorganisms or enzymes to biodegrade polyolefins (Santo et al., 2013; Gilan and Sivan, 2013).

A wide range of filler materials and polymer matrices are available for obtaining polyolefins/biomass composites (Table 10.1).

The main disadvantage of natural fibres is their hydrophilic nature, which lowers the compatibility with hydrophobic polymeric matrices like polyolefins during composite fabrication. Due to their hydrophilic character, swelling by water uptake can lead to microcracking of the composite and degradation of mechanical

Table 10.1 Natural fillers for thermoplastic composites

Natural filler	Matrix polyolefin	References
Wood flour/fibre	PE, PP, PVC, PS, PU	Nabi Saheb and Jog (1999)
Seaweed	PP, PE	Ferrero et al. (2014) and Jang et al. (2013)
Jute	PP, Epoxy, Polyester, Phenol–formaldehyde	Nabi Saheb and Jog (1999)
Doum fibres	PP	Essabir et al. (2013)
Sisal	PP	Jayaraman (2003)
Chicken eggshell	PP	Toro et al. (2007)
Pineapple	PP	Arib et al. (2006)
Cellulose	LDPE	Soulestin et al. (2007)
Flax, Kenaf	PE, PP	Chen and Porter (1994) and John and Anandjiwala (2009)
Lignin	PP, PP	Glasser et al. (2000, pp. 2–99) and Kharade and Kale (1999)
Chitosan	LDPE, PP	Vasile et al. (2013) and Suwannamek and Srikulkit (2013)
Corn starch	HDPE, LDPE	Oragwu and Igwe (2013)
Banana	PP, Polyester	Pothan et al. (2006b) and Paul et al. (2008)

properties (Arbelaiz et al., 2005; Premalal et al., 2002). The mechanical properties of biomass-reinforced polymer composite depend not only on the properties of constituents, but also on the properties of the region surrounding the filler known as the *interphase*. Separation of fibre–matrices coupling cause cracks and leak bonding under the effect of external load. Debonding depends on the shape and size of filler, on the stiffness of the matrix, and on interfacial adhesion.

Tensile, flexural, and impact properties are the most commonly investigated mechanical properties of natural fillers or synthetic fillers-reinforced plastic composites. Impact strength is one of the undesirable weak points of these materials in terms of mechanical performance. Some methods could be used in order to enhance the adhesion between hydrophobic matrix of polyolefins and the hydrophilic filler; namely, pretreatments of the filler surface by the addition of active agents as water soluble binder (hydrophilic polymers) together with an association of debonding agent (nonionic surfactant) and lubricants (silicone oil, fatty amides) (Barlow et al., 2004); and polymer modification by mixing with a more hydrophilic coupling agent in order to make it compatible with the filler. Several studies showed that maleated polypropylene (MAPP), maleated styrene-ethylene/butylene-styrene (SEBS-MA) or styrene-maleic anhydride (SMA), might be used as coupling agents to react with the hydroxyl groups of lignocellulosic fibre and to positively affect the dispersion (Olsen, 1991; Oksman et al., 1998). Properties like Young's modulus were improved due to entanglements and interdiffusion within the matrix (Gauthier et al., 1998). Other coupling agents including titanates, silanes, zirconates, isocyanates, and others, were used (Korta et al., 1990).

10.2 Recent progress in mechanical performance and design of polyolefin/biomass composites

10.2.1 Polyolefin/wood composites

As plastic/wood composites undergo degradation at high processing temperatures, it is important to select a suitable polymer matrix that can be melted and processed at relatively low temperatures. Thermoplastic polymers are preferred, as they allow easy processing of plastic/wood composites. The most common polymer matrices used are the polyolefins, such as PE and PP (Ashori, 2008; Xanthos, 2005).

As can be seen from Table 10.1, the majority of the work was done on wood flour/fibre and polyolefins. If compared with inorganic fillers, such as calcium carbonate (the most abundantly used filler globally), which have proven effective at enhancing mechanical and physical properties of composites materials, wood fillers has the advantage of being significantly less dense than inorganic materials. Therefore, wood fillers offer weight reduction to composites materials, which allows them to be used in a wider variety of new applications (Xanthos, 2010). When compared with synthetic fillers and reinforcements traditionally used in polymer composites, wood is considered a low-cost and more environmentally friendly raw material.

10.2.1.1 Processing of polyolefin/wood composites

The processing of wood composites has a crucial influence on the properties of the final products. Generally, the selection of processing techniques is based on the matrices, the feed form of the raw materials, the application of the final product, and costs. The most popular methods used to produce plastic/wood composites are compression (Umer et al., 2008), extrusion (Schirp and Stender, 2010; Hietala et al., 2011), and injection moulding (Bledzki and Faruk, 2006; Frisk and Schwendemann, 2004; Umer et al., 2008). The compression method can be used as one component of the manufacturing process or can be connected with extrusion or injection processes (Dai and Fan, 2014). Extrusion and injection moulding are two other important methods to manufacture plastic/wood composites. Although different, these processes follow the same basic steps: melting, shaping, and cooling (Michaeli and Menges, 1989). Stark et al. (2004) studied 50% pine flour-filled HDPE composites manufactured by injection and extrusion moulding. The authors suggest that the higher density of injection-moulded samples resulted in more intimate contact between the HDPE matrix and wood flour and, consequently, higher mechanical properties than extruded samples. The improvement in interfacial quality led to more efficient stress transfer between the matrix and fibre, resulting in increased strength. In a study conducted by Bledzki and Faruk (2004), these methods were also compared. In this case, composite materials were obtained based on polyethylene matrix (70%) and hardwood particles. In this study, the composite density was similar for both processes and higher mechanical properties that resulted by the injection moulding process.

Effective mixing is important for achieving good dispersion of wood filler and to optimize the final properties of the composite materials. Bledzki and Faruk (2004) studied the effect of the compounding method of softwood and hardwood fibre content on composite melt flow behaviour. The results showed that if it is used at a high content (50%) of wood fibre, all the compounding processes meet several problems such as agglomeration of wood filler as a result of poor dispersion and increase of viscosity of composite, which results in further processing difficulties and limited processing temperature of composite due to limited thermal stability of the wood filler. Moreover, higher values of MFI were obtained for composites filled with softwood fibre. The authors consider that waxes and other extractives may diffuse to the surface and increase flow ability of the composite.

10.2.1.2 Properties of polyolefin/wood composites

Thermoplastic/wood composites are typically described as materials that combine the properties of thermoplastics and wood. Composites made with wood and polymer matrix are used for the purpose of obtaining materials with some characteristics of these two resources, such as the appearance of wood and the performance of plastic in a humid area. In general, adding wood to a thermoplastic matrix should provide increased stiffness of the polymer, and the polymer, in turn, should ensure the protection of wood fibres against moisture absorption.

There are several aspects to take into account when making a wood polymer composite material. First, it is important to consider the effect of fibre geometry

(Bledzki and Faruk, 2003) and size (Migneault et al., 2009), as well as fibre distribution and content (Ayrilmis and Kaymakci, 2013) in order to control the physical properties of the composite material. In a study performed by Lomelí-Ramírez et al. (2014) it was demonstrated that better mechanical properties can be obtained for biocomposites with 30% content of wood filler and small particle size (4 mm), because of a better adhesion between the matrix and wood particles.

The type of wood filler is another aspect that should be considered. Bledzki and Faruk (2003) found that composites with hardwood fibres have better impact strength than softwood fibres. Ashori and Nourbakhsh (2010) found that composites containing oak fibres had inferior strength compared to pine composites. In the same study, it was reported that the higher the amount of extractible material in the wood, the lower the water absorption of the wood fibres. This was explained in terms of the extractible material interacting with the hydroxyl groups of the cellulose and lignin, making these inaccessible for interaction with water molecules. The same authors showed that wood fillers as fibres or flour could be used to prepare plastic/wood composites.

Blending of the plastics with the wood materials may require the use of many different types of additives. Compatibilizers may play a useful role in improving plastic/wood composites properties, such as interfacial adhesion, flow, and mechanical properties. Various techniques are used for the improvement of interfacial adhesion including the coating of wood flour with stearic acid (Raj and Kokta, 1991) and coupling with functional silanes (Xie et al., 2010) or aluminium ester (Qin et al., 2005). However, polymers functionalized with maleic anhydride (MA) or acrylic acid (AA) are introduced the most often into polyolefin composites (Chen et al., 2013). It was demonstrated that the functional groups of these two polymers strongly physically interact or chemically react with the surface of wood. The stiffness of the composites is not significantly influenced by the presence of the functionalized polymer, but usually the mechanical properties (tensile and flexural strength) increases considerably. The filler content can be correlated with the coupling agent as it was shown that the mechanical properties usually decreases with increasing filler content in the absence of coupling, but the correlation can be turned around and considerable increase can be achieved with the help of functionalized polymers (Danyadi et al., 2007).

An impressive amount of research work on plastic/wood composites has been published; most of these studies have investigated the mechanical, thermal, or morphological properties and how they are affected by the addition of the natural filler. Părpăriţă et al. (2014a) investigated the possibility of using different biomass materials as fillers and established their effect on the polypropylene matrix properties. The results showed an increase in the trend of crystallinity for composites, compared to the polymer matrix (Figure 10.1).

The crystallinity degree of the polypropylene was found to be 62.8% and the highest values were evidenced for composites based on polypropylene and *Eucalyptus globulus* and Norway spruce biomass, while the lowest values were for the composites based on polypropylene and pinecones and grape seeds biomass. The authors explained that this behaviour may be due to the fact that the biomass fillers introduced in the PP matrix settle in the amorphous region and act as nucleation centres, especially because the dimensions of the particles was small (160 μm).

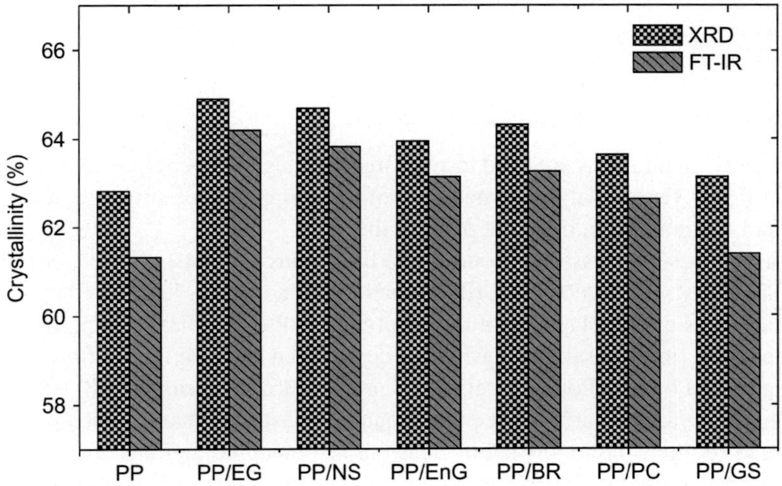

Figure 10.1 The crystallinity indices (%) for the PP and PP/lignocellulosic composites evaluated by X-ray diffraction and infrared spectroscopy.
Reproduced from Părpăriţă et al. (2014a), with permission from Elsevier.

The type of biomass also influences the degradation and pyrolysis behaviour of polypropylene due to interactions between partners (Părpăriţă et al., 2014b,c,d).

The proof of any material is in its performance. Most of the applications for wood–plastic composites have been for outdoor materials and, consequently, the durability of these products becomes a concern. These applications require the materials to stand up to a wide variety of environmental factors including moisture, UV radiation, high temperatures, freeze-thaw cycles, mechanical action, and biological invasion. These factors lead to colour degradation, which affects their aesthetic appearance and mechanical performance. Serious efforts have been made to develop photostabilized plastic/wood composites (Muasher and Sain, 2006). Li et al. (2014) investigated the UV degradation process on the durability of HDPE-based plastic–wood composites, comparing the roles played by hindered amine light stabilizers (HALS), ultraviolet absorbers (UVA), and ZnB. The results showed that compared with HALS and ZnB, UVA was more effective regarding the improvement of UV resistance.

10.2.2 Polyolefin/natural fibres composites

Natural fibres composites increased over the last few years owing to the following advantages:

1. Environmental friendliness
2. Easy availability
3. Intrinsic properties
4. Relatively low cost compared to conventional materials such as glass fibre or aramid fibre
5. Recyclability
6. Biodegradability

7. Competitive in terms of strength per weight of material (mechanical performance comparable with that of synthetic fibres)
8. Lower density
9. Easy processability
10. Satisfactory good thermal properties, a.s.o.

Generally, plant fibres are used to reinforce plastics in the composites industry.

Plant fibres vary widely in terms of chemical composition, structure, and dimensions, and originate from different parts of the plant.

Natural fibres are classified according to their source: plants, animals, or minerals. There are many varieties of plant fibres such as hairs (cotton, kapok), fibre sheafs of dicolytic plants or vessel sheafs of monocotylic plants (e.g., flax, hemp, jute, and ramie), and hard fibres (sisal, henequen, and coir), and a large variety of fibres obtained from trees, and others (Pothan et al., 2007, pp. 44–86; Pickering, 2008). Plant fibres obtained from various parts of the plant—namely, bast (jute, banana, hemp and flax), stem, seed (cotton, ramie, oil palm), leaf (sisal, pineapple and abaca), and coconut husk—are being used in the preparation of composites to achieve desired properties or to reduce the price of the finished article (Westrup et al., 2014).

A major disadvantage of natural fibres is their highly polar nature, which makes them incompatible with nonpolar polymers such as polyolefins. To modify the surface of natural fibres in order to reduce the hydrophilicity, several procedures have been applied (Belgacem and Gandini, 2005): coating with waxy substances; physical treatments, such as solvent extraction; physicochemical treatments, such as the use of corona and plasma discharges or laser, X-ray, and ultraviolet (UV) γ-ray irradiation; surface fibrillation; and chemical modifications, both by the direct condensation of the coupling agents onto the cellulose surface and by various modification/grafting strategies with suitable reagents including polycondensations and free radical or ionic polymerisations.

The physical methods used for surface modification involve methods such as stretching, calendaring, thermal treatment, production of hybrid yarns, and treatment with a solvent or extraction with a solvent of low molecular weight compounds. These methods do not change the chemical composition of the fibre. However, they change the structural and surface properties of the fibre and, thereby, influence the mechanical bonding to polymers.

These physicochemical treatments change the structural and surface properties of the fibre and, thereby, influence the mechanical bonding with the polymer. Surface activation by corona and plasma treatment of natural fibres increases their hydrophilicity by an increased number of aldehyde groups (Gassan and Gutowski, 2000; Gassan et al., 2000) and other modifications depending on the discharge gas used (Bledzki and Gassan, 1999; Yuan et al., 2004). Low-temperature plasma treatment causes mainly chemical implantation, etching, cross-linking, polymerisation, free radical formation, crystallisation, and surface roughness, which leads to better fibre–matrix interaction (Sdrobiş et al., 2012a,b,c).

Laser beam, UV radiations, and γ-irradiation also were applied on natural fibres to enhance surface properties and interaction with matrix (Belgacem and Gandini, 2005). Chemical treatment with alkali or mercerisation is one of the oldest methods used to modify natural fibres. The natural and artificial impurities from the surface of the fibre are removed on treatment with alkali, which brings about improved surface tension, improved wettability, and thereby improved interface bonding (Owolabi et al., 1985; Pothan et al., 2002; Ray and Sarkar, 2001).

The chemical coupling reactions are applied to attach the hydrophobic grafts to surface of the cellulosic fibres by reaction of surface –OH groups by etherification using chlorotriazine derivative (Pothan et al., 2000), esterification using acetic anhydride, alkyl ketene dimer, alkenyl succinic anhydride, and different fatty acids (with *n* varying from 6 to 22) or their chlorides (Joseph et al., 2002a, 2003a), urethane (Joseph et al., 2003a,b), or siloxane formation (Pothan et al., 2006a,b), which protects the fibre surface from water uptake and makes it more compatible with nonpolar matrices like polyolefins, hence ensuring interfacial adhesion. The most suitable is graft copolymerisation (Popescu et al., 2011; Sdrobiş et al., 2011, 2012b). For graft copolymerisation, vinylic monomers such as acrylonitrile (AN) (Mishra et al., 2001), methyl methacrylate, propionic, crotonic, and methacrylic anhydrides (Cetin and Hill, 1999) have been used.

10.2.2.1 Processing of polyolefin/natural fibres composites

As already mentioned, the hydrophilic natural fibres are generally incompatible with hydrophobic polyolefins. Therefore, a pretreatment such those already mentioned is applied before the processing step (George et al., 1998; Joseph et al., 2002a,b, 2003a,b; Kalaprasad and Thomas, 2003; Kalaprasad et al., 2004) but in certain studies where peroxide treatment was used, the peroxide was added to the molten polymer–fibre mixture (Malunka et al., 2006).

Natural fibre/polyolefin composites are normally processed through melt mixing, extrusion, compression moulding, and/or injection moulding (Arbelaiz et al., 2005; Joseph et al., 2003a,b; Keener et al., 2004; Mokoena et al., 2004; Malunka et al., 2006) or by solution mixing to avoid fibre deterioration (Li et al., 2000).

10.2.2.2 Properties of polyolefin/natural fibres composites

Mechanical performance and properties of natural fibre/polyolefin composites are summarized in Table 10.2.

Generally, natural fibre/polyolefin composites containing treated fibres show improved tensile strength and modulus, while the elongation at break decreased (Joseph et al., 1996a). Isocyanate-treated LDPE composites exhibited superior mechanical properties, and dicumylperoxide (DCP)-treated LLDPE-sisal showed higher tensile strength values (Mokoena et al., 2004), but because of the combined effects of cross-linking and grafting leading also to an increase in the elongation at break was also observed. Silane treatment of sisal fibres also strongly improves the tensile properties of the composites because the vinyl groups present in the silane undergo polymerisation in the presence of DCP and form a long hydrophobic polymeric chain on the fibre surface, which interacts with the polyethylene matrix through van der Waals type of bonding. It was found that the efficiency of the different treatments varies in the following order: alkali $<$ acetylation $<$ stearic acid $<$ $KMnO_4$ $<$ MA polymer treatment $<$ silane $<$ dicumyl peroxide $<$ dibenzoyl peroxide (Kalaprasad et al., 2004).

Table 10.2 **Mechanical performance and other properties of natural fibre/polyolefin composites**

Biocomposite	Treatment/processing	Mechanical performance and other properties	References
LDPE-sisal; LLDPE-sisal	Treatments using chemicals such as sodium hydroxide, isocyanate, permanganate and peroxide, maleic anhydride/solution mixing technique followed by extrusion. Efficiency of the different treatments varies in the following order: alkali < acetylation < stearic acid < $KMnO_4$ < maleic anhydride polymer treatment < silane < dicumyl peroxide < dibenzoyl peroxide	Improved tensile strength and modulus, while the elongation at break decreased; at higher sisal loadings (40% and 50%), poor wetting of fibres by the matrix enables easier conformational rearrangements of PE-chains during stress relaxation. Cardanol derivative of toluene diisocyanate (CTDIC) and dicumyl peroxide (DCP)-treated fibres showed the best properties; the melting temperatures of untreated composites increased with the amount of sisal fibre, while samples with 3% of DCP showed the opposite behaviour; untreated composites show higher dielectric constants at all frequencies than pure LDPE or its treated composites; show reduced water absorption	Joseph et al. (1996a), Kalaprasad et al. (2004), and Mokoena et al. (2004)
PE-sisal and sisal/glass LDPE hybrid	Fibre pretreatment with stearic acid	Improved interfacial shear strength by 23% with respect to untreated fibres; viscosity increases	Kalaprasad and Thomas (2003) and Torres and Cubillas (2005)
EVA-sisal	DCP-treated, grafting between the ethylene vinyl acetate copolymer (EVA) chains and sisal fibres		Joseph et al. (1996a,b)

Sisal/HDPE	Silane and permanganate treatments	Improvement in interfacial strength	Li and Mai (2006)
Short sisal fibres/ LDPE	Peroxide treatment	The presence of wax and sisal fibre, as well as peroxide treatment, has an effect on the thermal stability of the composites	Luyt and Malunka (2005)
Sisal fibres reinforce high-density polyethylene (HDPE) and low-density polyethylene (LDPE)	Nanocomposites were prepared by solution casting from toluene and melt mixing, both followed by melt pressing	Presence of whiskers, as well as nanocomposite preparation method, had an observable influence on the storage modulus of LDPE, but very little influence on that of HDPE; improvement in tensile modulus, and a decrease in elongation at break, but the stress at break only improved for the HDPE nanocomposites	Mokhena and Luyt (2014)
Flax-HDPE (10% fibre by volume)	Hand lay-up and compression moulding technique	*Significant* enhancement of *toughness* of the *composite*, ductility falls with increasing fibre reinforcement. Toughness of the composites by 50 kJ/ m^2 over that of recycled HDPE indicates that the flax fibre plays an important role in the toughening processes	Singleton et al. (2003)
HDPE/woven flax fibres/SiO_2	Hot press	Impact energy and stiffness value of high-density polyethylene/flax composites was markedly higher than that of high-density polyethylene but reflected the effects of composite structures and flax content. Incorporation of SiO_2 particles enhanced resistance to thermal degradation	Siengchin (2012)

Continued

Table 10.2 Continued

Biocomposite	Treatment/processing	Mechanical performance and other properties	References
Jute fibre-reinforced chemically functionalized high density polyethylene Jute fibre-reinforced–recycled polyethylene	Jute fibre-reinforced high-density polyethylene (HDPE) (50% virgin + 50% recycled) composites by injection moulding technique	The reinforcement of jute fibres into HDPE matrix reduces the tensile strength. The tensile strength of the composite decreases from 3.4% to 24.3% with the increase in fibre content from 10% to 30% when compared with specimen made of equal proportion of virgin and recycled HDPE; flexural strength increases with the increase in jute fibre content in the composite	Singh et al. (2013, 2014)
Kenaf bast powder-filled high-density polyethylene/ethylene propylene diene monomer	Mixing HDPE and EPDM in an internal mixer then fibres were added; treated kenaf bast powder using hydrochloric acid (HCl) hydrolysis	Treated filled HDPE/EPDM composites had higher tensile strength, elongation at break, and modulus than the untreated filled HDPE/EPDM composites	Safinas et al. (2013)
Coir and abaca fibre-reinforced linear low-density polyethylene (LLDPE) composites (30 wt.% fibre)	Compression moulding; Coir and abaca fibres were treated with methyl methacrylate (MMA) using ultraviolet radiation	Chemically treated fibre-reinforced specimens yielded better mechanical properties compared to the untreated composites, whereas coir fibre composites had better mechanical properties than abaca fibre-reinforced ones. For the improvement of the properties, optimized coir (coir fibre treated with 30% MMA) and abaca (abaca fibre treated with 40% MMA) fibres	Zaman et al. (2011)

Sisal/PP		The cutting mechanism of solid and hollow drill point geometry are substantially different	Bajpai and Singh (2013)
Flax fibre bundles/PP	Surface treatment with MAPP or matrix modification with MAPP/injection moulding	Stiffness of unmodified and MAPP-modified composites increases with increasing fibre bundle content, with the modulus values for the MAPP-modified composite slightly higher; water uptake increases with fibre bundle content	Arbelaiz et al. (2005)
PP–wood fibre	Ar-plasma treatment	The tensile strength decreases after adding 10 mass% of wood fibres, but then gradually increases with the addition of wood fibres until 30 mass% is reached, whereas the tensile modulus increases monotonically up to its maximum value at 30 mass% of wood fibres, storage moduli of plasma-treated wood fibre composites are higher than those for untreated fibre composites	Yuan et al. (2004)
Silane modified PP–wood pulp	Treatment with polymethylene polyphenyl isocyanate and MAH-PP coupling agent	Shear and tensile strengths increase by about 100% and 25%, respectively, for flax–polypropylene composites, when a maleic anhydrided (MAH) grafted PP coupling agent is added	Kokta et al. (1983), Maldas et al. (1989), and Raj et al. (2003)
Wheat straw and jute/PP	Maleic anhydride PP (MAPP) treatment improves wetting of filter paper, wheat straw, and jute fibres	Damaging initiations shifted to higher forces with a strong fibre–matrix adhesion	Gassan and Bledzki (1997)

Continued

Table 10.2 Continued

Biocomposite	Treatment/processing	Mechanical performance and other properties	References
Short sisal fibre reinforced PP	Maleic anhydride and $KMnO_4$ treated	Glass transition temperatures and dynamic moduli increase; sisal fibre degrades before the PP matrix; by the introduction of sisal fibre into the PP matrix, the extent of PP degradation during UV irradiation can be reduced	Joseph et al. (2002a,b, 2003a,b)
Untreated and treated jute and hemp fibre-reinforced polypropylene-maleic anhydride polypropylene copolymer (PP-MAPP)		Microfailure for jute occurred by splitting for jute, the microfailure modes for hemp fibre showed final fibre fracture	Park et al. (2006)
Jute-PP	With and without compatibilizer MAPP	Improved interaction in the case of the treated fibre composites	Doan et al. (2006)
Polypropylene/hemp shives	Melt mixing, poly(propylene)-co maleic anhydride (MAH-PP) 3 wt.% as compatibility agent. Brabender LabStation (Germany) 175 °C, 75 rpm for 10 min	Tensile strength increases up to an optimal value of hemp shive's content (40%); the flexural strength, flexural modulus, and Young's modulus increase with the increase in the content of hemp shives in the composites. The increase in the content of hemp shives in the composites produced the decrease in the impact strength. The stiffness of the composites increases with the increase in the artificial weathering time	Popa et al. (2013) and Niu et al. (2011)

25% and 40% (by weight) hemp fibre/PP	Idem above	Impact resistance was independent of temperature, whereas flexural and tensile properties were strongly affected	Tajvidi et al. (2010) and Dewidar et al. (2010)
Abaca and jute/PP	Coupling agent (maleated polypropylene-MAH-PP), Mixer-injection moulding	Tensile, flexural, and Charpy impact strengths increased for fibre loadings up to 40 wt.% and then decreased. Jute fibre composites provided best tensile properties but abaca fibre polypropylene composites	Bledzki et al. (2007)
Jute (20, 25, 30, and 35 wt.%)/PP	Single screw extruder and an injection moulding; raw jute fibre was oxidized and manufactured composites were post-treated with urea to increase the compatibility of the jute fibre with the polypropylene matrix	Post-treated jute fibre reinforced specimens yielded better mechanical properties compared to the oxidized and raw ones	Kabir et al. (2010) and Rahman et al. (2010)
Jute/PP	Chemical treatment by benzenediazonium salt in alkaline medium		
(10, 15, 20, 25, and 30 wt.%) Raw coir and abaca/fibre-reinforced PP	Raw coir and abaca were chemically treated with benzene diazonium salt	Chemically treated fibre-reinforced specimens yielded better mechanical properties compared to the raw composites, while coir fibre composites had better mechanical properties than abaca fibre-reinforced ones. Based on fibre loading, 30% fibre-reinforced composites had the optimum set of mechanical properties	Haque et al. (2010)

Continued

Table 10.2 Continued

Biocomposite	Treatment/processing	Mechanical performance and other properties	References
Short jute fibre (2–3 mm) reinforced polypropylene PP-based composites (20% fibre by weight)	Compression moulding	Short jute-based composites showed excellent mechanical properties and comparable to short E-glass-based composites. Soil degradation test of both types of composites indicated that jute/PP composites significantly lost much of its mechanical properties but E-glass/PP composites retained major portion of its original integrity	Khan et al. (2012)
Kenaf/PP fibres in a ratio of 1:1 in weight	Surface treatments to the comingled kenaf–polypropylene fibre mats were first carried out using different coupling agents to increase the compatibility of the natural fibre and the polypropylene matrix; composite laminates from the mats via compression moulding	Maleated polypropylene was able to promote strong adhesion between the kenaf fibres and the polypropylene matrix at their interfaces; excellent reinforcements	Deng et al. (2013)
Kenaf/PP fibres	Benkshire needle-punching machine in order to generate nonwoven felts and then thermal pressing	Optimum manufacture condition is at higher temperature (230 °C) but shorter time (60 s)	Hao et al. (2012)

Kenaf/recycled PP	Laccase-treated kenaf fibre, maleic anhydride grafted polypropylene (MAPP) was used with a ratio of 1:10 as coupling agent to fibre; extrusion and injection moulding	An optimum fibre loading of 40% gave the highest tensile properties. Flexural, impact, and thermal properties of the composites and crystallinity of the matrix were improved due to treatment	Islam et al. (2013)
30 wt.% Abaca/PP	Abaca fibres were modified by fungamix and natural enzyme and coupling agent (MA-PP), high-speed cascade mixer, melt compounding, then injection moulding process	The moisture absorption of the composites was found to be reduced 20–45% due to modification. Tensile strength found to be 5–45% and flexural strengths found to be 10–35% increased due to modification. Modified fibre composites found to better resistance in acid and base medium	Bledzki et al. (2010)

The length of fibres is also a critical parameter. In the case of LDPE-sisal composites, tensile strength increases with increasing fibre length and reaches a maximum at 6–8 mm fibre length, and then with further increase in fibre length the tensile strength decreases (Kalaprasad and Thomas, 2003).

LDPE-sisal composites ageing showed that the surface of the cardanol derivative of toluene diisocyanate (CTDIC)-treated sisal fibre is irregular or rough. This leads to strong physical interaction between the fibre and matrix. Therefore, treated natural fibres are likely to offer greater resistance to pullout from the matrix than synthetic fibres, and these result in superior mechanical properties (Joseph and Thomas, 1995). The change in the composites properties during ageing could be associated with increased water adsorption because of the fibres presence.

The lower thermal stability of up to 230 °C for natural fibres limits the number of thermoplastics to be considered as matrix material for natural fibre thermoplastic composites; namely, to thermoplastics whose processing temperature does not exceed 230 °C.

In hydrophilic fillers such as cellulose fibres, an increase in water absorption can be expected (George et al., 1998). As polyolefins are hydrophobic and natural fibres hydrophilic, the absorption of water is by virtue of the fibres alone. As the fibre loading increases, the cellulose content increases, which in turn results in the absorption of more water. The hydrophilicity of the fibre can be reduced to a considerable extent by suitable chemical treatment. In the case of treated-fibre composites, the water absorption rate decreases considerably compared to that of the untreated-fibre composites; the reduction depends on the nature of the treatment.

Natural fibre composites are a viable alternative to synthetic fibre-reinforced composites and they have an edge over conventional materials used in the building as well as automobile industries. Biofibre composites in the automotive industry both reduce material waste and increase fuel efficiency. If glass fibre composites are replaced by natural fibre composites, there will be a reduction in the weight of 10–20 kg (Bledzki et al., 2006). Natural fibre composites are also finding wide applications in construction.

10.2.3 Polyolefin/polysaccharide composites

10.2.3.1 Polyolefin/cellulose composites

As generally recognized, the most abundant, renewable polymer resource available worldwide today is *cellulose*. Krässig (1993) mentioned an estimation of 10^{11}–10^{12} tons of cellulose synthesized by photosynthesis, annually. Cellulose can be found also in agriculture residues, water plants, grasses, and other plant substances.

Cellulose is a polydisperse linear homopolymer, consisting of regio- and enantioselectively β-1,4-glycosidic linked D-glucopyranose units (so-called anhydroglucose units). Consequently, the hydroxyl groups are positioned in the ring plane (equatorial), while the hydrogen atoms are in the vertical position (axial). The polymer contains free hydroxyl groups at the C-2, C-3, and C-6 atoms. Based on the OH groups and the oxygen atoms of both the pyranose ring and the glycosidic bond, ordered hydrogen bond systems form various types of supramolecular semicrystalline structures (Klemm et al., 2002, pp. 275–287).

Processing of polyolefin/cellulose composites

Cellulose-based polymer composites are characterised by low cost, desirable fibre aspect ratio, low density, high specific stiffness and strength, biodegradability, flexibility during processing with no harm to the equipment, and good mechanical properties (Maya and Sabu, 2008).

The reduced thermal stability represents an important disadvantage of the cellulosic reinforcements, explaining the severe degradation of the material that occurs during the processing of the composites by extrusion and injection. Many studies have been conducted by choosing low-density polyethylene (LDPE) because it allows a low processing temperature due to relatively low melting point of the polymer (110 °C). Thus, a reasonable margin before thermal degradation of cellulose is available, which limits the thermal modification of cellulose during processing (Soulestin et al., 2007).

Properties of polyolefin/cellulose composites

Beside the abovementioned coupling agents, the use of polyethylenimines (PEIs) as coupling agents for cellulose-reinforced polypropylene was evaluated. The results reveal that the use of PEIs causes a significant improvement in the interfacial adhesion, the tensile properties, and the impact resistance of the composites (de la Orden et al., 2014). Yield strength for the cellulose/LLDPE system, where the cellulose had been pretreated with silane, showed a relatively small improvement as compared to the effect of the peroxides addition (Bataille et al., 1990).

The poor interfacial adhesion between the polar hydrophilic cellulosic fibre and the nonpolar hydrophobic polyolefins contributes dramatically to the lack of performance of cellulose–polymer composites. The fibre agglomerations are also detrimental to this interfacial adhesion.

Direct extraction of vegetal polymers such as cellulose, followed by a chemical modification in order to bring new properties, is one of the paths to produce a bioplastic.

Several different types of cellulose are used in biocomposites containing polyolefins, as mentioned in the following discussion.

Microcristalline cellulose (MCC) is a naturally occurring substance obtained from purified, partially depolymerised cellulose, prepared by treating alpha cellulose, obtained as a pulp from fibrous material with mineral acids (Azizi Samir et al., 2005). MCC is thermally stable until 300 °C, which is much higher than melting temperature of nylon 6 (225 °C) and polyethylene terephthalate (270 °C). Due to its high specific strength and modulus, low density, and thermally stable structure, MCC is a good candidate as reinforcement for engineering thermoplastics. MCC-filled polypropylene (PP) composites and cellulose nanofibre-filled composites were prepared by melt blending (Yang et al., 2011). Spoljaric et al. (2009) obtained PP–MCC composites containing poly(propylene-*graft*-maleic anhydride) (PP-*g*-MA) and MCC treated with silicone oil, stearic acid or alkyltitanate coupling agent to promote matrix–filler dispersion and compatibility. MCC content and PP-*g*-MA increased PP thermal stability and crystallisation temperature (T_c), though crystallinity was reduced due to cellulose II crystals. Due to effective interaction between PP and MCC, storage modulus, loss modulus, and glass transition temperature increased with MCC concentration. MCC and PP-*g*-MA presence reduced creep deformation and increased permanent strain.

Thermoplastic *cellulose esters* are now reconsidered for various materials applications. To improve mechanical weaknesses of cellulose esters-based materials, the technique of blending has been widely used. In order to obtain biodegradable materials, polyestercarbonates, polylactides, and polyhydroxyalcanoates were mostly used for blending with cellulose esters. Only a few studies were focused on obtaining durable cellulose ester-based blends containing cost-effective and easily available polyolefins as blending partners.

Blends of cellulose acetate butyrate (CAB) and various polyethylenes (PE) were prepared in the molten state in a batch mixer (Besson et al., 2012; Besson and Budtova, 2012). CAB was considered the matrix (85%) and PE was the rubbery dispersed phase. Mechanical results showed that, compared to low value of pure CAB (about 1 kJ/m^2), impact resistance is 5 to 6 times greater for all polyethylene/CAB studied blends.

Kosaka et al. (2007) obtained composites of linear low-density poly(ethylene-co-butene) (PE) or maleated linear low-density poly (ethylene-co-butene) (M-PE) and *cellulose* (CEL), *cellulose acetate* (CA), *cellulose acetate propionate* (CAP), *or* CAB by means of an internal laboratory mixer with 20 wt.% polysaccharide. They concluded that composites prepared with M-PE presented yield stress and elongation values higher than those of composites prepared with PE, showing the compatibilizer effect of MA.

A US patent (Mumpower and Touey, 1969) discusses the method of obtaining a filter material for cigarettes by blending continuous filaments of fine denier polyolefin fibres with coarser denier continuous filaments of *cellulose acetate* fibres into the form of a tow. In this combination, one has the advantage of a plasticizer bondable tow of cellulose acetate fibres and a tow of fine denier polyolefin fibres without their accompanying disadvantages. Partially hydrolyzed cellulose acetates with an acetyl content of 36–41% and the substantially completely acetylated cellulose triacetates with an acetyl content of 43% to about 44.5% are preferred for this purpose, while the polyolefin to be used are those with a molecular weight above 10,000 and below 1,000,000 and a softening point between 80 and 180 °C.

The research interest in *cellulose nanofibres* (CNFs) as reinforcement in both biodegradable and synthetic polymers has increased remarkably during the past decade (Jonoobi et al., 2010; Ljungberg et al., 2006), not only due to the abundant and renewable isolation sources, like wood or agricultural residues, but because of the large specific surface area and remarkable mechanical properties of CNFs as well.

The mechanical properties of cellulose nanocomposites largely depend on the processing methods. Compared to the solvent casting method, melt compounding, especially extrusion, is a more promising industrial-scale process method. Thus, the use of twin-screw extrusion for the development of cellulose nanocomposites has been a focus of research in the past several years. In this view, Lee et al. (2011) produced CNF by using a batch-type kneader with six twin-screw elements twisted with 22.5° to generate a high shearing force. CNFs were first distributed in polycaprolactone (PCL) with a low melting temperature to avoid the vaporization of water in a CNF–water solution and then this composite was further mixed with polypropylene (PP) at 180 °C. They used maleic anhydride-grafted-PP for improvement of distributions of the PCL domain and CNF in PP. Only 1% CNF was enough to increase the elastic modulus of the obtained hybrid composite.

Kenaf cellulose (KC) was used to obtain biocomposites with HDPE and polyethyleneglycol (PEG) by melt blending and compression moulding. The water content in the composite decreased due to HDPE presence that surrounded the KC fibres, preventing water from reaching the cellulose fibre particles. By increasing the KC content, the thermal stability was slightly improved (Behjat and Luqman Chuah, 2010; Behjat et al., 2009).

As concluded by Herrera-Francop et al. (Herrera-Francop and Aguilar-Vegam, 1997; Shumigin et al., 2011), by incorporation of *henequen cellulosic fibres* into LDPE, the tensile strength composite materials increased up to 50% compared to that of neat LDPE. The addition of the fibres decreases the ultimate strain values for the composite materials. Preimpregnation of the cellulosic fibres in a LDPE–xylene solution and the use of a silane coupling agent results in a small increment in the mechanical properties of the composites, attributed to an improvement in the interface between the fibres and the matrix. It was also noted that the fibre surface treatment improves fibre dispersion in the matrix.

Sdrobiş et al. (2012b) prepared LDPE/cellulose composites by melt mixing in a Brabender mixer and investigated the influence of *unbleached* (CUB) *and bleached* (CB) *cellulose kraft pulp fibres*, unmodified and modified with oleic acid by plasma treatment, on thermal, rheological and mechanical behaviour, by comparing them with those of the neat LDPE. The presence of oleic acid modified cellulose pulp fibres reduces the resistance of composites to the external forces, determining an improvement of flow for reinforced systems. In all treatments, the mechanical behaviour of biocomposites retained an acceptable level of strength. The impact strength of the composites was less modified, compared to pure LDPE, for LDPE/unbleached pulp cellulose. By using plasma-treated fibres as reinforcements, the impact strength is higher compared with untreated pulp fibres, proving that by plasma treatment the fibres become more compatible with the LDPE matrix (Figure 10.2). These materials could be advantageously used for the development of degradable composites with good physical properties. Natural

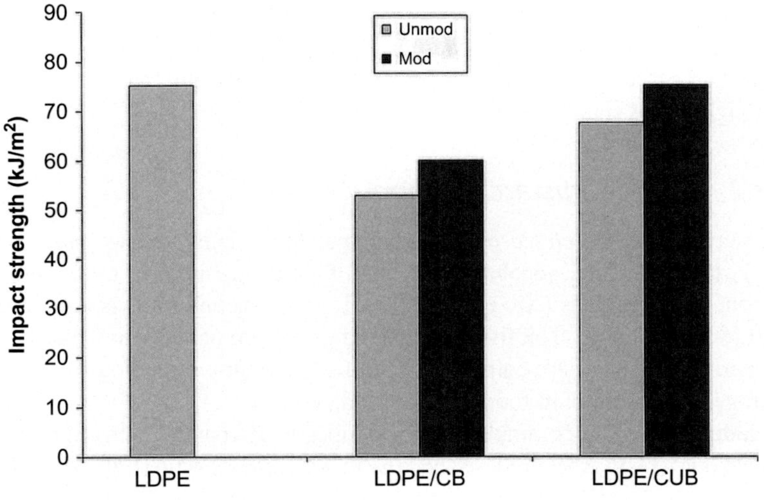

Figure 10.2 Charpy impact strength values for neat LDPE and LDPE/pulp fibres.

fibres addition and plasma treatment could change the properties of materials and control their mechanical, rheological, and thermal behaviour.

Due to their specific advantages (specific strength and modulus) over their conventional counterparts (glass and talc fibres), cellulose-based products are preferable for automotive applications (Ozen et al., 2013). Wood and pulp giant leader Weyerhaeuser replaced short glass fibres usually used for reinforcement in automotive applications with engineered cellulose fibre from trees and blended it with polypropylene with both high and low melt flow indices, obtaining a thermoplastic composite called THRIVET. In this case, cellulose requires less energy to produce and can reduce wear and tear on processing equipment when compared with composites containing abrasive short glass fibres. Ford Motor Company admitted that Weyerhaeuser's cellulose-based plastic composite materials meet the automaker's stringent requirements for stiffness, durability, and temperature resistance. Components weigh about 10% less and can be produced 20–40% faster with less energy when made with cellulose-based materials compared with fibreglass-based materials (PR Newswire, 2012). Cellulose-based fibres such as flax, sisal, abaca, hemp, jute, wood, coconut, banana, and wheat straw as reinforcement in plastic composites are reported to be used by many other automakers (General Motors, Daimler, Toyota, Fiat, Volkswagen, and BMW) in different parts and components, including spare-wheel compartment covers, underbody panels, seat backrests, headliners, engine and transmission covers, under-the-hood radiator end tanks, rear panel shelves, door trim panels, dashboard components and interior components such as the luggage compartment, speakers, carpeting, package shelves, luggage-trim upholstery, cowl-side trim, door scuff plate, tool box area, floor finish plate, and seat cushions (Kiziltas et al., n.d.; Ashori, 2008).

Nanoindentation with a continuous stiffness technique was used by Lee et al. (2007) to evaluate the hardness and elastic modulus in the interphase region, modified by maleic anhydride-grafted PP and γ-amino propyltrimethoxy silane (γ-APS) sizing in a *cellulose fibre-reinforced PP*. Depending on nanoindentation depth and spacing, the distinct properties of the transition zone were revealed by 1–4 indents. The authors assumed that the width of the property transition zone is less than 1 μm. However, three-dimensional finite element analysis shows that even a perfect interface without property transition has almost the same interphase width as that measured by nanoindentation. Using existing nanoindentation techniques, it will be difficult to calculate exact mechanical properties without the effect of neighbouring material property in at least an eight times smaller region than indent size.

10.2.3.2 Polyolefin/starch composites

Similar to cellulose, *starch* has received much attention lately due to such advantages as low cost, low density, nonabrasive nature, wide availability, or total compostability without toxic residues (Xie et al., 2013). It is the second most abundant biomass material found in nature. The five major starches of commerce (waxy maize, tapioca, potato, corn, and wheat) are commonly modified to improve rheological, processing, and storage performance, in food and nonfood systems.

Depending on the source, amylose and amylopectin percent in starch is different, ranging from about 10% to 20% amylase and 80% to 90% amylopectin (Ramesh et al., 1999).

Wheat starch (54–72% of the dry weight of wheat kernels) and wheat gluten are important and valuable coproducts of the wet processing of wheat flour (Maningat et al., 2009, pp. 441–510). Starch can be used either as extracted from the plant, found in literature as "native starch," or as "modified starch" because it undergoes chemical modifications in order to gain specific properties (Koshy et al., 2013, pp. 203–226).

Processing of polyolefin/starch composites

Starch-based degradable polymers are great candidates for bioplastics but, unfortunately, some properties like high gas permeability, brittleness, moisture sensitivity, low melt viscosity for processing, and low heat distortion temperature restrict their use in a large range of applications. As resulted from literature studies, starch can be used as filler to obtain reinforced plastics (Dufresne et al., 1996). Starch-based plastics with improved properties have been made by blending starch with other polymers, using moisture-resistant films, and by creating composite materials using fibres and inorganic materials. A small amount of starch (6–30%) induces biodegradability of the product if compounded with synthetic polymer. In order to obtain degradable composites with polyolefinic matrices, several attempts have been made to modify either starch or polyolefin to reduce the hydrophobic nature of the polyolefin or the hydrophilic character of the starch. In the presence of a plasticizer, starch becomes a thermoplastic material, as at high temperatures and shearing, it melts and fluidizes, enabling its use in injection, extrusion, and blowing equipment, such as those for synthetic plastics.

Properties of polyolefin/starch composites

In blends of starch with synthetic polymers such as polystyrene, polyethylene or polypropylene, the hydrophilic nature of starch leads to poor adhesion with the hydrophobic synthetic polymers. As mentioned also for cellulose, in order to obtain optimum properties, some compatibilizers are needed. For example, low-density polyethylene (LDPE-g-MAH) or HDPE functionalized with maleic anhydride (HDPE-g-MAH) were used as compatibilizer in a blend of PE with 30% natural corn starch (Matzinos et al., 2001) or HDPE/thermoplasticized starch (Sailaja and Chanda, 2001), emphasizing that the biodegradation of the compatible blends was only slightly lower than that of the noncompatible compounds. Another approach was to use poly(ethylene-co-acrylic), poly(ethylene-co-vinyl alcohol), or oxidized polyethylene as compatibilizer in polyethylene-starch composites, but the composites had unsatisfactory mechanical properties (Willet, 2004).

Evangelista et al. (1991) incorporated 25% starch previously modified with octenyl succinate into polyethylene (PE). They obtained blends with better mechanical properties than the unmodified polyethylene-starch blends. Another attempt was to incorporate oxidized polyethylene as a compatibilizer in blends of polyethylene and starch. Nevertheless, the modification was quite expensive, and the compounds still presented poor mechanical properties (Jane et al., 1991).

40–50% glycerol plasticized starch (GS) was compounded with LDPE or PP and monoethers of glycerol and fatty alcohols (GA), and in some cases epoxidized rubbers (ER) were used as compatibilizers. Mechanical properties and ageing susceptibility of blends depend strongly on their composition; that is, the content of plasticized starch in the blend and the content of glycerol in the starch. An increased susceptibility to biodegradation during soil or fungus ageing not only of the starch phase but also of the polymer phase was observed in some cases (Zuchowska et al., 1998).

Kaminsky (1984) patented a process for producing starch/polyolefin polymer compositions by polymerisation of at least one olefin in the presence of a Ziegler–Natta catalyst. Starch/polyolefin polymer compositions produced are useful in a variety of applications, including use in water-resistant adhesives and binders, paper sizes and coatings, textile sizes, and paint extenders.

The mechanical properties of LDPE/*thermoplastic starch* (TPS) blends and of linear LLDPE/TPS were studied by St-Pierre et al. (1997). They observed that these compounds had better processability and mechanical properties than those with the untreated starch. High percentages of thermoplastic starch led to a drastic reduction in the mechanical properties

Polyethylene–maleic anhydride terpolymers containing elastomeric segments provided excellent emulsification of PE/TPS blends with a fine morphology (volume diameter of 1.4 μm; number average diameter of 600 nm). The compatibilized materials exhibit a very high elongation at break of about 800%, the highest value ever reported for PE/TPS systems. Additionally, significant improvement in notched impact strength performance at interfacial saturation was found for these systems, with a performance similar to pure PE. Morphology is very important as an excellent correlation was found between the critical concentration for interfacial saturation and the mechanical properties (Cerclé et al., 2013).

Corn starch/HDPE and LDPE blends were prepared by injection moulding technique and PE-g-MAH was used as a compatibilizer (Oragwu and Igwe, 2013). The elongation at break of LDPE-corn starch blends decreased with increase in starch content, while the elongation at break of HDPE/corn starch blends first decreased with increase in starch content, up to 7.25 wt.%, and thereafter, decreased with further increase in corn starch contents. The addition of PE-g-MAH improved the tensile strength and elongation at break of the blends. Thakore et al. (1999) also reported a similar decrease in elongation at break of LDPE/potato starch blends with increase in potato starch content. This behaviour was attributed to the improved plasticity of starch acetate.

In another study, *tapioca starch* (TS) was used in order to improve the degradability of injection-moulded plastic packaging. TS and LDPE was melt-compounded using a twin-screw extruder at screw speed of 70–80 rpm, barrel temperature of 150, 130, 130 and 140 °C, followed by injection moulding process. The presence of high starch contents had an adverse effect on the tensile properties of LDPE/tapioca starch blends. Young's modulus and tensile strength were found to decrease with the addition of tapioca starch. However, elongation at break of the samples increased slightly when starch was added. The samples become ductile when increasing TS loadings. A growth of microbes colony was observed on the surface of LDPE/TS biocomposites, indicating that granular starch present on the surface of the samples was attacked by microorganisms, until most of it was assimilated as a carbon source (Ali et al., 2013).

Among all kinds of starches, *cassava starch* (CS) is a promising raw material in Brazil. It was found that the addition of CS to PP increases its thermal stability, especially when the CS is plasticized (PCS) and when PP-g-MAH is added. The Young's modulus and the tensile strength of the PP increase with 10 phr of CS and start to decrease at higher contents. However, when PP-g-MAH is added to the PP/CS blends, the mentioned properties decrease. The degradability of the PP/CS blend in an agricultural soil significantly affected the Young's modulus, whose values decreased from 1841 to 834 MPa (González et al., 2009).

10.2.3.3 Polyolefin/chitosan composites

Processing of polyolefin/chitosan composites

Numerous studies have been conducted on antimicrobial activities of chitosan in solutions or films (Dutta et al., 2009; Sunilkumar et al., 2012; Ziani et al., 2009; Yoksan and Chirachanchai, 2009; Zhong et al., 2011), but few published reports exist on testing the feasibility of chitosan-based antimicrobial packaging film through incorporation of chitosan into polyethylene film. Park et al. (2010) evaluated the antimicrobial activity of chitosan-incorporated LDPE films, measured their film properties, and determined that significant extensions of red colour shelf life were observed in sliced flesh red meats stored in LDPE/chitosan films.

In order to convert a commodity polymer, such as polyethylene (PE), to a functional polymer by using chitosan, functionalization of the surface of PE film with chitosan via γ-ray irradiation technique followed by copper ion immobilization on this surface was realized. Authors used solubilization of chitosan in water by forming an organic salt with hydroxybenzyl triazole as a key factor (Pasanphan and Chirachanchai, 2008). Surface functionalization of polyethylene via plasma source implantation using O_2 or CF_4 gas was performed in order to improve adhesion between linear LDPE and chitosan or corn zein (Shin et al., 2002) in view of using the mixture in food packaging applications due to the O_2 gas barrier property and easy recyclability of the multilayer film.

Antimicrobial materials based on chitosan and polyethylene composites have been obtained either by incorporation or by surface coating and electrospinning, which are destined for food packaging as poultry meat packaging (Munteanu et al., 2014; Paslaru et al., 2013; Vasile et al., 2013, 2014).

Properties of polyolefin/chitosan composites

The addition of the chitosan surfactant to a PE surface improves the blood compatibility of the PE (Sagnella and Mai-Ngam, 2005).

Mir et al. (2011) developed HDPE blends containing different percentages of chitosan (up to a maximum of 35 wt%) and vinyl triethyoxysilane as a cross-linking agent with improved antimicrobial and mechanical properties.

Biodegradable blends of LDPE (up to 60 wt.%) chitosan have been obtained by melt processing in the presence of an interfacial modifier as epoxy functionalized LDPE, and 0.1 wt.% of pro-oxidant zinc sterate (Zn $(C_{18}H_{35}O_2)_2$) was incorporated in order to accelerate UV degradation of LDPE (Prasanna and Sailaja, 2012). The obtained systems present the dual advantages of biodegradability and antibacterial characteristics, while the LDPE part would give strength and sealing capabilities to the packaging material. Nino et al. (1999, pp. 198–203) prepared a blend of LDPE with starch and chitin along with urea as plasticizer. Kurek et al. (2012) used casting method to obtain coated LDPE films with chitosan and found out that this process enhanced polyethylene gas barrier properties at low relative humidities.

10.2.4 Polyolefin/lignin composites

10.2.4.1 Processing of polyolefin/lignin composites

Blends of various polymer/lignin ratios were prepared by mechanical mixing followed by thermal extrusion using a mixing extruder with a single- or twin-screw configuration, rotation speed of 200–250 rpm, and a take-up speed of 100 m/min. Extrusion temperatures varied depending on the composite/blend composition, varying from 145 to 247 °C.

To increase the interfacial bonding, modification of either thermoplastics by reactive extrusion (Darie et al., 2011) or lignin (epoxidation, alkylation, etc.) is applied sometimes (Simionescu et al., 1993, 1994).

Most compounding processes require the addition of compatibilizers, which circumvent incompatibility problems relating to poor interfacial adhesion between filler and polymer. The addition of maleic anhydride-modified polymers (e.g., maleic anhydride-grafted polymers) represents a prominent method for compatibilizing fillers and polymers. During compounding, the MA-functionalized polymer reacts with the hydroxyl groups of the filler to form graft copolymers.

10.2.4.2 Properties of polyolefin/lignin composites

Thermoplastics with high lignin content have been produced by alkylated kraft lignin or blending kraft lignin with synthetic polymers, and it was found that the lignin-based materials possess mechanical properties comparable to polystyrene (Kadla and Kubo, 2004; Li et al., 1997). Although 100% kraft lignin and alkylated kraft lignin films and fibres are quite brittle, they can be effectively plasticized by blending with suitable polyesters and polyethers (Cazacu et al., 2002, 2004a,b; Macoveanu et al., 2001a,b; Pascu et al., 2001; Tudorachi et al., 2000b; Vasile et al., 1998, 2006; Vasile, 2002). The variation/improvement of the characteristics is strongly dependent upon the chemical functional groups of both lignins (alkaline lignin and lignosulfonate, epoxy modified lignin).

Many works of lignin-filled thermoplastics including polyethylene (Glasser et al., 2000, pp. 2–99) and polypropylene (Kharade and Kale, 1999; Pouteau et al., 2003, 2004) have already been reported. Due to its phenolpropanoidic structure similar to that of hindered phenols, lignin can act as stabilizer (Vasile et al., 1997) or initiator of PO degradation (Košiková et al., 1993a,b,c). Usually, the addition of lignin deteriorates the mechanical properties of plastics due to the poor miscibility and influences biodegradation characteristics.

The incorporation of kraft lignin (10–55 wt.%) into polypropylene was accompanied by the increase of the tensile modulus with lignin content, and a decrease of maximum tensile strength, indicating a poor adhesion between nonpolar structure of PP and more polar structure of lignin and a poor dispersion into the matrix (Sanchez and Esposito-Alvarez, 1999). A similar result was found with polyethylene (Tudorachi et al., 2000a,b).

(Kharade and Kale, 1999; Vasile and Cazacu, 2013) have incorporated up to 30 wt.% dry lignin powder in LDPE, HDPE, and polypropylene (PP), and they observed that the tensile strength and elongation at break values decreased with lignin content for all three polymer blends, indicating a poor compatibility or interaction between lignin and the matrix polymer. However, incorporation of lignin into a binary polyolefin mixture (70 PP/30 PE) in amounts of 2.5–40% (w/w) gave polymer blends with good thermal, physicomechanical, and surface properties (Feldman et al., 1995). The polyolefin blends containing kraft and aminated lignins present improved strength, elongation at break, and other mechanical indices (Wang et al., 1992). Canetti et al. (2006) observed an increase in the thermal degradation temperature of the blends as a function of lignin content, both in oxidative and nonoxidative conditions, because of the formation of a protective surface able to reduce the oxygen diffusion toward the polymer bulk. Gregorová et al. (2005) showed that the incorporation of lignin in polyolefin matrices improves the heat resistance of PP and PP plaques.

Compatibility between PO and lignin can be improved by introduction of reactive functional groups in each or in only one blending compound. Thus, Casenave et al. (1996) developed a new technique to improve the interfacial adhesion by establishing the chemical bonds between the matrix and the additive and to ensure a better dispersion and adhesion to the synthetic matrix (Wang et al., 1992) grafted by a catalytic reaction of the ethylene or propylene monomers on the lignin surface and then mixed with the polyethylene matrix. Young's modulus for PE/L-g-PE remains constant and closed to that of pure PE, and morphological studies indicate a better lignin-grafted/matrix adhesion.

Another way to improve the compatibility between lignin and polyolefins is utilization of a compatibilizing agent such as ethylene-vinyl acetate copolymer (EVA), ethylene-proylene rubber grafted with MA (EP-MA) (Darie et al., 2007), and glycidylmethacrylate-grafted-polypropylene (PP-g-GMA) (Simionescu et al., 1995, pp. 222–223). Alkylating lignin with bromododecane improves compatibility with polypropylene and enables up to 70 wt.% lignin to be incorporated into a thermally stable composite (Zhong, 2011).

Pascu et al. (2000, 2002) showed that by a combination of the chemical procedures of component modification (e.g., epoxidation of lignin) with physical treatments of surfaces (e.g., photooxidation, plasma, electron beam treatment, and UV-irradiation) the bio/environmental disintegrable polyolefin materials can be obtained.

Other effects of lignin incorporation in PO are an increase in conductivity, hydrophilicity, improved printability, and it imparts biodegradability characteristics (Košiková et al., 1993c, pp. 537–542). In some cases, lignin behaves as an antifungal (El-Aghoury et al., 2006) or improves antioxidant (Pouteau et al., 2003) properties of the products.

The lignin in the concentration of 1–2 wt.% possessed the ability to act as a processing stabilizer and as an antioxidant during thermo-oxidative ageing of polypropylene films (Košiková and Lábaj, 2009).

10.2.5 Other fillers

Various literature surveys show that polyolefin/biomass degradable composites can be realized with a wide range of other biomass materials. Table 10.3 contains a selection of such compositions and the most important findings regarding their mechanical performance.

Table 10.3 Biomass/polyolefins degradable composites containing other fillers

Biomass	Matrix	Method/compatibilizer	Important properties	References
Coir fibre (10, 15, and 20 wt.%)	PP	Hot press; raw coir fibre chemically treated with basic chromium sulphate and sodium bicarbonate salt in acidic media	During chemical treatment, hydrophilic −OH groups in the raw coir cellulose were converted to hydrophobic −OH–Cr groups; improvement in interfacial adhesion between the coir and PP upon treatment. Chemically treated specimens and 20% fibre-reinforced composites had the optimum mechanical properties	Mir et al. (2013)
Waste shell (S)	β-PP	Melt blending/(S) was modified by pimelic acid (PA) through solution method	Shell modified by PA was a β nucleating agent with high efficiency, resulting in the maximum K_β value of 0.56. PA modification promoted the dispersion and interfacial bonding of shell powder in PP. The addition of PA-modified shell powder largely increased the impact strength of PP	Lin et al. (2013)
Waste shellfish shell	PP	Melt processing/granules of shellfish shell (SS) were prepared from waste shellfish shell by removing cuticle, crushing, grinding, and shearing emulsification	PP/SS composite shows a higher yield strain, yield strength, tensile strength, and elongation at break than traditional commercial calcium carbonate (CC)-filled PP. Yield strength of PP/SS composite with 2% SS is improved by 11.1% due to the formation of β-crystalline PP phase	Li et al. (2012)

seaweed (SW)	PP	Compressing moulding/*Laminaria japonica* in the form of brown algae (BA) and *Enteromorpha crinite* in the form of green algae (GA) were characterised by pretreatments with ethanol, acetone, and sulfuric acid	GA/PP biocomposites showed higher thermal stability and greater impact strength than BA/PP biocomposites. Sulfuric acid-treated GA/PP biocomposites showed generally improved thermal and mechanical properties	Jang et al. (2013)
Durian skin fibre (DSF)	PP	0, 2.5, and 5.0 wt.% maleic anhydride polypropylene (MAPP); Haake 600P twin-screw extruder at 190 °C with screw speed of 250 rpm. 10, 30, and 50 wt.% DSF content. Fibre size ranges from 50 to 100 μm, 100 to 250 μm, and 250 to 500 μm	The optimized value for the maximum impact strength of PP–DSF composite was found at 50 wt.% DSF content and fibre size range of 250–500 μm with 0 wt.% MAPP. The maximum impact property for PP–DSF composite was 10.66 kJ/m^2. Fibre content significantly affects the strength of PP–DSF composites, followed by the fibre size and MAPP content	Nur Aimi et al. (2014)
Old newspapers (ONPs)—85% of thermomechanical pulp from hardwood and 15% of calcium carbonate mineral filler	PP	20–50 wt.% PP/ONP fibres, 0–8% of MAP, Brabender plastograph internal mixing at 80 rpm during 10 min at a temperature of 180 °C	Micromechanics of the fibres were obtained using Hirsch model, Bowyer–Bader methodology, and Kelly–Tyson equations. Due to the presence of a percentage of calcium carbonate in the obtained fibres (10%), the computed intrinsic characteristics were addressed as equivalent	Serrano et al. (2013)

Continued

Table 10.3 Continued

Biomass	Matrix	Method/compatibilizer	Important properties	References
Chicken eggshell (ES)—contains 95% calcium carbonate	PP and antioxidant (Irganox 1010/Irgafos 168 (50/50)); 20 and 40 wt.% ES	Brabender Plasticorder internal mixer, 75 rpm, temperature of 190 °C for 15 min	Improved Young's modulus (E) with the increment of ES content; ES was better than all types of commercial carbonate fillers used. Although ES composites showed lower E values than talc composites, talc filler could be replaced by up to 75% with ES while maintaining a similar stiffness and E compared to the talc composites; improvement in the mechanical properties was attributed to a better ES/matrix interface related to the geometric ratio of the ES particles similar to talc particles	Toro et al. (2007)

10.3 Conclusions and future trends

The use of polymer composites filled with natural-organic fillers, as replacement of mineral-inorganic fillers, is of great interest in view of the reduction in the use of petroleum-based, nonrenewable resources and, in general, in a more intelligent utilization of environmental and financial resources. There is still a great deal of research and development going on in many parts of the world in the area of composites materials based on polyolefins and natural fillers. Challenges are to find better extrusion/injecting/compression systems, enhancing mechanical properties, and the properties of the filler fraction through various modification methods in order to improve outdoor performance. Much work can still be done in order to find new applications and to increase the marketability of these materials.

Although there are plenty of studies referring to obtaining and characterization of biocomposites based on polyolefins, including their transformation in biodegradable materials by using suitable processes, microorganisms, and enzymes, which will open new ways for their applications and help to maintain a high level of production, there are still many research aspects to be followed. In efforts toward finding new applications where degradable/disintegrable composites containing synthetic polymers like polyolefins are still in demand, a wide variety of biocomposites processing techniques, as well as some factors like moisture content, fibre type, size, and amount, coupling agents, and their influence on composites properties affecting these processes are yet open for investigation.

Acknowledgments

The authors acknowledge the financial support from EC through the Marie Curie FP7 program, grant PIRSES GA-2009-247550 BIOFUEL, and from UEFISCDI through the research project BIONANOMED 164/2012.

References

Ali, R.R., Rahman, W.A.W.A., Kasmani, R.M., Ibrahim, N., Mustapha, S.N.H., Hasbullah, H., 2013. Tapioca starch biocomposite for disposable packaging ware. Chem. Eng. Trans. 32, 1711–1716. http://dx.doi.org/10.3303/CET1332286.

Arbelaiz, A., Fernandez, B., Ramos, J.A., Retegi, A., Llano-Ponte, R., Mondragon, I., 2005. Mechanical properties of short flax fiber bundle/polypropylene composites: influence of matrix/fiber modification, fibre content, water uptake and recycling. Compos. Sci. Technol. 65, 1582–1592. http://dx.doi.org/10.1016/j.compscitech.2005.01.008.

Arib, R.M.N., Sapuan, S.M., Ahmad, M.M.H.M., Paridah, M.T., Khairul Zaman, H.M.D., 2006. Mechanical properties of pineapple leaf fibre reinforced polypropylene composites. Mater. Des. 27 (5), 391–396. http://dx.doi.org/10.1016/j.matdes.2004.11.009.

Ashori, A., 2008. Wood–plastic composites as promising green-composites for automotive industries. Bioresour. Technol. 99, 4661–4667. http://dx.doi.org/10.1016/j.biortech.2007.09.043.

Ashori, A., Nourbakhsh, A., 2010. Reinforced polypropylene composites: effects of chemical compositions and particle size. Bioresour. Technol. 101, 2515–2519. http://dx.doi.org/10.1016/j.biortech.2009.11.022.

Ayrilmis, N., Kaymakci, A., 2013. Fast growing biomass as reinforcing filler in thermoplastic composites: *Paulownia elongata* wood. Ind. Crop. Prod. 43, 457–464. http://dx.doi.org/10.1016/j.indcrop.2012.07.050.

Azizi Samir, M.A., Alloin, F., Dufresne, A., 2005. Review of recent research into cellulosic whiskers, their properties and their application in nanocomposite field. Biomacromolecules 6 (2), 612–626. http://dx.doi.org/10.1016/j.indcrop.2012.07.050.

Bajpai, P.K., Singh, I., 2013. Drilling behavior of sisal fiber-reinforced polypropylene composite laminates. J. Reinf. Plast. Compos. 32, 1569–1576. http://dx.doi.org/10.1177/0731684413492866.

Barlow, F., Khanna, Y., Pietsch, D.B., Underwood, J., 2004. Melt Blending Compositions that Include Cellulosic Pulp Fibers Having an Alpha-Cellulose Purity of greater than 80% by weight, Water Soluble Binder, Lubricant, Compatibilizer, Matrix Polymer. US Patent US6743507 B2.

Bataille, P., Allard, P., Cousin, P., Sapieha, S., 1990. Interfacial phenomena in cellulose/polyethylene composites. Polym. Compos. 11 (5), 301–304. http://dx.doi.org/10.1002/pc.750110508.

Behjat, T., Luqman Chuah, A., 2010. Thermal properties of high density polyethylene-kenaf cellulose composites. Polym. Polym. Compos. 18 (5), 257–261.

Behjat, T., Russly, A.R., Luqman Chuah, A., 2009. Mechanical and morphological properties of kenaf cellulose/LDPE biocomposites. Am. Eurasian J. Agric. Environ. Sci. 5 (6), 777–785.

Belgacem, M.N., Gandini, A., 2005. The surface modification of cellulose fibres for use as reinforcing elements in composite materials. Compos. Interfaces 12 (1–2), 41–75. http://dx.doi.org/10.1163/1568554053542188.

Besson, F., Budtova, T., 2012. Cellulose ester-polyolefine binary blend: Morphological, rheological and mechanical properties. Eur. Polym. J. 48 (5), 981–989. http://dx.doi.org/10.1016/j.eurpolymj.2012.02.010.

Besson, F., Vanhille, A., Budtova, T., 2012. Cellulose ester/polyolefin binary blends: rheology, morphology and impact properties. In: Paper Presented at the 11th European Symposium on Polymer Blends – Polyblends, Donostia – San Sebastian, Spain.

Bledzki, A.K., Faruk, O., 2003. Wood fiber reinforced polypropylene composites: effect of fiber geometry and coupling agent on physic-mechanical properties. Appl. Compos. Mater. 10, 365–379. http://dx.doi.org/10.1023/A:1025741100628.

Bledzki, A.K., Faruk, O., 2004. Extrusion and injection moulded microcellular wood fiber reinforced polypropylene composites. Cell. Polym. 23 (4), 211–227.

Bledzki, A.K., Faruk, O., 2006. Microcellular wood fiber reinforced PP composites – A comparative study between extrusion, injection moulding and compression moulding. Int. Polym. Process. 21, 256–262. http://dx.doi.org/10.3139/217.0085.

Bledzki, A.K., Gassan, J., 1999. Composites reinforced with cellulose based fibres. Prog. Polym. Sci. 24, 221–274.

Bledzki, A.K., Faruk, O., Sperber, V.E., 2006. Cars from bio-fibres. Macromol. Mater. Eng. 291 (5), 449–457. http://dx.doi.org/10.1002/mame.200600113.

Bledzki, A.K., Mamun, A.A., Faruk, O., 2007. Abaca fibre reinforced PP composites and comparison with jute and flax fiber PP composites. Express Polym. Lett. 1 (11), 755–762. http://dx.doi.org/10.3144/expresspolymlett.2007.104.

Bledzki, A.K., Mamun, A.A., Jaszkiewicz, A., Erdmann, K., 2010. Polypropylene composites with enzyme modified abaca fiber. Compos. Sci. Technol. 70 (5), 854–860. http://dx.doi.org/10.1016/j.compscitech.2010.02.003.

Canetti, M., Bertini, F., De Chirico, A., Audiso, G., 2006. Thermal degradation behaviour of isotactic polypropylene blended with lignin. Polym. Degrad. Stab. 91, 494–498. http://dx.doi.org/10.1016/j.polymdegradstab.2005.01.052.

Casenave, S., Aït-Kadfi, A., Riedl, B., 1996. Mechanical behaviour of highly filled lignin/polyethylene composites made by catalytic grafting. Can. J. Chem. Eng. 74 (2), 308–315. http://dx.doi.org/10.1002/cjce.5450740216.

Cazacu, G., Pascu, M.C., Profire, L., Vasile, C., 2002. Environmental friendly polymer materials I. Polyolefins-lignin based materials. J. Environ. Prot. Ecol. 3 (1), 242–248.

Cazacu, G., Mihaies, M., Pascu, M.C., Profire, L., Kowarski, A.I., Vasile, C., 2004a. Polyolefins/lignosulfonate blends. IX. Functionalized polyolefins/lignin blends. Macromol. Mater. Eng. 289, 880–889. http://dx.doi.org/10.1002/mame.200300378.

Cazacu, G., Pascu, M.C., Profire, L., Kowarski, A.I., Vasile, C., 2004b. Lignin role in a complex polyolefin blend. Ind. Crop. Prod. 20 (204), 205–221. http://dx.doi.org/10.1016/j.indcrop.2004.04.030.

Cerclé, C., Sarazin, P., Favis, B.D., 2013. High performance polyethylene/thermoplastic starch blends through controlled emulsification phenomena. Carbohydr. Polym. 92 (1), 138–148. http://dx.doi.org/10.1016/j.carbpol.2012.08.107.

Cetin, N.S., Hill, C.A.S., 1999. An investigation of the reaction of epoxides with wood. J. Wood Chem. Technol. 19, 247–264. http://dx.doi.org/10.1080/02773819909349611.

Chandra, R., Rustgi, R., 1997. Biodegradation of maleated linear low-density polyethylene and starch blends. Polym. Degrad. Stab. 56 (2), 185–202. http://dx.doi.org/10.1016/S0141-3910(96)00212-1.

Chen, H.L., Porter, R.S., 1994. Composite of polyethylene and kenaf, a natural cellulose fiber. J. Appl. Polym. Sci. 54 (11), 1781–1783. http://dx.doi.org/10.1002/app.1994.070541121.

Chen, J., Wang, Y., Gu, C., Liu, J., Liu, Y., Li, M., Lu, Y., 2013. Enhancement of the mechanical properties of basalt fiber–wood–plastic composites via maleic anhydride grafted high-density polyethylene (MAPE) addition. Materials 6, 2483–2496. http://dx.doi.org/10.3390/ma6062483.

Dai, D., Fan, M., 2014. Wood fibers as reinforcements in natural fiber composites: structure, properties, processing and applications. In: Hodzic, A., Shanks, R. (Eds.), Natural Fibre Composites: Materials, Processes and Properties. Woodhead Publishing, UK, pp. 3–65. http://dx.doi.org/10.1533/9780857099228.1.3.

Danyadi, L., Janecska, T., Szabo, Z., Nagy, G., Moczo, J., Pukanszky, B., 2007. Wood flour filled PP composites: compatibilization and adhesion. Compos. Sci. Technol. 67, 2838–2846. http://dx.doi.org/10.1016/j.compscitech.2007.01.024.

Darie, R., Cazacu, G., Vasile, C., Kozlowski, M., 2007. Blends with polypropylene matrix and lignin additive. In: Paper Presented at the Italic 4 – Science & Technology of Biomass: Advances and Challenges, Rome, Italy.

Darie, R., Vasile, C., Kozlowski, M., 2011. Compatibilization of Complex Polymeric Systems, Reactive Compatibilization-cum-impact Modification Applied to Virgin Polymer and Waste. LAP Lambert Academic Publishing, Saarbrucken, Germany.

de la Orden, M.U., González Sánchez, C., González Quesada, M., Martínez Urreaga, J., 2014. Effect of different coupling agents on the browning of cellulose polypropylene composites during melt processing. Polym. Degrad. Stab. 95, 201–206. http://dx.doi.org/10.1016/j.polymdegradstab.2009.11.024.

Deng, S., Beehag, A., Hillier, W., Zhang, D., Ye, L., 2013. Kenaf–polypropylene composites manufactured from blended fiber mats. J. Reinf. Plast. Compos. 32, 1198–1210. http://dx.doi.org/10.1177/0731684413485652.

Dewidar, M., Bakrey, M., Hashim, A.M., Abdel-Haleem, A., Diab, Kh., 2010. Mechanical properties of polypropylene reinforced hemp fiber composite. Available from: http://www.ipme.ru/e-journals/MPM/no_21512/MPM215_04_dewidar.pdf 2010 (28 June 2014).

Doan, T.-T.-L., Gao, S.-L., Mäder, E., 2006. Jute/polypropylene composites I. Effect of matrix modification. Compos. Sci. Technol. 66, 952–963. http://dx.doi.org/10.1016/j.compscitech.2005.08.009.

Dufresne, A., Cavaille, J.Y., Helbert, W., 1996. New nanocomposite materials: microcrystalline starch reinforced thermoplastic. Macromolecules 29, 7624–7626. http://dx.doi.org/10.1021/ma9602738.

Dutta, P.K., Tripathi, S., Mehrotra, G.K., Dutta, J., 2009. Perspectives for chitosan based antimicrobial films in food applications. Food Chem. 114 (4), 1173–1182. http://dx.doi.org/10.1016/j.foodchem.2008.11.047.

El-Aghoury, A., Vasudeva, R., Banu, D., Elektorowicz, M., Feldman, D., 2006. Contribution to the study of fungal attack on some plasticized vinyl formulations. J. Polym. Environ. 14 (2), 135–147. http://dx.doi.org/10.1007/s10924-006-0004-9.

Essabir, H., Elkhaoulani, A., Benmoussa, K., Bouhfid, R., Arrakhiz, F.Z., Qaiss, A., 2013. Dynamic mechanical thermal behavior analysis of doum fibers reinforced polypropylene composites. Mater. Des. 51, 780–788. http://dx.doi.org/10.1016/j.matdes.2013.04.092.

Evangelista, R.L., Sung, W., Jane, J.L., Gelina, R.J., Nikolov, Z.L., 1991. Effect of compounding and starch modification on properties of starch-filled low-density polyethylene. Ind. Eng. Chem. Res. 30 (8), 1841–1846. http://dx.doi.org/10.1021/ie00056a025.

Feldman, D., Lacasse, M.A., Wang, J., Luchian, C., 1995. Lignin and its polyblends. J. Macromol. Sci. Part A: Pure Appl. Chem. 32 (8–9), 1613–1619. http://dx.doi.org/10.1007/978-1-4615-0643-0_5.

Ferrero, B., Fombuena, V., Fenollar, O., Boronat, T., Balart, R., 2014. Development of natural fiber-reinforced plastics (NFRP) based on biobased polyethylene and waste fibers from Posidonia oceanica seaweed. Polym. Compos. http://dx.doi.org/10.1002/pc.23042.

Frisk, H., Schwendemann, D., 2004. Injection moulding: compounding wood fibers with plastic. Kunststoffe Plast Eur. 94, 76–80.

Gassan, J., Bledzki, A.K., 1997. The influence of fiber-surface treatment on the mechanical properties of jute-polypropylene composites. Compos. A: Appl. Sci. Manuf. 28 (12), 1001–1005. http://dx.doi.org/10.1016/S1359-835X(97)00042-0.

Gassan, J., Gutowski, V.S., 2000. Effects of corona discharge and UV treatment on the properties of jute–fibre epoxy composites. Compos. Sci. Technol. 60, 2857–2863. http://dx.doi.org/10.1016/S0266-3538(00)00168-8.

Gassan, J., Gutowski, V.S., Bledzki, A.K., 2000. About the surface characteristics of natural fibres. Macromol. Mater. Eng. 283, 132–139. http://dx.doi.org/10.1002/1439-2054(20001101)283:1<132::AID-MAME132>3.0.CO;2-B.

Gauthier, R., Joly, C., Coupas, A.C., Gauthier, H., Escoubes, M., 1998. Interfaces in polyolefin/cellulosic fiber composites: chemical coupling, morphology, correlation with adhesion and aging in moisture. Polym. Compos. 19, 287–300. http://dx.doi.org/10.1002/pc.10102.

George, J., Bhagawan, S.S., Thomas, S., 1998. Effects of environment on the properties of low-density polyethylene composites reinforced with pineapple-leaf fibre. Compos. Sci. Technol. 58, 1471–1485. http://dx.doi.org/10.1016/S0266-3538(97)00161-9.

Gilan, I., Sivan, A., 2013. Effect of proteases on biofilm formation of the plastic-degrading actinomycete Rhodococcus ruber C208. FEMS Microbiol. Lett. 342 (1), 18–23. http://dx.doi.org/10.1111/1574-6968.12114.

Glasser, W.G., Northey, R.A., Schultz, T.P., 2000. Lignin chemistry, technology, and utilization: a brief history. In: Glasser, W.G., Northey, R.A., Schultz, T.P. (Eds.), Lignin: Historical, Biological and Materials Perspectives. ACS Symposium Series, vol. 742. ACS, Washington, DC, pp. 2–99.

González, J., Albano, C., Hernández, M., Ichazo, M., Lavady, M., 2009. Characterization of polypropylene/cassava starch compounds obtained in an internal mixer, vol. 24, no. 1, Revista de la Facultad de Ingeniería Universidad Central de Venezuela. Available from: http://www.scielo.org.ve/scielo.php?pid=S0798-40652009000100003&script=sci_arttext (15 June 2014).

Gregorová, A., Cibulková, Z., Košiková, B., Šimon, P., 2005. Stabilization effect of lignin in polypropylene and recycled polypropylene. Polym. Degrad. Stab. 89, 553–558. http://dx.doi.org/10.1016/j.polymdegradstab.2005.02.007.

Hao, A., Zhao, H., Jiang, W., Yuan, L., Chen, J.Y., 2012. Mechanical properties of kenaf/polypropylene nonwoven composites. J. Polym. Environ. 20, 959–966. http://dx.doi.org/10.1007/s10924-012-0484-8.

Haque, M., Rahman, R., Islam, N., Huque, M., Hasan, M., 2010. Mechanical properties of polypropylene composites reinforced with chemically treated coir and abaca fiber. J. Reinf. Plast. Compos. 29, 2253–2261. http://dx.doi.org/10.1177/0731684409343324.

Herrera-Francop, J., Aguilar-Vegam, J., 1997. Effect of fiber treatment on the mechanical properties of LDPE–henequen cellulosic fiber composites. J. Appl. Polym. Sci. 65 (1), 197–207. http://dx.doi.org/10.1002/(SICI)1097-4628(19970705)65:1<197::AID-APP24>3.0.CO;2-#.

Hietala, M., Samuelsson, E., Niinimäki, J., Oksman, K., 2011. The effect of pre-softened wood chips on wood fiber aspect ratio and mechanical properties of wood–polymer composites. Compos. A: Appl. Sci. Manuf. 42, 2110–2116. http://dx.doi.org/10.1016/j.compositesa.2011.09.021.

Islam, M.R., Beg, M.D.H., Gupta, A., 2013. Characterization of laccase-treated kenaf fiber reinforced recycled polypropylene composites. BioResources 8 (3), 3741–3752.

Jane, J.L., Gelina, R.J., Nikolov, Z., Evangelista, R.L., 1991. Degradable Plastics from Octenyl Succinate Starch. US Patent 5059642.

Jang, Y.H., Han, S.O., Sim, I.N., Kim, H.I., 2013. Pretreatment effects of seaweed on the thermal and mechanical properties of seaweed/polypropylene biocomposite. Compos. A: Appl. Sci. Manuf. 47, 83–90. http://dx.doi.org/10.1016/j.compositesa.2012.11.016.

Jayaraman, K., 2003. Manufacturing sisal–polypropylene composites with minimum fibre degradation. Compos. Sci. Technol. 63 (3–4), 367–374. http://dx.doi.org/10.1016/S0266-3538(02)00217-8.

John, M.J., Anandjiwala, R.D., 2009. Chemical modification of flax reinforced polypropylene composites. Compos. A: Appl. Sci. Manuf. 40 (4), 442–448. http://dx.doi.org/10.1016/j.compositesa.2009.01.007.

Jonoobi, M., Harun, J., Mathew, A.P., Oksman, K., 2010. Mechanical properties of cellulose nanofiber (CNF) reinforced polylactic acid (PLA) prepared by twin screw extrusion. Compos. Sci. Technol. 70, 1742–1747. http://dx.doi.org/10.1016/j.compscitech.2010.07.005.

Joseph, K., Thomas, S., 1995. Effect of ageing on the physical and mechanical properties of sisal-fiber-reinforced polyethylene composites. Compos. Sci. Technol. 53 (1), 99–110. http://dx.doi.org/10.1016/0266-3538(94)00074-3.

Joseph, K., Thomas, S., Pavithran, C., 1996a. Effect of chemical treatment on the tensile properties of short sisal fibre-reinforced polyethylene composites. Polymer 37 (23), 5139–5149. http://dx.doi.org/10.1016/0032-3861(96)00144-9.

Joseph, K., Varghese, S., Kalaprasad, G., Thomas, S., Prasannakumari, L., Koshy, P., Pavithran, C., 1996b. Influence of interfacial adhesion on the mechanical properties and fracture behaviour of short sisal fibre reinforced polymer composites. Eur. Polym. J. 32 (10), 1243–1250. http://dx.doi.org/10.1016/S0014-3057(96)00051-1.

Joseph, P.V., Joseph, K., Thomas, S., 2002a. Short sisal fiber reinforced polypropylene composites: the role of interface modification on ultimate properties. Compos. Interfaces 9 (2), 171–205. http://dx.doi.org/10.1163/156855402760116094.

Joseph, P.V., Rabello, M.S., Mattoso, L.H.C., Joseph, K., Thomas, S., 2002b. Environmental effects on the degradation behaviour of sisal fibre reinforced polypropylene composites. Compos. Sci. Technol. 62 (10–11), 1357–1372. http://dx.doi.org/10.1016/S0266-3538(02)00080-5.

Joseph, P.V., Joseph, K., Thomas, S., Pillai, C.K.S., Prasad, V.S., Groeninckx, G., Sarkissova, M., 2003a. The thermal and crystallisation studies of short sisal fibre reinforced polypropylene composites. Compos. A: Appl. Sci. Manuf. 34 (3), 253–266. http://dx.doi.org/10.1016/S1359-835X(02)00185-9.

Joseph, P.V., Mathew, G., Joseph, K., Groeninckx, G., Thomas, S., 2003b. Dynamic mechanical properties of short sisal fibre reinforced polypropylene composites. Compos. A: Appl. Sci. Manuf. 34 (3), 275–290. http://dx.doi.org/10.1016/S1359-835X(02)00020-9.

Kabir, A.M., Huque, M.M., Islam, M.R., Bledzki, A.K., 2010. Mechanical properties of jute fiber reinforced polypropylene composite; effect of chemical treatment by benzenediazonium salt in alkaline medium. BioResources 5 (3), 1618–1625. Available from: http://ojs.cnr.ncsu.edu/index.php/BioRes/issue/view/30 (28 August 2014).

Kadla, J.F., Kubo, S., 2004. Lignin-based polymer blends: analysis of intermolecular interactions in lignin-synthetic polymer blends. Compos. A: Appl. Sci. Manuf. 35 (3), 395–400. http://dx.doi.org/10.1016/j.compositesa.2003.09.019.

Kalaprasad, G., Thomas, S., 2003. Melt rheological behavior of intimately mixed short sisal–glass hybrid fiber-reinforced low-density polyethylene composites. II. Chemical modification. J. Appl. Polym. Sci. 89 (2), 443–450. http://dx.doi.org/10.1002/app.11976.

Kalaprasad, G., Francis, B., Thomas, S., Kumar, C.R., Pavithran, C., Groeninckx, G., Thomas, S., 2004. Effect of fibre length and chemical modifications on the tensile properties of intimately mixed short sisal/glass hybrid fibre reinforced low density polyethylene composites. Polym. Int. 53 (11), 1624–1638. http://dx.doi.org/10.1002/pi.1453.

Kaminsky, W., 1984. Process for Producing Starch/Polyolefin Polymer Compositions. US Patent 4431788 A.

Keener, T.J., Stuart, R.K., Brown, T.K., 2004. Maleated coupling agents for natural fibre composites. Compos. A: Appl. Sci. Manuf. 35 (3), 357–362. http://dx.doi.org/10.1016/j.compositesa.2003.09.014.

Khan, MdN, Roy, J.K., Akter, N., Zaman, H.U., Islam, T., Khan, R.A., 2012. Production and properties of short jute and short e-glass fiber reinforced polypropylene-based composites. Open J. Compos. Mater. 2, 40–47. http://dx.doi.org/10.4236/ojcm.2012.22006.

Kharade, A.Y., Kale, D.D., 1999. Lignin-filled polyolefins. J. Appl. Polym. Sci. 72 (10), 1321–1326. http://dx.doi.org/10.1002/(SICI)1097-4628(19990606)72:10<1321::AID-APP12>3.0.CO;2-9.

Kiziltas, A., Kiziltas, E.E., Boran, S., Gardner, D.J. Micro- and nanocellulose composites for automotive applications. Available from: http://www.speautomotive.com/SPEA_CD/SPEA2013/pdf/BNF/BNF3.pdf (14 June 2014).

Klemm, D., Schmauder, H.P., Heinze, T., 2002. Polysaccharides II: Polysaccharides from eukaryotes. In: De Baets, S., Vandamme, E., Steinbüchel, A. (Eds.), Biopolymers. Wiley-VCH, Weinheim, pp. 275–287.

Kokta, B.V., Chen, R., Daneault, C., Valade, J.L., 1983. Use of wood fibers in thermoplastics composites. Polym. Compos. 4 (4), 229–232. http://dx.doi.org/10.1002/pc.750040407.

Korta, B.V., Maldas, D., Daneault, C., Beland, P., 1990. Composites of polyvinyl chloride-wood fibers. I. Effect of isocyanate as a bonding agent. Polym.-Plast. Technol. Eng. 29, 87–118. http://dx.doi.org/10.1080/03602559008049836.

Kosaka, P.M., Kawano, Y., Petri, H.M., Fantini, M.C.A., Petri, D.F.S., 2007. Structure and properties of composites of polyethylene or maleated polyethylene and cellulose or cellulose esters. J. Appl. Polym. Sci. 103 (1), 402–411. http://dx.doi.org/10.1002/app.24836.

Koshy, R.R., Pothan, L.A., Thomas, S., 2013. Starch-based bionanocomposite: processing techniques. In: Dufresne, A., Thomas, S., Pothan, L.A. (Eds.), Biopolymer Nanocomposites, Processing, Properties and Applications. John Wiley & Sons, Inc, Hoboken, New Jersey, US, pp. 203–226.

Košiková, B., Lábaj, J., 2009. Lignin-stimulated protection of polypropylene films and DNA in cells of mice against oxidation damage. BioResources 4 (2), 805–815.

Košiková, B., Damianova, V., Kacurokova, M.J., 1993a. Sulfur-free as composites of polypropylene films. J. Appl. Polym. Sci. 47 (6), 1065–1073. http://dx.doi.org/10.1002/app.1993.070470613.

Košiková, B., Demianova, V., Kacurakova, M.J., 1993b. Lignin biopolymers in the composite polypropylene films. In: Kennedy, J., Phillips, G.O., Williams, D.A. (Eds.), Cellulosics Chemical, Biochemical Material Aspects. Ellis Horwood, NY, pp. 537–542.

Košiková, B., Miklesova, K., Damianova, V., 1993c. Characteristics of free radicals in composite lignin/polypropylene films studied by the EPR method. Eur. Polym. J. 29 (11), 1495–1497. http://dx.doi.org/10.1016/0014-3057(93)90065-N.

Krässig, H.A., 1993. Cellulose, Structure, Accessibility and Reactivity. Gordon and Breach Publishers, Philadelphia, PA.

Kurek, M., Scetar, M., Voilley, A., Galic, K., Debeaufort, F., 2012. Barrier properties of chitosan coated polyethylene. J. Membr. Sci. 403–404, 162–168. http://dx.doi.org/10.1016/j.memsci.2012.02.037.

Lee, S.H., Wang, S., Pharr, G.M., Xu, H., 2007. Evaluation of interphase properties in a cellulose fiber-reinforced polypropylene composite by nanoindentation and finite element analysis. Compos. A: Appl. Sci. Manuf. 38 (6), 1517–1524. http://dx.doi.org/10.1016/j.compositesa.2007.01.007.

Lee, S.H., Teramoto, Y., Endo, T., 2011. Cellulose nanofiber-reinforced polycaprolactone/polypropylene hybrid nanocomposite. Compos. A: Appl. Sci. Manuf. 42 (2), 151–156. http://dx.doi.org/10.1016/j.compositesa.2010.10.014.

Li, Y., Mai, Y.-W., 2006. Interfacial characteristics of sisal fiber and polymeric matrices. J. Adhes. 82 (5), 527–554. http://dx.doi.org/10.1080/00218460600713840.

Li, Y., Mlynár, J., Sarkanen, S., 1997. The first 85% kraft lignin-based thermoplastics. J. Polym. Sci. B 35 (12), 1899–1910. http://dx.doi.org/10.1002/(SICI)1099-0488(19970915)35:12<1899::AID-POLB5>3.0.CO;2-L.

Li, Y., Mai, Y.-W., Ye, L., 2000. Sisal fibre and its composites: a review of recent developments. Compos. Sci. Technol. 60 (11), 2037–2055. http://dx.doi.org/10.1016/S0266-3538(00)00101-9.

Li, H.Y., Tan, Y.Q., Zhang, L., Zhang, Y.X., Song, Y.H., Ye, Y., Xia, M.S., 2012. Bio-filler from waste shellfish shell: preparation, characterization, and its effect on the mechanical properties on polypropylene composites. J. Hazard. Mater. 217–218, 256–262. http://dx.doi.org/10.1016/j.jhazmat.2012.03.028.

Li, H., Zhang, Z., Song, K., Lee, S., Chun, S.-J., Zhou, D., Wu, Q., 2014. Effect of durability treatment on ultraviolet resistance, strength, and surface wettability of wood plastic composite. BioResources 9 (2), 3591–3601.

Lin, Z., Guan, Z., Chen, C., Cao, L., Wang, Y., Gao, S., Xu, B., Li, W., 2013. Preparation, structures and properties of shell/polypropylene biocomposites. Thermochim. Acta 551, 149–154. http://dx.doi.org/10.1016/j.tca.2012.10.009.

Ljungberg, N., Cavaille, J.Y., Heux, L., 2006. Nanocomposites of isotactic polypropylene reinforced with rod-like cellulose whiskers. Polymer 47 (18), 6285–6292. http://dx.doi.org/10.1016/j.polymer.2006.07.013.

Lomelí-Ramírez, M.G., Barrios-Guzmán, A.J., García-Enriquez, S., Rivera-Prado, J.J., Manríquez-González, R., 2014. Chemical and mechanical evaluation of bio-composites based on thermoplastic starch and wood particles prepared by thermal compression. BioResources 9 (2), 2960–2974.

Lucas, N., Bienaime, C., Belloy, C., Queneudec, M., Silvestre, F., Nava-Saucedo, J.-E., 2008. Polymer biodegradation: mechanisms and estimation techniques – a review. Chemosphere 73 (4), 429–442. http://dx.doi.org/10.1016/j.chemosphere.2008.06.064.

Luyt, A.S., Malunka, M.E., 2005. Composites of low density polyethylene and short sisal fibers: the effect of wax addition and peroxide treatment on thermal properties. Thermochim. Acta 426 (1–2), 101–107. http://dx.doi.org/10.1016/j.tca.2004.07.010.

Macoveanu, M.M., Constantin, L., Manoliu, Al, Pascu, M.C., Profire, L., Cazacu, G., Vasile, C., 2001a. Polyolefins/lignosulfonates blends. IV. Bio-and environmental degradation testing of polyolefins/lignosulfonates blends. Cellul. Chem. Technol. 35 (1–2), 197–203.

Macoveanu, M.M., Georgescu-Buruntea, N., Pascu, M.C., Casariu, M., Ioanid, A., Vidrascu, P., Vasile, C., 2001b. Polyolefins/lignosulfonates blends. V. Bio and environmental degradation. Cellul. Chem. Technol. 35 (3–4), 263–273.

Maldas, D., Kokta, B.V., Daneault, C., 1989. Composites of polyvinyl chloride–wood fibers: IV. Effect of the nature of fibers. J. Vinyl Technol. 11 (2), 90–99. http://dx.doi.org/10.1002/vnl.730110209.

Malunka, M.E., Luyt, A.S., Krump, H., 2006. Preparation and characterization of EVA–sisal fiber composites. J. Appl. Polym. Sci. 100, 1607–1617. http://dx.doi.org/10.1002/app.23650.

Maningat, C.C., Seib, P.A., Bassi, S.D., Woo, K.S., Lasater, G.D., 2009. Wheat starch: production, properties, modification and uses. In: BeMiller, J., Whistler, R. (Eds.), Starch: Chemistry and Technology, a volume in Food Science and Technology, 3rd ed. Elsevier Inc., USA, pp. 441–510.

Matzinos, P., Bikiaris, D., Kokkou, S., Panayiotou, C., 2001. Processing and characterization of LDPE/starch products. J. Appl. Polym. Sci. 79, 2548–2557. http://dx.doi.org/10.1002/1097-4628(20010401)79:14<2548:AID-APP1064>3.0.CO;2-3.

Maya, J.J., Sabu, T., 2008. Biofibers and biocomposites. Carbohydr. Polym. 71, 343–364. http://dx.doi.org/10.1016/j.carbpol.2007.05.040.

Michaeli, W., Menges, G., 1989. Prediction of product properties in extrusion and injection molding. Adv. Polym. Technol. 9 (1), 69–85. http://dx.doi.org/10.1002/adv.1989.060090108.

Migneault, S., Koubaa, A., Erchiqui, F., Chaala, A., Englund, K., Wolcott, M.P., 2009. Effects of processing method and fiber size on the structure and properties of wood–plastic composites. Compos. A: Appl. Sci. Manuf. 40, 80–85. http://dx.doi.org/10.1016/j.compositesa.2008.10.004.

Mir, S., Yasin, T., Halley, P.J., Siddiqi, H.M., Nich, T., 2011. Thermal, rheological, mechanical and morphological behavior of HDPE/chitosan blend. Carbohydr. Polym. 83, 414–421. http://dx.doi.org/10.1016/j.carbpol.2010.07.069.

Mir, S.S., Nafsin, N., Hasan, M., Hasan, N., Hassan, A., 2013. Improvement of physico-mechanical properties of coir-polypropylene biocomposites by fiber chemical treatment. Mater. Des. 52, 251–257. http://dx.doi.org/10.1016/j.matdes.2013.05.062.

Mishra, S., Misra, M., Tripathy, S.S., Nayak, S.K., Mohanty, A.K., 2001. Graft copolymerization of acrylonitrile on chemically modified sisal fibers. Macromol. Mater. Eng. 286, 107–113. http://dx.doi.org/10.1002/1439-2054(20010201)286:2<107::AID-MAME107>3.0.CO;2-0.

Mokhena, T.C., Luyt, A.S., 2014. Investigation of polyethylene/sisal whiskers nanocomposites prepared under different conditions. Polym. Compos. 35 (11), 2221–2233. http://dx-.doi.org/10.1002/pc.22887. Available from: http://onlinelibrary.wiley.com/doi/10.1002/pc.22887/pdf (28 August 2014).

Mokoena, M.A., Djoković, V., Luyt, A.S., 2004. Composites of linear low density polyethylene and short sisal fibres: the effects of peroxide treatment. J. Mater. Sci. 39 (10), 3403–3412. http://dx.doi.org/10.1023/B:JMSC.0000026943.47803.0b.

Muasher, M., Sain, M., 2006. The efficacy of photostabilizers on the color change of wood filled plastic composites. Polym. Degrad. Stab. 91 (5), 1156–1165. http://dx.doi.org/10.1016/j.polymdegradstab.2005.06.024.

Mumpower, R.C., Touey, G.P., 1969. Blends of Cellulose Acetate and Polyolefin Fibers in Tow Form. US Patent 3451887 A.

Munteanu, B.S., Pâslaru, E., Fras Zemljic, L., Sdrobiş, A., Pricope, G.M., Vasile, C., 2014. Chitosan coatings applied to polyethylene surface to obtain food-packaging materials. Cellulose Chem. Technol. 48 (5-6), 565–575.

Muthukumar, T., Aravinthana, A., Dineshram, R., Venkatesan, R., Doble, M., 2014. Biodegradation of starch blended high density polyethylene using marine bacteria associated with biofilm formation and its isolation characterization. J. Microbial Biochem. Technol. 6, 116–122. http://dx.doi.org/10.4172/1948-5948.1000131.

Nabi Saheb, D., Jog, J.P., 1999. Natural fiber polymer composites: a review. Adv. Polym. Technol. 18 (4), 351–363. http://dx.doi.org/10.1002/(SICI)1098-2329(199924)18:4<351::AID-ADV6>3.0.CO;2-X.

Netravali, A.N., Chabba, S., 2003. Composites get greener. Mater. Today 6 (4), 22–29. http://dx.doi.org/10.1016/S1369-7021(03)00427-9.

Newswire, P.R., 2012. Weyerhaeuser Makes Breakthrough in Thermoplastic Composites. Available from: http://www.prnewswire.com/news-releases/weyerhaeuser-makes-breakthrough-in-thermoplastic-composites-171547381.html (15 June 2014).

Nino, K.A., Imam, S.H., Gordon, S.H., Wong, L.J.G., 1999. Extruded plastics containing starch and chitin: physical properties and evaluation of biodegradability. In: Imam, S.H., Greene, R.V., Zaidi, B.R. (Eds.), Biopolymers – Utilizing Nature's Advanced Materials. ACS Symposium Series 723, Washington DC. pp. 198–203.

Niu, P., Liu, B., Wei, X., Wang, X., Yang, J., 2011. Study on mechanical properties and thermal stability of polypropylene/hemp fiber composites. J. Reinf. Plast. Compos. 30 (1), 36–44. http://dx.doi.org/10.1177/0731684410383067.

Nur Aimi, N., Anuar, H., Manshor, M.R., Wan Nazri, W.B., Sapuan, S.M., 2014. Optimizing the parameters in durian skin fiber reinforced polypropylene composites by response surface methodology. Ind. Crop. Prod. 54, 291–295. http://dx.doi.org/10.1016/j.indcrop.2014.01.016.

Oksman, K., Lindberg, H., Holmgren, A., 1998. The nature and location of SEBS-MA compatibilizer in polyethylene-wood flour composites. J. Appl. Polym. Sci. 69, 201–209. http://dx.doi.org/10.1002/(SICI)1097-4628(19980705)69:1<201::AID-APP23>3.0.CO;2-0.

Olsen, D.J., 1991. Effectiveness of maleated polypropylenes as coupling agents for wood flour/polypropylene composites. In: ANTEC: Proceedings of the 49th Annual Technical Conference. Society of Plastics Engineers, Brookfield, CT, Montreal, Canada, pp. 1886–1891.

Oragwu, I.P., Igwe, I.O., 2013. Studies on the mechanical and water uptake properties of some polyolefins/corn starch blends (1). Am. J. Eng. Res. 2 (7), 22–27.

Owolabi, O., Ozvikovzki, T., Kovacs, I., 1985. Coconut fibre reinforced thermosetting plastics. J. Appl. Polym. Sci. 30, 1827–1836. http://dx.doi.org/10.1002/app.1985.070300504.

Ozen, E., Kiziltas, A., Kiziltas, E.E., Gardner, D.J., 2013. Natural fiber blend-nylon 6 composites. Polym. Compos. 34, 544–553. http://dx.doi.org/10.1002/pc.22463.

Park, J.-M., Quang, S.T., Hwang, B.-S., DeVries, K.L., 2006. Interfacial evaluation of modified jute and hemp fibers/polypropylene (PP)-maleic anhydride polypropylene copolymers (PP-MAPP) composites using micromechanical technique and non-destructive acoustic emission. Compos. Sci. Technol. 66, 2686–2699. http://dx.doi.org/10.1016/j.compscitech.2006.03.014.

Park, S.I., Marsh, K.S., Dawson, P., 2010. Application of chitosan-incorporated LDPE film to sliced fresh red meats for shelf life extension. Meat Sci. 85, 493–499. http://dx.doi.org/10.1016/j.meatsci.2010.02.022.

Părpăriţă, E., Darie, R.N., Popescu, C.-M., Uddin, MdA, Vasile, C., 2014a. Structure–morphology–mechanical properties relationship of some polypropylene/lignocellulosic composites. Mater. Des. 56, 763–772. http://dx.doi.org/10.1016/j.matdes.2013.12.033.

Părpăriţă, E., Brebu, M., Uddin, MdA, Yanik, J., Vasile, V., 2014b. Pyrolysis behaviors of various biomasses. Polym. Degrad. Stab. 100, 1–9. http://dx.doi.org/10.1016/j.polymdegradstab.2014.01.005.

Părpăriţă, E., Nistor, M.T., Popescu, M.-C., Vasile, C., 2014c. TG/FT–IR/MS study on thermal decomposition of polypropylene/biomass composites. Polym. Degrad. Stab. 109, 13–20. http://dx.doi.org/10.1016/j.polymdegradstab.2014.06.001.

Părpăriţă, E., Uddin, Md.A., Watanabe, T., Kato, Y., Yanik, J., Vasile, C., 2014d. Gas production by steam gasification of polypropylene/biomass waste composites in a dual-bed reactor. J. Mater. Cycles Waste Manage. http://dx.doi.org/10.1007/s10163-014-0308-0.

Pasanphan, W., Chirachanchai, S., 2008. Polyethylene film surface functionalized with chitosan via γ-ray irradiation in aqueous system: an approach to induce copper (II) ion adsorptivity on PE. React. Funct. Polym. 68, 1231–1238. http://dx.doi.org/10.1016/j.reactfunctpolym.2008.05.006.

Pascu, M.C., Macoveanu, M.M., Vasile, C., Ioanid, A., Oghina, R.C., 2000. Polyolefins/lignosulfonates blends. III. Photo-oxidation of isotactic polypropylene/epoxidized lignosulfonate blends. Cellul. Chem. Technol. 34, 399–420.

Pascu, M.C., Macoveanu, M.M., Ioanid, A., Vasile, C., 2001. Polyolefins/lignosulfonates blends VI. Bio and environmental behaviour of photo-oxidized blends. Cellul. Chem. Technol. 35 (5–6), 513–522.

Pascu, M., Vasile, C., Popa, G., Mihaila, I., Pohoata, V., 2002. Modification of polymer blends properties by plasma/electron beam treatment. I. Plasma diagnosis and bulk properties of plasma treated blends. Int. J. Polym. Mater. 51, 181–192. http://dx.doi.org/10.1080/00914030213031.

Paslaru, E., Fras Zemljic, L., Bracic, M., Vesel, A., Petrinic, I., Vasile, C., 2013. Stability of a chitosan layer deposited onto a polyethylene surface. J. Appl. Polym. Sci. 130 (4), 2444–2457. http://dx.doi.org/10.1002/app.39329.

Paul, S.A., Boudenne, A., Ibos, L., Candau, Y., Joseph, K., Thomas, S., 2008. Effect of fiber loading and chemical treatments on thermophysical properties of banana fiber/polypropylene commingled composite materials. Compos. A: Appl. Sci. Manuf. 39 (9), 1582–1588. http://dx.doi.org/10.1016/j.compositesa.2008.06.004.

Pickering, K. (Ed.), 2008. Properties and Performance of Natural-Fiber Composites. Woodhead Publishing Limited and CRC Press LLC, UK and USA.

St-Pierre, N., Favis, B.D., Ramsay, B.A., Ramsay, J.A., Vergoogt, H., 1997. Processing and characterization of starch/polyethylene blends. Polymer 38, 647–656. http://dx.doi.org/10.1016/S0032-3861(97)81176-7.

Popa, M.I., Pernevan, S., Sirghie, C., Spiridon, I., Chambre, D., Copolovici, D.M., Popa, N., 2013. Mechanical properties and weathering behavior of polypropylene-hemp shives composites. J. Chem. 2013 (343068), 8. http://dx.doi.org/10.1155/2013/343068 (28 August 2014).

Popescu, M.C., Totolin, M., Tibirna, C.M., Sdrobiş, A., Stevanovic, T., Vasile, C., 2011. Grafting of softwood kraft pulps fibers with fatty acids under cold plasma conditions. Int. J. Biol. Macromol. 48, 326–335. http://dx.doi.org/10.1016/j.ijbiomac.2010.12.011.

Pothan, L.A., Zimmermann, Y., Thomas, S., Spange, S., 2000. Determination of polarity parameters of chemically modified cellulose fibers by means of the solvatochromic technique. J. Polym. Sci. B 38, 2546–2553. http://dx.doi.org/10.1002/1099-0488(20001001)38:19<2546::AID-POLB60>3.0.CO;2-W.

Pothan, L.A., Bellmann, C., Kailas, L., Thomas, S., 2002. Influence of chemical treatments on the electrokinetic properties of cellulose fibres. J. Adhes. Sci. Technol. 16, 157–178. http://dx.doi.org/10.1163/156856102317293687.

Pothan, L.A., Simon, F., Spange, S., Thomas, S., 2006a. XPS studies of chemically modified banana fibers. Biomacromolecules 7, 892–898. http://dx.doi.org/10.1021/bm050462a.

Pothan, L.A., Thomas, S., Groeninckx, G., 2006b. The role of fiber/matrix interaction on the dynamic mechanical properties of chemically modified banana fiber/polyester composites. Compos. A: Appl. Sci. Manuf. 37 (9), 1260–1269. http://dx.doi.org/10.1016/j.compositesa.2005.09.001.

Pothan, L.A., Luyt, A.S., Thomas, S., 2007. Polyolefin-natural fiber composites. In: Nwabunma, D., Kyu, T. (Eds.), Polyolefin Composites. John Wiley, New Jersey, pp. 44–86.

Pouteau, C., Dole, P., Cathala, B., Averous, L., Boquillon, N., 2003. Antioxidant properties of lignin in polypropylene. Polym. Degrad. Stab. 81 (1), 9–18. http://dx.doi.org/10.1016/S0141-3910(03)00057-0.

Pouteau, C., Baumberger, S., Cathala, B., Dole, P., 2004. Lignin-polymer blends: evaluation of compatibility by image analysis. C. R. Biol. 327 (9–10), 935–943. http://dx.doi.org/10.1016/j.crvi.2004.08.008.

Prasanna, K., Sailaja, R.R.N., 2012. Blends of LDPE/chitosan using epoxy-functionalized LDPE as compatibilizer. J. Appl. Polym. Sci. 124, 3264–3275. http://dx.doi.org/10.1002/app.35262.

Premalal, H.G.B., Ismail, H., Baharin, A., 2002. Comparison of the mechanical properties of rice husk powder filled polypropylene composites with talc filled polypropylene composites. Polym. Test. 21, 833–839. http://dx.doi.org/10.1016/S0142-9418(02)00018-1.

Qin, Te-fu, Huang, Luo-hua, Li, Gai-yun, 2005. Effect of chemical modification on the properties of wood/polypropylene composites. J. For. Res. 16 (3), 241–244.

Rahman, MdR, Hasan, M., Monimul Huque, Md, Islam, MdN, 2010. Physico-mechanical properties of jute fiber reinforced polypropylene composites. J. Reinf. Plast. Compos. 29, 445–455. http://dx.doi.org/10.1177/0731684408098008.

Raj, R.G., Kokta, B.V., 1991. Reinforcing high density polyethylene with cellulosic fibers I: the effect of additives on fiber dispersion and mechanical properties. Polym. Eng. Sci. 31 (18), 1358–1362. http://dx.doi.org/10.1002/pen.760311811.

Raj, R.G., Kokta, B.V., Daneault, C., 2003. A comparative study on the effect of aging on mechanical properties of LLDPE–glass fiber, mica, and wood fiber composites. J. Appl. Polym. Sci. 40, 645–655. http://dx.doi.org/10.1002/app.1990.070400502.

Ramesh, M., Mitchell, J.R., Jumel, K., Harding, S.E., 1999. Amylose content of rice starch. Starch 51, 311–313. http://dx.doi.org/10.1002/(SICI)1521-379X(199909)51:8/9<311::AID-STAR311>3.0.CO;2-E.

Ray, D., Sarkar, B.K., 2001. Characterization of alkali-treated jute fibers for physical and mechanical properties. J. Appl. Polym. Sci. 80, 1013–1020.

Ross, H., 2012. Plastic Waste Degradation Options. Available from: http://plasticwastesolutions.com/plastic-waste-degradationoptions/ (9 September 2014).

Safinas, MdSA, Bakar, A.A., Ismail, H., 2013. Properties of kenaf bast powder-filled high density polyethylene/ethylene propylene diene monomer Composites. BioResources 8 (2), 2386–2397.

Sagnella, S., Mai-Ngam, K., 2005. Chitosan based surfactant polymers designed to improve blood compatibility on biomaterials. Colloids Surf. B Biointerfaces 42, 147–155. http://dx.doi.org/10.1016/j.colsurfb.2004.07.001.

Sailaja, R.R.N., Chanda, M., 2001. Use of maleic anhydride-grafted polyethylene as compatibilizer for HDPE-tapioca starch blends: effects on mechanical properties. J. Appl. Polym. Sci. 80, 863–872. http://dx.doi.org/10.1002/1097-4628(20010509)80:6<863::AID-APP1164>3.0.CO;2-R.

Sanchez, C.G., Esposito-Alvarez, L.A., 1999. Micromechanics of lignin/polypropylene composites suitable for industrial applications. Die Angew. Makromol. Chem. 272 (1), 65–70. http://dx.doi.org/10.1002/(SICI)1522-9505(19991201)272:1<65::AID-APMC65>3.0.CO;2-5.

Santo, M., Weitsman, R., Sivan, A., 2013. The role of the copper-binding enzyme – laccase – in the biodegradation of polyethylene by the actinomycete *Rhodococcus ruber*. Int. Biodeter. Biodegr. 84, 204–210. http://dx.doi.org/10.1016/j.ibiod.2012.03.001.

Schirp, A., Stender, J., 2010. Properties of extruded wood-plastic composites based on refiner wood fibers (TMP fibers) and hemp fibers. Eur. J. Wood. Wood. Prod. 68, 219–231. http://dx.doi.org/10.1007/s00107-009-0372-7.

Sdrobiş, A., Cazacu, G., Totolin, M., Vasile, C., 2011. Alkaline solution swelling of fatty acids modified with softwood kraft pulps fibers under cold plasma conditions. Cellul. Chem. Technol. 45 (5–6), 329–338.

Sdrobiş, A., Biederman, H., Kylian, O., Vasile, C., 2012a. Modification of cellulose/chitin mix fibers under different cold plasma conditions. Cellulose 20, 509–524. http://dx.doi.org/10.1007/s10570-012-9842-x.

Sdrobiş, A., Darie, R.N., Totolin, M., Cazacu, G., Vasile, C., 2012b. Low density polyethylene composites containing cellulose pulp fibers. Compos. Part B 43, 1873–1880. http://dx.doi.org/10.1016/j.compositesb.2012.01.064.

Sdrobiş, A., Ioanid, G.E., Stevanovic, T., Vasile, C., 2012c. Modification of cellulose/chitin mix fibers with n-isopropylacrylamide and poly (n-isopropylacrylamide) under cold plasma conditions. Polym. Int. 61 (12), 1767–1777. http://dx.doi.org/10.1002/pi.4268.

Serrano, A., Espinach, F.X., Julian, F., del Rey, R., Mendez, J.A., Mutje, P., 2013. Estimation of the interfacial shears strength, orientation factor and mean equivalent intrinsic tensile strength in old newspaper fiber/polypropylene composites. Compos. Part B 50, 232–238. http://dx.doi.org/10.1016/j.compositesb.2013.02.018.

Shin, G.H., Lee, Y.H., Lee, J.S., Kim, Y.S., Choi, W.S., Park, H.J., 2002. Preparation of plastic and biopolymer multilayer films by plasma source ion implantation. J. Agric. Food Chem. 50 (16), 4608–4614. http://dx.doi.org/10.1021/jf020169r.

Shumigin, D., Tarasova, E., Krumme, A., Meier, P., 2011. Rheological and mechanical properties of poly(lactic) acid/cellulose and LDPE/cellulose composites. Mater. Sci. (MEDŽIAGOTYRA) 17 (1), 32–38.

Siengchin, S., 2012. Impact, thermal and mechanical properties of high density polyethylene/flax/SiO_2 composites: effect of flax reinforcing structures. J. Reinf. Plast. Compos. 31, 959–966. http://dx.doi.org/10.1177/0731684412451333.

Simionescu, C.I., Rusan, V., Macoveanu, M.M., Cazacu, G., Lipsa, R., Vasile, C., Stoleriu, A., Ioanid, A., 1993. Lignin–epoxy composites. Compos. Sci. Technol. 48, 317–323. http://dx.doi.org/10.1016/0266-3538(93)90149-B.

Simionescu, C.I., Turta, C.I., Bucova, S.A., Macoveanu, M.M., Cazacu, G., Vasile, C., 1994. Sinteza, caracterizarea fizico-chimica si spectrala a unor rasini lignin epoxidice. Buletinul Acad. St. a Republicii Moldova. Stiinte Biologice si Chimice 3 (270), 62–73.

Simionescu, C.I., Macoveanu, M.M., Vasile, C., Ciobanu, Fl, Cazacu, G., 1995. In: Paipetas, S.A., Youtsos, A.G. (Eds.), High Technology Composites in Modern Applications. University of Patras, Greece, pp. 222–223.

Singh, A.A., Palsule, S., Singh, A., 2013. Jute fiber-reinforced chemically functionalized high density polyethylene (JF/CF-HDPE) composites with in situ fiber/matrix interfacial adhesion by Palsule Process. Compos. Interfaces 20 (8), 553–573. http://dx.doi.org/10.1080/15685543.2013.813192.

Singh, E., Deepak, D., Aggarwal, L., Gupta, V.K., 2014. Fabrication and characterization of jute fiber-reinforced–recycled polyethylene composites. In: Proceedings of the International Conference on Research and Innovations in Mechanical Engineering Lecture Notes in Mechanical Engineering, pp. 315–322.

Singleton, A.C.N., Baillie, C.A., Beaumont, P.W.R., Peijs, T., 2003. On the mechanical properties deformation and fracture of a natural fibre/recycled polymer composite. Compos. Part B 34, 519–526. http://dx.doi.org/10.1016/S1359-8368(03)00042-8.

Soulestin, J., Quiévy, N., Sclavons, M., Devaux, J., 2007. Polyolefins–biofiber composites: a new way for an industrial production. Polym. Eng. Sci. 47 (4), 467–476. http://dx.doi.org/10.1002/pen.20706.

Spoljaric, S., Genovese, A., Shanks, R.A., 2009. Polypropylene–microcrystalline cellulose composites with enhanced compatibility and properties. Compos. A: Appl. Sci. Manuf. 40 (6–7), 791–799. http://dx.doi.org/10.1016/j.compositesa.2009.03.011.

Stark, N.M., Matuana, L.M., Clemons, C.M., 2004. Effect of processing method on surface and weathering characteristics of wood-flour/HDPE composites. J. Appl. Polym. Sci. 93, 1021–1030. http://dx.doi.org/10.1002/app.20529.

Sunilkumar, M., Francis, T., Thachil, Eth, Sujith, A., 2012. Low density polyethylene–chitosan composites: a study based on biodegradation. Chem. Eng. J. 204–206, 114–124. http://dx.doi.org/10.1016/j.cej.2012.07.058.

Suwannamek, N., Srikulkit, K., 2013. Effect of chitosan compounds on composite fibers properties. Adv. Mater. Res. 701, 93–97. http://dx.doi.org/10.4028/www.scientific.net/AMR.701.93.

Tajvidi, M., Motie, N., Rassam, G., Rajaee, S., Falk, R.H., Felton, C., 2010. Mechanical performance of hemp fiber polypropylene composites at different operating temperatures. J. Reinf. Plast. Compos. 29 (5), 664–674. http://dx.doi.org/10.1177/0731684408100266.

Thakore, L.M., Iyer, S., Desai, A., Lele, A., Desvi, S., 1999. Morphology thermomechanical properties, and biodegradability of low density polyethylene/starch blends. J. Appl. Polym. Sci. 74, 2791–2802. http://dx.doi.org/10.1002/(SICI)1097-4628(19991213)7.

Toro, P., Quijada, R., Yazdani-Pedram, M., Arias, J.L., 2007. Eggshell, a new bio-filler for polypropylene composites. Mater. Lett. 61 (22), 4347–4350. http://dx.doi.org/10.1016/j.matlet.2007.01.102.

Torres, F.G., Cubillas, M.L., 2005. Study of the interfacial properties of natural fibre reinforced polyethylene. Polym. Test. 24, 694–698. http://dx.doi.org/10.1016/j.polymertesting.2005.05.004.

Tudorachi, N., Cascaval, C.N., Rusu, M., 2000a. Biodegradable polymer blends based on polyethylene and natural polymers. Degradation in soil. J. Polym. Eng. 20 (4), 287–304. http://dx.doi.org/10.1515/POLYENG.2000.20.4.287.

Tudorachi, N., Rusu, M., Cascaval, N.C., Constantin, L., Rugina, V., 2000b. Biodegradable polymeric materials I. Polyethylene–natural polymer blends. Cellul. Chem. Technol. 34, 101–111.

U.S. Geological Survey Science in the Decade 2007–2017, 2007. Available from: http://pubs.usgs.gov/circ/2007/1309/pdf/C1309.pdf (10 September 2014).

Umer, R., Bickerton, S., Fernyhough, A., 2008. Modelling the application of wood fiber reinforcements within liquid composite moulding processes. Compos. A: Appl. Sci. Manuf. 39, 624–639. http://dx.doi.org/10.1016/j.compositesa.2007.08.023.

Vasile, C., 2002. Role of polymer degradation processes in environmental pollution and protection. Polymery (Poland) XLVII (7/8), 517–522.

Vasile, C., 2009. Environmentally degradable polymeric materials: definitions and background. In: Vasile, C., Zaikov, G.E. (Eds.), Environmentally Degradable Materials Based on Multicomponent Polymeric Systems. Brill, Netherlands, pp. 1–34.

Vasile, C., Cazacu, G., 2013. Bio and nanocomposites containing lignin. In: Dufresne, A., Thomas, S., Pothan, L.A. (Eds.), Biopolymer Nanocomposites. Processing, Properties and Applications. Wiley Series on Polymer Engineering and Technology, New York, pp. 565–599.

Vasile, C., Macoveanu, M.M., Cazacu, G., Popa, V.I., 1997. In: Martuscelli, E., Musto, P., Ragosta, G. (Eds.), 5th Mediterranean School on Science and Technology of Advanced Polymer-Based Materials. UNESCO OFFICE, Capri, Italy, pp. 185–209, October 4–9.

Vasile, C., Downey, M., Wong, B., Macoveanu, M.M., Pascu, M.C., Choi, J.H., Sung, C., Baker, W., 1998. Polyolefins/lignosulfonates blends. II. Isotactic polypropylene/epoxi-modified lignin. Cellul. Chem. Technol. 32 (1–2), 61–88.

Vasile, C., Iwanczuk, A., Frackowiak, S., Cazacu, G., Constantinescu, G., Kozlowski, M., 2006. Modified lignin/polyethylene blends. Cellul. Chem. Technol. 40 (5), 345–351.

Vasile, C., Darie, R.N., Cheaburu-Yilmaz, C.N., Pricope, G.M., Bračič, M., Pamfil, D., Hitruc, G.E., Duraccio, D., 2013. Low density polyethylene – chitosan composites. Compos. Part B 55, 314–323. http://dx.doi.org/10.1016/j.compositesb.2013.06.008.

Vasile, C., Darie, R.N., Sdrobiş, A., Pâslaru, E., Pricope, G., Baklavaridis, A., Munteanu, S.B., Zuburtikudis, I., 2014. Effectiveness of chitosan as antimicrobial agent in LDPE/CS composite films as poultry minced meat packaging materials. Cellul. Chem. Technol. 48 (3–4), 325–336.

Wang, Q., Aıt-Kadfi, A., Kaliaguine, S., 1992. Catalytic grafting: a new technique for polymer–fiber composites. J. Polym. Sci. 44, 1107–1119. http://dx.doi.org/10.1002/app.1993.070480113.

Westrup, J.L., Da Silva, P.M.M., Dal-Bó, A.G., Benavides, R., Caldart, C.A., Coronetti, J.C., Doy, T.A., Frizon, T., Da Silva, L., 2014. Effect of chemical treatments on the properties of HDPE composites with Luffa cylindrical fiber. Cellul. Chem. Technol. 48 (3–4), 337–344.

Wiles, D.M., Scott, G., 2006. Polyolefins with controlled environmental degradability. Polym. Degrad. Stab. 91, 1581–1592. http://dx.doi.org/10.1016/j.polymdegradstab.2005.09.010.

Willet, J.L., 2004. Water sorption and diffusion in starch/polyolefins blends. J. Polym. Eng. Sci. 35 (14), 1184–1190. http://dx.doi.org/10.1002/pen.760351409.

Xanthos, M. (Ed.), 2005. Functional Fillers for Plastics. first ed. Wiley-VCH Verlag, Germany.

Xanthos, M. (Ed.), 2010. Functional Fillers for Plastics, enlarged ed. Wiley-VCH Verlag, Weinheim.

Xie, Y.J., Hill, C.A.S., Xiao, Z.F., Militz, H., Mai, C., 2010. Silane coupling agents used for natural fiber/polymer composites: a review. Compos. A: Appl. Sci. Manuf. 41, 806–819. http://dx.doi.org/10.1016/j.compositesa.2010.03.005.

Xie, F., Pollet, E., Halley, P.J., Avérous, L., 2013. Starch-based nano-biocomposites. Prog. Polym. Sci. 38 (10–11), 1590–1628. http://dx.doi.org/10.1016/j.progpolymsci.2013.05.002.

Yang, H.S., Gardner, D.J., Nader, J.W., 2011. Dispersion evaluation of microcrystalline cellulose/cellulose nanofibril-filled polypropylene composites using thermogravimetric analysis. J. Therm. Anal. Calorim. 103 (3), 1007–1015. http://dx.doi.org/10.1007/s10973-010-1147-4.

Yoksan, R., Chirachanchai, S., 2009. Silver nanoparticles dispersing in chitosan solution: preparation by γ-ray irradiation and their antimicrobial activities. Mater. Chem. Phys. 115 (1), 296–302. http://dx.doi.org/10.1016/j.matchemphys.2008.12.001.

Yuan, X., Jayaraman, K., Bhattacharyya, D., 2004. Effects of plasma treatment in enhancing the performance of wood fibre–polypropylene composites. Compos. A: Appl. Sci. Manuf. 35, 1363–1374. http://dx.doi.org/10.1016/j.compositesa.2004.06.023.

Zaman, H.U., Khan, M.A., Khan, R.A., 2011. A comparative study on the mechanical and degradation properties of plant fibers reinforced polyethylene composites. Polym. Compos. 32 (10), 1552–1560. http://dx.doi.org/10.1002/pc.21168.

Zhong, M., 2011. Strong, Flexible High-Lignin Polypropylene Blends. Available from http://www.4spepro.org/pdf/003642/003642.pdf (15 June 2014).

Zhong, Y., Song, X., Li, Y., 2011. Antimicrobial, physical and mechanical properties of kudzu starch–chitosan composite films as a function of acid solvent types. Carbohydr. Polym. 84, 335–342. http://dx.doi.org/10.1016/j.carbpol.2010.11.041.

Ziani, K., Fernández-Pan, I., Royo, M., Maté, J.I., 2009. Antifungal activity of films and solutions based on chitosan against typical seed fungi. Food Hydrocoll. 23 (8), 2309–2314. http://dx.doi.org/10.1016/j.foodhyd.2009.06.005.

Zuchowska, D., Steller, R., Meissner, W., 1998. Structure and properties of degradable polyolefin-starch blends. Polym. Degrad. Stab. 60 (2–3), 471–480. http://dx.doi.org/10.1016/S0141-3910(97)00110-9.

Mechanical performance of PC-based biocomposites

11

H.N. Dhakal
University of Portsmouth, Portsmouth, UK

11.1 Introduction

Glass and carbon fibre-reinforced polymer composites have several advantages over their metal–matrix counterparts, most commonly recognised being lightweight, high strength, and having an increased "strength-to-weight ratio." However, the major drawback of such composites is that they are difficult to recycle after the end of their useful life. Environmental legislation as well as consumer pressure for the adaptation of sustainable and environmentally friendly materials use has triggered a paradigm shift toward using natural materials as a substitute for nonrenewable conventional fibres like glass and carbon (Faruk et al., 2012; Bledzki et al., 2002). An example of environmental legislation can be considered the European Union's directive on end-of-life vehicles, which requires all new vehicles should be 85% reusable and recycled by weight, 10% can be used for energy recovery, and only 5% can be used in landfills from 2015 onward (Paul et al., 2015).

Over the last decade a number of researchers have been involved in investigating the properties of lignocellulosic biofibres such as hemp, kenaf, flax, and jute for composite and biocomposites as load-bearing constituents. The use of these renewable reinforcements in composites has increased significantly due to their relative cheapness, acceptable specific strength and modulus, good thermal insulation properties, CO_2 neutrality, reduced tool wear, and recycling and biodegradable possibility with enhanced energy recovery compared to their conventional counterparts such as glass fibres, which are non-renewable and nonbiodegradable and energy intensive (Satyanarayana et al., 2009).

Composites that use biodegradable or nonbiodegradable polymer resin as the matrix and biodegradable natural biofibres as the reinforcements are called biocomposites. For composites to be fully biocomposites, according to Mohanty et al. (2000), both the matrices and the reinforcements need to be biodegradable. Biopolymers are an important class of materials and the use of these matrices helps in the preservation of petroleum-based polymers and help to reduce the volume of landfill wastes and eventually help in protecting the environmental damage by the reduction of CO_2 emissions and contribute toward sustainable development.

Due to their lack of adequate mechanical and thermal properties, biodegradable matrices are not fully exploited yet in the creation of load-bearing and fully biodegradable biocomposites (Madsen and Lilholt, 2003). On the other hand, biocomposites based on nonbiodegradable matrices such as epoxy, vinyl ester, and unsaturated polyester reinforced with natural biofibres including hemp, sisal, flax, and jute are already produced in

Biocomposites: Design and Mechanical Performance. http://dx.doi.org/10.1016/B978-1-78242-373-7.00004-4

an industrial scale and have drawn a great deal of interest as these materials can be used for demanding product applications in automotive, building, furniture, and packaging applications (Bolton, 1995). However, the application of these biocomposites is mainly limited to interior parts and nonstructural components of automotive (Gayer and Schuh, 1996). The main reasons for this scenario are the inadequate mechanical properties of these materials required for structural and semistructural applications. This is due to a weak interface between the reinforcement and the matrix due to inherent moisture absorption behaviour of these biofibres and their structure and the damage mechanisms not well understood. Moreover, a main disadvantage of natural fibres is their natural origin, which implies an inherent large variation in fibre properties leading to a lower durability and a lower strength than glass fibres (Peijs, 2000; Stamboulis et al., 2001).

Bisphenol A polycarbonate (PC), an amorphous thermoplastic polymer, is used for a varieties of applications because of its several useful properties such as light weight and good impact resistance, optical transparency, dimensional stability, and high heat distortion temperature. PC is widely used in the manufacture of different components in the automotive industry, packaging, medical devices, and aerospace apparatus (Lu et al., 2014; Antonakou et al., 2014). PC, however, is not biodegradable and it can be a challenging task to use PC as matrix material in biofibre-reinforced composites due to its amorphous morphology, causing it to be difficult to impregnate into the reinforcements during the fabrication of composites.

Mechanical properties such as tensile, flexural, impact, and fracture toughness of the composites depend mainly on the fibre–matrix interface adhesion. Good adhesion between fibre and matrix leads to an efficient transfer of applied load from the matrix to the reinforcement. However, biofibres are hydrophilic and have low moisture resistance and such incompatibility between hydrophilic fibres and hydrophobic matrices can lead to a poor interfacial bonding between fibres and matrix and, thus, critically influences the strength and the modulus (Dhakal et al., 2013).

In order to produce structural biocomposites with reduced weight and improved mechanical and thermal properties, the problems associated with natural biofibre-reinforced composites and biocomposites, such as incompatibility between hydrophilic fibres and hydrophobic matrices and issues of moisture absorption and their effects on the properties, needs to be investigated and understood. Considering these facts, this chapter aims at highlighting the development of PC-based biocomposites and important parameters relating to their mechanical performance (Nayak et al., 2000; Yan et al., 2014).

11.2 Advantages of biofibres as composite reinforcements

Biofibres have been used for a long time in different industries and have the potential to be used increasingly as reinforcements in polymer engineering composites because they offer several advantages compared to conventional glass fibres, which are listed as follows (Peijs, 2000; Bledzki et al., 1997; Joshi et al., 2004; Eichhorn and Young, 2004):

- unlike glass and carbon fibre, biofibres are environmentally friendly;
- lower density than E-glass and carbon fibre;
- some biofibres have a higher specific modulus than E-glass;
- lower cost than E-glass and carbon fibres;
- CO_2 neutral;
- reduced tool wear during the processing;
- natural appearance;
- available abundantly and are renewable;
- low processing temperature;
- provide good insulation against heat and sound;
- reduced dermal and respiratory irritation;
- natural biofibres are more ductile than glass and carbon fibres;
- nontoxic;
- good thermal and insulation properties;
- enhanced energy recovery;
- good toughness properties;
- easily degradable.

11.3 Disadvantages of biofibres

Biofibres have some disadvantages, which are as follows (Peijs, 2000; Bledzki et al., 1997; Eichhorn et al., 2001; Summerscales et al., 2010):

- more easily inflammable;
- high moisture absorption and swelling properties;
- large scatter of parameters;
- incompatible with some polymeric matrices;
- due to hydrophilic nature, poor fibre–matrix interaction (decreases the mechanical properties);
- fibres may swell due to moisture;
- susceptibility to fungal and insect attack;
- relatively low degradation temperature;
- the variable quality of fibres, which depend on unpredictable influences such as weather and growing conditions;
- biofibre composite material shows much less thermal conductivity when compared to a glass fibre-reinforced composite;
- structure inhomogeneous;
- crop yield may cause price to vary.

11.4 Characterisation and mechanical performance of PC-based biofibre-reinforced biocomposites

11.4.1 PC as matrix materials in biocomposites: structures and properties

Thermoset resins such as epoxy, unsaturated polyester, and vinyl ester have been used extensively in the composite as matrix materials. However, these resins are brittle in nature and have poor resistance to crack propagation. PC is an important class of thermoplastic materials. Due to its outstanding mechanical and optical properties, many research studies have been conducted investigating the effect of PC (Figure 11.1) in composite materials as matrix.

A comprehensive study on the properties of different polymers along with PC has been undertaken by Velde and Kiekens (2001). The physical and thermal properties of PC adapted from their work are presented in Tables 11.1 and 11.2. As far as lightweight and strong biocomposites development is concerned, low density and higher mechanical properties of PC matrix can be a very attractive characteristic for biofibres as these fibres inherently absorb a large amount of water, which weakens the fibre–matrix interface and reduces mechanical properties. Despite its other benefits, PC has some drawbacks, including poor solvent resistance and relative softness making it liable to scratching and abrasion, which requires the application of some form of coating. PC also has high notch sensitivity leading to low fatigue strength (Khan et al., 2005). Compared to other polymers such as PP, LDPE, HDPE, and PET, PC absorbs significantly higher water (PC 1.19% vs. PET 0.07% for 24 h). But compared to PA6 1.8% and PA-66 1.6% of water absorption for 24 h, PC absorbs less water (Velde and Kiekens, 2001). However, the aging characteristic can be crucial when choosing PC as matrix material for biocomposites.

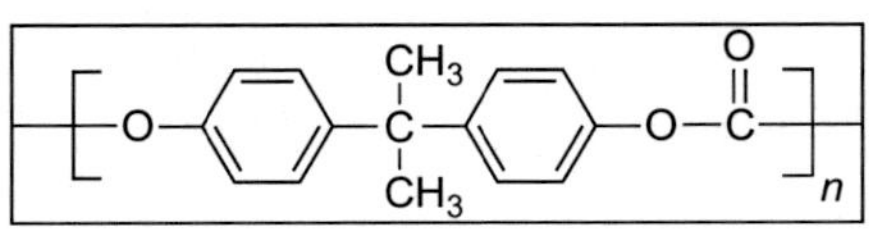

Figure 11.1 Molecular structure of PC.
Source: http://www.lg-dow.com/tech/tech.htm.

Table 11.1 Important physical and thermal properties of PC

PC	Limits	Density (g/cm^3)	Water content (% 24 h)	T_g (°C)	T_d, 1.8 MPa (°C)	T_p (°C)	αT (10_5°C_1)	Thermal conductivity (W/m/°C)
	Lower	1.19	0.12	140.50	127	260	6.5	0.19
	Upper	1.24	1.19	150	140	330	7.0	

Source: Velde and Kiekens (2001).

Table 11.2 Important mechanical properties of PC

PC	Limits	σ_{max} (MPa)	E (GPa)	σ_f (MPa)	E_f (GPa)	ε (%)	Izod, 1/8″ (J/m)
	Lower	53	2.3	140.50	81.4	90	534
	Upper	72	3.0	150	93.2	125	908

Source: Velde and Kiekens (2001).

Mechanical properties such as tensile, flexural, and impact values of PC are presented in Table 11.2. These properties are important in using PC as matrix material in biocomposites.

Tensile properties listed in Table 11.2 for PC are higher compared to other polymers such as PP and HDPE. However, these properties can be significantly affected under elevated temperatures. The work carried out by Lu et al. (2014) for PC at four different temperatures (i.e., 0, 30, 60, and 90 °C) suggests that the tensile stress depends greatly on test temperatures. Their results show that the higher the temperature, the lower the tensile stress. Similarly, the strain was also temperature dependent; the higher the test temperature, the larger the tensile strain. This is attributed to the molecular chains of the PC sliding easily at higher temperature.

11.4.2 Biofibres as reinforcing materials in biocomposites: structures and properties

Natural biofibres are largely divided into two categories based on their origin; namely, plant- or vegetable-based and animal-based. The mechanical performance of composites depends on the properties of its reinforcements. Reinforcements (individual constituents) improve the mechanical properties such as strength, stiffness, and creep resistance of polymers and their strength-to-weight and stiffness-to-weight ratios. In a fibre-reinforced polymer, the fibres serve as a reinforcement and, therefore, have to show a high tensile strength and stiffness, whereas the tasks of the matrix are to hold the fibres together, to transmit the shear forces, and to work as a coating. The chemical composition, diameter, crystallinity, surface properties, and fibre geometry contribute to the strength of the reinforcement. But unlike conventional man-made fibres, these properties vary significantly for biofibres due to their natural origin (Figure 11.2). It is evident that the mechanical properties of the fibres determine the stiffness and tensile strength of the composites decisively.

Another important factor that significantly affects the properties of biofibres is fibre defects, including kink bands and crack running along the fibre bundles. Kink bands are folds, or bends in the fibre walls. These are areas of low strength and when the fibre is loaded, kink starts to extend, leading to the failure of the fibre. Under the tensile strength, for example, the kink bands and crack running along the fibre act as stress concentration factors and as a result, these points can be sites for the initiation of delamination and fibre–matrix deboning. SEM of hemp biofibre showing kink band and fibre split along its longitudinal axis are shown in Figure 11.3.

The processing temperature of biofibres in thermoplastic matrices is limited due to the possibility of fibre degradation at elevated temperatures. The thermoplastic matrices that can be used at lower temperatures are limited due to their higher melting point and glass transition temperatures. For example, PC has a glass transition temperature of 150 °C, processing temperature of PC needs to be above 150 °C, and if biofibre-reinforced biocomposites are to be fabricated using PC as matrix, then the processing temperature must be monitored closely so that it does not exceed their degradation temperature.

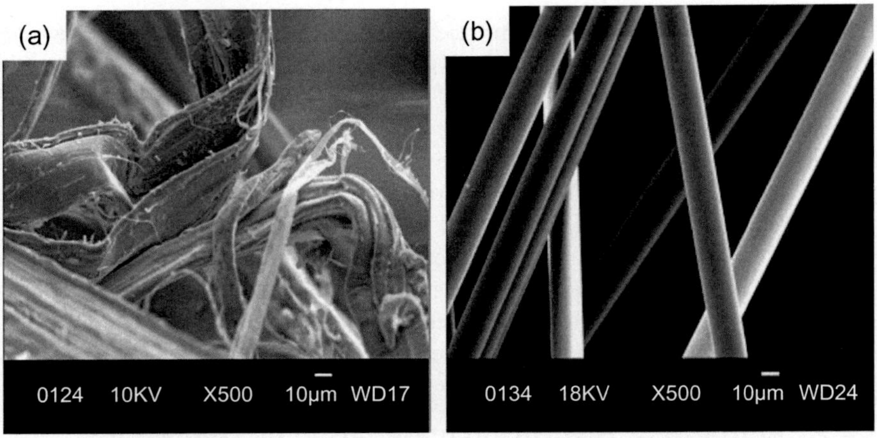

Figure 11.2 Structure of natural oil palm biofibre (a) showing nonuniform diameter in its structure and conventional glass fibre showing uniform fibre (b) diameter and length.
Panel b: Reprinted from Dhakal et al. (2007b), Copyright (2007), with permission from Elsevier.

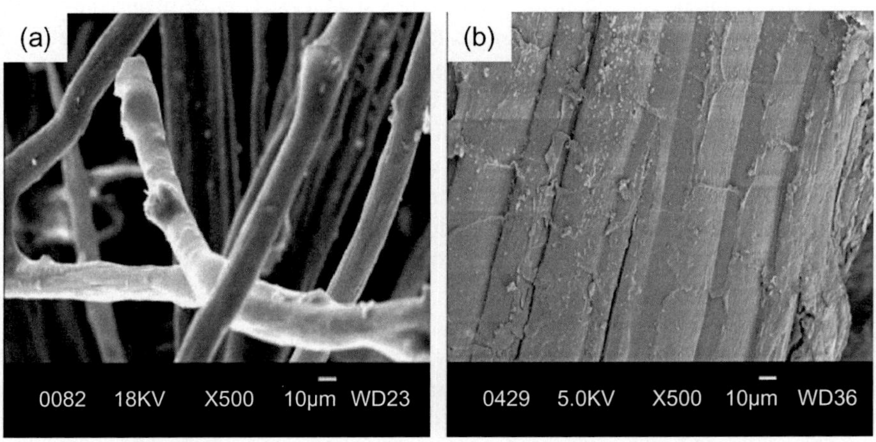

Figure 11.3 SEM micrographs of an example of (a) "kink bands" and (b) fibre split/crack observed in hemp fibres.
Panel a: Reprinted from Dhakal et al. (2007b), Copyright (2007), with permission from Elsevier.

The structural reliability of biocomposites mainly depends on the interface between the matrix and the fibres. It is obvious that the reinforcement must be the stronger component if it is to give good mechanical properties. The bond between the matrix and the reinforcement (fibres) is critical, as good interfacial adhesion between the matrix and the fibres transfers the stress from the matrix to the fibres, improving the mechanical properties of the composites (Bisinda and Ansell, 1991).

11.4.3 Tensile properties of PC-based biocomposites

Tensile testing of biocomposites is widely employed in order to assess their properties and performance. The testing is an important part in the process of developing this type of composite material.

In order to predict the tensile behaviour of natural fibre-reinforced composites, several models have already been developed (Zeidman and Sawhney, 2002; Gassan et al., 2001; Davies and Bruce, 1997). These models have been concerned about important parameters such as cellulose content, primary and secondary walls, and spiral angle to calculate the elastic properties of natural fibre-reinforced polymeric composites.

In Gassan et al. (2001), two structural models are presented to predict elastic properties of natural fibres. In model A, three cell wall layers are taken into account by using a laminated plate model where a natural fibre cell is considered as antisymmetrical laminated structure. In model B, the elliptic geometry and the hollow structure of the cross-section of the fibre cell (central lumen) are furthermore taken into account by using a thick laminated tube model (Gassan et al., 2001). The report suggests based on both models that the elastic properties are dependent also on spiral angle, cellulose content and fibre cross-section. The report concludes that, in general, modulus in fibre axis decreases with increasing spiral angle as well as the degree of anisotropy, while shear modulus reaches a maximum value for a spiral angle of 45° and fibre cell modulus increases with increasing cellulose content. It is worth noting though that the results suggested by these models need to be compared with the experimental results. It is well known that along with the parameters mentioned by the models described above, the mechanical and thermal properties of biocomposites are dependent on the different parameters such as fibre volume fraction and aspect ratio, microvoid contents, fibre structure and geometry, matrix properties, and manufacturing processes used.

The work of Khan et al. (2005) involving jute fibre-reinforced PC composites produced by the compression-moulding method indicates that the tensile properties of jute-reinforced biocomposites depend on parameters such as processing methods and temperatures, fibre thickness, and fibre volume fraction. The tensile strength for 13 and 32 wt.% of randomly oriented jute-PC composites is recorded approximately as 49 and 63 MPa, respectively. The reasons for the improvement on these properties are suggested as a result of fibre treatment and good fibre and matrix adhesion. The work further suggested that above fibre wt.% of 32, there was less impregnation of PC matrix and onto the jute reinforcement and, as a result, the fibre–matrix adhesion was reduced. As a consequence, for fibre wt.% of 35 and 42, the tensile strength was reduced.

Fibre treatment is one of the key factors for property enhancement. The fibre treatment enhances the adhesion capability of fibre and the PC matrix and as a result, fibre pull out is prevented whereas for untreated composites, fibre tends to agglomerate into the bundle and become unevenly distributed through the matrix. A better dispersion of fibre in the matrix can be expected for the treated composites (Khan et al., 2005). The tensile properties of biofibre-reinforced composites (both thermoplastics and thermosets) are mainly influenced by the interfacial adhesion between the matrix and the fibres. Several chemical modifications are employed to improve the matrix–fibre interfacial bonding resulting in the enhancement of tensile properties of the biocomposites.

Scanning electron microscopy provides an excellent technique for examination of surface morphology of natural and treated biofibres. Micrographs of untreated biofibres, such as the one presented in Figure 11.4a, indicates the presence of impurities on the surface, mostly waxes and oil. Waxes and oils provide a protective layer to the surface of the fibres. However, after alkali treatment, as presented in Figure 11.4b, a surface free of impurities is obtained.

In general, the tensile strength of the natural biofibre-reinforced polymer biocomposites increase with fibre content, up to a threshold or optimum value; the strength value then drops.

The properties of pineapple leaf fibre-reinforced PC biocomposites fabricated by the injection-moulding method were investigated by Threepopnatkul et al. (2009). The surface of pineapple leaf fibre (PALF) was treated with sodium hydroxide (PALF/NaOH) and modified with two different functionalities such as c-aminopropyl trimethoxy silane (PALF/Z-6011) and c-methacryloxy propyl trimethoxy silane (PALF/Z-6030). The results showed that the improvement in tensile properties were strongly dependent upon the surface treatment. Results from their study suggest that tensile strength of PALF/NaOH composites modified with different functionalities contributed toward the increase of tensile strength. However, the failure of PC/PALF was changed from ductile to brittle. The Young's modulus of PC composites significantly increased with increasing PALF content; from 1.120 GPa for neat PC to 1.960 GPa (75% increase) for NaOH-treated composites. The increase in modulus of composite was attributed to improved adhesion between the PALF and the PC matrix as well as the effect of reinforcement. Improved adhesion between PALF and PC makes it feasible for stress transfer to take place from the PC matrix to the PALF, thereby improving the strength and the modulus of the composites.

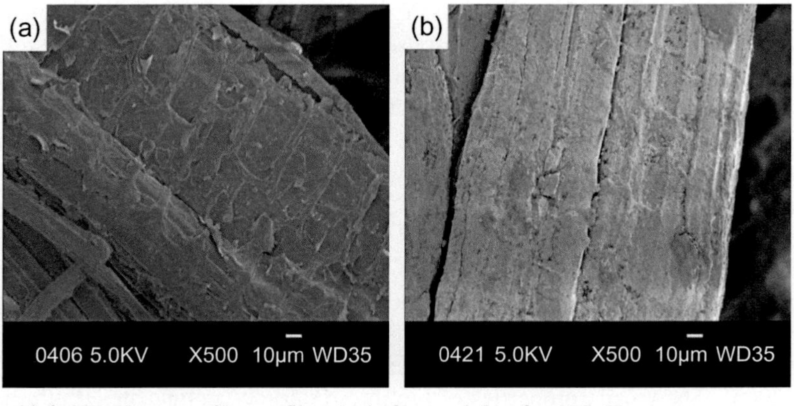

Figure 11.4 SEM images of hemp fibre (a) before and (b) after NOaH treatment.

11.4.4 Flexural properties

Flexural properties of jute/PC biocomposites reported by Khan et al. (2005) showed that both flexural strength and modulus of neat PC increased with various concentrations of jute reinforcement. The study reports that flexural strength at break and flexural modulus showed an increasing trend with the introduction of 32 wt.% of jute reinforcement into PC matrix. The flexural strength recorded for 13 and 32 wt.% of jute-PC composites is 67 and 87 MPa, respectively. A similar observation was made on the flexural modulus where 13 and 32 wt.% of fibre content PC composite exhibited flexural modulus of 3 and 4 GPa, respectively. A strong interfacial adhesion between jute and PC was reported due to chemical modification of jute fibre.

11.4.5 Impact properties

Low velocity impact of fibre-reinforced plastics has been the subject of many experimental and analytical investigations (Bogdanovich and Friedrich, 1994; Naik and Sekher, 1998). Susceptibility to low-velocity impact damage of composite materials has been well documented (Richardson and Wisheart, 1996).

One of the problems with the structural applications of fibre-reinforced composites, particularly natural biofibre-reinforced biocomposites, is the response to impact damage. Such damage may be caused by bumps or crashes and falling objects and debris. Many natural fibre-reinforced composite materials have been found to be very sensitive to impact loading (Santulli and Caruso, 2009; De Rosa et al., 2012).

Low-velocity impacts occur at a velocity below 10 m/s, which is the most effective method used for natural fibre composites and biocomposites as this technique resembles the real, practical situation. The impact results depend on various parameters such as impactor geometry and mass, laminate thickness, and impact velocity (Dhakal et al., 2014a,b).

Evaluation of impact and postimpact damage is important. When a projectile strikes a laminated composite, fracture processes such as delamination, matrix cracking, and fibre fracture frequently occur (Dhakal et al., 2012a,b). Therefore, the residual impact damage, especially in the case of barely visible impact damage (BVID), can cause serious structural failure in the useful life of composite laminates, which needs to be well understood.

The effects of PALF content and chemical treatment on the impact properties of pineapple leaf/PC composites were investigated by Threepopnatkul et al. (2009). Their results suggest that impact strength dropped with the increase of fibre content. It can be assumed that at the lower content of reinforcement, PC, as a tough polymer, influences the impact toughness. When the fibre content is increased, at higher concentration, then the influence of PC gets reduced. This lower value of impact strength might be explained by the presence of strong fibre–matrix interface resulting from the chemical treatment, which reduced the impact energy absorption.

11.4.6 Water absorption and its effects on the properties of biocomposites

Most polymers and composites absorb water at both room and elevated temperatures, and it is well known that moisture induces property changes in polymers and their composites. One of the main concerns for the use of natural fibre-reinforced composite materials, which contains the strong polar groups, and its structure is susceptible to moisture absorption as these fibres favour the bonding water molecules by hydrogen bridges and its effects on the physical, mechanical, and thermal properties (Dhakal et al., 2007a).

In order to understand the water sorption characteristics of biofibre-reinforced biocomposites, it is important to achieve an understanding of the water sorption mechanism in the plant fibres themselves. Water absorption behaviour of natural fibre-reinforced composites has been widely reported. However, there is not much information on water sorption behaviour of plant fibres themselves. The physics of water in relation to water sorption in plant fibres has been described in great detail by Madsen (2004). Various models have been proposed to view the charge distribution within a water molecule (Israelachhvili, 1991; Skaar, 1972).

Cellulosic fibres are hydrophilic and absorb moisture. Lignocellulosics are three-dimensional, polymeric composites made up primarily of cellulolse, hemicelluloses, and lignin. While all types of lignocellulosic fibres differ in chemical composition, within certain limits, all lignocellulosics have very similar properties; that is, they all swell and shrink as the moisture content of the cell wall changes, burn, decay, and degraded by acids, bases, and ultraviolet radiation.

Lignocellulosic structure of plant fibres change dimensions with changing moisture content because the cell wall polymers contain hydroxyl and other oxygen-containing groups that attract moisture through hydrogen bonding (Dhakal et al., 2014a,b). The hemicelluloses are mainly responsible for moisture sorption, but the accessible cellulose, noncrystalline cellulose, lignin, and surface of crystalline cellulose also play major roles. Moisture swells the cell wall, and the fibre expands until the cell wall is saturated with water (fibre saturation point, FSP). Beyond this saturation point, moisture exists as free water in the void structure and does not contribute to further expansion. This process is reversible, and the fibre shrinks as it loses moisture below the FSP.

Biofibres as lignocelluloses materials are inherently prone to moisture absorption. As a result, composites made from reinforcing these fibres also uptake moisture. The porous structure of these fibres make the water flow path easy. In biocomposites, additional factors such as void contents, humidity and temperature, fibre orientation and types, fibre volume fraction, as well as type of resin system, further promote moisture uptake. Moisture penetration into composite materials is conducted by three different mechanisms (Lin et al., 2002):

1. The first involves the ingress of water molecules inside the microgaps between polymer chains.
2. The second involves capillary transport into the gaps and indentations at the fibre–matrix interfaces.
3. The third involves propagation of microcracks in the matrix arising from the swelling of wet fibres (particularly in the case of natural biofibre-reinforced biocomposites).

Based on the abovementioned mechanisms, the diffusion behaviour can be related to either Fickian, non-Fickian or anomalous, and an intermediate between Fickian and non-Fickian (Shen and Springer, 1999; Errajhi et al., 2005). These three cases of diffusion can be distinguished theoretically by the shape of the sorption curve represented by the coefficient of diffusion

$$D = \pi \left(\frac{kh}{4M_{\mathrm{m}}} \right)^2 \tag{4.1}$$

where k is the initial slope of a plot of $M(t)$ versus $t^{1/2}$, $M(t)$ is the mass of the specimen during aging, $t^{1/2}$ is the square root of time, M_{m} is the maximum weight gain, and h is the thickness of the composites.

In the course of their service life, biocomposites are exposed to various environmental conditions such as acids, alkaline, UV rays, and high temperatures. In such exposures, the strength and the modulus of biocomposites are reduced. Moreover, the water diffusion through the matrix, fibre, and fibre–matrix interfaces causes swelling, plasticization, and hydrolysis, strongly affecting the physical, mechanical, and thermal properties as a result of weaker interfacial bonding between fibres and matrix caused by prolonged exposure to water and humidity (Chow et al., 2007; George et al., 1998).

It is well reported in the literature that the reduction of moisture absorption in biofibres and their composites can be achieved through the optimisation of the interface between the fibre and the matrix using various treatments. The following section attempts to explain the phenomenon of obtaining optimum fibre–matrix strength and increasing the water repellence behaviour of biocomposites.

11.5 Optimisation of fibre and matrix

11.5.1 Fibre modification

From the early development of polymer composites, the optimisation of the interface has been of pivotal importance. In conventional man-made fibre composites such as glass and carbon, various methods of modifying the fibre surface, like silane sizings, electrochemical oxidiation, and so on, have been found to be very successful in controlling the interface (Thomason and Adzima, 2001).

Due to heterogeneous and inconsistent chemical structure of biofibres, it is always difficult to control fibre/matrix compatibility of biofibre-based composites. However, a whole range of methods have been developed to enhance fibre/matrix compatibility and these methods are useful in these biocomposites (Zafeiropoulos et al., 2002). The thermal characteristics, crystallinity index, reactivity, and surface morphology of untreated and chemically modified and some natural fibres were investigated by Mwaikambo and Ansell (2004). In their study, hemp, sisal, jute, and kapok fibres were subjected to alkalisation by using sodium hydroxide. The result showed that alkalisation modifies biofibres, promoting the development of fibre–matrix adhesion, which then will result in increased interfacial energy and, hence, improvement in the mechanical and thermal stability of the natural fibre-reinforced biocomposites.

The alkalisation effects on biofibres are of particular importance for fibre–matrix adhesion and the creation of high fibre surface area required for the optimisation of fibre–resin reinforcement. The modification of cellulose fibres, therefore, develops into hydroxyl groups. These changes will effectively result in improved surface tension, wetting ability, swelling, adhesion, and compatibility with polymeric materials. The work of Eichhorn et al. (2001) has revealed that hemp fibre appears to have the highest crystallinity index and thermal stability when alkalised and acetylated with and without acid catalyst followed by jute, sisal, and then kopak fibres. In their report, the authors have outlined that kapok fibre has the highest reaction affinity to chemicals followed by jute, sisal, and hemp fibres. The chemical modification of hemp fibre has been investigated by Sebe et al. (2000). In their work, they have found that the interfacial properties of hemp and polyester were enhanced by the chemical modification of fibre.

11.6 Future for biofibre-reinforced PC-based biocomposites

The increased use and acceptance of biofibre-reinforced PC-based biocomposites in structural and semistructural applications will depend on compatibility between biofibre and PC matrix and matching their processing characteristics so that optimal mechanical performance is realised. It is important that the development of the surface treatment technique that enhances the fibre–matrix adhesion between PC and biofibres is established. Another important factor for the future of biocomposites is that it also depends on a competitive price in comparison to the conventional composites.

Development of biofibre-reinforced hybrid composites with the incorporation of nanoparticulates into PC matrix would play an important role in improving the mechanical performance of biocomposites. Another area that needs attention is the investigation of impact and the fracture toughness of biofibre-reinforced PC biocomposites. Low-velocity impact damage causes delamination in biocomposites and the fracture toughness tests (mode I and mode II) can characterize the delamination behaviour. If biofibre-reinforced PC-based biocomposites are to be used as structural materials, mechanical properties including impact and postimpact delamination behaviours as well as fatigue strength need to be analysed and understood.

References

Antonakou, E.V., Kalogiannis, K.G., Stefanidis, S.D., Karakoulia, S.A., Triantafyllidis, K.S., Lappas, A.A., Achilias, A.S., 2014. Catalytic and thermal pyrolysis of polycarbonate in a fixed-bed reactor: the effect of catalysts on products yields and composition. Polym. Degrad. Stab. 110, 482–491.

Bisinda, E.T.N., Ansell, M.P., 1991. The effect of silane treatment on the mechanical and physical properties of sisal-epoxy composites. Compos. Sci. Technol. 41, 165–178.

Bledzki, A.K., Gassan, J., Nichilas, P., Cheremisinoff, E.D., 1997. Natural Fibre Reinforced Plastics, Handbook of Engineering Polymeric Materials. Marcel Dekker, New York.

Bledzki, A.K., Sperber, V.E., Faruk, O., 2002. Natural and wood fibre reinforcement in polymer. Rapra Review Reports.

Bogdanovich, A.E., Friedrich, K., 1994. Initial and progressive failure analysis of laminated composites structures under dynamic loading. Compos. Struct. 27, 439–456.

Bolton, J., 1995. The potential of plant fibres as crops for industrial use. Outlook Agr. 24, 85–89.

Chow, C.P.L., Xing, X.S., Li, R.K.Y., 2007. Moisture absorption studies of sisal fibre reinforced polypropylene composites. Compos. Sci. Technol. 67, 306–313.

Davies, G.C., Bruce, D.M., 1997. A stress analysis model for composite coaxial cylinders. J. Mater. Sci. 32, 5425–5447.

De Rosa, I.M., Dhakal, H.N., Santulli, C., Sarasini, F., Zhang, Z.Y., 2012. Post-impact static and cyclic flexural characterisation of hemp fibre reinforced laminates. Compos. Part B 43, 1382–1396.

Dhakal, H.N., Zhang, Z.Y., Richardson, M.O.W., 2007a. Effect of water absorption on the mechanical properties of hemp fibre reinforced unsaturated polyester composites. Compos. Sci. Technol. 67, 1674–1683.

Dhakal, H.N., Zhang, Z.Y., Richardson, M.O.W., Errajhi, O.A.Z., 2007b. The low velocity impact response of non-woven hemp fibre reinforced unsaturated polyester composites. Composite Struct 81, 559–567.

Dhakal, H.N., Zhang, Z.Y., Bennett, N., 2012a. Influence of fibre treatment and glass fibre hybridisation on thermal degradation and surface energy characteristics of hemp/unsaturated polyester composites. Compos. Part B 43, 2757–2761.

Dhakal, H.N., Zhang, Z.Y., Bennett, N., Reis, P.N.B., 2012b. Low-velocity impact response of non-woven hemp fibre reinforced unsaturated polyester composites: influence of impactor geometry and impact velocity. Compos. Struct. 94, 2756–2763.

Dhakal, H.N., Zhang, Z.Y., Guthrie, R., Bennett, N., 2013. Development of flax/carbon fibre hybrid composites for enhanced properties. Carbohydr. Polym. 96 (1), 1–8.

Dhakal, H.N., Skrifvars, M., Adekunle, A., Zhang, Z.Y., 2014a. Falling weight impact response of jute/methacrylated soybean oil bio-composites under low velocity impact loading. Compos. Sci. Technol. 92, 134–141.

Dhakal, H.N., Zhang, Z.Y., Bennett, N., Arraiza, A.L., Vallejo, F.J., 2014b. Effect of water immersion ageing on the mechanical properties of flax and jute fibre biocomposites evaluated by nanoindentation and flexural testing. J. Compos. Mater. 48 (11), 1399–1406.

Eichhorn, S.J., Young, R.J., 2004. Composite micromechanics of hemp fibres and epoxy resin microdroplets. Compos. Sci. Technol. 64, 767–772.

Eichhorn, S.J., Baillie, C.A., Zafeiropoulos, N., Maikambo, L.Y., Ansell, M.P., Dufresne, A., Entwistle, K.M., Herrera-Franco, P.J., Escamilla, G.C., Groom, L., Hughes, M., Hill, C., Rials, T.G., Wild, P.M., 2001. Review: current international research into cellulosic fibres and composites. J. Mater. Sci. 36, 2107–2131.

Errajhi, O.A.Z., Osborne, J.R.F., Richardson, M.O.W., Dhakal, H.N., 2005. Water absorption characteristics of aluminised E-glass fibre reinforced unsaturated polyester composites. Compos. Struct. 71, 333–336.

Faruk, O., Bledzki, A.K., Fink, H.P., Sain, M., 2012. Biocomposites reinforced with natural fibers: 2000–2010. Prog. Polym. Sci. 37, 1552–1596.

Gassan, J., Chate, A., Bledzki, A.K., 2001. Calculation of elastic properties of natural fibres. J. Mater. Sci. 36, 3715–3720.

Gayer, U., Schuh, T.G., 1996. Automotive application of natural fibres composites. In: Proceedings of the First International Symposium on Lignocellulosic Composites. UNESP-Sao Paulo State University, Brazil.

George, J., Bhagawan, S.S., Thomas, S., 1998. Effects of environment on the properties of LDPE composites reinforced with pineapple-leaf fibre. Compos. Sci. Technol. 58, 1471–1484.

Israelachhvili, J.N., 1991. Intermolecular and Surface Forces, second ed. Academic Press, London, UK, Chapter 8.
Joshi, S.V., Drazal, L.T., Mohanty, A.K., Arora, S., 2004. Are natural fibre composites environmentally superior to glass fibre reinforced composites? Compos. A: Appl. Sci. Manuf. 35, 371–376.
Khan, M.A., Hassan, M.M., Drazal, L.T., 2005. Effect of 2-hydroxyethyl methacrylate (HEMA) on the mechanical and thermal properties of jute-polycarbonate composite. Compos. A: Appl. Sci. Manuf. 36, 71–81.
Lin, Q., Zhou, X., Dai, G., 2002. Effect of hydrothermal environment on moisture absorption and mechanical properties of wood flour–filled polypropylene composites. J. Appl. Polym. Sci. 85, 2824–2832.
Lu, F., Kang, G., Jiang, H., Zhang, J., Liu, Y., 2014. Experimental studies on the uniaxial ratchetting of polycarbonate polymer at different temperatures. Polym. Test. 39, 92–100.
Madsen, B. 2004. Properties of plant fibre yarn polymer composites – an experimental study. Technical University of Denmark, Report BYG, DTU-R082, 2004. ISSN 1601-2917, ISBN 87-7877-145-5.
Madsen, B., Lilholt, H., 2003. Physical and mechanical properties of unidirectional plant fibre composites – an evaluation of the influence of porosity. Compos. Sci. Technol. 63, 1265–1272.
Mohanty, A.K., Misra, M., Hinrichsen, G., 2000. Biofibres, biodegradable polymers and biocomposites: an overview. Macromol. Mater. Eng. 276/277, 1–24.
Mwaikambo, L.Y., Ansell, M.P., 2004. Chemical modification of hemp, sisal, jute and kopok fibres by alkalisation. J. Appl. Polym. Sci. 84, 2222–2234.
Naik, N.K., Sekher, Y.C., 1998. Damage in laminated composites due to low velocity impact. J. Reinf. Plast. Compos. 17, 1232–1263.
Nayak, S.K., Tripahy, S.S., Rout, J., Mohanty, A.K., 2000. Coir–polyester composites: effect on fibre surface treatment on mechanical properties of composites. Inter. Plast. Eng. Technol. 4, 79–86.
Paul, V., Kanny, K., Redhi, G.G., 2015. Mechanical, thermal and morphological pproperteis of a biobased composite derived from banana source. Compos. A: Appl. Sci. Manuf. 68, 90–100.
Peijs, T., 2000. Composites turn green. E-polymers 2002, no. T_002. http://www.e-polymers.org, pp. 1–12.
Richardson, M.O.W., Wisheart, M.J., 1996. Review of the low velocity impact properties of composites materials. J. Compos. A 27, 1123–1131.
Santulli, C., Caruso, A.P., 2009. Effect of fibre architecture on the falling weight impact properties of hemp/epoxy composites. J. Biobased Mater. Bioenergy 3, 291–297.
Satyanarayana, K.G., Arizaga, G.G.C., Wypych, F., 2009. Biogradable composites based on lignocellulosic fibres – an overview. Prog. Polym. Sci. 34, 982–1021.
Sebe, G., Cetin, N.S., Hill, C.A.S., Hughes, M., 2000. RTM Hemp fibre reinforced polyester composites. Appl. Compos. Mater. 7, 341–349.
Shen, C.H., Springer, G., 1999. Moisture absorption and disorption of composite materials. J. Compos. Mater. 10, 2–20.
Skaar, C., 1972. Water in wood. Syracuse University Press, New York, USA.
Stamboulis, A., Baillie, C.A., Peijs, T., 2001. Effects of environmental conditions on mechanical and physical properties of flax fibers. Compos. A: Appl. Sci. Manuf. 32, 1105–1114.
Summerscales, J., Dissanayake, N.P.J., Virk, A.S., Hall, W., 2010. A review of bast fibres and their composites. Part 1 – fibres as reinforcements. Compos. A 41, 1329–1334.

Thomason, J.L., Adzima, L.J., 2001. Sizing up the interphase: an insider's guide to the science of sizing. Compos. Part A 32, 313–321.

Threepopnatkul, P., Kaerkitcha, N., Athipongarporn, N., 2009. Effect of surface treatment on performance of pineapple leaf fibre-pllycarbonate composites. Compos. Part B 40, 628–632.

Velde, K.V., Kiekens, P., 2001. Thermoplastic polymers: overview of several properties and their consequences in flax fibre reinforced composites. Polym. Test. 20, 885–893.

Yan, L., Chouw, N., Jayaraman, K., 2014. Flax and its composites – a review. Compos. Part B 56, 296–317.

Zafeiropoulos, N.E., Baillie, C.A., Hodgkinson, J.M., 2002. Engineering and characterisation of the interface in flax fibre/polypropylene composite materials. Part II. The effect of surface treatments on the interface. Compos. A: Appl. Sci. Manuf. 33, 1185–1190.

Zeidman, M., Sawhney, P.S., 2002. Influence of fibre length distribution on strength efficiency of fibres in yarn. Text. Res. J. 72, 216–220.

Nylon uses in biotechnology

12

A. Tomasini, H.H. León-Santiesteban
Department of Biotechnology, Universidad Autónoma Metropolitana-Iztapalapa, Mexico, Mexico, D.F.

12.1 Introduction

Polymers are molecules with high molecular weights, greater than 5000. Synthetic polymers such as polystyrene, polyurethane, and polyamides are widely used and have multiple applications in the modern world.

Polyamides also called "nylons" are used in a wide variety of applications; mainly as nylon fibers. Their high tensile strength and good dye ability make them one of the best. A wide variety of polyamides with different properties can be synthesized varying the diamine and the diacid in aliphatic compounds. These polyamides present differences in melting points, glass transition temperatures, water absorption, and mechanical properties, that is, very specific sets of properties (Long and Turner, 2000).

The indiscriminate use of these fibers and their strong resistance make nylon fibers a recalcitrant compound causing, nowadays, serious environmental pollution problems. For this reason, it is important to propose physical, chemical, and/or biological processes to degrade polyamides. Some microorganisms capable of degrading nylon fibers have been reported (Deguchi et al., 1997, 1998).

The goal of this chapter is to describe some applications of nylon fibers (materials) within the biotechnology field. First, the chemical characteristics of polyamides will be briefly described, then some thermal and mechanical properties and the advances attained in the biodegradation of nylon fibers. Finally, some applications of nylon-immobilized microorganisms and enzymes will be presented.

12.2 Chemical characteristics of polyamides (nylon fiber)

The term "nylon" is accepted as the generic name for the family of polyamides, which are characterized by an amide linkage (–CONH–). Nylon was the first commercial polymer. Nylon-6,6 development began at DuPont with the polymer experiments of Wallace Carothers in 1928, and made its commercial debut as a fiber in 1938 and as a molding compound in 1941. Nylon-6 was marketed in 1946 throughout Europe. The thermoplastic properties of nylon have increased its applications. A global annual consumption of around 6.8 million tons of nylon-6 and nylon-6,6 was reported in 2011 (Press Information, PCI nylon Gmbh, 27 June 2011, http://www.konsens.de/44.html). This marked increase is due to the combination of toughness, rigidity, and its "lubrication-free" performance, peculiarly suited for mechanical bearing and gear applications. The properties of nylon, such as fast crystallization, fast molding cycling, great strength, good elasticity, resistance

Biocomposites: Design and Mechanical Performance. http://dx.doi.org/10.1016/B978-1-78242-373-7.00006-8

to abrasion, high degree of solvent resistance (it can only be solved into phenol, cresol, and formic acid at ambient temperature), toughness, lubricity, and water resistance have extended its use and applications in modern life (Peters and Arisman, 2000).

The most commercialized polyamides are nylon-6 (polycaprolactam) and nylon-6,6 (hexamethylene adipamide). Generally, the highest production of nylon-6 occurs in the United States, while nylon-6,6 is produced mainly in European countries.

The general structure of polyamides (nylon) is shown in Figure 12.1; the hydrogen-bonding interactions provide the nylon fiber with more mechanical strength, at lower degrees of polymerization, than that exhibited by hydrocarbon polymers, such as polyethylene.

The reaction mechanism of nylon-6 synthesis is shown in Figure 12.2. The polymerization of caprolactam is carried out by H_2O and heating at high temperature to produces a linear polymer based on aminocarboxilic acids (Carraher, 2012; Rudin, 1999). Polycaprolactam is crystalline, has a melting point (T_m) of 228 °C, and has been successfully applied as a molding compound.

Nylon-6,6 synthesis, as well as nylon-6 synthesis, does not require a catalyst; the adipic acid plays the role of a catalyst in the reaction at high temperature, between 270 and 280 °C, in the presence of acetic acid. Nylon-6,6 derives from the condensation of a six-carbon diamine (hexamethylenediamine) and a six-carbon dibasic acid (adipic acid) and, normally, the chain ends with monofunctional reactants (Figure 12.3) (Peters and Arisman, 2000). The melting point of this nylon-6,6 is higher than that of nylon-6, (T_m of 269 °C), and the polymer at this temperature collapses forming a low-viscosity fluid (Peters and Arisman, 2000).

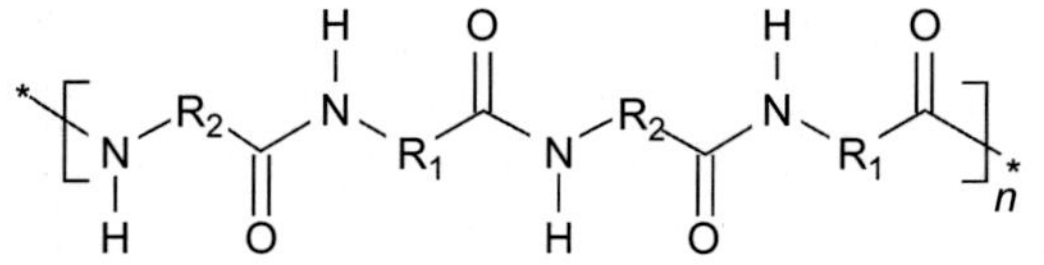

Figure 12.1 General structure of polyamides where R_1 and R_2 represent either aliphatic or aromatic compounds.

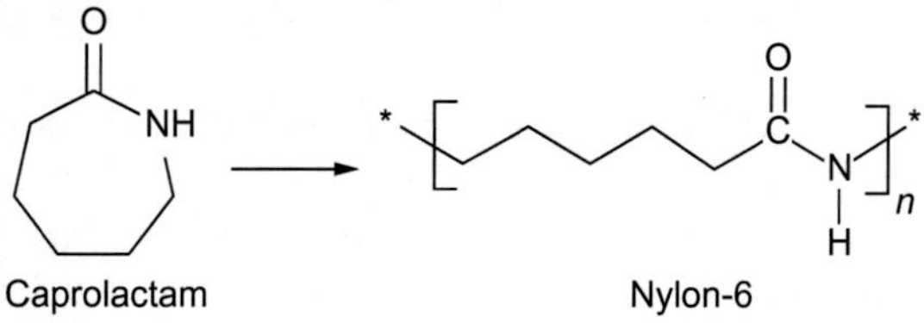

Figure 12.2 Synthesis of nylon-6.

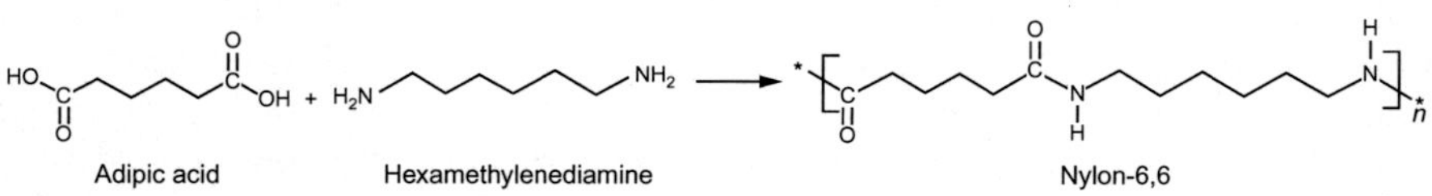

Figure 12.3 Synthesis of nylon-6,6.

There are other varieties of nylon such as nylon-4,6; nylon-9,9; nylon-6,10; and nylon-6,12 made from di-carboxylic acid, respectively, and nylon-6,11 made by self-condensation of 11-aminoundecanoic acid. The nylon nomenclature is based on the number of carbon atoms in the diacid and the diamine chains; for example, nylon produced from reaction of 1,6-hexanediamine (six carbons) and sebacic acid (ten carbons) is called nylon-6,10. These varieties of nylon present lower moisture absorption than nylon-6 and -6,6. Also, there are aromatic, semiaromatic polyamides with different characteristics, but the most used are nylon-6 and -6,6 (Sperling et al., 2000; Peters and Arisman, 2000).

Originally, nylons were used in fiber applications and this still constitutes a large market for nylons. Their high tensile strength and good dye ability make them superior in fiber performance. Approximately 60% of nylon fibers are used to manufacture home products, such as ropes, bands, abrasive cloths, wires, and hoses; 20% is used to manufacture articles of clothing; 10% in vulcanization products; and 10% in other products, as in automobile manufacturing.

In this chapter, first we will describe briefly the principal chemical, thermal, and mechanical characteristics, the various types of nylon fibers, and the structural proprieties. Thereafter, the pollution problems caused by nylon fiber accumulation in the environment and its biodegradation will be discussed. Finally, we describe the use of the nylon fiber in biotechnology applications.

12.3 Nylon structure

Nylons are classified as semicrystalline thermoplastics with a high degree of crystallinity and, therefore, high melting points (Olivares et al., 1996). The degree of crystallinity depends largely on the processing conditions; for instance, thermal and pressure treatments (Gogolewski and Pennings, 1973). Nylon-6, one of the most economically important polyamides, presents mainly two different crystalline forms. The γ crystalline form is obtained when nylon chains are oriented on parallel form and are joined by hydrogen bonds (Figure 12.4a). The α crystalline form is more stable than the γ

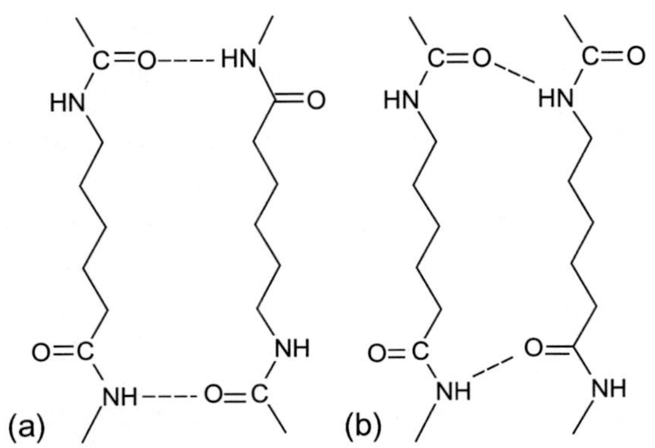

Figure 12.4 Crystalline forms of nylon-6: (a) γ crystalline form; (b) α crystalline form.

crystalline form and it is characterized for having hydrogen bonds among nylon chains oriented on antiparallel form (Figure 12.4b).

D'Alò et al. (1974) found that nylon-6, which was obtained from a cold roll and left one week with about 5% of water, had α crystalline form with two X-ray diffraction peaks at 20.3–20.7° (α_I) and at 23.3–23.9° (α_{II}). Meanwhile, when nylon-6 was spun and drawn between two thermoregulated rolls (slow at room temperature and fast at 150 °C) with a draw ratio of 3.8, a strong diffraction peak at 21.3 °C was only found, which corresponds to γ form. However, when the draw roll temperature increased to 195 °C, the transition of $\gamma \rightarrow \alpha$ transformations was reached, producing both α_I and α_{II}; although, Dasgupta et al. (1996) computed that the γ crystalline form is thermodynamically more stable for nylon-*n* with $n \geq 8$.

Greco and Nicolais (1976) calculated that the crystallinity of nylon-6, nylon-6,6, nylon-6,10, nylon-11, and nylon 12 were 27%, 27%, 19%, 19%, and 26%, respectively.

12.4 Thermal properties of nylons

One of the most important differences between plastics and rubber materials is the localization of plastics' glass transition temperature (T_g), which is the temperature where the amorphous region of the polymer undergoes a glassy-rubbery change. Rubber materials have typically T_g below room temperature, while plastics have T_g above room temperature. At low temperatures, polymers can be considered as glassy solids with a high modulus and as purely elastic solids (Charles, 2002).

Generally, T_g of aliphatic polyamides is less than T_g of aromatic polyamides, which are considered as a high-performance material due to their thermal resistance and/or mechanical strength. Commercial aliphatic nylons have T_g between 40 and 60 °C (Table 12.1). Meanwhile, aromatic polyamides as nylon-4,1 presents at less 2-times more T_g values than an aliphatic nylon (Table 12.1). Nylon-4,1 is a completely amorphous polymer with rather low crystalline structures. Among aromatic polyamides, poly(*m*-phenylene isophthalamide) (PMPI or Nomex®), poly(*p*-phenylene terephthalamide) (PPTA or Kevlar®), and 3,4′-diaminodiphenyl ether (ODA)/PPTA (Technora®) are the most important from an economic point of view. PMPI and PPTA are highly crystalline polymers (>60%) and possess T_g of 275 and ~345 °C, respectively (Table 12.1).

If temperature is increased farther than T_g, nylon polymers will undergo a transition between glassy state and rubbery state. This transition is named *leathery region* and is characterized by a decrease on elastic modulus (*E*). After leathery region, the *E* value will become more stable (but with less value than in glassy state) when temperature is increased; this stable zone is named *rubbery plateau* and is so characterized because the polymeric materials do not undergo permanent deformations. As temperature is increased farther, nylon polymers will melt, changing their state of aggregation from solid to liquid. The temperature necessary to achieve this change is denominated the melting temperature (T_m). At T_m, crystalline structures will be completely broken down (Aklonis, 1981; Charles, 2002).

Table 12.1 shows the T_m values of some important nylon polymers. Usually, aromatic polyamides melt at temperatures above 350 °C, while aliphatic polyamides melt

Table 12.1 **Thermal properties of the most economically important nylons**

Polymer	Abbreviations	T_g (°C)	T_m (°C)	References
Poly(imino-(1-oxohexamethylene))	Nylon-6[a]	53	220	Greco and Nicolais (1976) and Galanty (1999)
Poly(iminoadipoyliminohexamethylene)	Nylon-6,6[a]	57	269	Greco and Nicolais (1976) and Viers (1999)
Poly(iminohexamethyleneiminoazelaoyl)	Nylon-6,9[a]	58	220	Wen (2007a,b)
Poly(iminohexamethyleneiminosebacoyl)	Nylon-6,10[a]	45	227	Greco and Nicolais (1976) and Kohan (1999)
Poly(iminoadipoyliminododecamethylene)	Nylon-6,12[a]	46	207–247	Peterson and Zhao (1999)
Poly(imino-1-oxoundecamethylene)	Nylon-11[a]	42	184–188	Greco and Nicolais (1976) and Apgar (1999)
Poly(imino-1-oxododecamethylene)	Nylon-12[a]	43	260	Rohde-Liebenau (1999) and Greco and Nicolais (1976)
Poly(methylene isophthalamides)	Nylon-4,1[b]	138-146	262–270	Gaymans and Van der Ham (1984)
Poly(*m*-phenylene isophthalamide)	PMPI[b]	275	365	García et al. (2010)
Poly(*p*-phenylene terephthalamide)	PPTA[b]	~345	>500	García et al. (2010)
Co-poly-(*p*-phenylene/3,4′-oxy diphenylene terephthalamide)	ODA/PPTA[b]	–	>500	García et al. (2010)

[a]Aliphatic polyamide.
[b]Aromatic polyamide.

at temperatures below 300 °C (Gallini, 2002). Gaymans and Harkema (1977) demonstrated that the T_m was directly related to aromaticity; if aromaticity is increased, T_m will be increased, too.

12.5 Mechanical properties of nylons

As a result of their semicrystalline nature, nylons are polymers with high strength and good toughness. Their mechanical properties are strongly influenced by their preparation procedure, their chemical structure, temperature, moisture, and the fillers added. Table 12.2 summarizes some of the most relevant mechanical properties of nylons at room temperature.

Elastic modulus or modulus (*E*) is a mechanical property that indicates the stiffness of materials to elastic deformation under a load. Nylon polymers can be regarded as either stiff or flexible material, depending on the value of elastic modulus. Commercial aliphatic nylons have elastic modulus values between 1.2 and 2.9 GPa, which make them flexible materials (Table 12.2). Gogolewski and Pennings (1985) improved the value of elastic modulus of nylon-6 fibers up to 19 GPa by dry-spinning from a formic acid 75/chloroform 25 cosolvent mixture and hot-drawn at 234 °C. On the other hand, Zussman et al. (2006) prepared nylon-6,6 nanofibers by the electrospinning process with values of elastic modulus between 0.45 and 0.95 GPa. These same researchers prepared nylon-6,6 by melt extraction followed by cold drawing and calculated an elastic modulus of 1.2 GPa. Ehrenstein et al. (2000) synthetized amphiphilic nylon-6.24 and nylon-6.34, by melt polycondensation, using hexamethylenediamine and long alkane carboxylic diacids (tetracosanedioic acid and tetratriacontanedioic acid, respectively). They found that those nylons showed thermal properties very similar to conventional nylons. Although, their mechanical properties were lower than commercial nylons. Nylon-6.24 and nylon-6.34 exhibited elastic modulus of 1.3 and 0.7 GPa, respectively.

Nevertheless, when nylons are reinforced with mineral fillers (such as halloysite, organoclay, wollastonite, recycled carbon fiber, short glass fiber, and mica) or elastomers (such as polyvinyl butyral (PVB), styrene-ethylene-butylene-styrene (SEBS), and maleated styrene-ethylene-butylene-styrene (SEBS-g-MA)) their value of elastic modulus can be considerably enhanced, which produce inexpensive stiff composites (Table 12.2). Handge et al. (2010) observed that mineral halloysite promote the crystallization (γ modification) of nylon-6 but when the filler fraction is above 10 wt.%, the nylon-6/halloysite composite becomes less ductile at room temperature. This contrasts with what was observed at 80 °C, where nylon composites behaved with more ductility because the nylon was no longer in the glass state. Yoo et al. (2011) prepared nylon-6 composite containing both montmorillonite (MMT) and glass fiber (GF) by melting processing. They found that when the proportion of MMT is constantly maintained at 5 wt.% and the proportion of GF is increased from 0 to 30 wt.%, elastic modulus will increase from approximately 6 to 12 GPa. Feng et al. (2013) prepared nylon-6 composite, by melt extraction process, using recycled carbon fiber (RCF) chemically modified. These researchers observed that increasing the proportion of modified RCF into a blend from 0 to 25 wt.%, elastic modulus was increased from

Table 12.2 **Some mechanical properties of nylon polymers**

Polyamide	Modulus (*E*, GPa)	Yield strength (MPa)	Elongation at break (%)	Tensile strength (MPa)	References
Nylon-6	2.86	80.9	272	79–80.83	Majumdar et al. (1994), Wen (2007a,b), and Galanty (1999)
2% Halloysite	3.64	80.2	186	80.2	Hedicke-Höchstötter et al. (2009)
5% Halloysite	3.79	81.8	96	81.8	Hedicke-Höchstötter et al. (2009)
2% Organoclay	4.30	84.8	65.9	~86	Hedicke-Höchstötter et al. (2009)
5% Organoclay	4.36	87.2	63.8	~87	Hedicke-Höchstötter et al. (2009)
30% Wollastonite	~5	UR	UR	~74	Unal and Mimaroglu (2004)
Nylon-6	2.0	UR	105	69	Unal and Mimaroglu (2012)
30% Mica	4.5	UR	3.3	63.2	Unal and Mimaroglu (2012)
30% SGF	8.2	UR	3.6	165	Unal and Mimaroglu (2012)
10% Mica + 20% SGF	8.2	UR	3.0	133	Unal and Mimaroglu (2012)
15% Mica + 15% SGF	7.4	UR	2.7	115	Unal and Mimaroglu (2012)
20% Mica + 10% SGF	5.8	UR	2.8	93.9	Unal and Mimaroglu (2012)

Continued

Table 12.2 Continued

Polyamide	Modulus (E, GPa)	Yield strength (MPa)	Elongation at break (%)	Tensile strength (MPa)	References
Nylon-6, 6	2.65	80.1	210	82.7–90.3	Majumdar et al. (1994) and Viers (1999)
Nylon-6, 6	1.98	UR	34.5	37.1	Lai and Lin (2006)
15% PVB	2.38	UR	74.7	53.6	Lai and Lin (2006)
15% SEBS	2.45	UR	49.1	43.9	Lai and Lin (2006)
15% SEBS-g-MA	2.16	UR	72.9	43.3	Lai and Lin (2006)
Nylon-6, 12	1.86	52.4	240	60.7	Majumdar et al. (1994) and Peterson and Zhao (1999)
Nylon-11	1.22	41.4	230–250	47	Majumdar et al. (1994) and Hu et al. (2004)
Nylon-12	1.34	40.8	300	50	Majumdar et al. (1994)
Nylon-12, 12	1.58	44.8	400	56	Majumdar et al. (1994)
PMPI	7.8–12.1	UR	20–45	590	García et al. (2010) and Gallini (2002)
PPTA	70–112	UR	2.4–3.6	2900–3000	García et al. (2010) and Gallini (2002a,b)
ODA/PPTA	72	UR	4.6	3400	García et al. (2010) and Gallini (2002a,b)

UR, unreported; SGF, short glass fiber; PVB, polyvinyl butyral; SEBS, styrene-ethylene-butylene-styrene; SEBS-g-MA, maleated styrene-ethylene-butylene-styrene; PMPI, poly(*m*-phenylene isophthalamide); PPTA, poly(*p*-phenylene terephthalamide); ODA/PPTA, 3,4′-diaminodiphenyl ether/poly(*p*-phenylene terephthalamide).

~2.5 to ~20 GPa. Furthermore, the presence of modified RCF improved the nucleation capacity and crystallinity of nylon-6.

On the other hand, aromatic nylons are very stiff materials, their elastic modulus is one or two orders of magnitude bigger than aliphatic polyamides (Table 12.2).

Besides elastic deformation, polymers are deformed plastically. The transition point between plastic and elastic behaviors is known as *yield strength*. This yield can be defined as the stress where polymers begin to deform plastically. As can be seen from Table 12.2, nylon-6 and nylon-6,6 are the aliphatic polyamides with the highest resistance to plastic deformation. In the particular case of nylon-6, the yield strength does not decrease when nylon is blended with silicates; in fact, the yield strength increases slightly.

Elongation at break is a percentage measure of plastic deformation at the moment of rupture. Generally, commercial aliphatic nylons are thermoplastics more ductile than aromatic nylons; however, the latter are more impact resistant (Table 12.2). Nylon-6 mineral composites are less ductile than their counterpart without fillers. Contrary to what was observed when nylon-6,6 was blended with elastomers, in this case the blends were more ductile than its counterpart without elastomers (Table 12.2).

In some cases, nylon polymers need to resist high stresses before breaking. Tensile stress is the mechanical property that describes the maximum stress that it is necessary to apply to break the nylon sample. Nylon-6 and nylon-6,6 are more ductile than other commercial aliphatic nylons, which is reflected in their tensile strength value (Table 12.2). However, inorganic fillers are capable of reinforcing the nylon structure, which leads to an increase in the magnitude of tensile strength. Yoo et al. (2011) increased the tensile strength value of nylon-6 composite (MMT/GF) from ~90 to ~145 MPa, when the proportion of GF was increased from 0 to 30 wt.% and the proportion of MMT was constantly maintained. Feng et al. (2013) improved the tensile strength value of nylon-6/modified RCF composite from ~55 to 170 MPa, increasing the proportion of modified RFC into the blend from 0 to 25 wt.%.

Commercial aromatic nylons are polyamides that need the highest stress to break them; their tensile strength lies between 0.59 and 3.4 GPa (Table 12.2).

12.6 Biodegradation of nylon

Nylon fiber is extensively used due to its high tensile strength. Polyamides (nylon) present high resistance to biodegradation owing mainly to the symmetry of their molecular structures and their strong intermolecular hydrogen bonds, resulting in a crystalline morphology with excellent mechanical and thermal properties (Tokiwa et al., 2009).

During the polymerization reaction to produce nylon-6, some molecules are not polymerized and remain as oligomers (i.e., as nylon-4) and others undergo head-to-tail condensation to form cyclic oligomers. The characteristics of polyamides, such as their strong intermolecular cohesive force by the hydrogen bond makes the rate of abiotic degradation very slow compared to the accumulation rate in the environment.

For this reason, it is a priority to propose degradation processes with faster degradation rates than the accumulation rate (Sudhakar et al., 2007).

The degradation of polyamides using chemical processes has been proposed, for example, degradation by solvent and thermal degradation (Nam et al., 2011; Holland and Hay, 2000). Biodegradation processes involve the presence of live organisms, such as microorganisms or plants. These processes have the advantage that they are, in general, friendlier to the environment than the chemical processes. To our knowledge, there is no information about the use of plants for polyamide biodegradation. Several attempts have been made to use microorganisms, principally bacteria and fungi, to degrade polyamide fibers. The first studies indicated that recalcitrant compounds, such as polyamides, were not biodegradable because of the nature of molecules and the scarce knowledge at that time (Alexander, 1975); fortunately, scientific development has demonstrated that it is possible.

Due to the nature of nylon, its degradation is not easy to determine. The weight loss and the determination of molecular weight with a viscometer are the most used methods. Other methods used to determine nylon degradation require sophisticated equipment, such as scanning electron microscopy (SEM) and epi-fluorescence microscopy, Fourier transformation infrared spectroscopy (FT-IR), and differential scanning calorimetry; these methods are based principally on the observation or determination of the structure of the nylon fiber. Another method used to assess biodegradation is through determination of growth of microorganisms in nylon as the sole carbon source.

Weight loss is an easily used technique that does not require specialized equipment: degradation is determined by the difference in weight between nylon-6 (initial value) and nylon-6 degraded. The relative viscosity of nylon (control and degraded) is determined using a viscometer and the average molecular weight is calculated using the viscosity values. A decrease in the viscosity value and molecular weight is observed in the degraded nylon.

Changes presented in the surface of degraded nylon fiber are observed using electron microscopy or epi-fluorescent microscopy; the images must be analyzed with the aid of the indicated software.

The spectra of nylon and degraded nylon are obtained using a spectrometer FT-IR and are compared to find the functional groups of both fibers. Formation of new groups could occur in the degraded nylon. These new groups may be due to hydrolysis and oxidation processes through chemical or enzymatic reactions.

Thermal transitions of nylon fiber are determined by differential scanning calorimetry, obtaining the melting and glass transition temperatures that indicate the nylon fiber degradation.

Finally, an indirect method to determine nylon fiber degradation is to obtain the growth of the used microorganisms in a culture medium using nylon fiber as the sole carbon source.

Later, we will mention some of the studies on biodegradation of nylon fiber using bacteria and fungi. Several of these works describe the possible reactions carried out during degradation by microorganisms, hydrolysis, and oxidation. Other studies mention also the probable enzymes involved in the degradation processes.

The oligomer 6-aminohexanoate is one of those oligomers produced during polymerization, and it is a wastewater compound from the nylon factories, contributing to increased environmental pollution (Negoro, 2000). *Flavobacterium* sp. K1725 and *Pseudomona aeuruginosa* (NK87) were isolated from soil and wastewater from nylon factories. This last strain grew in a culture medium containing 6-aminohexanoate as the sole carbon and nitrogen source. Furthermore, a *Pseudomonas aeuruginosa* MCMB-407 was isolated from a wastewater treatment plant of a nylon-6 factory by Kulkarni and Kanekar (1998); this strain was able to remove ϵ-caprolactam, a compound used in nylon-6 production, and a reduction in the chemical oxygen demand was also observed. The plasmid-dependence of nylon oligomer-degrading enzymes from *Flavobacterium* sp. K172 has been identified (Negoro et al., 1983; Negoro, 2000). White rot fungi degrade nylon-6,6 by an oxidation process.

Polyamide 4 or nylon-4, a linear polymer of γ-aminobutyric acid, is synthesized from 2-pyrrolidine by ring-opening polymerization. The branched polyamide showed higher tensile strength than the linear polymer; this polyamide could become a biobased polymer with the advantage of being readily biodegradable. *Pseudomonas* sp. strains ND-10 and ND-11 were able to degrade nylon-4 in activated sludge. These bacteria produce an extracellular hydrolytic enzyme that can degrade nylon-4 producing γ-aminobutyric acid (Yamano et al., 2008).

Hashimoto et al. (2002) studied the degradation of nylon-4, nylon-6, and their blend in both composted soil and activated sludge. Both nylons and their blend were buried in the soil composted for more than 10 years. They observed that nylon-6 weight remained almost constant and practically all nylon-4 weight decreased at 6 months. The same experiment, but in an activated sludge, showed similar results, indicating than nylon-4 can be degraded by microorganisms in 28 days but nylon-6 was not modified in the same time. They showed that the pH is also an important factor in nylon-4 degradation. From the activated sludge, they isolated three strains, *Fusarium solani*, *F. oxysporum*, and *Clonostachys rosea*, and suggest that probably *F. solani* is involved in the nylon-4 degradation process. It is known that *F. solani* and *F. oxysporum*, common microorganisms found in soil, are fungi capable of degrading lignin.

Tachibana et al. (2010) isolated a bacterium and a fungus and studied their role in nylon-4 degradation. They identified the bacteria as a *Stenotrophomonas* sp. and the fungus as *Fusarium* sp. Nylon-4, film and power, was decomposed when both strains were grown in a mineral medium for 2 months; probably an exotype extracellular enzyme is involved in nylon degradation.

Temperature could modify the structural characteristics of nylon fiber. Nylon-12 presents glass transition at less than 60 °C, but at higher temperatures the degree of crystallinity increases. Thermal degradation of nylon has been studied and it was found that at 280 °C crystallinity structure was lost in nylon-6,6, but in nylon-6 a small increase in crystallinity was observed (Holland and Hay, 2000; Tomita et al. (2003)). A thermophile bacterium, strain 26, was identified as a closely neighboring species to *Bacillus pallidus*. Strain 26 was capable of degrading nylon-12 as well as nylon-6 at 60 °C, probably by the random cleavage of the amide bonds in the nylon chain (Tomita et al., 2003).

Many studies and efforts have shown that some microorganisms are able to degrade nylon-6. The white rot fungi, *Phanerochaete chrysosporium* and *Trametes versicolor*, degrade nylon-6 and nylon-6,6 membrane.

A white rot fungus, strain IZU-154, degrades the membrane of nylon-6,6 in 20 days, and manganese addition accelerates degradation activity, suggesting that ligninolytic enzymes are involved (Deguchi et al., 1997). Managanese peroxidase, a ligninolytic enzyme produced by the strain IZU-154, was purified and it was shown that this enzyme is involved in the degradation of nylon-6,6 membrane (Deguchi et al., 1998). Later, two ligninolytic white rot fungi were reported (Friedrich et al., 2007) as degrading nylon-6. These fungi were *P. chrysosoporium*, a known ligninolytic fungus, and *Bjerkandera adusta*. The former had been reported before by Klun et al. (2003). They showed that the molecular mass decreased 50% after 3 months of exposure to the fungus. On the other hand, it was shown that *B. adusta* presented a higher ability to degrade nylon-6 than *P. chrysosporium*. Molecular mass of nylon decreased 66% after 2 months, when cultivated with *B. adusta*. It was reported also that manganese peroxidase is the responsible enzyme for nylon-6 degradation by *B. adusta* presenting high nylon degradation ability (Friedrich et al., 2007).

Some fungi were isolated from samples of nylon-6 from a factory producing nylon-6 and none was capable of degrading this fiber. Marine bacteria have been shown to degrade nylon-6 and nylon-6,6. *Bacillus cereus*, *Bacillus sphericus*, *Vibrio furnisii*, and *Brevundimonas versiculars*, all marine bacteria, isolated from the Indian Ocean (Sudhakar et al., 2007), were tested to observe their ability for nylon degradation. Bacteria grew in mineral salt medium indicating that they utilized carbon and nitrogen from nylon-6 and -6,6 as carbon and nitrogen sources. *B. cereus* presented maximum growth at 60 days of culture in both substrates, nylon-6 and -6,6. A 42% decrease in molecular weight of nylon-6,6 and 31% of nylon-6 were observed after 3 months of *B. cereus* cultivation. Oxidation and hydrolysis reactions were found in the nylon-6 and nylon-6,6 degradation process exerted by *B. cereaus* (Sudhakar et al., 2007).

12.7 Immobilization of microorganisms

The attachment of any microorganism on a support or carrier is defined as immobilization. From a biotechnological point of view, the microbial immobilization is a cultivation strategy to decrease some stress factors (mainly expressed in cultures with free cells) that directly impact on the microbial metabolism and limit the production of some metabolites or the metabolic ability for removing some toxics. Carriers can be classified as two kinds, organic and synthetic supports, both support are a source of nutrients or provide only a surface for growth. Among the synthetic supports used to immobilize microorganisms are plastic polymers, such as polyurethanes, polyesters, polyamides, and so on. However, polyamides (nylons) have been historically used less as carriers compared to other plastic polymers. Even so, nylons have been demonstrated to be biodegradable by some microorganisms.

The immobilization of microorganisms on nylon materials began in the 1980s and was mainly focused on the production of enzymes. However, over time, nylon-immobilized microorganisms have been used to produce other kind of metabolites; for instance, organic acids, which have been used as biocatalysts for degrading some hazardous chemicals. Microorganisms immobilized on nylon have been successfully cultivated in liquid, solid, and semisolid media. Some representative examples of this technology are described below.

12.7.1 Liquid cultures

The liquid culture condition has been widely used to develop applications with biomass immobilized on nylon materials. *P. chrysosporium*, a ligninolytic fungus, has been immobilized in nylon fiber to increase peroxidases activities produced by this fungus. *P. chrysosporium* ATCC 24725 was immobilized on nylon-web cubes to produce lignin peroxidase (LiP), in batch and repeated batch shake cultures. The production of LiP in batch with immobilized *P. chrysosporium* ATCC 24725 on nylon-web cubes was greater than the production of LiP with free biomass, even with biomass immobilized on agar and agarose gel beads. Besides, LiP (~72 UL^{-1}) was produced continuously in several repeated batch shake cultures (Zhong et al., 1988). In parallel, Linko (1988) immobilized *P. chrysosporium* ATCC 24725 on nylon fiber to produce LiP, but using a mechanically agitated bioreactor with two types of bioreactor configurations. In the first one (F1), the carrier was a nylon-web sheet wrapped around the temperature control loop of the tank; in the second configuration (F2), carriers were nylon-web cubes placed between a stainless steel cylinder and the outer wall of the bioreactor. The maximum LiP activity attained in F1 and F2 were 375 and 660 UL^{-1} at 95 and 109 h, respectively. Later, Linko and Zhong (1991) optimized the production of LiP in shake cultures by *P. chrysosporium* ATCC 24725 biomass immobilized on nylon-web cubes.

Also, manganese peroxidase (MnP) has been produced using immobilized fungi. *P. chrysosporium* I-1512 was immobilized on a sheet of nylon net; the surface had a hydrophobic roughness (R_a) of 21.05 ± 4.82 μm. SEM showed that the attachment of mycelia was by adhesion to the carrier rather than by entrapment into the network, probably due to the hydrophobic nature of the carrier. The immobilized fungus was cultured in a bubble-column bioreactor. Two phases of growth were observed: phase I and phase II. In phase I (0–40 h), the bioreactor was fed with an air flow, fungal biomass was mainly produced, and high-rate oxygen consumption was observed. At the beginning of phase II (40–120 h), the medium was supplemented with a phospholipid (NAT89) and veratryl alcohol and air was replaced by pure oxygen. The maximum LiP and MnP activities during phase II were 600 and 3600 UL^{-1}, respectively (Laugero et al., 1996).

Domínguez et al. (2001) studied the production of ligninolytic enzymes by *P. chrysosporium* BKM-F-1767 immobilized on nylon sponge cubes inside a rotating drum bioreactor operated under both batch and continuous mode. They assayed two aeration fluxes (0.5 and 1 vvm). When the bioreactor was operated at 0.5 vvm, the maximum LiP activity was reached at 7 days and was of 144 UL^{-1}, whereas the maximum MnP activity was of 570 U L^{-1} at 10 days. When the bioreactor was operated at 1 vvm, the maximum LiP and MnP activities were of 364 and 1350 U L^{-1} at 6 and 11 days, respectively. These enzymatic activities were threefold higher than those attained at 0.5 vvm.

On the other hand, *Aspergillus* has been immobilized also in nylon fiber to improve enzymes and metabolites production. Gawande and Kamat (1998) immobilized two strains of *Aspergillus* (sp. 5 and sp. 44) on 400-mesh nylon-6 bolting cloth (squares) to produce xylanase in shake cultures. The fungal biomasses were successfully attached on mesh nylon squares and very little mycelium remained free at the end of the cultures. The maximum xylanase activity obtained using *Aspergillus* sp. 5 was of 28.2 IU mL^{-1}, which was 1.68-times higher xylanase activity than that produced with free mycelia. The maximum xylanase production by *Aspergillus* sp. 44 was of 11.5 IU mL^{-1} in both immobilized and free mycelium cultures. Finally, the authors indicated that immobilized *Aspergillus* sp. 5 and sp. 44 could be used in three and four successive batch cultivations to produce xylanases, respectively.

A. niger PM1 was immobilized on a circular nylon-6 mesh (pore size of 50 μm) for citric acid production (Papagianni and Mattey, 2004).

Hui et al. (2010) achieved production of cellulose by *A. terreus* immobilized on woven nylon pads using cellulose as the sole carbon source. Results showed that immobilized *A. terreus* biomass yields 4.5-fold more cellulolytic enzymes than free biomass. The immobilized biomass was more active than the free biomass and cellulolytic enzyme was produced for more than 100 days (eight medium changes) versus two cycles of medium change using free biomass.

Moreover, microbial biomass immobilized on nylon fibers has been used for other purposes completely opposite to the production of biotechnological relevant products; for instance, as a biocatalyst to remove and/or degrade toxic compounds. León-Santiesteban et al. (2011) attained immobilization of *Rhizopus oryzae* ENHE (CDBB-H-1877) on nylon-web cubes for the removal of pentachlorophenol (PCP). Authors showed that both the nylon-immobilized biomass and carrier contribute to the removal of PCP. The nylon-web cubes sorbed up to 3 mg PCP g^{-1} nylon, whereas the fungal biomass immobilized on nylon-web cubes impregnated with PCP (3 mg PCP g^{-1} nylon) attained a removal of up to 88.6% and 92% (12.5 and 25 mg PCP L^{-1}) at 48 and 72 h, respectively.

Heitkamp and Stewart (1996) developed a new porous nylon material based on nylon-6,6 (67%) and glass fiber (33%) to improve bacterial immobilizing for bioremediation applications. The prototype named R533 showed very few pores with a size < 1 μm. However, by SEM, it was found that there were some cavities with a size between 800 and 1200 μm in length, and some openings of up to 3.9 mm in length. The authors estimated that the total pore volume and microbe-accessible pore volume were 0.487 and 0.445 mL g^{-1}, respectively, which indicated that 91% of the total pores were accessible to bacteria. *Pseudomona putida* PNP1 was immobilized in R533 and a *p*-nitrophenol (PNP) removal rate of 5.95 kg of PNP m^{-3} day^{-1} was achieved for wastes containing 1200 mg PNP L^{-1}, using a bench-scale packed-bed bioreactor. The immobilized *Pseudomona* strains in nylon mesh technique was used to degrade some volatile organic compounds (VOC) inside a trickle-bed bioreactor (Kozliak et al., 2000).

12.7.2 Solid- and semisolid-state fermentations

Solid-state cultures are defined as the growth of microorganisms on a solid substrate (organic or inorganic) in the absence or near-absence of free water. In this culture, the available water for microorganism is absorbed within the solid matrix (Pandey, 1992). In semisolid-state cultures, the microorganisms also grow within a solid substrate, but in the presence of free water, regardless of the absorbed water by the carrier (Rodriguez et al., 1997).

Nylon as a carrier in solid- and semisolid-state cultures rarely has been taken into account, and very few studies on this topic are available. Cabaleiro et al. (2002) cultivated two ligninolytic fungi, *P. chrysosporium* BKM-F 1767 and *Phlebia radiata* VTT-D-84236, on nylon-web cubes to study the production of ligninolytic enzymes and the effect of proteases on these extracellular enzymes. Results showed that nylon-immobilized *P. chrysosporium* BKM-F 1767 only produced MnP, reaching its maximum activity at day 6 ($510\,U\,L^{-1}$). Whereas, nylon-immobilized *P. radiata* VTT-D-84236 mainly produced LiP and laccase and their maximum activities were of 23 and $13\,U\,L^{-1}$ at day 7 and 5, respectively. Both strains produced proteases, apparently *P. chrysosporium* BKM-F 1767 secreted proteases mainly during primary metabolism, whereas *P. radiata* VTT-D-84236 produced them at the onset of secondary metabolism.

Laccase, MnP, and LiP were produced by *P. chrysosporium* BKM-F 1767 immobilized on fibrous nylon cubes, using semisolid-state cultivation supplemented with solid manganese (IV) oxide, as a promoter and stabilizer of the ligninolytic enzymes (Rodriguez et al., 1997)

12.8 Immobilization of enzymes

Immobilization, from an enzymatic standpoint, is a procedure where an enzyme is chemically linked to a support to improve its reusability, stability, and storage. Nylon materials are a good support for enzymatic immobilization because they are nonporous materials that permit the liquid to flow through them, they have mechanical strength, and are relatively hydrophobic. In most cases, nylons should be chemically modified to improve their interaction with enzymes. When nylon materials are modified to link enzymes on their surfaces or within them, this is typically known as "activation." Many procedures have been described for this purpose, and many of them have followed the same theoretical bases but with some modifications; for instance, the treatment used to increase the reactivity of functional groups of nylons and of the molecules linked to nylon materials to increase the distance between the enzymes and the support.

The state of the art of enzyme immobilization on nylon supports is presented in the following. Immobilization of enzymes has been classified based on the method applied to improve the reactivity of functional groups of nylon materials.

12.8.1 Partial hydrolysis of nylon polymers

As is well known, nylon polymers are linked by means of peptide bonds that provide stability and rigidness. When these covalent bonds are completely hydrolyzed, the rigidness of polymers is compromised and their integrity is lost. However, if the hydrolysis is achieved under controlled conditions, only some peptide bonds will be broken, releasing both carboxyl and amino groups. These functional groups by themselves are not capable of forming covalent linkages with enzymes, but they can react covalently with other molecules that, in turn, can covalently react with enzymes. The procedure to form these chemical bridges between the enzyme and the support is known as activation. Historically, glutaraldehyde (GTA) was the first reagent used to activate nylon because this bifunctional reagent is capable of linking free amino groups on the enzyme to amino groups on the surface of nylon supports by means of covalent bonds (Sundaram and Hornby, 1970). Table 12.3 describes some examples of enzymatic immobilization carried out on nylon supports partially hydrolyzed with HCl and activated with GTA.

Salleh (1982) used polyethylenimine (PEI) as a spacer molecule to increase the distance between the enzymes and the hydrophobic surface of nylon. In this case, PEI formed covalent linkage with free aldehyde groups of activated nylon forming PEI-nylon, which was reactivated with GTA forming activated PEI-nylon (Figure 12.5).

García and Galindo (1990) used activated PEI-nylon net to immobilize glucose oxidase (type V from *A. niger*). On the other hand, Pialis et al. (1996) used 3,3′,5,5′-tetramethylbenzidine as spacer molecules that were activated with GTA to immobilize a tyrosinase (mushroom source) on nylon-6,6 membranes to produce L-DOPA in a 1-L stirred batch reactor. These authors found an L-DOPA production rate of 1.70 mg $L^{-1}h^{-1}$ over a 170-h period. After 14 days of operation, the immobilized tyrosinase lost 20% of its initial activity. Pialis and Saville (1998) enhanced the L-DOPA production with tyrosinase immobilized on nylon-6,6 membranes with 3,3′,5,5′-tetramethylbenzidine as spacers in a semibatch operation mode.

The HCl treatment is not the only alternative to hydrolyze nylon supports. Silva et al. (2007) cleaved the peptide bonds of woven nylon-6,6 with a protease from *Bacillus* sp. The hydrolyzed nylon-6,6 was activated with GTA, it was reacted 1,6 hexadiamine (Hex, spacer molecule), and reactivated with GTA. The activated Hex-nylon was used to immobilize a laccase (from *T. hirsuta*).

12.8.2 Alkylation of nylon polymers

Alkylation is another strategy to break down the peptide bonds in nylon polymers. Sundaram (1979) found that nylon polymers can be activated by means of *ortho*-alkylation reactions with dimethyl sulfate (DMS) yielding a reactive imidate salt of nylon. This imidate salt can quickly react with amino groups on enzymes and form nylon amidine. Both the activated nylon (nylon imidate) and the immobilized enzyme (nylon amidine) are stable. However, in the presence of small nucleophiles or other enzymes, this stability is compromised. The immobilized enzyme undergoes a facile exchange reaction in the presence of nucleophiles. Vaillant et al. (2000) applied *o*-alkylation with DMS for activating nylon to immobilize pectinlyase and endocellulase (from *A. niger*). They observed that the thermal stability of enzymes enhanced after immobilization.

Table 12.3 Enzymatic immobilization carried out on nylon supports partially hydrolyzed with HCl and activated with GTA and their application

Enzyme	Support	Shape	Application	System	References
Urease	Nylon-6	Tubes (inside)	Urea hydrolysis	Continuous mode[a]	Sundaram and Hornby (1970)
Urease Urate oxidase	Nylon-6,6	Tubes (inside) powder	Urea hydrolysis Uric acid oxidation	Continuous mode[a]	Filippusson (1972)
Glucose oxidase Urease	Nylon-6	tubes (inside) powder	Urea hydrolysis Glucose oxidation	Continuous[a]	Inman and Hornby (1972)
Cellulase	Nylon and nylon with glass	blocks	Carboxymethylcellulose and sawdust hydrolysis	Stirred jacketed reactor	Jain and Wilkins (1987)
Lipase	Nylon-6	Pellets	Ester synthesis	Batch and packed-bed reactor (continuous mode)	Carta et al. (1992)
Pectin lyase	Nylon-6	–	Pectin degradation	batch	Alkorta et al. (1996)
Phenol oxidase	–	Membranes	L-DOPA oxidation	Capillary-membrane reactor (semi-continuous mode)	Boshoff et al. (1998)
Alkaline protease	Nylon-6	–	Protein hydrolysis	Batch	Chellapandian and Sastry (1994)

[a] Automated analysis.

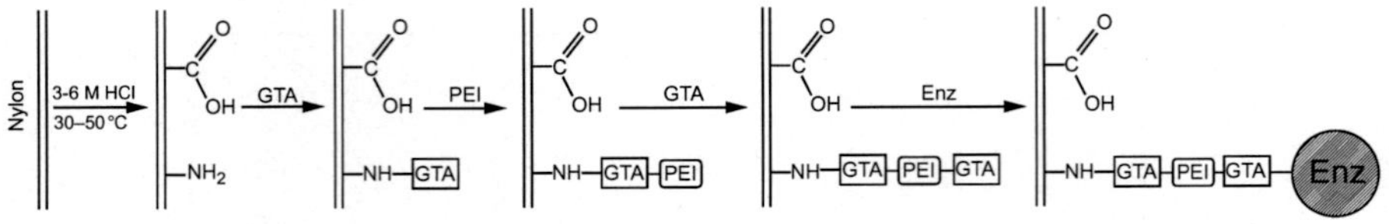

Figure 12.5 Schematic representation of enzyme immobilization on nylon partially hydrolyzed with HCl. Glutaraldehyde (GTA) and polyethylenimine (PEI) were used as activation reagent and spacer molecule, respectively. Enz represents to any enzyme covalently linked to activated nylon.

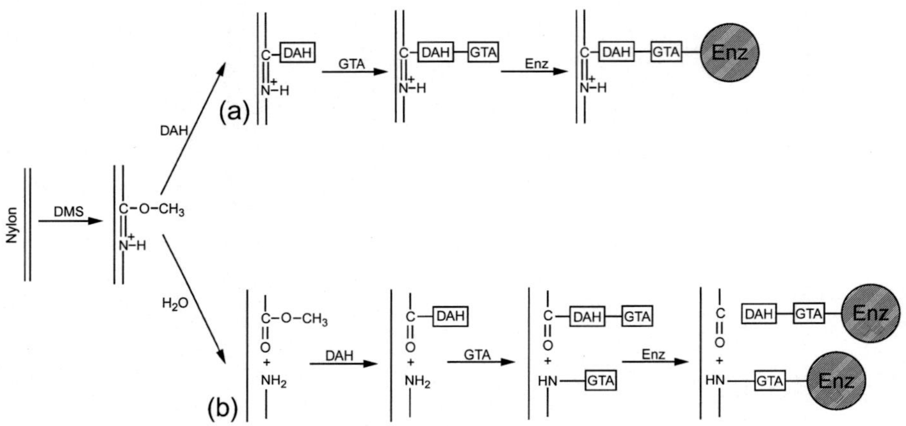

Figure 12.6 Reaction scheme of immobilization of any enzyme (Enz) on nylon alkylated with dimethyl sulfate (DMS). DAH and GTA represent to diaminohexane and glutaraldehyde which were used as spacer molecule and activation reagent, respectively. (a) Mechanism reaction proposed by Onyezili and Onitiri (1981) to describe the immobilization of urease. (b) Mechanism reaction proposed by Onyezili (1987) to describe the immobilization of invertase.

Onyezili and Onitiri (1981) increased the distance between the imidate nylon and the enzyme with diaminohexane (DAH) forming amidates after carrying out *o*-alkylation with DMS. Then, these amidates were activated with GTA through classical Schiff's base and were used to immobilize covalently a urease (from Jack bean) (Figure 12.6a). Later, Onyezili (1987), during the immobilization of invertase (from yeast source) inside nylon-6 tubes, observed that GTA by itself had the capacity to react with the *o*-alkylated nylon before the introduction of the polyamide spacer molecule (Figure 12.6b). Moisture, inevitable during the high-temperature alkylation, caused the hydrolysis of nylon and the resultant amine groups providing binding sites for GTA without prior introduction of polyamine spacer molecules. These additional binding sites enhanced the invertase immobilization.

Onyezili (1988) immobilized urease (from Jack bean) inside *o*-alkylated nylon-6 tubes with DMS but using lysine (Lys) as spacer molecules, which were then reactivated with GTA. A similar activation procedure was applied by Leoni et al. (1998) to covalently immobilize a sulfatase (from *Helix pomatia*) on granular nylon-6,6 for the production of some desulfo-glucosinolates (DS-*epi*-progoitrin, DS-glucotropaeolin, and DS-glucoraphenin) in a column bioreactor.

Obón et al. (1990) immobilized acylase I, an enzyme that catalyzes the hydrolysis of *N*-acetyl-L-amino acids, on *o*-alkylated nylon-6 and -6.6 membranes. These authors used both PEI and diethylaminoethylamine (DEA) as spacer molecules during the immobilization procedure. They first linked the PEI molecules to the imidate salt of nylon and then they reactivated the nylon amidine with GTA. The activated nylon amidine reacted with DEA to link the chemically modified nylon to the enzyme. On the other hand, Manjón et al. (1991) used PEI and β-mercaptoethylamine as spacers for the immobilization of a lipase (from *Mucor miehei*) on *o*-alkylated nylon net to synthesize low molecular esters from acid and alcohols in organic media.

Michalon et al. (1990) immobilized trypsin on GTA-activated nylon-6,6 spheres; however, the spheres were previously *o*-alkylated with H_2SO_4 in methanol. This nylon methylation was also carried out by Appendini and Hotchkiss (1997), but for the immobilization of lysozyme (hen egg white).

12.8.3 Direct activation of nylon with GTA

In these cases, nylon supports were not hydrolyzed or *o*-alkylated prior to the activation. Nylon supports were directly activated with GTA. After that, activated nylon supports were covalently linked to spacer molecules, GTA, and enzyme. Table 12.4 summarizes some examples of this activation procedure.

When the aldehyde groups of GTA react with amino groups of the nylon, a Schiff base is formed (condensation reaction), but this bond is unstable. Braun et al. (1996) resolved that problem by reducing the bond with $NaCNBH_3$ to get a more stable secondary amine without reducing the terminal aldehyde group (Figure 12.7). After reduction of the Schiff base, they covalently linked the terminal aldehyde group of the activated nylon to amino groups of adipic acid dihydrazide (spacer molecule) forming

Table 12.4 Enzymatic immobilization carried out on nylon supports without nylon hydrolysis and alkylation and their application

Support	Activation	Enzyme	Application	System	References
Nylon	GTA-PEI-GTA	Glucose oxidase	Oxidation of glucose	Continuous mode	Salleh (1982)
Nylon-6,6		Lysozyme	Breakage of bacterial cell walls	Batch	Appendini and Hotchkiss (1997)
Nylon-6		Invertase	Hydrolysis of sucrose	Packed-bed reactor	Amaya-Delgado et al. (2006)
Nylon-6		Invertase INVA	Hydrolysis of sucrose	Batch	Calixto-Romo et al. (2008)

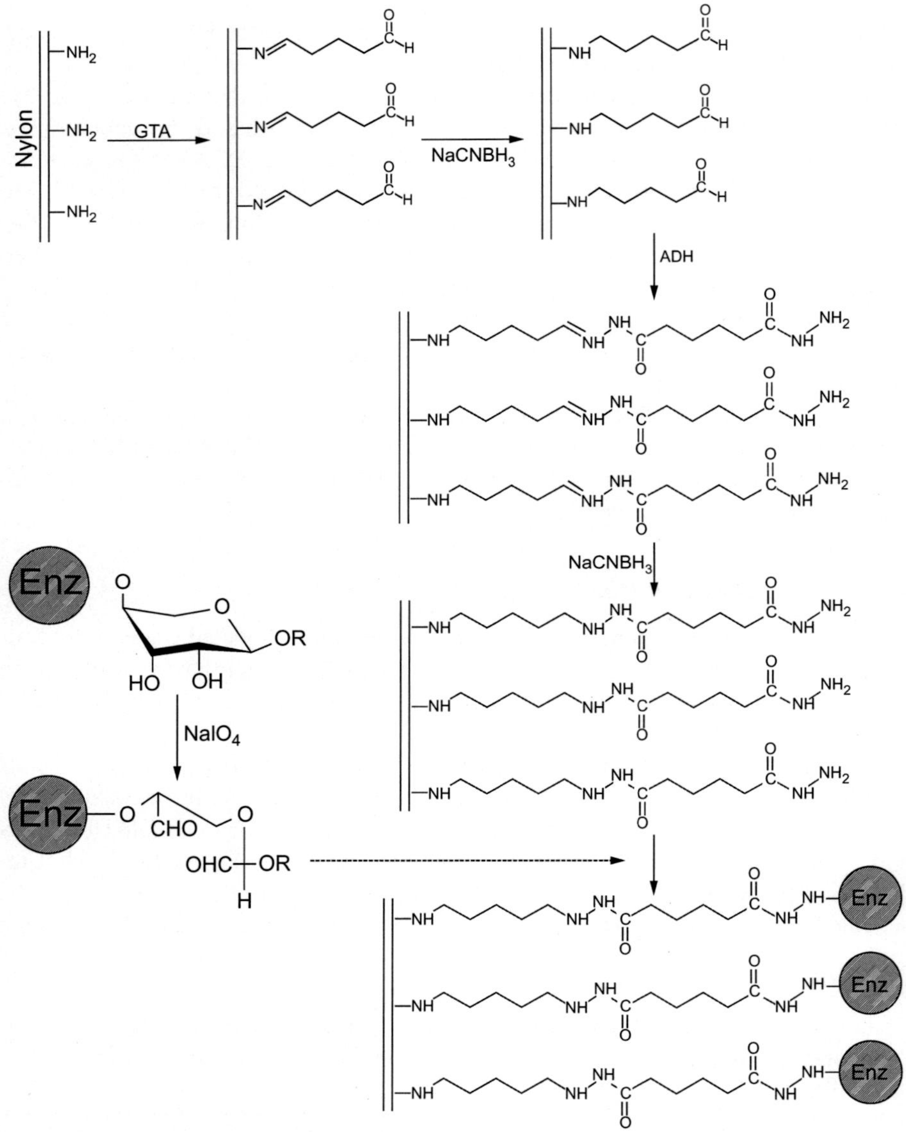

Figure 12.7 Reaction mechanism stated by Braun et al. (1996) for the immobilization of lipase (Enz) on nylon directly activated with glutaraldehyde (GTA).

hydrazones. The terminal amino groups of adipic acid dihydrazide were used as linkage between enzymes and the activated nylon (Figure 12.7).

Braun et al. (1996) immobilized a lipase from the yeast *Candida rugosa*, but they first oxidized the carbohydrate moieties in the enzyme with $NaIO_4$ to form aldehyde groups that facilitated the covalent interaction with the amino groups of the activated nylon (Figure 12.7).

12.8.4 Quaternization of nylon

Miyama et al. (1982) developed a methodology to activate nylon to immobilize urease by ionic bonds through alkylation reactions between the tertiary amines of AQ-nylons (Toray Industries, Inc) and alkyl bromide reagents. These authors observed that when the alkylation reagents were more hydrophilic, the immobilized urease showed more affinity to the substrate (but less than nonimmobilized enzyme) and more water was absorbed onto the support. Similar results were reported for the immobilization of uricase on quaternized AQ-nylon (Miyama et al., 1984).

12.8.5 Grafting copolymerization

El-Masry et al. (2001) immobilized a β-galactosidase (from *A. oryzae*) on grafted nylon Hydrolon membranes (Pall Italia, srl-Milano, Italy) by two methods: diazotization and condensation (Figure 12.8). Glycidyl metacrylate (GMA) was used as a grafting monomer, whereas phenylenediamine (PDA) and GTA were used as spacer molecules. Grafting copolymerization was carried out by using a $K_2S_2O_8/Na_2S_2O_3$/GMA system in the presence of copper acetate, which yielded nylon/polyGMA membranes. These nylon/polyGMA membranes were activated with *p*-phenylenediamine forming aminoaryl derivatives.

For the diazotization method (Figure 12.8a), the terminal amino groups of aminoaryl derivatives were oxidized in the presence of $NaNO_2$/HCl. Then, these groups reacted with aromatic rings of tyrosine residues of the β-galactosidase, thus forming the diazo group between enzymes and grafting nylon membranes.

On the other hand, in the immobilization by condensation (Figure 12.8b), the terminal amino groups of aminoaryl derivatives were reacted with GTA to form

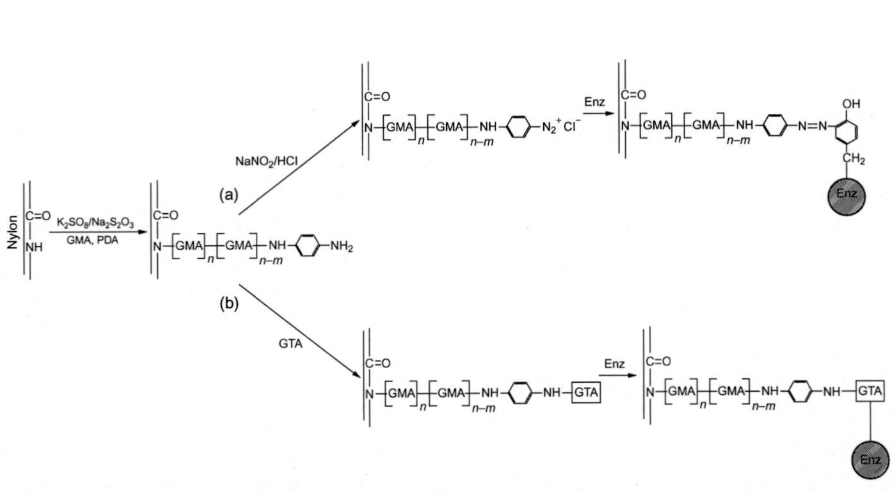

Figure 12.8 Diagram where is shown the grafting copolymerization of nylon membranes using glycidyl metacrylate (GMA) as a grafting monomer. PDA, GTA, and Enz represent to phenylenediamine, glutaraldehyde, and enzyme, respectively. (a) Grafting copolymerization by diazotization method; (b) grafting copolymerization by condensation method.

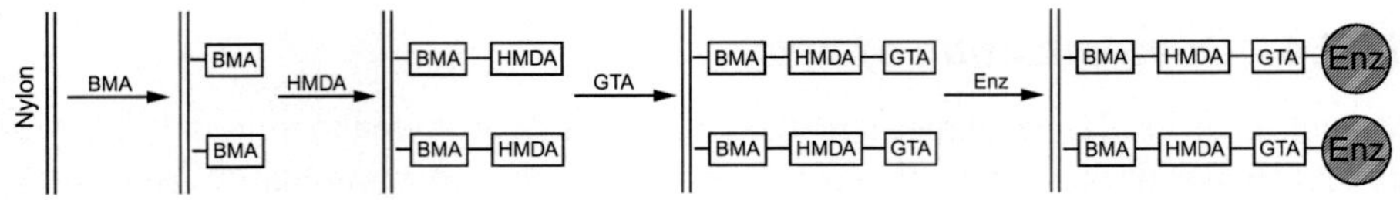

Figure 12.9 Simplified scheme of grafting copolymerization of nylon membranes. Butyl methacrylate (BMA) was used as a grafting monomer. HMDA, GTA, and Enz symbolize hexamethylenediamine, glutaraldehyde, and enzyme, respectively.

imines. The terminal aldehyde groups of these imines were used to immobilize the β-galactosidase, principally on the amino groups of arginine residues.

El-Sherif et al. (2001) also grafted nylon Hydrolon membranes, but using butyl methacrylate (BMA) as hydrophobic monomer (Figure 12.9). They immobilized urease (type III from Jack beans) using hexamethylenediamine (HMDA) and GTA as spacer molecule and coupling agent, respectively. Grafting was achieved inside an ethanolic solution of BMA containing potassium persulfate. Then, the grafted membranes were reacted with HMDA aqueous solution and they were finally reactivated with GTA. To immobilize the urease on activated HMDA-poly(BMA)-nylon membranes, it was necessary to form Schiff bases between the terminal aldehyde groups of activated membranes and the amino groups of arginine and lysine residues of the enzyme (Figure 12.9).

However, because urease is an enzyme with high content of thiol groups (−SH) whose oxidation can change the structure of the enzyme and thus its activity, it was necessary to add β-mercaptoethanol to the reaction medium to prevent the oxidation of these functional groups; without this antioxidant, the activity of the immobilized urease decreased considerably. In contrast, when β-mercaptoethanol was added to the reaction medium, the enzymatic activity was constant during about 1 month.

12.8.6 Cross-linking

The immobilization by cross-linking is when an enzyme is trapped within a matrix. Inman and Hornby (1972) immobilized, by this method, both glucose oxidase and urease (*A. niger*, Jack bean, respectively) into nylon-6 membranes. The enzymes were perfused through the nylon membranes. Although, it was necessary to link the enzymes to GTA instead of activating the membranes, because the membranes did not resist a partial hydrolysis with HCl and they did not have sufficient free amino groups to form a Schiff base with GTA; thus, direct activation was impossible too.

An exceptional case of enzymatic immobilization by cross-linking was described by Burdick and Schaeffer (1987). They did not really immobilize enzymes on nylon material. However, they used a woven nylon fabric as physical support to maintain a gelatin matrix that contained both cross-linker agents and enzymes. In that work, bisvinylsulfone methyl ether, GTA, or oxidized dextran were used as cross-linkers and the immobilized enzymes were both glucose oxidase-catalase and lactate oxidase-catalase.

12.8.7 Adsorption

Adsorption is, perhaps, the least useful procedure to attach an enzyme on nylon support because the interaction between the enzyme and the nylon support is very weak. However, Ruiz et al. (2000) adsorbed a recombinant laccase (from *T. versicolor* produced in *Aspergillus* sp.) on nylon-6,6 membranes by drying an enzyme solution under dry N_2 gas. Laccase adsorbed on nylon-6,6 membranes presented high activity in diethyl ether and ethyl acetate and it did not decrease its activity over a period of 3 days.

This review allows us to conclude that nylon fiber has a great use potential. The advantages of using this bicomposite are that it could be used in several cycles of production or biodegradation processes and that the nylon fiber, after use, could be also biodegraded. Therefore, the use of this biocomposite will not cause pollution problems. Many studies are still needed to improve the use of nylon with enzymes or microorganisms at an industrial level.

References

Aklonis, J.J., 1981. Mechanical properties of polymers. J. Chem. Educ. 58 (11), 892. http://dx.doi.org/10.1021/ed058p892.

Alexander, M., 1975. Environmental and microbiological problems arising from recalcitrant molecules. Microb. Ecol. 2 (1), 17–27. http://dx.doi.org/10.1007/bf02010378.

Alkorta, I., Garbisu, C., María Llama, J., Serra, J.L., 1996. Immobilization of pectin lyase from Penicillium italicum by covalent binding to nylon. Enzyme Microb. Technol. 18 (2), 141–146. http://dx.doi.org/10.1016/0141-0229(95)00094-1.

Amaya-Delgado, L., Hidalgo-Lara, M.E., Montes-Horcasitas, M.C., 2006. Hydrolysis of sucrose by invertase immobilized on nylon-6 microbeads. Food Chem. 99 (2), 299–304. http://dx.doi.org/10.1016/j.foodchem.2005.07.048.

Apgar, G., 1999. Nylon 11. Polymer Data Handbook, second ed. Oxford University Press, Inc, New York.

Appendini, P., Hotchkiss, J.H., 1997. Immobilization of lysozyme on food contact polymers as potential antimicrobial films. Packag. Technol. Sci. 10 (5), 271–279. http://dx.doi.org/10.1002/(sici)1099-1522(199709/10)10:5<271::aid-pts412>3.0.co;2-r.

Boshoff, A., Edwards, W., Leukes, W.D., Rose, P.D., Burton, S.G., 1998. Immobilisation of polyphenol oxidase on nylon and polyethersulphone membranes: Effect on product formation. Desalination 115 (3), 307–312. http://dx.doi.org/10.1016/S0011-9164(98)00050-2.

Braun, B., Klein, E., López, J.L., 1996. Immobilization of *Candida rugosa* lipase to nylon fibers using its carbohydrate groups as the chemical link. Biotechnol. Bioeng. 51 (3), 327–341. http://dx.doi.org/10.1002/(sici)1097-0290(19960805)51:3<327::aid-bit8>3.0.co;2-f.

Burdick, B., Schaeffer, J., 1987. Co-immobilized coupled enzyme systems on nylon mesh capable of gluconic and pyruvic acid production. Biotechnol. Lett. 9 (4), 253–258. http://dx.doi.org/10.1007/bf01027159.

Cabaleiro, D.R., Rodriguez-Couto, S., Sanroman, A., Longo, M.A., 2002. Comparison between the protease production ability of ligninolytic fungi cultivated in solid state media. Process Biochem. 37 (9), 1017–1023. http://dx.doi.org/10.1016/s0032-9592(01)00307-7.

Calixto-Romo, M., Santiago-Hernández, J.A., Vallejo-Becerra, V., Amaya-Delgado, L., Montes-Horcasitas, M., Hidalgo-Lara, M., 2008. Expression, purification and immobilization of the intracellular invertase INVA, from Zymomonas mobilis on crystalline cellulose and Nylon-6. J. Ind. Microbiol. Biotechnol. 35 (11), 1455–1463. http://dx.doi.org/10.1007/s10295-008-0447-1.

Carraher Jr., C.E., 2012. Step-reaction polymerization (polycondensation reactions). In: Introduction to Polymer Chemistry, third ed. CRC Press Taylor & Francis Group, Florida, USA.

Carta, G., Gainer, J.L., Gibson, M.E., 1992. Synthesis of esters using a nylon-immobilized lipase in batch and continuous reactor. Enzyme Microb. Technol. 14 (11), 904–910. http://dx.doi.org/10.1016/0141-0229(92)90054-r.

Charles, A.H., 2002. Thermoplastics. Handbook of Plastics, Elastomers, and Composites, fourth ed. McGraw Hill Professional, Access Engineering, New York, USA.

Chellapandian, M., Sastry, C.A., 1994. Immobilization of alkaline protease on nylon. Bioprocess. Eng. 11 (1), 17–21. http://dx.doi.org/10.1007/bf00369610.

D'Alò, B., Coppola, G., Pallesi, B., 1974. Studies of crystalline forms of nylon-6 by X-ray and i.r. spectrophotometry. Polymer 15 (3), 130–132. http://dx.doi.org/10.1016/0032-3861(74)90189-X.

Dasgupta, S., Hammond, W.B., Goddard, W.A., 1996. Crystal structures and properties of nylon polymers from theory. J. Am. Chem. Soc. 118 (49), 12291–12301. http://dx.doi.org/10.1021/ja944125d.

Deguchi, T., Kakezawa, M., Nishida, T., 1997. Nylon biodegradation by lignin-degrading fungi. Appl. Environ. Microbiol. 63 (1), 329–331.

Deguchi, T., Kitaoka, Y., Kakezawa, M., Nishida, T., 1998. Purification and characterization of a nylon-degrading enzyme. Appl. Environ. Microbiol. 64 (4), 1366–1371.

Domínguez, A., Rivela, I., Couto, S.R., Sanroman, M.A., 2001. Design of a new rotating drum bioreactor for ligninolytic enzyme production by *Phanerochaete chrysosporium* grown on an inert support. Process Biochem. 37 (5), 549–554. http://dx.doi.org/10.1016/s0032-9592(01)00233-3.

Ehrenstein, M., Dellsperger, S., Kocher, C., Stutzmann, N., Weder, C., Smith, P., 2000. New polyamides with long alkane segments: nylon 6.24 and 6.34. Polymer 41 (10), 3531–3539. http://dx.doi.org/10.1016/S0032-3861(99)00590-X.

El-Masry, M.M., De Maio, A., Martelli, P.L., Casadio, R., Moustafa, A.B., Rossi, S., Mita, D.G., 2001. Influence of the immobilization process on the activity of β-galactosidase bound to nylon membranes grafted with glycidyl methacrylate: Part 1. Isothermal behavior. J. Mol. Catal. B Enzym. 16 (3–4), 175–189. http://dx.doi.org/10.1016/S1381-1177(01)00061-3.

El-Sherif, H., Martelli, P.L., Casadio, R., Portaccio, M., Bencivenga, U., Mita, D.G., 2001. Urease immobilisation on chemically grafted nylon membranes: Part 1: isothermal characterisation. J. Mol. Catal. B Enzym. 14 (1–3), 15–29. http://dx.doi.org/10.1016/S1381-1177(00)00237-X.

Feng, N., Wang, X., Wu, D., 2013. Surface modification of recycled carbon fiber and its reinforcement effect on nylon 6 composites: mechanical properties, morphology and crystallization behaviors. Curr. Appl. Phys. 13 (9), 2038–2050. http://dx.doi.org/10.1016/j.cap.2013.09.009.

Filippusson, H., Hornby, W.E., McDonald, A., 1972. The use of immobilised derivatives of urease and urate oxidase in automated analysis. FEBS Lett. 20 (3), 291–293. http://dx.doi.org/10.1016/0014-5793(72)80089-9.

Friedrich, J., Zalar, P., Mohorcic, M., Klun, U., Krzan, A., 2007. Ability of fungi to degrade synthetic polymer nylon-6. Chemosphere 67 (10), 2089–2095. http://dx.doi.org/10.1016/j.chemosphere.2006.09.038.

Galanty, P.G., 1999. Nylon 6. Polymer Data Handbook, second ed. Oxford University Press, Inc, New York.

Gallini, J., 2002. Polyamides, aromatic. Encyclopedia of Polymer Science and Technology. John Wiley & Sons, Inc., New York, USA. http://dx.doi.org/10.1002/0471440264.pst249.

García, J., Galindo, E., 1990. An immobilization technique yielding high enzymatic load on nylon nets. Biotechnol. Tech. 4 (6), 425–428. http://dx.doi.org/10.1007/bf00159390.

García, J.M., García, F.C., Serna, F., de la Peña, J.L., 2010. High-performance aromatic polyamides. Prog. Polym. Sci. 35 (5), 623–686, 10.1016/j.progpolymsci.2009.09.002.

Gawande, P.V., Kamat, M.Y., 1998. Immobilization of *Aspergillus* sp. on nylon bolting cloth for production of xylanase. J. Ferment. Bioeng. 86 (2), 243–246. http://dx.doi.org/10.1016/s0922-338x(98)80122-0.

Gaymans, R.J., Harkema, S., 1977. Melting behavior of aliphatic and aromatic diamides. J. Polym. Sci. Polym. Phys. Ed. 15 (3), 587–590. http://dx.doi.org/10.1002/pol.1977.180150315.

Gaymans, R.J., Van der Ham, A.G.J., 1984. Nylon 4, I: an amorphous polyamide. Polymer 25 (12), 1755–1758, 10.1016/0032-3861(84)90247-7.

Gogolewski, S., Pennings, A.J., 1973. Crystallization of polyamides under elevated pressure: nylon-6 (polycapramide). Polymer 14 (9), 463–464. http://dx.doi.org/10.1016/0032-3861(73)90015-3.

Gogolewski, S., Pennings, A.J., 1985. High-modulus fibres of nylon-6 prepared by a dry-spinning method. Polymer 26 (9), 1394–1400. http://dx.doi.org/10.1016/0032-3861(85)90317-9.

Greco, R., Nicolais, L., 1976. Glass transition temperature in nylons. Polymer 17 (12), 1049–1053. http://dx.doi.org/10.1016/0032-3861(76)90005-7.

Handge, U.A., Hedicke-Höchstötter, K., Altstädt, V., 2010. Composites of polyamide 6 and silicate nanotubes of the mineral halloysite: Influence of molecular weight on thermal, mechanical and rheological properties. Polymer 51 (12), 2690–2699. http://dx.doi.org/10.1016/j.polymer.2010.04.041.

Hashimoto, K., Sudo, M., Ohta, K., Sugimura, T., Yamada, H., Aoki, T., 2002. Biodegradation of nylon4 and its blend with nylon6. J. Appl. Polym. Sci. 86 (9), 2307–2311. http://dx.doi.org/10.1002/app.11235.

Hedicke-Höchstötter, K., Lim, G.T., Altstädt, V., 2009. Novel polyamide nanocomposites based on silicate nanotubes of the mineral halloysite. Compos. Sci. Technol. 69 (3–4), 330–334, 10.1016/j.compscitech.2008.10.011.

Heitkamp, M.A., Stewart, W.P., 1996. A novel porous nylon biocarrier for immobilized bacteria. Appl. Environ. Microbiol. 62 (12), 4659–4662.

Holland, B.J., Hay, J.N., 2000. Thermal degradation of nylon polymers. Polym. Int. 49 (9), 943–948. http://dx.doi.org/10.1002/1097-0126(200009)49:9<943::aid-pi400>3.0.co;2-5.

Hu, G., Wang, B., Zhou, X., 2004. Effect of EPDM-MAH compatibilizer on the mechanical properties and morphology of nylon 11/PE blends. Mater. Lett. 58 (27–28), 3457–3460 10.1016/j.matlet.2004.07.002.

Hui, Y.S., Amirul, A.A., Yahya, A.R.M., Azizan, M.N.M., 2010. Cellulase production by free and immobilized *Aspergillus terreus*. World J. Microbiol. Biotechnol. 26 (1), 79–84. http://dx.doi.org/10.1007/s11274-009-0145-9.

Inman, D.J., Hornby, W.E., 1972. Immobilization of enzymes on nylon structures and their use in automated analysis. Biochem. J. 129 (2), 255.

Jain, P., Wilkins, E.S., 1987. Cellulase immobilized on modified nylon for saccharification of cellulose. Biotech. Bioeng. 30 (9), 1057–1062. http://dx.doi.org/10.1002/bit.260300908.

Klun, U., Friedrich, J., Krzan, A., 2003. Polyamide-6 fibre degradation by a lignolytic fungus. Polym. Degrad. Stab. 79 (1), 99–104. http://dx.doi.org/10.1016/s0141-3910(02)00260-4.

Kohan, M.I., 1999. Nylon 6,10. Polymer Data Handbook, second ed. Oxford University Press, Inc, New York.

Kozliak, E.I., Ostlie-Dunn, T.L., Jacobson, M.L., Mattson, S.R., Domack, R.T., 2000. Efficient steady-state volatile organic compound removal from air by live bacteria immobilized on fiber supports. Biorem. J. 4 (1), 81–96. http://dx.doi.org/10.1080/10588330008951134.

Kulkarni, R.S., Kanekar, P.P., 1998. Bioremediation of ε-caprolactam from nylon-6 waste water by use of *Pseudomonas aeruginosa* MCM B-407. Curr. Microbiol. 37 (3), 191–194. http://dx.doi.org/10.1007/s002849900362.

Lai, S.M., Lin, Y.C., 2006. Mechanical properties of nylon 6,6/polyvinyl butyral blends. Polym.-Plast. Technol. Eng 45 (3), 421–428. http://dx.doi.org/10.1080/03602550600553978.

Laugero, C., Sigoillot, J.C., Moukha, S., Frasse, P., BellonFontaine, M.N., Bonnarme, P., Mougin, C., Asther, M., 1996. Selective hyperproduction of manganese peroxidases by *Phanerochaete chrysosporium* I-1512 immobilized on nylon net in a bubble column reactor. Appl. Microbiol. Biotechnol. 44 (6), 717–723.

Leoni, O., Iori, R., Haddoum, T., Marlier, M., Wathelet, J.P., Rollin, P., Palmieri, S., 1998. Approach to the use of immobilized sulfatase for analytical purposes and for the production of desulfo-glucosinolates. Ind. Crop. Prod. 7 (2–3), 335–343. http://dx.doi.org/10.1016/S0926-6690(97)00066-6.

León-Santiesteban, H., Meraz, M., Wrobel, K., Tomasini, A., 2011. Pentachlorophenol sorption in nylon fiber and removal by immobilized *Rhizopus oryzae* ENHE. J. Hazard. Mater. 190 (1–3), 707–712. http://dx.doi.org/10.1016/j.jhazmat.2011.03.101, S0304-3894(11)00417-1 [pii].

Linko, S., 1988. Production of lignin peroxidase by immobilized *phanerochaete chrysosporium* in an agitated bioreactor. Ann. NY Acad. Sci. 542 (1), 195–203. http://dx.doi.org/10.1111/j.1749-6632.1988.tb25828.x.

Linko, S., Zhong, L.C., 1991. Central composite experimental design in the optimization of lignin peroxidase production in shake cultures by free and immobilized *Phanerochaete chrysosporium*. Bioprocess Eng. 6 (1–2), 43–48.

Long, T.E., Turner, S.R., 2000. Step-growth polymerization in: Styrene Polymers and Copolymers. In: Craver, C.D., Carraher, C.E. (Eds.), Applied Polymer Science: 21st Century. Elsevier Science Ltd, UK, pp. 979–997.

Majumdar, B., Keskkula, H., Paul, D.R., 1994. Mechanical behaviour and morphology of toughened aliphatic polyamides. Polymer 35 (7), 1399–1408, 10.1016/0032-3861(94)90339-5.

Manjón, A., Iborra, J.L., Arocas, A., 1991. Short-chain flavour ester synthesis by immobilized lipase in organic media. Biotechnol. Lett. 13 (5), 339–344. http://dx.doi.org/10.1007/bf01027679.

Michalon, P., Couturier, R., Hacques, M.-F., Favre-Bonvin, G., Ville, A., Marion, C., 1990. Enzyme coupling method on calibrated nylon spheres: application to the selective trypsinization of histones in chromatin. Biochem. Biophys. Res. Commun. 167 (1), 9–15. http://dx.doi.org/10.1016/0006-291X(90)91722-5.

Miyama, H., Kobaya, T., Nosaka, Y., 1982. Immobilization of urease on synthetic polymers. Biotechnol. Bioeng. 24 (12), 2757–2763. http://dx.doi.org/10.1002/bit.260241216.

Miyama, H., Kobayashi, T., Nosaka, Y., 1984. Immobilization of enzyme on nylon containing pendant quaternized amine groups. Biotechnol. Bioeng. 26 (11), 1390–1392. http://dx.doi.org/10.1002/bit.260261117.

Nam, K.T., Pant, H.R., Jeong, J.W., Pant, B., Kim, B.I., Kim, H.Y., 2011. Solvent degradation of nylon-6 and its effect on fiber morphology of electrospun mats. Polym. Degrad. Stab. 96 (11), 1984–1988. http://dx.doi.org/10.1016/j.polymdegradstab.2011.08.012.

Negoro, S., 2000. Biodegradation of nylon oligomers. Appl. Microbiol. Biotechnol. 54 (4), 461–466. http://dx.doi.org/10.1007/s002530000434.

Negoro, S., Taniguchi, T., Kanaoka, M., Kimura, H., Okada, H., 1983. Plasmid-determined enzymatic degradation of nylon oligomers. J. Bacteriol. 155 (1), 22–31.

Obón, J.M., Manjón, A., Cánovas, M., Iborra, J.L., 1990. Anionic exchange nylon laminated membranes for enzyme immobilization. Biotechnol. Tech. 4 (5), 357–362. http://dx.doi.org/10.1007/bf00157436.

Olivares, M., López-Valdivia, H., Vázquez-Polo, G., Carrasco, H., Alvarez-Castillo, A., Oliva, E., Castaño, V.M., 1996. Studies on the effects of γ-radiation on the mechanical properties of Nylon 6–12 fibers. Polym. Bull. 36 (5), 629–636. http://dx.doi.org/10.1007/bf00342456.

Onyezili, F.N., 1987. Glutaraldehyde activation step in enzyme immobilization on nylon. Biotechnol. Bioeng. 29 (3), 399–402. http://dx.doi.org/10.1002/bit.260290318.

Onyezili, F.N., 1988. The enzyme coupling process in urease immobilization on *o*-alkylated nylon tubes. J. Biochem. Biophys. Methods 16 (4), 255–262. http://dx.doi.org/10.1016/0165-022x(88)90059-0.

Onyezili, F.N., Onitiri, A.C., 1981. Polysaccharide derivatives as coats for nylon tube urease. Anal. Biochem. 117 (1), 121–125. http://dx.doi.org/10.1016/0003-2697(81)90701-6.

Pandey, A., 1992. Recent process developments in solid-state fermentation. Process Biochem. 27 (2), 109–117. http://dx.doi.org/10.1016/0032-9592(92)80017-w.

Papagianni, M., Mattey, M., 2004. Physiological aspects of free and immobilized *Aspergillus niger* cultures producing citric acid under various glucose concentrations. Process Biochem. 39 (12), 1963–1970. http://dx.doi.org/10.1016/j.procbio.2003.09.027.

Peters, E.N., Arisman, R.K., 2000. Engineering thermoplastic in styrene polymers and copolymers. In Craver, C.D., Carraher, C.E. (Eds.), Applied Polymer Science: 21st Century. Elsevier Science Ltd, UK, pp. 177–196.

Peterson, G.G., Zhao, W.B., 1999. Nylon 6,12. Polymer Data Handbook, second ed. Oxford University Press, Inc, New York.

Pialis, P., Saville, B.A., 1998. Production of L-DOPA from tyrosinase immobilized on nylon 6,6: enzyme stability and scaleup. Enzym. Microb. Technol. 22 (4), 261–268, 10.1016/S0141-0229(97)00195-6.

Pialis, P., Jimenez Hamann, M.C., Saville, B.A., 1996. L-DOPA production from tyrosinase immobilized on nylon 6,6. Biotechnol. Bioeng. 51 (2), 141–147. http://dx.doi.org/10.1002/(sici)1097-0290(19960720)51:2<141::aid-bit2>3.0.co;2-j.

Rodriguez, C.S., Santoro, R., Cameselle, C., Sanroman, A., 1997. Laccase production in semi-solid cultures of *Phanerochaete chrysosporium*. Biotechnol. Lett. 19 (10), 995–998. http://dx.doi.org/10.1023/a:1018495216946.

Rohde-Liebenau, H.U.W., 1999. Nylon 12. Polymer Data Handbook, second ed. Oxford University Press, Inc, New York.

Rudin, A., 1999. Chapter 4 – Effects of polymer isomerism and conformational changes. In: Rudin, A. (Ed.), Elements of Polymer Science and Engineering, second ed. Academic Press, San Diego, pp. 121–153. http://dx.doi.org/10.1016/B978-012601685-7/50004-5.

Ruiz, A., Malavé, A., Felby, C., Griebenow, K., 2000. Improved activity and stability of an immobilized recombinant laccase in organic solvents. Biotechnol. Lett. 22 (3), 229–233. http://dx.doi.org/10.1023/a:1005698301681.

Salleh, A., 1982. Activation of nylon by alkaline glutaraldehyde solution for enzyme immobilisation. Biotechnol. Lett. 4 (12), 769–774. http://dx.doi.org/10.1007/bf00131150.

Silva, C., Silva, C.J., Zille, A., Guebitz, G.M., Cavaco-Paulo, A., 2007. Laccase immobilization on enzymatically functionalized polyamide 6,6 fibres. Enzym. Microb. Technol. 41 (6–7), 867–875, 10.1016/j.enzmictec.2007.07.010.

Sperling, L.H., Metanomski, W.V., Carraher Jr., C.E., 2000. Polymer nomenclature in: styrene polymers and copolymers. In: Craver, C.D., Carraher, C.E. (Eds.), Applied Polymer Science: 21st Century. Elsevier Science Ltd, UK, pp. 49–54.

Sudhakar, M., Priyadarshini, C., Doble, M., Murthy, P.S., Venkatesan, R., 2007. Marine bacteria mediated degradation of nylon 66 and 6. Int. Biodeterior. Biodegrad. 60 (3), 144–151. http://dx.doi.org/10.1016/j.ibiod.2007.02.002.

Sundaram, P.V., 1979. The reversible immobilization of proteins on nylon activated through the formation of a substituted imidoester, and its unusual properties. Biochem. J. 183 (2), 445–451.

Sundaram, P.V., Hornby, W.E., 1970. Preparation and properties of urease chemically attached to nylon tube. FEBS Lett. 10 (5), 325–327, 10.1016/0014-5793(70)80464-1.

Tachibana, K., Hashimoto, K., Yoshikawa, M., Okawa, H., 2010. Isolation and characterization of microorganisms degrading nylon 4 in the composted soil. Polym. Degrad. Stab. 95 (6), 912–917. http://dx.doi.org/10.1016/j.polymdegradstab.2010.03.031.

Tokiwa, Y., Calabia, B.P., Ugwu, C.U., Aiba, S., 2009. Biodegradability of plastics. Int. J. Mol. Sci. 10 (9), 3722–3742. http://dx.doi.org/10.3390/ijms10093722.

Tomita, K., Hayashi, N., Ikeda, N., Kikuchi, Y., 2003. Isolation of a thermophilic bacterium degrading some nylons. Polym. Degrad. Stab. 81 (3), 511–514. http://dx.doi.org/10.1016/s0141-3910(03)00151-4.

Unal, H., Mimaroglu, A., 2004. Influence of filler addition on the mechanical properties of nylon-6 polymer. J. Reinf. Plast. Compos. 23 (5), 461–469. http://dx.doi.org/10.1177/0731684404031977.

Unal, H., Mimaroglu, A., 2012. Mechanical and morphological properties of mica and short glass fiber reinforced polyamide 6 composites. Int. J. Polym. Mater. Polym. Biomater. 61 (11), 834–846. http://dx.doi.org/10.1080/00914037.2011.610053.

Vaillant, F., Millan, A., Millan, P., Dornier, M., Decloux, M., Reynes, M., 2000. Co-immobilized pectinlyase and endocellulase on chitin and nylon supports. Process Biochem. 35 (9), 989–996, 10.1016/S0032-9592(00)00131-X.

Viers, B.D., 1999. Nylon 6,6. Polymer Data Handbook, second ed. Oxford University Press, Inc, New York.

Wen, J., 2007a. Heat capacities of polymers. In: Mark, J. (Ed.), Physical Properties of Polymers Handbook. Springer, New York, pp. 145–154. http://dx.doi.org/10.1007/978-0-387-69002-5_9.

Wen, J., 2007b. Some mechanical properties of typical polymer-based composites. In: Mark, J. (Ed.), Physical Properties of Polymers Handbook. Springer, New York, pp. 487–495. http://dx.doi.org/10.1007/978-0-387-69002-5_28.

Yamano, N., Nakayama, A., Kawasaki, N., Yamamoto, N., Aiba, S., 2008. Mechanism and characterization of polyamide 4 degradation by *pseudomonas* sp. J. Polym. Environ. 16 (2), 141–146. http://dx.doi.org/10.1007/s10924-008-0090-y.

Yoo, Y., Spencer, M.W., Paul, D.R., 2011. Morphology and mechanical properties of glass fiber reinforced Nylon 6 nanocomposites. Polymer 52 (1), 180–190. http://dx.doi.org/10.1016/j.polymer.2010.10.059.

Zhong, L.-C., Linko, S., Lindholm, N., Linko, Y.-Y., 1988. Lignin peroxidase production by immobilized *phanerochaete chrysosporium*. Ann. NY Acad. Sci. 542 (1), 153–158. http://dx.doi.org/10.1111/j.1749-6632.1988.tb25820.x.

Zussman, E., Burman, M., Yarin, A.L., Khalfin, R., Cohen, Y., 2006. Tensile deformation of electrospun nylon-6,6 nanofibers. J. Polym. Sci. B Polym. Phys. 44 (10), 1482–1489. http://dx.doi.org/10.1002/polb.20803.

Mechanical performance of polyvinyl acetate (PVA)-based biocomposites

13

A. Kaboorani, B. Riedl,
Université Laval, Québec, QC, Canada

13.1 Introduction

13.1.1 Green chemistry

Green chemistry, also known as sustainable chemistry is, among other things, the design of chemical products and processes to prevent potential pollution sources and use renewable materials. This also includes avoiding potential environmental contamination via spills, leaks, and other types of accidents. By using environmentally friendly materials, the risk for contaminating the environment is substantially reduced if an accident should occur. It is for this reason that it is important to develop a green ideology and methodology when pursuing the synthesis of compounds (Lech, 2008).

Wood transformation is inherently applying several principles of green chemistry and green engineering, one of which is the use of renewable rather than depleting processes. Material and energy inputs should be renewable rather than depleting. Dealing with a renewable material such as wood, industry is applying this principle to a large extent and it should secure this position in the future.

Although wood is known as a sustainable material, adhesives and coatings that are used with wood are not sustainable and are a source of concern. In this chapter, principles of green chemistry and green engineering are considered in developing polyvinyl acetate (PVA) nanocomposites with an application in wood adhesives.

13.1.2 A brief history of PVA

PVA is a colorless, usually nontoxic thermoplastic adhesive prepared by the polymerization of vinyl acetate. PVA was discovered in 1912 by Dr. Fritz Klatte in Germany. It is one of the most widely used water-dispersed adhesives. PVA is made up of a water-based emulsion of a widely used type of glue, referred to variously as wood glue, white glue, carpenter's glue, school glue, or PVA glue. PVA is largely used in glass fiber-reinforced plastics to improve the stress and antishrink properties. It is also used in automobile headlights to promote their gloss performance. In addition, PVA may be added to cement/concrete where it can improve the water-resistance properties. Because PVA is a emulsion, not a true water solution, once the film is dry, it is rather hydrophobic. PVA setting and curing is accomplished by the removal of water due to evaporation or absorption into a substrate. PVA adhesives produce clear, hard films

Biocomposites: Design and Mechanical Performance. http://dx.doi.org/10.1016/B978-1-78242-373-7.00009-3

that have good weather resistance and withstand water, grease, oil, and petroleum fuels. Additional properties are high initial tack, almost invisible bond line, softening at 30–45 °C, good biodegradation resistance, poor resistance to creep under load, and low cost. PVA adhesives and copolymers are also used as hot-melt adhesives, sealants, in fabric finishing, plastic wood, and inks.

13.1.3 Case study: PVA as a wood adhesive

Recently there has been a strong tendency to use so-called green materials for industry. The tendency has forced industry to look for renewable materials with the least impact on the environment. One of the industries that can benefit from this trend and subsequently experience significant and constant growth is the wood industry. Wood can provide the market with the properties that cannot be offered by other materials. Wood has a great potential to become the dominant material for the construction industry. As a material for construction, wood has unique properties, such as high strength, flexibility, fire resistance, durability, insulation, and a low carbon footprint. Introduction and development of engineered wood products (EWPs), which can be produced in different shapes and sizes, has opened new markets for the wood industry and has expanded current markets.

Adhesives play an important role in the performance of EWPs. Production of EWPs with high performance needs durable adhesives. Commercial durable wood adhesives contain formaldehyde, causing concerns about health during the production and service of EWPs. The wood industry is under increasing pressure to eliminate formaldehyde from its products; hence, the wood industry is looking for alternatives.

PVA is a good alternative to replace some wood adhesives containing formaldehyde. PVA is a linear and thermoplastic polymer. It is water-soluble, biodegradable with excellent chemical resistance, and has no toxic effect on humans. As a wood adhesive, utilization of PVA is very simple and its curing does not need high temperatures.

PVA has three main drawbacks: low resistance to water and humidity, poor performance at elevated temperatures, and high susceptibility to creep. The drawbacks stem from the structure of PVA adhesive. PVA is a linear amorphous polymer with a weak polar interaction among the molecular chains, which results in a relatively low glass transition temperature (T_g). Some research projects have been targeted at modifications of PVA in order to improve its performance. The modifications can be divided in two main groups: (1) Copolymerizing vinyl acetate with more hydrophobic monomers or functional monomers (Zhou, 1991; Cai, 1997; Chen, 1996) and (2) blending PVA with other adhesives or hardeners (Lu, 1996; Comyn, 1997; Wang, 1999; Qiao et al., 2000; Huang et al., 2002; López-Suevos and Frazier, 2006; Kim and Kim, 2005a,b).

Using the abovementioned strategies could increase some properties of PVA at the expense of reducing some other properties. Moreover, some additives are so acidic that they can damage wood subtracts, finally affecting the overall performance of wood joints. For instance, copolymers of vinyl acetate and butyl acrylate or ethylene can increase water resistance and toughness of the adhesive, but will reduce its tensile modulus and stiffness, especially at elevated temperatures (Qiao and Easteal, 2001).

As the cost of adhesives covers a substantial part of the total production cost of plants producing finger-jointed lumber, any attempt to boost performance of relatively cheap PVA adhesive at high temperatures could enhance the use of wood components in buildings.

The properties of adhesives can be modified or adjusted by adding several compounds. The inclusion of tackifiers, adhesion promoters, and fillers to adhesives is commonly done to alter the performance of the adhesives with respect to specific applications (FerrandizGomez et al., 1996; Pastor-Sempere et al., 1995; Poh and Firdaus, 2010; Macia-Agullo et al., 1992). Conventional compounds sometimes modify or adjust some properties of the adhesives at the expense of the other properties. Meanwhile the extent of improvements achieved by conventional compounds is limited in some cases.

The introduction of nanotechnology has opened an opportunity for the adhesives industry to develop a new generation of adhesives. The incorporation of nanoparticles into a polymer matrix can lead to a simultaneous improvement of different material properties (Wetzel et al., 2003; Cho et al., 2006). The development of nanoparticle-reinforced adhesive materials is presently one of the most explored areas in materials science and engineering. The exceptional properties of nanoparticles have led to widespread research in this area. Nanofillers provide many advantages over classical microreinforcements for adhesive materials; for example, they allow thin bond lines and consequently lower the risk of embrittlement within the bulk adhesive material, resulting in improved adhesive tensile strength (FerrandizGomez et al., 1996).

Nanocomposites, with dispersed nanoparticles, have been studied extensively due to their capability to improve mechanical, physical, thermal, and barrier properties with very low nanoparticles loading of 1–5 wt.% (Ji et al., 2003; Zhang and Archer, 2004; Ash et al., 2004). For such composites, montmorillonite (MMT) and other clay nanoparticles have been used by many researchers (Pegoretti et al., 2004; Lee and Han, 2003; Chang et al., 2002; Ma et al., 2004; Fornes and Paul, 2004).With a structure of stacked platelets and one dimension of the platelet in the nanometer scale, MMT has a high aspect ratio and specific surface when exfoliated. If the platelets are dispersed properly, its nanosize can provide a significant amount of interface between the clay and the matrix resin with only a small weight percentage of MMT, thus contributing to the excellent mechanical and physical properties of the nanocomposites.

A new generation of polymeric-based materials have been introduced by using nanotechnology. During the infancy period of nanotechnology, inorganic nanomaterials like nanoclay and metal oxide nanoparticles (ZnO, ZrO_2, TiO_2, SiO_2, and Al_2O_3) were used. As the renewability and, more importantly, the health risk of these materials were questioned, a great deal of research was directed toward exploring the use of alternate organic nanoparticles. The result of the research was the introduction of nanocelluloses. Cellulose has a natural nanostructure. Different methods are used to explore the nanostructure of cellulose, leading to the introduction of various nanocelluloses; namely, nanocrystalline cellulose (NCC) and microfibrillated cellulose (MFC). NCCs, also known as whiskers, consist of rod-like cellulose crystals with widths of 5–70 nm and lengths between 100 nm and several micrometers. NCC is obtained by the acid hydrolysis of cellulose under conditions where the amorphous regions are selectively hydrolyzed. NCC is abundant, renewable, and nontoxic. Compared to mineral fillers, NCC has a lower density, a high form factor of about 70 and a high specific area of 150 m^2/g. Among all advantages, surface reactivity of NCC provides a chance to graft a wide variety of chemical functions to the surface of NCC, resulting in NCC with different properties.

In this chapter, the feasibility of improvement in bonding strength of PVA in dry and wet states, and also at elevated temperatures through using different nanomaterials are discussed. Nanomaterials include nanoclay and NCC. Moreover, the effects of dispersion methods on mechanical properties and the morphology of PVA nanocomposites are studied.

13.2 Experimental analysis of PVA based bio-composites

13.2.1 Materials

A commercial PVA was received in liquid form. Two types of hydrophilic nanoclay, namely Nanofil® 116 and Lit.G-105 (polymer grade (PG) MMT) were supplied by Southern Clay Products, Inc., Gonzales, TX, and Nanocor, Inc., Arlington Heights, IL, respectively. Nanoaluminum oxide particles (Sasol Dispal 25 F4) were provided by Sasol Germany GmbH. NCC, provided kindly by Forest Products Laboratory in Madison, WI, was used as reinforcing material (in 5.5% suspension). Two wood species were used to make adhesive joints—sugar maple (*Acer saccharum*) and black spruce (*Picea mariana*)—which were obtained from trees grown in Quebec, Canada.

13.2.2 Methods

Various blends of PVA with nanoparticles were prepared. Two methods were used to mix PVA with nanoparticle: high-speed mixer and ultrasonication. A high-speed mixer, Gast MFG Corp. 4AMFRV13C, with the addition of small glass beads was used to mix PVA with nanoclay. The formulations having different amounts of nanoclay were mixed at 1500 rpm for 15 min. For the ultrasonication method, a high-intensity ultrasonic probe (60 kHz, maximum amplitude 100 nm from tip to tip, Branson PG) was used to disperse the nanoparticles. The amounts of nanoparticles were 1%, 2%, and 4% wt/wt.% (solid content). The nanoparticles were added to PVA on a base of solid content of PVA in the emulsion of PVA adhesive. As ultrasonication generates a lot of energy in the form of heat and shear, the temperature of the solution increases rapidly, causing some interference with the results. In order to keep the temperature of solution under control ($<50\,^\circ C$), the vessel containing the solution was cooled by means of a recirculating ethylene glycol bath. Sonication experiments were carried out with 50% amplitude and a volume of 20 ml of the nanoparticle–water mixture. During ultrasonication, the sonication power was gradually raised while maintaining the temperature of the mixture below 50 °C. The ultrasonicated solution was added to PVA. New formulated blends of PVA and nanoparticles were mixed for 30 min. For NCC, a simple mixer was used to blend PVA with NCC, as NCC had been already dispersed in water by the supplier.

For transmission electron microscopy (TEM) and atomic force microscopy (AFM), samples of nanocomposites were prepared by casting the NCC–PVA composites on Teflon sheets. Prior to further analyses, the sheets of nanocomposites were allowed to dry at room temperature for at least 2 weeks and conditioned to about 60% RH.

13.2.2.1 Fabrications and tests of wood joints

Mixed solutions of PVA and different nanomaterials were used to bond wood samples. Prior to gluing, the moisture content of wood was fixed at 12% by conditioning wood to 20 °C and 60% relative humidity for two months. After applying glue on the surface of the wood, the samples were pressed in an MTS hydraulic test machine with 50 kN capacity at 2.46 kg/cm^2 pressure for 2 h. Before testing, glued samples were conditioned to 20 °C and 60% relative humidity for 2 weeks. Twenty samples were tested for each set of formulation.

To evaluate the impact of nanoparticles on the performance of wood joints, the shear strength of wood joints was measured in dry and wet states, and at an elevated temperature. An MTS hydraulic test machine with 50 kN capacity was used for load application, and the data was acquired by a computer. Wood failure and maximum load were recorded for each test. The block shear tests were carried out according to ASTM D905-98. The sizes of samples for "wet state" tests were the same as those for dry state tests. For "wet state" tests, the samples were taken directly out of the water after being immersed in water for 24 h. Before the tests, excess water was wiped off the samples. During the water immersion period, temperature of water was maintained at 23 ± 1 °C.

Block shear tests at the elevated temperature were carried out according to ASTM 7247-07. Samples made of black spruce were heated in an oven, having a temperature controller with an integral and derivative (PID) control algorithm, until the temperature of the middle of the samples reached 100 °C. On average, it took 30 min to reach 100 °C in the middle of the samples. After reaching 100 °C in the middle of the samples, the samples were kept at 100 °C for 15 min more, followed by immediate block shear tests. The shear strength of the sample was measured by an MTS hydraulic test machine.

13.2.2.2 Transmission electron microscopy

TEM allows a qualitative understanding of the internal structure, spatial distribution, and dispersion of the nanoparticles within the polymer matrix, and views of the defect structure through direct visualization. Analyses were performed on a JEOL JEM-1230, transmission electron microscope at 80 kV. TEM specimens, having 50–70 nm thickness, were prepared by ultramicrotoming the nanocomposite samples encapsulated in an epoxy matrix.

13.2.2.3 Atomic force microscopy

AFM observations were carried out using a NanoScope IIIa, an atomic force microscope (Veeco Instruments Inc.). AFM measurements were done under ambient air conditions in tapping mode. The sensitivity of the tip deviation and the scanner resolution was 0.1 nm. The resolution was set to 512 lines by 512 pixels for all observations. Two topographic and phase images were obtained within each sample for scan areas of 10 × 10 and 50 × 50 μm. Surface roughness was calculated in 50 ×50 μm scan areas, using the classical mean surface roughness parameters Ra and Rq (RMS). The parameters were calculated by the Nanoscope 5.30r3sr3 software.

13.3 Results of adding nanoclay and NCC to PVA based bio-composites

The results of adding nanoclay and NCC to the PVA to bonding strength of PVA are presented in two separate sections.

13.3.1 Nanoclay

Two types of nanoclay (at different loading levels of 1%, 2%, and 4%), coming from different companies (Southern Clay Products, Inc., Gonzales, TX, and Nanocor, Inc., Arlington Heights, IL), were first mixed with PVA by using a high-speed mixer. In the following, the effects of adding nanoclay on the bonding strength measured on sugar maple are discussed.

The values of bonding strength for PVA and its nanocomposites prepared by a high-speed mixer are given in Figure 13.1. Adding nanoclay to PVA clearly did not increase significantly shear strength of joints bonded by PVA, although all values of strength of formulations with added nanoparticles were slightly above those of pure PVA. Optimal loading level of nanoclay depended on the type of nanoclay. Adding only

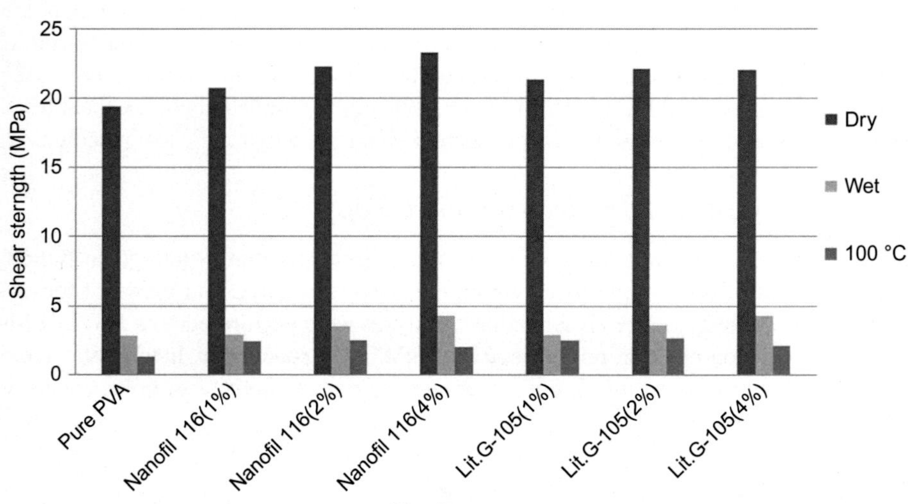

Figure 13.1 Values of bonding strength for PVA/nanoclay nanocomposites (mixed by a high-speed mixer) in dry and wet states, and at the elevated temperature.

1% of Lit.G-105 enhanced shear strength by 10% and further increase did not change the strength remarkably. In contrast, a significant improvement on shear strength was achieved for Nanofil 116 as nanoclay loading increased. The highest gain in strength was obtained at 4% loading of Nanofil 116. The constant increase in the strength in the case of Nanofil 116 could be related to its compatibility with the matrix. Such constant increase was not seen in Lit.G-105, possibly because Lit.G-105 cannot be mixed well with PVA at higher loadings due to its incompatibility with the matrix. Further morphological study is required to draw any clear conclusion regarding this phenomenon.

Results of shear strength measurements performed on samples after spending 24 h in water are presented in Figure 13.1. Effects of nanoclay on the strength were significant. Nanoclay enhanced the resistance of wood joints toward water. At 1% loading, nanoclay did not change the strength considerably. But as loading nanoclay was increased, marked improvements in shear strength were achieved; as high as 53%. Increase in the barrier properties of glue line against water could explain the increase. Lit.G-105 and Nanofil 116 improved the water resistance of PVA almost in the same degree and no significant difference was detected.

Shear strength of samples after being exposed to elevated temperature is given in Figure 13.1. Joints bonded with pure PVA lost their integrity at a low load. Effects of adding nanoclay on shear strength was significant. One percent nanoclay improved shear strength of PVA by 83%. Such important improvement stems from the fact that nanoclay changed the response of the adhesives toward elevated temperatures. In terms of the model for the origin of properties, 2% exfoliated clay theoretically has enough surface area to interact with about 60% of the polymer chains, confining them to a lower free volume. This may account for the increase in cohesive strength of the adhesive. The other 40% of the polymer chains reside in a higher free volume state than do the bulk, unmodified polymer chains. These chains have less constricted motion and exhibit more rubbery behavior than in the unmodified polymer (Krywko et al., 2002; Giannelis et al., 1999).

Effects of nanoclay loading on shear strength at the elevated temperature were significant. At 1% and 2%, the strength was improved as nanoclay loading increased in the formulations. But at 4% loading, a reduction in strength was detected in comparison with the joints bonded with 2% nanoclay. Difficulty involved in optimizing dispersion of nanoclay in the matrix at high loading could be a reason for the reduction. Lit.G-105 increased the shear strength of wood joints a little bit more than Nanofil 116. No wood failure occurred in the tests and all failure occurred in the glue line.

As the high-speed mixer could not disperse well one type of nanoclay (Lit.G-105) in PVA, we decided to use ultrasonication to disperse Lit.G-105 in PVA. Such an application allowed us to measure the efficiency of ultrasonication in dispersing nanoclay in PVA. The results are presented in the following discussion.

Results of measuring shear strength of wood joints in dry tests are shown in Figure 13.2. Although fluctuations caused by varying loading of nanoclay on bonding strength were not found significant, all joints with nanoclay in their formulations had improved shear strength. The increase in wood joints strength was between 2% and 7%. The strength of joints showed improvements as nanoclay content increased in the matrix. An increase in bonding strength of wood joints not only could be measured in

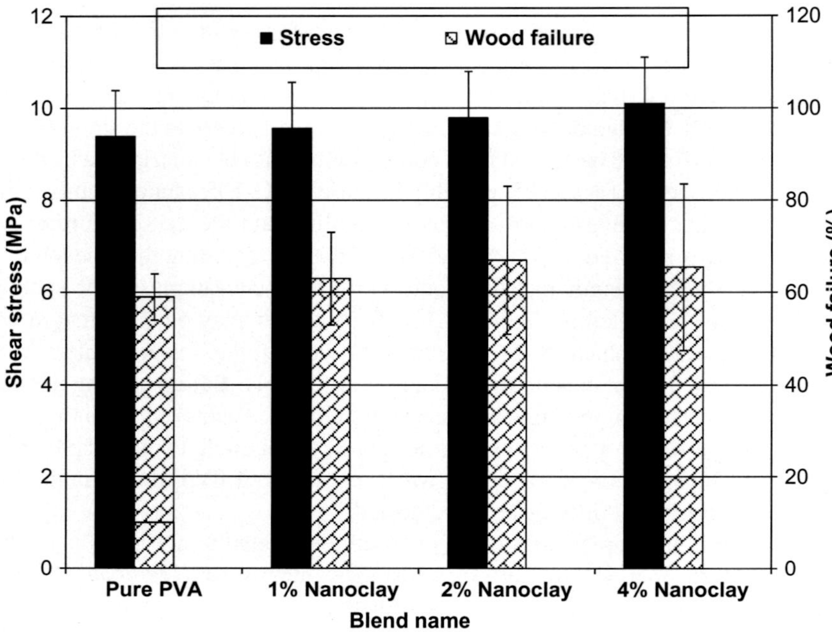

Figure 13.2 Values of bonding strength for PVA/nanoclay (Lit.G-105) nanocomposites (mixed by ultrasonication technique) in dry state at room temperature (Kaboorani and Riedl, 2013).

terms of shear strength, but also in terms of wood failure under shear load. Inclusion of nanoclay to PVA increased wood failure of wood joints under shear load (Figure 13.2). Higher wood failure in joints having nanoclay in their adhesives means that nanoclay increased the strength of the glue line to a level that the strength of the glue line surpassed the strength of wood.

Figure 13.3 presents the values of shear strength after 24-h exposure of wood joints to water. Water exposure decreased the shear strength of wood joints. The shear strength of joints made of pure PVA decreased drastically, dropping down to one-sixth of their values in a dry state. Adding nanoclay to PVA improved the resistance of the glue line to water. The extent of improvement was between 25% and 64%. Any increase in nanoclay loading in the matrix gave a boost to water resistance of wood joints (at all levels of loading). Positive effects of nanoclay on water resistance can be attributed to better barrier properties of the glue line.

Past research shows that nanomaterials can improve the barrier properties of polymers (Nielsen, 1967; Bharadwaj, 2001; Falla et al., 1996; Plate and Yampol'skii, 1994; Thran et al., 1999; Pethrick, 1997; Wang et al., 2004; Becker et al., 2003; Winberg et al., 2005). The better barrier properties are associated with the fact that permeate molecules are forced to follow tortuous pathways, reducing the diffusion coefficients (Fischer, 2003). In addition, the inclusion of nanoclay with layered structures and their adhesion to the polymer generate additional free volume, more likely affecting the polymeric chains located near interfacial regions (Plate and Yampol'skii, 1994;

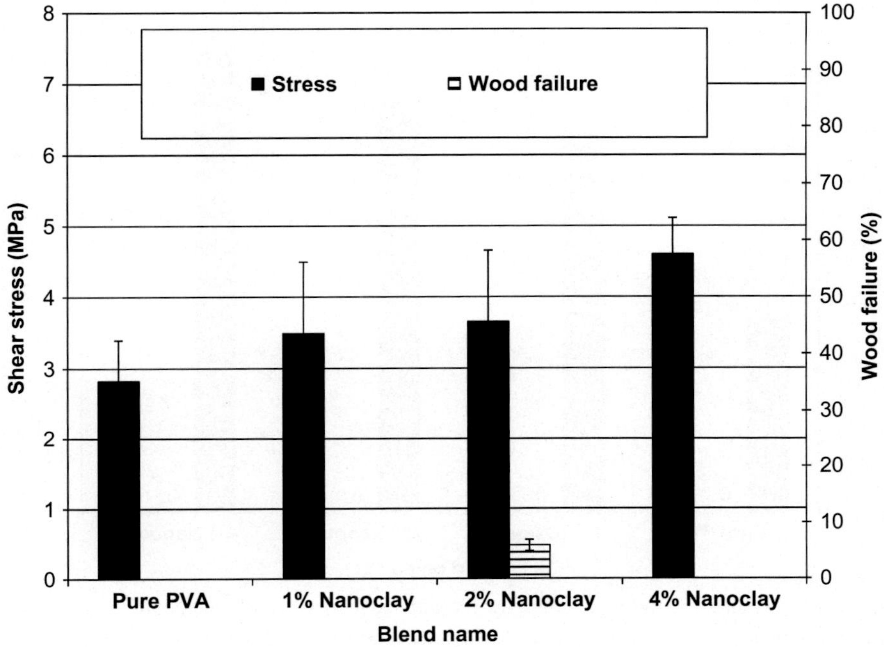

Figure 13.3 Values of bonding strength for PVA/nanoclay (Lit.G-105) nanocomposites (mixed by ultrasonication technique) after 24 h of water exposure (Kaboorani and Riedl, 2013).

Thran et al., 1999; Pethrick, 1997). Wood failure of joints was affected by adding nanoclay as well. Wood joints bonded with PVA and 2% nanoclay showed a small percentage of wood failure under the shear load. The wood failure could be related to good dispersion of nanoclay in PVA matrix.

Values of shear strength of wood joints at 100 °C are given in Figure 13.4. Nanoclay had a positive effect on heat resistance of wood joints. The effects were more pronounced at 1% loading (resulting in a 96% increase in shear strength at 100 °C). As nanoclay content in the PVA matrix increased, shear strength of wood joints improved at the elevated temperature. The extent of improvement was very significant at 4% nanoclay loading. The PVA with 4% nanoclay had 1.40 times higher shear strength. In our past experience (Kaboorani and Riedl, 2011) with a high-shear mixer, a decrease in shear strength of wood occurred at 4% nanoclay loading because of poor dispersion of nanoclay in the PVA matrix. Such a reduction did not occur in this study where an ultrasonic technique was used to disperse nanoclay particles in PVA. Thus, the results showed that using ultrasonic technique to disperse nanoclay particles in the matrix can enhance the effectiveness of nanoclay particles in the matrix. No wood failure was observed in the joints at the elevated temperature.

13.3.1.1 Transmission electron microscopy

Figure 13.5 shows the TEM images for nanocomposites containing nanoclay. Adding nanoclay at 1% and 2% loadings led to good dispersion of nanoclay in

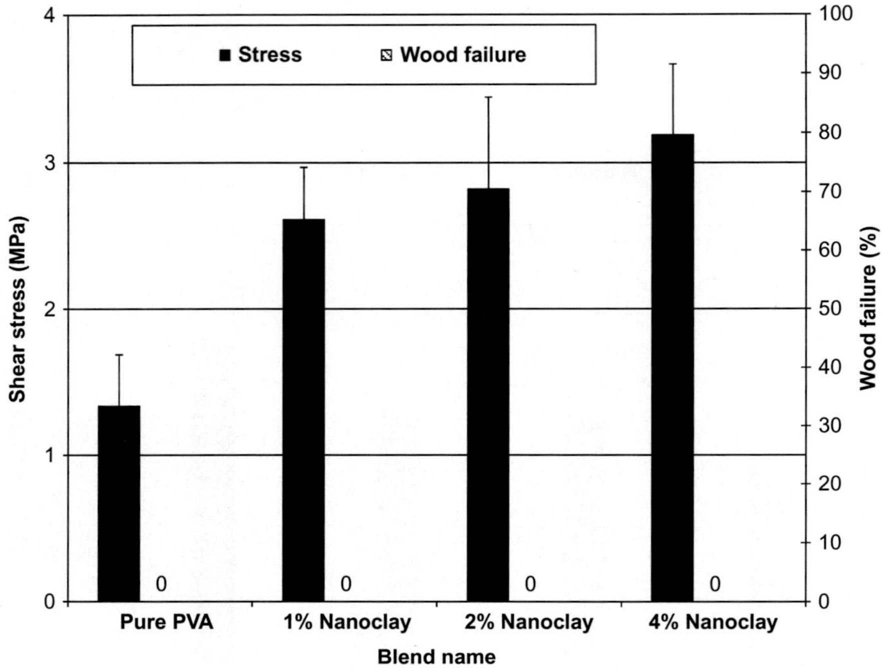

Figure 13.4 Values of bonding strength for PVA/nanoclay (Lit.G-105) nanocomposites (mixed by ultrasonication technique) at elevated temperature (100 °C) (Kaboorani and Riedl, 2013).

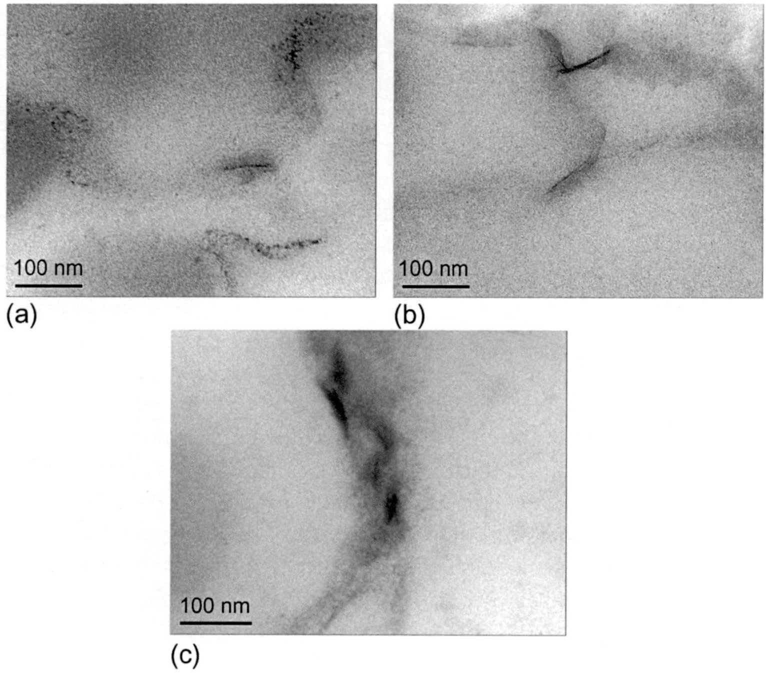

Figure 13.5 Images of PVA/nanoclay composites with different nanoclay (Lit.G-105) loadings: (a) 1%, (b) 2%, and (c) 4% (Kaboorani and Riedl, 2013).

matrix. The distance between nanoclay platelets was increased and the polymer chains entered between the platelets space. This nanostructure is referred to as exfoliated structure, giving superior properties to nanocomposites. As can be seen in Figure 13.5c, nanocomposite with 4% is composed mostly of intercalated structure with very large aggregates or tactoids in the order of several tens of silicate layers. Intercalated structure is not considered an ideal structure for a nanocomposite and it does not grant the nanocomposites the superior properties that exfoliated structure does. As observed in measuring bonding strength (in dry condition and at elevated temperature), adding nanoclay at 1% and 2% loadings gave a significant boost to the properties, but improvement of the properties at 4% content was not much different than with nanocomposites with 1% and 2% nanoclay. This phenomenon should be related to difficulty of dispersing nanoclay in matrix at high loadings. Past research has shown that there is a direct linkage between properties of nanocomposites and quality of nanoclay dispersion (Macia-Agullo et al., 1992; Teresa et al., 2008; Mirzataheri et al., 2010; Zilg et al., 1999; Ranade et al., 2002; Fu and Qutubuddin, 2001). In fact, the extent of improvement, as a result of adding nanoclay, cannot be solely proportional to nanoclay loading because of the difficulty of obtaining good dispersion at high loading.

13.3.1.2 Atomic force microscopy

Although TEM is used to study the quality of nanoclay dispersion in the polymer, subjectivity of TEM observations raises some questions regarding the results obtained by TEM. In order to draw a firm conclusion on the structure of nanocomposites a quantitative technique should be used. In this study, AFM was used to determine the effects of adding nanoclay on PVA surface structure. The AFM images for dried surfaces of pure PVA and its composites are presented in Figure 13.6. Pure PVA had a smooth surface as it had low roughness values. As nanoclay was added to the matrix, a reorganized surface was observed. The reorganization became more notable as nanoclay loading in the matrix increased. At 4% nanoclay loading, the surface of PVA film totally reorganized as a big increase in roughness values was detected, although this roughness is still too small to be detected by the human eye or resulted from light diffusion.

The results of this study show that the ultrasonication technique is an effective way to disperse nanoclay in the PVA matrix. The extent of improvement on PVA performance as a wood adhesive was superior or at least similar to that obtained by high-speed mixing. When a high-speed mixer was used to disperse nanoclay at a high loading (4%) in the matrix, a remarkable reduction was observed in the improvement gained by adding nanoclay. Such a reduction was not observed in the case of the ultrasonic technique. Although adding 4% nanoclay to the PVA matrix resulted in an intercalated structure, nanoclay did increase the shear strength of wood joints in humid conditions and at the elevated temperature. The results show that the ultrasonic technique is very efficient in dispersing nanoclay, especially at high loadings, in contrast to the high-shear speed mixer. High-speed mixing could disperse nanoclay in the PVA only at low loadings and increased bonding strength of PVA in different

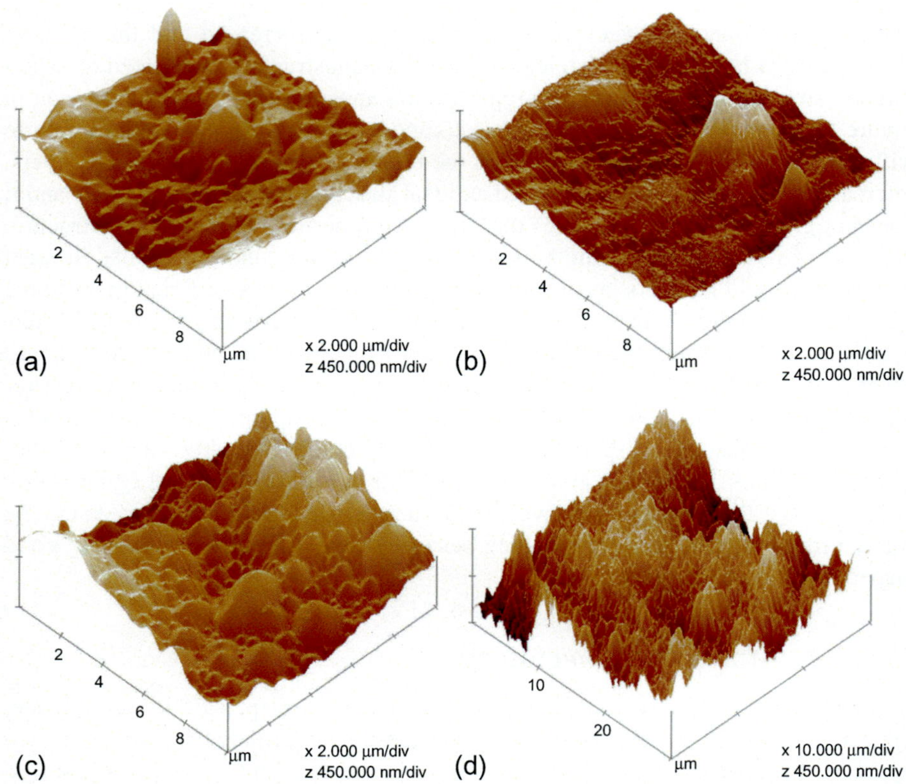

Figure 13.6 AFM images of pure PVA and its composites with different nanoclay (Lit.G-105) loadings: (a) pure PVA, (b) 1% nanoclay, (c) 2% nanoclay, and (d) 4% nanoclay (Kaboorani and Riedl, 2013).

conditions. High-speed mixing has some disadvantages: possible damage to PVA emulsion (because of strong shear force used during the mixing), high cost, and high energy consumption. By contrast, the ultrasonication technique has minimum negative impact on PVA emulsion. Moreover, the ultrasonication technique is economical, as ultrasonic mixing could take place before production of PVA and the solution containing nanoclay can be added to PVA during the production process. By considering the results obtained from this research and our previous work (Kaboorani and Riedl, 2011), and the advantages of the ultrasonication technique over high-speed mixing, adding nanoclay to PVA in an industrial scale seems feasible and can be recommended to wood adhesive manufacturers.

13.3.2 Nanocrystalline cellulose

NCC was added to PVA at 1%, 2%, and 3%. Figure 13.7 shows shear stress at failure of wood joints bonded by pure PVA and its composites with NCC. Adding NCC to PVA did increase the stress but not in a significant way. Although not being significant

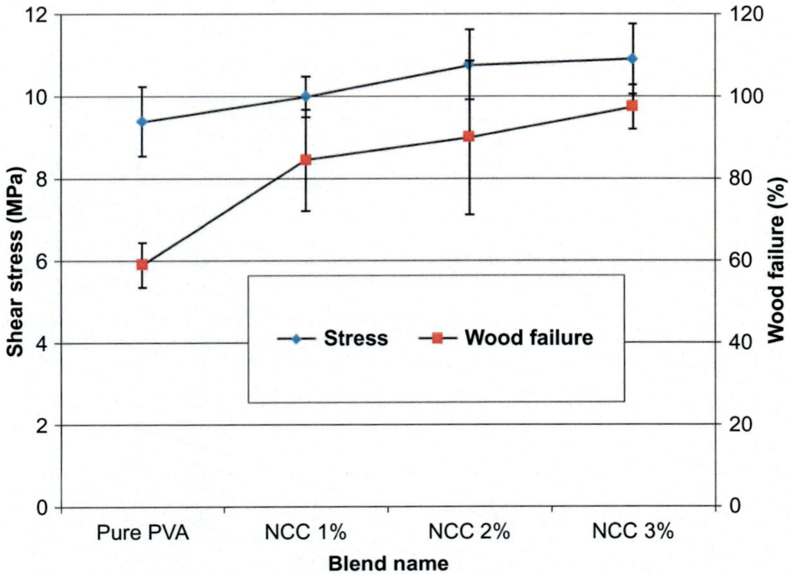

Figure 13.7 Values of bonding strength for PVA/NCC nanocomposites in the dry state.

in terms of the stress at failure, adding NCC to PVA significantly increased the wood failure, proving its positive effects on bonding strength of PVA glue line. Wood failure value increased from 59% for pure PVA to 84% for adhesive with 1% NCC and finally to 97% for adhesive with 3% NCC. Such an increase shows that NCC made the glue line of PVA stronger than wood, causing failure in the wood rather than in the glue line. The loading level of NCC affected the performance of PVA. Wood failure percentage increased as more NCC was added to PVA.

The values of shear stress at failure of wood joints after spending 24 h in water are shown in Figure 13.8. Nanocellulose significantly affected shear strength of wood joints after spending 24 h in water. Adding only 1% NCC to PVA increased the strength by 63%. Further increase in NCC (from 1% to 2%) increased shear strength of wood joints but on a much smaller scale. An increase in NCC loading from 2% to 3% did not change the strength much. The improvement in water resistance of PVA was so high that some failure occurred in the glue line instead of the wood. As more NCC was inserted in PVA, the percentage of wood failure increased, reaching its maximum value at 3% NCC level (21%).

Figure 13.9 shows the values of shear strength at 100 °C. NCC significantly affected PVA performance at this high temperature. Adding NCC to PVA doubled the shear strength of PVA. There was no significant difference between different loading levels of NCC. Adding only 1% NCC to PVA resulted in a 100% increase in shear strength and further addition did not change the strength much.

At 1% loading, NCC improved bonding strength of PVA significantly. But adding more NCC (2% and 3%) did not increase the properties beyond the extent obtained by 1% addition. This trend is related to the quality of NCC dispersion. At 1% NCC

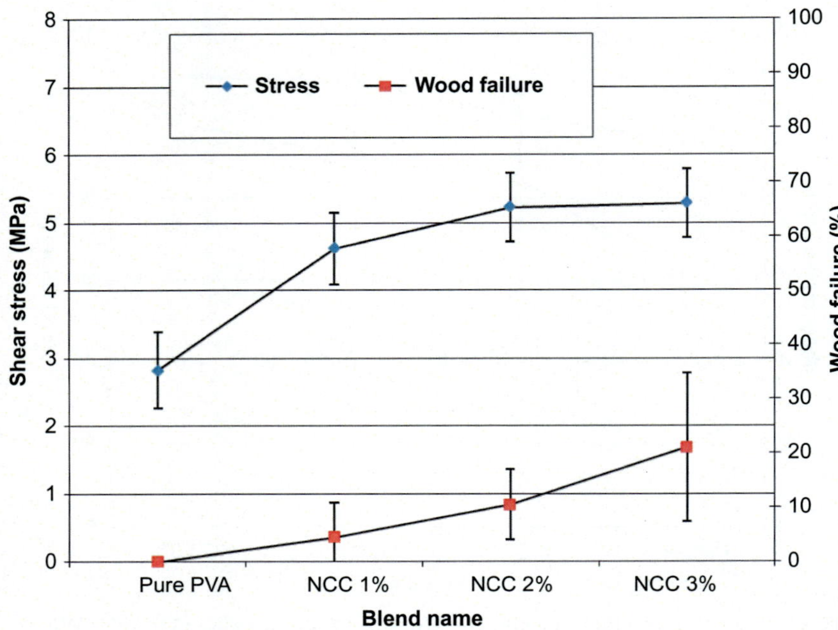

Figure 13.8 Values of bonding strength for PVA/NCC nanocomposites after spending 24 h in water.

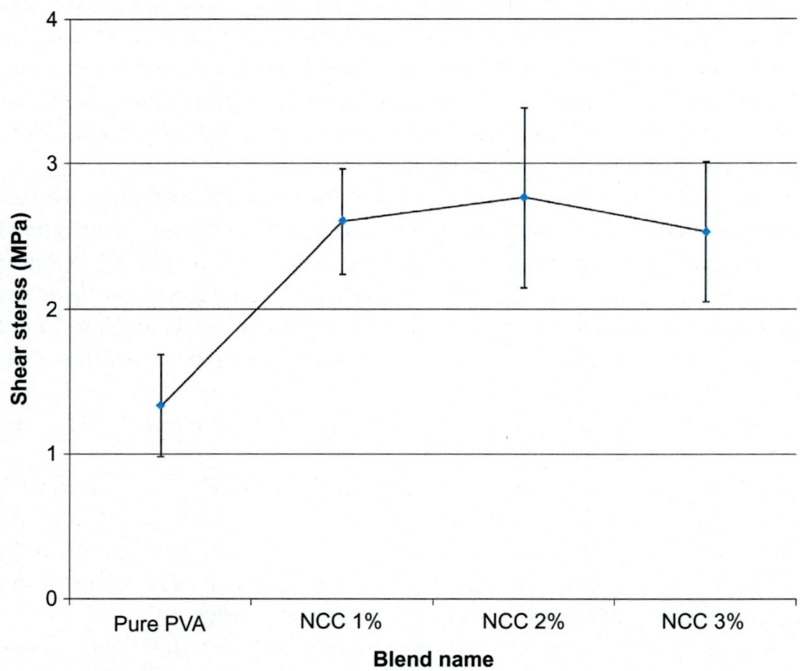

Figure 13.9 Values of bonding strength for PVA/NCC nanocomposites at elevated temperature (100 °C).

loading, significant improvement on bonding strength was achieved as the quality of NCC dispersion was good. At 2% and 3% NCC loadings, dispersion of NCC in PVA encountered difficulty, as shown by the AFM results (Kaboorani et al., 2012), and a poor dispersion of NCC in the matrix was obtained leading to a leveling-off of the improvement. In the context of the other studies conducted on the reinforcement of PVA with inorganic nanomaterials, the extent of improvement achievable by NCC is higher than that of inorganic nanomaterials; namely, nanoclay and aluminum oxide. The improvement is especially much higher for bonding strength in wet conditions.

In this study, unmodified and hydrophilic NCC was used to reinforce PVA. Using such unmodified and hydrophilic NCC led to agglomeration of NCC and difficulty in dispersion of NCC in PVA at high loadings (2% and 3%). The agglomeration limits the extent of improvement that be achieved by NCC. Modifying the surface of NCC a little bit and imposing a slightly hydrophobic character on the surface can help to prevent agglomeration of NCCs and to have a better NCC dispersion in the matrix at a high loading. The modification of surface should be done in controlled conditions without making NCC too hydrophobic or degrading its reinforcing effects.

13.4 Conclusion

The results of this study showed that ultrasonication technique is efficient in dispersing nanoclay in PVA at low (1% and 2%) and high (4%) loadings. Bonding strength of newly formulated adhesives measured on block shear samples increased in wet conditions and at the elevated temperature. In the dry state, the positive effects of nanoclay on strength of the glue line could be detected in terms of wood failure percentage. The strength of the glue line was so high that some failure occurred in the wood rather than in the glue line. Improved barrier properties strengthened the resistance of the glue line to water and, subsequently, a significant increase was observed in bonding strength in the wet state.

In contrast to the results obtained from the dry condition, the extent of improvement on bonding strength in the wet state was proportional to nanoclay loading. Bonding strength of PVA at the elevated temperature was also affected by adding nanoclay. As nanoclay loading in the PVA matrix increased, the shear strength of wood joints at the elevated temperature improved, despite the fact that the biggest gain in the shear strength was observed at 1% nanoclay loading. The morphological studies of nanocomposites revealed that the fluctuations observed in bonding strength tests were related to dispersion quality of nanoclay in the matrix (PVA). AFM proved that it is a credible technique to examine the quality of dispersion, as the results of AFM fully conformed to the TEM observations. At low loadings (1% and 2%), an exfoliated structure is achieved, causing a significant improvement in PVA properties. At high loading (4%), a coexistence of exfoliated and intercalated was observed and so improvements on the shear strength of wood joints were achieved to an extent.

NCC was found to be an effective nanoreinforcing material for PVA. Bonding strength of wood joints was improved significantly as NCC was inserted to PVA. The improvement was measurable in terms of wood failure percentage in the dry condition and in terms of shear strength in the wet condition and at the elevated temperature. Fluctuations in properties of PVA as a wood adhesive and a polymer at different loadings of NCC could be explained by the quality of NCC dispersion in the matrix.

Acknowledgments

The authors acknowledge financial support from the Canadian Forest Nano Products Network (ArboraNano). The authors would like to express their thanks to Forest Products Laboratory-USDA Forest Service for so kindly providing nanocrystalline cellulose (NCC) for this research.

References

Ash, B.J., Siegel, R.W., Schadler, L.S., 2004. Mechanical behavior of alumina/poly(methyl methacrylate) nanocomposites. Macromolecules 37 (4), 1358–1369.

Becker, O., Cheng, Y.B., Varley, R.J., Simon, G.P., 2003. Layered silicate nanocomposites based on various high-functionality epoxy resins: the influence of cure temperature on morphology, mechanical properties, and free volume. Macromolecules 36 (5), 1616–1625.

Bharadwaj, R.K., 2001. Modeling the barrier properties of polymer-layered silicate nanocomposites. Macromolecules 34 (26), 9189–9192.

Cai, Y., 1997. Synthesis of melamine-formaldehyde resin modified poly(VAc-MMA) emulsion. China Adhes. 6 (2), 43–45.

Chang, J.H., Seo, B.S., Hwang, D.H., 2002. An exfoliation of organoclay in thermotropic liquid crystalline polyester nanocomposites. Polymer 43 (10), 2969–2974.

Chen, Y., 1996. Researches of wood adhesives with high strength, water and creep resistance. Nianjie 17 (1), 25–27.

Cho, J., Joshi, M.S., Sun, C.T., 2006. Effect of inclusion size on mechanical properties of polymeric composites with micro and nano particles. Compos. Sci. Technol. 66 (13), 1941–1952.

Comyn, J., 1997. Adhesion Science. The Royal Society of Chemistry, London, pp. 36–39.

Falla, W.R., Mulski, M., Cussler, E.L., 1996. Estimating diffusion through flake-filled membranes. J. Membr. Sci. 119 (1–2), 129–138.

FerrandizGomez, T.D., FernandezGarcia, J.C., OrgilesBarcelo, A.C., MartinMartinez, J.M., 1996. Effects of hydrocarbon tackifiers on the adhesive properties of contact adhesives based on polychloroprene. I. Influence of the amount of hydrocarbon tackifiers. J. Adhes. Sci. Technol. 11 (10), 1303–1319.

Fischer, H., 2003. Polymer nanocomposites: from fundamental research to specific applications. Mater. Sci. Eng. C 23 (6–8), 763–772.

Fornes, T.D., Paul, D.R., 2004. Structure and properties of nanocomposites based on nylon-11 and -12 compared with those based on nylon-6. Macromolecules 37 (20), 7698–7709.

Fu, X., Qutubuddin, S., 2001. Polymer–clay composite: exfoliation of organophilic montmorillonite nanolayers in polystyrene. Polymer 42 (2), 807–813.

Giannelis, E.P., Krishnamoorti, R., Manias, E., 1999. Polymer-silicate nanocomposites: model systems for confined polymers and polymer brushes. Adv. Polym. Sci. 138, 107–147.

Huang, M., Kuo, S., Wu, H., Chang, F., Fang, S., 2002. Miscibility and hydrogen bonding in blends of poly(vinyl acetate) with phenolic resin. Polymer 43 (8), 2479–2487.

Ji, X.L., Hampsey, J.E., Hu, Q.Y., He, J.B., Yang, Z.Z., Lu, Y.F., 2003. Mesoporous silica-reinforced polymer nanocomposites. Chem. Mater. 15 (19), 3656–3662.

Kaboorani, A., Riedl, B., 2011. Effects of addition of nano-clay on performance of polyvinyl acetate (PVA) as a wood adhesive. Compos. A: Appl. Sci. Manuf. 42, 1031–1039.

Kaboorani, A., Riedl, B., Blanchet P., 2013. Ultrasonication technique: a method for dispersing nanoclay in wood adhesives. J. Nanomater. 2013, 1–9, http://dx.doi.org/10.1155/2013/341897.

Kaboorani, A., Riedl, B., Blanchet, P., Fellin, M., Hosseinaei, O., Wang, S., 2012. Nanocrystalline cellulose (NCC): a renewable nano material for polyvinyl acetate (PVA) adhesive. Eur. Polym. J. 48 (11), 1829–1837.

Kim, S., Kim, H., 2005a. Thermal stability and viscoelastic properties of MF/PVAc hybrid resins on the adhesion for engineered flooring in under heating system; ONDOL. Thermochim. Acta 444 (2), 134–140.

Kim, S., Kim, H., 2005b. Effect of addition of polyvinyl acetate to melamineformaldehyde resin on the adhesion and formaldehyde mission in engineered flooring. Int. J. Adhes. Adhes. 25 (5), 456–461.

Krywko, W.P., McAuley, K.B., Cunningham, M.F., 2002. Mathematical modeling of particle morphology development induces by radical concentration gradients in seeded styrene homopolymerization. Polym. React. Eng. 10 (3), 135–161.

Lech, J., 2008. Green Chemistry: Nitration of Tyrosine. CHMC42: Synthetic Chemistry. University of Toronto Press, Toronto, Ontario.

Lee, K.M., Han, C.D., 2003. Linear dynamic viscoelastic properties of functionalized block copolymer/organoclay nanocomposites. Macromolecules 36 (3), 804–815.

López-Suevos, F., Frazier, C.E., 2006. Fracture cleavage analysis of PVAc latex adhesives: influence of phenolic additives. Holzforschung 60 (3), 313–317.

Lu, G., 1996. Improvement of polyvinyl acetate emulsions. Huagong Shikan 10 (6), 17.

Ma, J., Xiang, P., Mai, Y.W., Zhang, L.Q., 2004. A novel approach to high performance elastomer by using clay. Macromol. Rapid Commun. 25 (19), 1692–1696.

Macia-Agullo, T.G., Fernindez-Garcia, J.C., Torro-Palau, A., Orgiles-Barcelo, A.C., Martin-Martinez, J.M., 1992. Addition of silica to polyurethane adhesives. J. Adhes. 38 (1–2), 31–53.

Mirzataheri, M., Atai, M., Mahdavian, A., 2010. Physical and mechanical properties of nanocomposite barrier film containing encapsulated nanoclay. J. Appl. Polym. Sci. 118 (6), 3284–3291.

Nielsen, L., 1967. Models for the permeability of filled polymer systems. J. Macromol. Sci. (Chem.) 1 (5), 929–942.

Pastor-Sempere, N., Fernandez-Garcia, J.C., Orgiles-Barcelo, A.C., Torregrosa-Macia, R., Martin-Martinez, J.M., 1995. Fumaric acid as a promoter of adhesion in vulcanized synthetic rubbers. J. Adhes. 50 (1), 25–42.

Pegoretti, A., Kolarik, J., Peroni, C., Migliaresi, C., 2004. Recycled poly(ethylene terephthalate)/layered silicate nanocomposites: morphology and tensile mechanical properties. Polymer 45 (8), 2751–2759.

Pethrick, R.A., 1997. Positron annihilation—a probe for nanoscale voids and free volume? Prog. Polym. Sci. 22, 1–47.

Plate, N.A., Yampol'skii, Y.P., 1994. Relationship between structure and transport properties for high free volume polymeric materials. In: Paul, D.R., Yampol'skii, Y.P. (Eds.), Polymeric Gas Separation Membranes. CRC Press, Boca Raton, FL, pp. 155–207 (Chapter 5).

Poh, P.T., Firdaus, S.Z., 2010. Effect of hybrid tackifiers on adhesion properties of epoxidized natural rubber-based pressure-sensitive adhesives. J. Polym. Environ. 18 (3), 335–338.

Qiao, L., Easteal, A., 2001. Aspects of the performance of PVAc adhesives in wood joints. Pigm. Resin Technol. 30 (20), 79–87.

Qiao, L., Easteal, A.J., Bolt, C.J., Coveny, P.K., Franich, A., 2000. Improvement of the water resistance of poly(vinyl acetate) emulsion wood adhesive. Pigm. Resin Technol. 29 (3), 152–158.

Ranade, A., D'Souza, N., Gnade, B., 2002. Exfoliated and intercalated polyamide-imide composite with montmorillonite. Polymer 43 (13), 3759–3766.

Teresa, V.G., Carlos, G.S., Javier, R.D.R., Virgilio, G.G., 2008. Nylon 6/organoclay nanocomposites by extrusion. J. Appl. Polym. Sci. 08 (5), 2923–2933.

Thran, A., Kroll, G., Faupel, F., 1999. Correlation between fractional free volume and diffusivity of gas molecules in glassy polymers. J. Polym. Sci. B Polym. Phys. 37 (23), 3344–3358.

Wang, J., 1999. Study on improving properties of polyvinyl acetate emulsion. Zhanjie 20 (3), 16–19.

Wang, Y.Q., Wu, Y.P., Zhang, H.F., Zhang, L.Q., Wang, B., Wang, Z.F., 2004. Free volume of montmorillonite/styrene-butadiene rubber nanocomposites estimated by positron annihilation lifetime spectroscopy. Macromol. Rapid Commun. 25 (23), 1973–1978.

Wetzel, B., Haupert, F., Zhang, M.Q., 2003. Epoxy nanocomposites with high mechanical and tribological performance. Compos. Sci. Technol. 63 (14), 2055–2067.

Winberg, P., Eldrup, M., Pedersen, N.J., van Es, M.A., Maurer, F.J.H., 2005. Free volume sizes in intercalated polyamide clay nanocomposites. Polymer 46 (19), 8329–8349.

Zhang, Q., Archer, L.A., 2004. Optical polarimetry and mechanical rheometry of poly(ethylene oxide)–silica dispersions. Macromolecules 37 (5), 1928–1936.

Zhou, H., 1991. Research on the improvement of water resistance of polyvinyl acetate emulsion. Nianjie 12 (4), 11–12.

Zilg, C., Mülhaupt, R., Finter, J., 1999. Morphology and toughness/stiffness balance of nanocomposites based upon anhydride-cured epoxy resins and layered silicates. Macromol. Chem. Phys. 200 (3), 661–670.

Mechanical performance of flax-based biocomposites

14

A. Bourmaud, A. Le Duigou, C. Baley
Université de Bretagne Sud, Lorient Cedex, France

14.1 Introduction

In recent years, many research and industrial developments deal with the integration of plant fibers as composite polymer reinforcement. In Europe, the most interesting are flax or hemp fibers due to their high-performance specific mechanical properties (Beaugrand and Berzin, 2013; Bourmaud et al., 2010; Placet, 2009), their environmental and ecological benefits (Joshi et al., 2004; Le Duigou et al., 2011; Pervaiz and Sain, 2003), and their low cost.

With the same fiber volume fraction, and despite similar fibers mechanical properties, plant fiber composites exhibit mechanical performances lower than those of glass fiber ones (Bourmaud and Baley, 2007, 2009). This phenomenon is mainly due to the conjugated effects of a moderate fiber/matrix adhesion and fiber dispersion or individualization into the composites. In spite of the retting and mechanical separation operations, the cohesive nature of these bundles results in a certain amount persisting through to the final composite component. Moreover, the flow conditions during processing can also result in the formation of clusters of fibers. As a result, the bundles may cause a significant drop in the reinforcement aspect ratio and, hence, a drop in their reinforcing capacity. Bos et al. (2002) showed how flax fiber bundles can affect the mechanical properties of composites. Rask et al. (2012) studied the damages processes of PP/flax UD by using X-ray diffraction; they concluded that well-separated fibers are recommended for composite reinforcements. Andersons and Joffe (2011) enounced similar conclusions by demonstrating that a probabilistic model, assuming perfect separation and regular spacing of fibers, yields an upper limit of strength for UD flax composites.

Composites materials reinforced by natural fibers can be manufactured using classic processes; Zhou et al. (2005) and Baley et al. (2005a) show examples for different fiber/matrix combinations. During processing the reinforcing fibers are subjected to high temperatures, which affect their properties. Velde and Baetens (2001) showed that water removal and degradation of waxes occur from 120 °C, while pectin degradation starts around 180 °C. For Gassan and Bledzki (2001), the degradation of flax fibers, and a reduction in their toughness, appears from 170 °C due to depolymerization and chain breakage. Nanoindentation tests performed *in situ* on flax fiber cross-sections before and after manufacture (Bourmaud and Baley, 2010) revealed a strong drop in fiber stiffness after injection molding. A transformation by film stacking was shown to be less severe for the plant walls due to lower shear levels associated with this process.

Biocomposites: Design and Mechanical Performance. http://dx.doi.org/10.1016/B978-1-78242-373-7.00013-5

Similar conclusions on process-mechanical property relationships were reached by Le Duigou et al. (2008) for flax/PLA biocomposites.

Thus, the use of aggressive processes involving high shear stresses has a strong influence on the fiber morphology (Ausias et al., 2013; Mano et al., 2010), resulting in a reduction in the length and diameter of the fibers bundles. The reduction in diameters is interesting for the aspect ratio but the drop in lengths must be limited; thus, high rotation speeds induce an improved fiber dispersion but lower aspect ratios. Technologies such as Buss comixers can be used to reduce fiber changes (Shon and White, 1999) thanks to adapted profiles and delayed fiber introduction. The residence time is an important parameter, too short a duration results in poor dispersion, while long residence can favor matrix and fiber degradation (Joseph et al., 1999). Increasing the rotation speed of the extrusion screw reduces residence time but can in some cases induce drops in modulus and flexural and tensile strengths of composites (Mano et al., 2010). Finally, the temperature of extruders and injection processes is an important parameter. A high temperature favors fiber dispersion through higher matrix fluidity (Siaotong et al., 2006). However, as shown by Li et al. (2009) on flax/PE composites, a compromise must be found as high temperatures can cause permanent damage to plant fibers.

Consequently, the use of different processing methods, due to variations in shear rate or residence times, could induce important differences into fiber morphology or composite microstructure. The aim of this chapter is to provide an overview of plant fiber composites processing with an important focus on flax-reinforced thermoplastics. First, we will describe plant fiber structure and main properties. The second part will be dedicated to the influence of the process on plant fiber properties, considering the isolated fiber or within a polymer matrix. The third section focuses on the influence of the processing method on composite properties; we'll detail the impact of the fiber morphology or volume fraction as well as the processing method on the composite mechanical performances or on its water uptake. Finally, the last section presents the impact of the process, especially injection molding, on the plant fiber composite microstructure; the skin-core effect will be presented as well as the impact of the fiber individualization (due to retting or mechanical fiber extraction) and the consequences on the composite performances.

14.2 Plant fibers for composite reinforcement: structure and properties

14.2.1 Plant fiber structure and mechanical behavior

The structure–function relationship must be considered when selecting plant fiber. For example, coir fiber or flax fiber do not have a similar function in the wild and, consequently, structure. Coir fiber is fruit-wall fiber that should help the fruit to resist breakage during drooping. Thus, it exhibits high interest for bioinspired impact-resistant structure. In contrast, flax fiber acts as a reinforcing element within the stem. Its stiffness and strength could be used for light and high-performance composites structures.

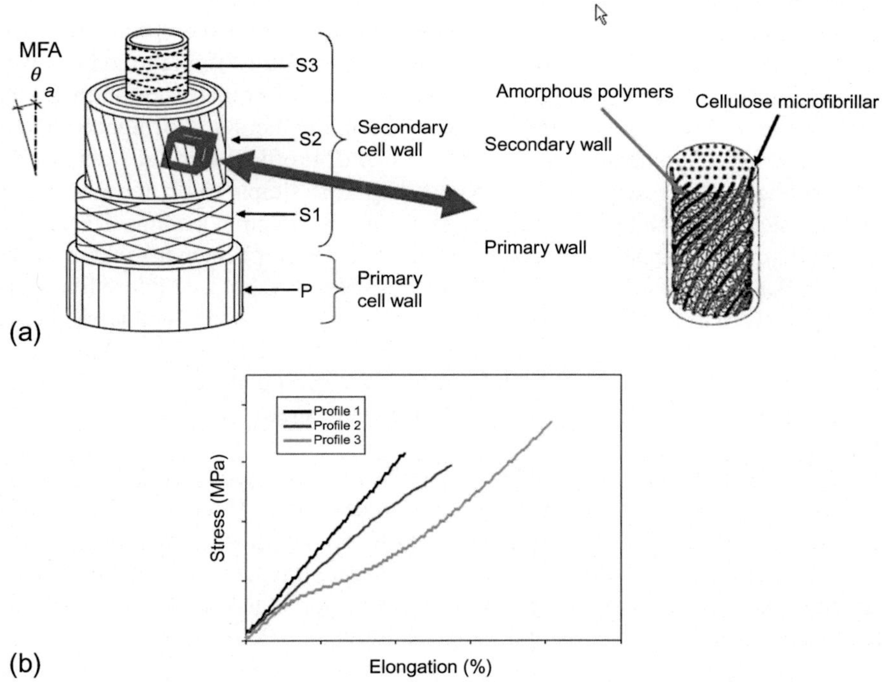

Figure 14.1 (a) Schematic representation of flax fiber showing helix arrangement of cellulose fibrils. Diagram on right adapted from Hearle (1963). (b) Typical stress–strain curves for flax fibers (Gourier et al., 2014).

Plant fibers consist of composite layers reinforced by cellulose fibrils (Hearle, 1963; Morvan et al., 2003; Figure 14.1) whose axial Young's modulus is in the range 134–160 GPa (Fratzl et al., 2004; Mohanty et al., 2000). The hierarchical architecture of flax is composed of an outer layer with a primary wall (0.1–0.5 μm thick), which is flexible enough to allow wall expansion during growth (Burgert, 2006). The primary layer is coated with a polymer matrix (mainly pectins of middle lamella) that ensures the intercellular cohesion in a fiber bundle. A secondary wall represents around 80% of the fiber cross-section and is composed of three layers, S1 (0.5–2 μm thick), S2 (5–10 μm), and S3 (0.5–1 μm) and provides the reinforcement of the plant structure (Morvan et al., 2003). The S2 layer is reinforced by cellulose fibrils with a helix angle of 10°, the microfibrillar angle, to the plant axis (Baley, 2002). In the center of the fiber there may be a cavity called a lumen.

Thus, plant fibers are composite nanostructures that are renewable, durable, and biodegradable. Their hierarchical architecture results in complex mechanical behavior (Charlet et al., 2010) and they exhibit various tensile behavior (Figure 14.1).

The elementary flax fiber tensile characterization is governed by the XPT 25-501-2 standard (AFNOR, 2009), which details the method parameters. The first behavior (Profile 1, Figure 14.1b) is truly elastic; tensile stress–strain plot is linear. The second behavior (Profile 2) includes a loss of linearity that could be due to the damage mechanism. Clear distinction between Profiles 1 and 2 is made by measuring Young's modulus either at the beginning (0.1% of deformation) or at the end of the curve ($E_{standard}$ according to XPT 25-501-2). Therefore, for Profile 1, Young's modulus variation between the beginning and the end of the curve is lower than 10%. Profile 2 has higher variation. Profile 3 is abundantly observed in the literature for lignocellulosic fibers and drastically changes from Profiles 1 and 2. Profile 3 has three distinct domains corresponding, respectively, to fiber loading, with simple strain due to test setup and fiber microfibrils and pectic matrix in stress length/direction loading. Then a nonlinear part appears, explained by complex accommodation phenomena within the S2 layer. Similar to wood and hemp fibers, beyond a stress threshold, cellulose fibrils and matrix polysaccharide in S2 network disrupt their hydrogen bonds and leads; hence, flowing in stress direction inducing aligning of fibrils in fiber length (Baley et al., 2012; Placet et al., 2014). The second linear area corresponds to a quasi-elastic response of the fiber. A meticulous analysis of the curve shows that the tensile modulus increases with the strain.

The tensile properties of flax fibers have been widely studied and their specific stiffness is comparable to that of glass fibers (Baley, 2002; Hughes et al., 2007; Le Duigou et al., 2008). The compression strength of single fibers, measured by Bos et al. (2004) is also high, with values up to 1200 MPa. Moreover, the structure of these plant fibers is highly anisotropic (Baley et al., 2006; Bourmaud and Baley, 2009) with a transverse modulus for flax around 7 GPa (Baley et al., 2006).

14.2.2 Influence of the microfibrillar angle, cellulose rate, and biochemical composition on fiber mechanical properties

According to the flax variety, cellulose represents between 65% and 75% of the total weight of flax fibers (Altaner and Jarvis, 2008). The hemicellulose (glucomananes) not only acts as a matrix but also, thanks to a branched structure similar to that of cellulose (Charlet et al., 2009a), may act as a coupling agent (Baley, 2002; Morvan et al., 2003); its nature varies according to whether it is in the primary layer of fibers (xyloglucanes or arabinoxylanes) or in the secondary layer (xylanes and glucomananes) (Baley et al., 2006; Burgert, 2006; Charlet et al., 2009b).

Two types of pectins are present in flax fibers (Morvan et al., 2003): those that are found in the adjoining and the primary layers, which form large macromolecules and ensure cohesion between cells, and those in the secondary layers, which are part of the matrix covering the cellulose mesofibrils (Zykwinska et al., 2008); these are assemblies of elementary sugars (oses). The most common pectins in flax fibers are homogalacturonanes and rhamnogalacturonanes (Girault et al., 1997; Gorshkova et al., 1996; Morvan et al., 2003; His et al., 2001).

At room temperature flax fibers contain (depending on the variety, the retting, and growth conditions) between 6% and 10% by weight of water, which acts as a cell wall plasticizer (Baley et al., 2005b). The presence of absorbed water is frequently invoked when plant fibers are being considered for polymer reinforcement. Most thermoplastic manufacturing processes involve steps at high temperatures, and the amount of absorbed water in the fibers will evolve according to the temperature and duration of

Table 14.1 Biochemical composition of flax fiber (Bourmaud, 2011)

Component	(%)	References
Cellulose	64–84	Bismark et al. (2002), Khalil et al. (2000), and Alix et al. (2008)
Hemicellulose	10–21	Wardrop and Bland (1959), and Lapierre and Belmont (1994)
Pectin	2–15	Batra (1998) and Buschle-Diller and Zeronian (1993)
Lignin	1–5	Lapierre and Belmont (1994) and Gorshkova et al. (2000)
Wax	1.5	Batra (1998)
Water	6–10	Batra (1998) and Bos (2004)

exposure. Consequently, the fiber water content could be considered as a key parameter for thermoplastic composites processing.

Table 14.1 (Bourmaud, 2011) summarizes the flax fiber biochemical composition according to the literature.

Due to the high mechanical properties of cellulose, its rate has a direct impact on the fiber properties (Table 14.2) as well as the structure and the conformation of the noncellulosic polymers (hemicelluloses and pectins) as detailed in the following section.

Moreover, the microfibrillar angle (MFA) has a great influence on the mechanical behavior of the elementary fibers and, consequently, on their mechanical performances as evidenced for wood (Salmén, 2004). Thus, flax or hemp, which exhibit low MFA values (Table 14.2), have interesting mechanical properties, in opposition to sisal or cotton. Nevertheless, this condition is not sufficient; in the case of bamboo, despite a low MFA, the fiber properties are weak due to a low cellulose rate.

Table 14.2 (Bourmaud, 2011) shows average mechanical properties for various vegetal fibers. Moreover, it is important to know exactly the main parameters influencing the plant fiber mechanical properties: the variety (Baley and Bourmaud, 2014), the climatic conditions (Lefeuvre et al., 2014), the soil land cultural practices (Bert, 2013), the retting (Martin et al., 2013), the scutching (Martin et al., 2014), and the water absorbed fraction (Baley et al., 2005b; Gourier et al., 2014).

14.2.3 Schematic representations of secondary cell wall structure

As described previously, the cell wall stiffness is mainly induced by the cellulose mesofibrils as well as by the structure and biochemical composition of the polysaccharidic network where these fibrils are embedded. Several analyses of this system exist in the literature. According to Hearle (1963), the cellulose mesofibrils are strongly linked to the amorphous polysaccharides, giving a network with galactans and xyloglucans. This network could flow above a threshold shear stress due to the breakage of the components' hydrogen bonds. Other studies on wood have revealed a stick-slip (Velcro®) mechanism in the secondary cell walls (Keckes et al., 2003; Fratzl et al., 2004; Altaner

Table 14.2 Cellulose rate and microfibrillar angle of main vegetal fiber (Bourmaud, 2011)

Fiber	Cellulose rate (%)	Microfibrillar angle (°)	Young's modulus (GPa)	Strength at break (MPa)	References
Flax	64–85	10	46–85	600–2000	John et al. (2008), Bismark et al. (2002), Khalil et al. (2000), Baley (2002), Batra (1998), Charlet et al. (2007, 2009c), Davies and Bruce (1998), Ganster and Fink (1999), and Tröger et al. (1998)
Hemp	58–85	11	17–44	368–800	Trivaudey and Placet (2010), Beckermann and Pickering (2008), Bourmaud and Baley (2009), Duval et al. (2011), Pickering et al. (2007), Saheb and Jog (1999), Sedan et al. (2007), and Le Troedec et al. (2008)
Jute	61–75	8	26.5	393–773	Bisanda and Ansell (1992), Roe and Ansell (1985), Shah and Lakkad (1981), Liu et al. (2009), and Varma et al. (1984)
Ramie	68	7.5	24.5–87	560–1600	Angelini et al. (2000), Bodros and Baley (2008), Goda et al. (2006), Paiva Júnior et al. (2004), and Mohanty et al. (2000)
Sisal	65	20	7–25	530–700	Idicula et al. (2005), Bourmaud and Baley (2009), Chand et al. (1988), Kumar et al. (1995), and Satyanarayana et al. (1990)
Cotton	82–99	30–40	3.5–8	287–597	Bisanda and Ansell (1992), Chand et al. (1988), Kumar et al. (1995), and Satyanarayana et al. (1990)
Bamboo	42–50	2–10	10–40	340–510	Rao and Rao (2007), Amada et al. (1997), Amada and Untao (2001), and Jain et al. (1992)

and Jarvis, 2008). According to Burgert (2006) and Keckes et al. (2003), the Velcro® mechanism is partly reversible after loading and other hydrogen bonds between cellulose and hemicellulose will be built elsewhere. Altaner and Jarvis (2008) propose a slightly different model in which, in addition to hydrogen bonds with cellulose, the hemicellulose chains can make bridges or loops with cellulose mesofibrils.

More structural models are detailed by biologists' groups. Comparative works on the fiber structure of different varieties of flax (Alix et al., 2008) have shown stiffness differences related to cellulose content in the cell walls; the varieties with the highest cellulose content were represented with cellulose fibrils being the closest together and a less well-developed matrix between fibrils. The bridges between glucomananes chains have been proposed to explain the stiffness differences observed (Alix et al., 2008).

In flax fibers, other components such as arabinogalactan proteins rich in glycine (GRP) and β-1–4 galactans have also been proposed to play the role of hemicelluloses, that is, of compatibilizers between the cellulose mesofibrils and the pectic chains (RG-I) to which they are linked, representing a true interphase between the matrix and the cellulose and guaranteeing the cohesion of the system (Alix et al., 2008; Gorshkova and Morvan, 2006; Morvan et al., 2003). This configuration can be reinforced by the presence of homogalacturonans that will consolidate the interphase between the galactans and the pectic matrix (Alix et al., 2008; Morvan, 2010). This analysis was based on models of *in vitro* interactions studied by Zykwinska et al. (2007) and primary cell wall models elaborated by Cosgrove (2005).

Recent studies (Bourmaud et al., 2013b; Lefeuvre et al., 2013) have shown the prime importance of the organization of the plant cell wall components and their structure in the mechanical performance. More than the quantity of cellulose, which is present in sufficient amounts to ensure good performance, the ratio between the structural polysaccharides and matrix within the S2 layer present a high correlation with the stiffness of flax fibers.

14.3 Influence of the process on the fiber properties

14.3.1 Impact of a thermal cycle on the fiber performances

High temperatures have an effect on fiber mechanical properties. As evidenced by Baley et al. (2005b) or Stamboulis et al. (2001), a temperature exposure induces a decrease in the fiber strength and the apparition of defaults inside the cell walls. Cichocki and Thomason (2002) highlighted a decrease of the transverse and shear modulus of jute fibers with the temperature increase. Gassan and Bledzki (2001) showed that the mechanical properties of jute or flax are affected by the temperature around 170 °C; in another work, Placet (2009), by using DMA on hemp bundles, evidenced a decrease of the storage modulus and an increase of the loss factor during a heating ramp between 20 and 200 °C. Ho and Ngo (2005) performed tensile tests on coir and hemp bundles between 23 and 80 °C; they also showed an important decrease of the Young's modulus and of the strength and the strain at break with the increase of the temperature. Manufacture of flax-reinforced thermoplastics at 150 °C should not modify the fiber

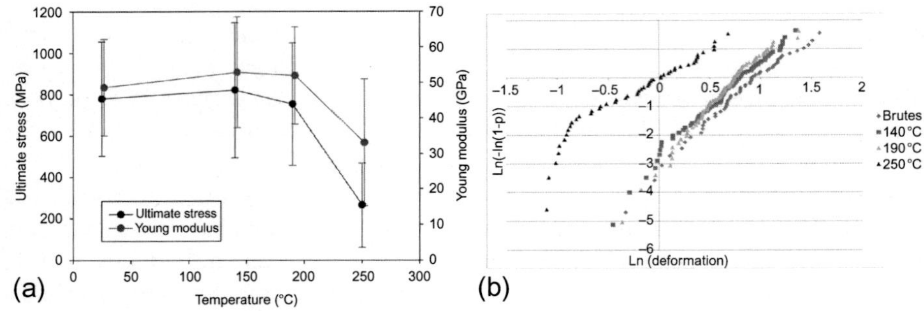

Figure 14.2 (a) Stress at break and Young's modulus for the different temperature of thermal treatment (values at 20 °C for the raw fibers). (b) Weibull coordinates for tensile strength and tensile strain at break (Gourier et al., 2014).

structure but will affect the absorbed water content, depending on the temperature and exposure time, and can modify the interfaces between the constituents.

As shown in Figure 14.2a, drying fibers (14 h at 105 °C) results in significant loss of strength (–44% on average) and failure strain (–39%), together with a change in the modulus versus strain behavior (Baley et al., 2012). If we treat the flax fiber structure as a layered composite reinforced by helical cellulose fibrils, when this is loaded in tension the microfibril angle to the fiber axis decreases and stiffness increases. Assuming that the knee on the tensile modulus–strain plot corresponds to the onset of mesofibril sliding, a damage threshold can be defined in terms of the local shear stress in the S2 layer (secondary cell wall). This shear stress drops after drying from 45 to 9 MPa.

Raw fibers as well as fiber cycled at 140 and 190 °C follow Weibull distribution (Figure 14.2a) with a quasi-linear relationship between $\ln(\ln(1-p)$ versus $\ln(\varepsilon)$ in Weibull coordinates) and relatively close Weibull modulus (generally around 3.0) to the literature's data. The behavior of flax fiber at 250 °C is different with rupture in the slope typical of bimodal behavior. Reduction of m Weibull modulus at 250 °C indicates that thermal cycle induces a dramatic structural change within the fiber responsible for heterogeneous strength distribution and defect generation. In addition, bimodal behavior highlights that defects should be distributed in two populations.

14.3.2 Impact of a thermal cycle on fiber surface and adhesion

The interface in natural fiber composites is often described as weak, due to the supposed incompatibility between a hydrophilic fiber and hydrophobic matrix. However, this does not take into account the complexity of the phenomena involved. The quality of the fiber–matrix bond influences the properties of composites. However, at a macroscopic scale there are many other factors, so it is not easy to attribute changes in behavior directly to the interface. For this reason, it is simpler to use a direct interfacial measurement such as tests on microcomposites (microdroplet debonding, fragmentation, pull-out, etc.). Figure 14.3a shows a PLA microdroplet on single flax fiber and Figures 14.3b and c the typical mechanical behaviors.

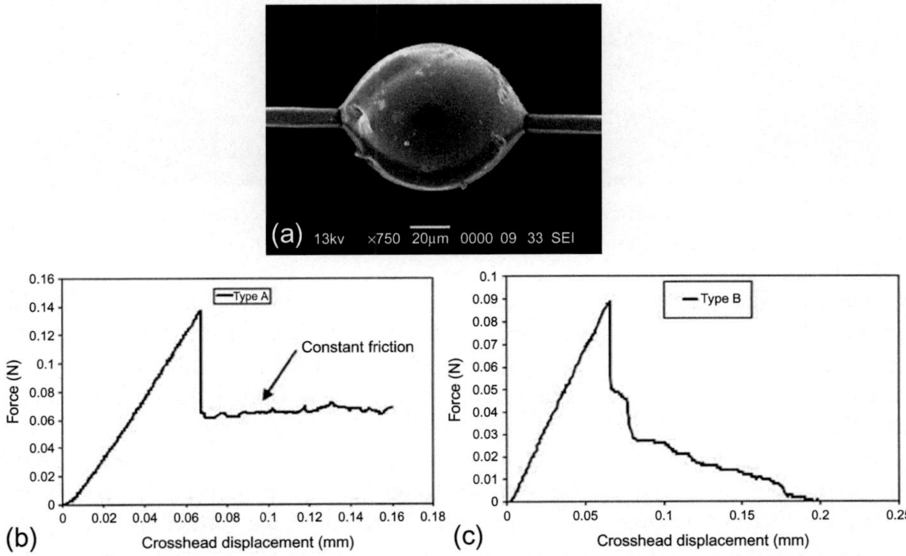

Figure 14.3 (a) PLA microdroplet on single flax fiber. (b and c) Typical mechanical behaviors recorded during debonding test of flax/PLLA (Le Duigou et al., 2012).

When natural fibers are used to reinforce thermoplastic matrix composites they are exposed to high temperatures that affect their properties (Bourmaud and Baley, 2010). However, although several studies have examined the influence of drying on mechanical properties of fibers, few authors have examined how the removal of water affects surface properties and adhesion between natural fibers and biopolymers. This aspect has been studied in some detail for wood, however (Christiansen, 1990, 1991; Nussbaum, 1999; Brito et al., 2008; Gérardin et al., 2007).

A fiber drying cycle of 14 h at 105 °C reduces flax/PLLA interface shear strength (–20%) and friction stress (–30%) evaluated by microbond tests (Table 14.3) (Le Duigou et al., 2012).

The multilayer nature of these fibers influences debonding mechanisms and partial peeling of fibers can occur. Drying treatment increases the proportion of peeling failures, suggesting that fiber transverse properties are also reduced (Figure 14.4a).

The average roughness (RMS) of untreated fibers is comprised of between 10.8 and 14.7 nm over an area of 1 µm^2 (Figure 14.4b), whereas after the drying cycle values vary from 3.3 to 4.8 nm for the same area. The drying cycle tends to reduce fiber

Table 14.3 Microbond test results on PLLA/flax and dried flax systems (Le Duigou et al., 2012)

Material	Number of tests	L_e/D_g	τ_{app} (MPa)	$\tau_{friction}$ (MPa)
PLLA/flax	60	1.32 ± 0.10	16.4 ± 3.8	5.7 ± 1.2
PLLA/dried flax	56	1.33 ± 0.07	13.6 ± 3.3	4.3 ± 1.2

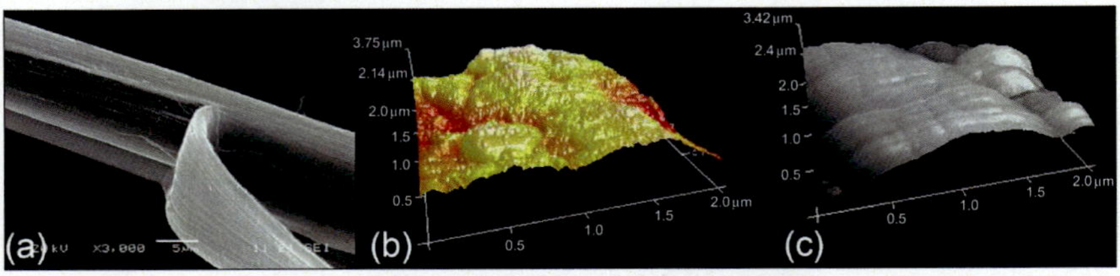

Figure 14.4 Fiber surfaces after microdroplet sliding: (a) peeling mechanism; (b) TM-AFM phase of raw flax fiber; (c) Dried flax fibers (Le Duigou et al., 2012).

roughness. Indeed, surface components such as hydrophilic pectins can lose water and shrink (volume reduction) during thermal treatment (Basu et al., 2011). There has been extensive work on the drying of wood that provides indications on the mechanisms acting during drying of natural fibers. According to Christiansen (1990, 1991), different phenomena occur at the interface between wood and phenolic resin following excessive drying of the wood. First, the surface of the wood can be deactivated around 160 °C due to a modification that reduces wettability (Christiansen, 1990). This is caused by different mechanisms such as the movement to the surface of nonwetting species. Gérardin et al. (2007) showed by XPS that these species (low-electron donors), originate from the degradation of hemicellulose. Chemical degradation of surface components also occurs (Christiansen, 1991). In the presence of oxygen, an oxidation of carbonyl and hydroxyl groups occurs at the wood surface, which leads to a reduction in the number of sites available for bonding to resins (Gérardin et al., 2007). A reorientation of surface molecules may also occur, particularly at temperatures above the T_g of surface components such as hemicellulose and pectins (Tissaoui, 1996). For example, the T_g of hemicellulose varies from 200 °C for a residual humidity (RH) of 0% to 70 °C for a RH of 10% (similar to storage conditions) (Tissaoui, 1996). Finally, drying of wood can cause closure of micropores on the surface, which limits impregnation by a resin (Christiansen, 1990). This phenomenon can also affect mechanical anchoring mechanisms (Sernek, 2002).

In addition to the loss in adhesion properties, the fiber drying cycle has a strong influence on the behavior of the fiber–matrix interface after debonding (Figure 14.3b with friction). Thus, an increase of 20% in type B behavior (nonconstant friction) is noted compared to undried fibers. As shown in Figure 14.4a, this type of behavior corresponds to partial peeling of fibers.

14.3.3 Impact of the process on composite reinforcement morphology (diameter, length, and aspect ratio)

The compounding or injection molding shear rate could achieve high values and induce significant decrease in fiber length. Figure 14.5 (Bourmaud and Baley, 2010) shows the evolution of the fibers' morphology before and after compounding and injection molding for a PP–flax 30 wt.%.

An important decrease of fiber length can be noticed during reprocessing. The main cause of this decrease is a fiber fracture principally attributed to the compounding step. Antich et al. (2006) have investigated the aim of reprocessing with an internal

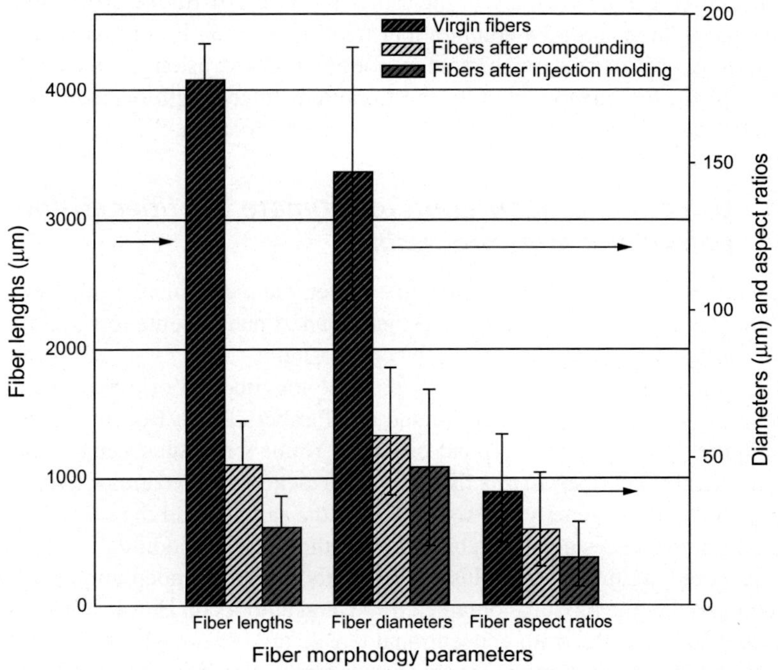

Figure 14.5 Evolution of fiber morphology after the compounding and injection molding process (Bourmaud and Baley, 2010).

mixer on sisal fiber lengths and they have confirmed a similar result (decrease of 60% in length after mixing). In Bourmaud and Baley (2010), the decrease in fiber length after compounding and injection molding is 73.1% and 85.1%, respectively.

These important changes in fiber morphology evidences the important shear rate suffered by fibers during the process inducing important modifications into the structure and the organization of the flax fibers. We can see that the processing induces a slightly greater decrease of aspect ratio due to the conjugated decrease of fiber length and diameter as evidenced in a previous work (Bourmaud and Baley, 2007). Aspect ratio (*L*/*D*) is calculated by dividing fiber length (*L*) by its diameter (*D*). The aspect ratio is the most influential parameter on mechanical properties of the composite for short fiber-reinforced polymers (Kelly and Tyson, 1965).

Thus, for plant fibers, we can notice an important diminution of fiber lengths but also an important reduction of the diameter of bundles caused by the shearing during the process. In initial pellets, fibers are aggregated into bundles and the recycling induces bundle divisions. These bundles into the composite are mainly due to the fiber organization into the plant where the fibers are located at the periphery of the stem, and arranged into bundles, each one having several tens of fibers. In some cases, moreover, we can notice an important decrease of the flax fibers length with the fiber content after extrusion as well as injection molding (Ausias et al., 2013). The addition of fibers in the matrix increases the viscosity of the material and then increases shear stress during processing, which involves more fiber ruptures. In contrast, we can notice that, after a slight decrease of the aspect ratio for low fiber content, a threshold could be highlighted around a volume fiber fraction of 30%. From this value, the decrease of the aspect ratio is much more important after extrusion as well as injection. We can explain this phenomenon by the important fiber length decrease due to the important shear rate induced by the high fiber volume fraction.

14.3.4 Use of nanoindentation to estimate the fiber stiffness before and after processing

During a process cycle, in opposition to glass fiber, the mechanical properties of plant fiber could change. Only a local measurement such as nanoindentation could estimate the fiber mechanical properties within the composite.

Nanoindentation can be used to determine the longitudinal or transverse mechanical properties of natural fibers (Bourmaud and Pimbert, 2008; Bourmaud and Baley, 2009). Figure 14.6 presents a typical curve of Young's modulus versus indentation depth for a virgin flax fiber. In this figure, we can clearly see two areas: the first one is attributed to the heterogeneity and roughness of the sample, and the second one, after around 10 nm, evidences a good stabilization of the Young's modulus.

The average longitudinal modulus of flax fibers from nanoindentation experiments is low compared to the modulus obtained by several authors on Hermes fibers (Charlet et al., 2007; Baley, 2002) with conventional tensile tests (from 46.9 to 59.1 GPa), but the scales and the solicitation modes are very different. Gindl and Schöberl (2004) and Gindl et al. (2008) indicated that longitudinal modulus of wood cell walls obtained by nanoindentation is considerably lower than the tensile Young's modulus or model

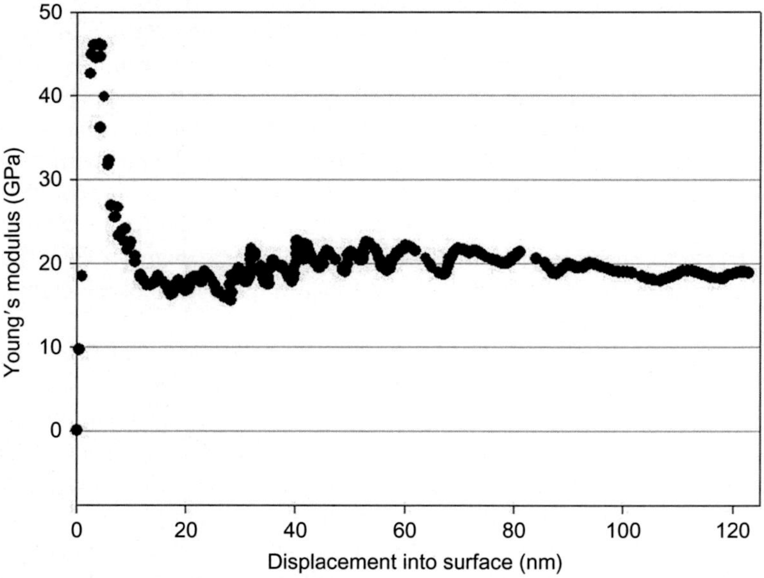

Figure 14.6 Typical curve of Young's modulus versus indentation depth for a virgin flax fiber (Bourmaud and Baley, 2010).

Table 14.4 Nanoindentation longitudinal modulus and hardness of Hermes fibers before and after processing (Bourmaud and Baley, 2010)

Materials	Number of indents	Longitudinal modulus E_L (GPa)	Hardness H (MPa)
Virgin Hermes fibers	304	18.4 ± 1.9	527 ± 79
Hermes fibers after film stacking	368	12.65 ± 1.3	280 ± 46
Hermes fibers after injection molding	241	9.78 ± 2.37	246 ± 82

calculations. Because of the inclination on faces of the Berkovich-type indenter, the wall is loaded at an angle of approximately 25°. Consequently, the resulting three-dimensional stress is not only governed by the longitudinal modulus but also affected by the transverse modulus, inducing an underestimation of the longitudinal modulus.

Table 14.4 shows the longitudinal moduli and hardness values obtained for Hermes fibers before and after two processing cycles (Bourmaud and Baley, 2010). A significant decrease of mechanical properties can be noticed after film-stacking and injection-molding processes. The modulus and hardness decreases are 31.25% and 46.9%, respectively, and 46.8% and 53.3% for the film-stacked and injected-molded sample.

In the case of injection molding, this drop in mechanical properties could be induced by the process parameters and especially by the combined effect of shear rate and temperature. The film-stacking process also induces a decrease in the mechanical properties of flax fibers mainly induced by the high-temperature exposure and to a lesser extent by the elimination of water causing a diminution of the fiber softness.

14.4 Plant fiber composites properties: relationship between the processing method and final properties

14.4.1 Impact of the loading rate on plant fiber composites mechanical properties

The effect of fiber content on the mechanical properties could be investigated for PP–flax composites. As evidenced in Figure 14.7, the Young's modulus (a) and the tensile strength (b) of polypropylene composites reinforced by flax fibers are higher than the values obtained for the polypropylene matrix (Ausias et al., 2013), highlighting the strengthening mechanisms induced by plant fibers. The values of elastic moduli increase with fiber volume fraction. This trend has also been confirmed by John and Anandjiwala (2009), Baiardo et al. (2004), and Joseph et al. (1999). In contrast, the tensile strength increases with the fiber until it reaches a maximum value at a mass fraction of about 32% (threshold on Figure 14.7b). From this value, the tensile strength stabilizes, due to the concentration of strain within the matrix between fibers. Arbelaiz et al. (2006) also show that resistance increases with the fiber up to 30–40% by mass.

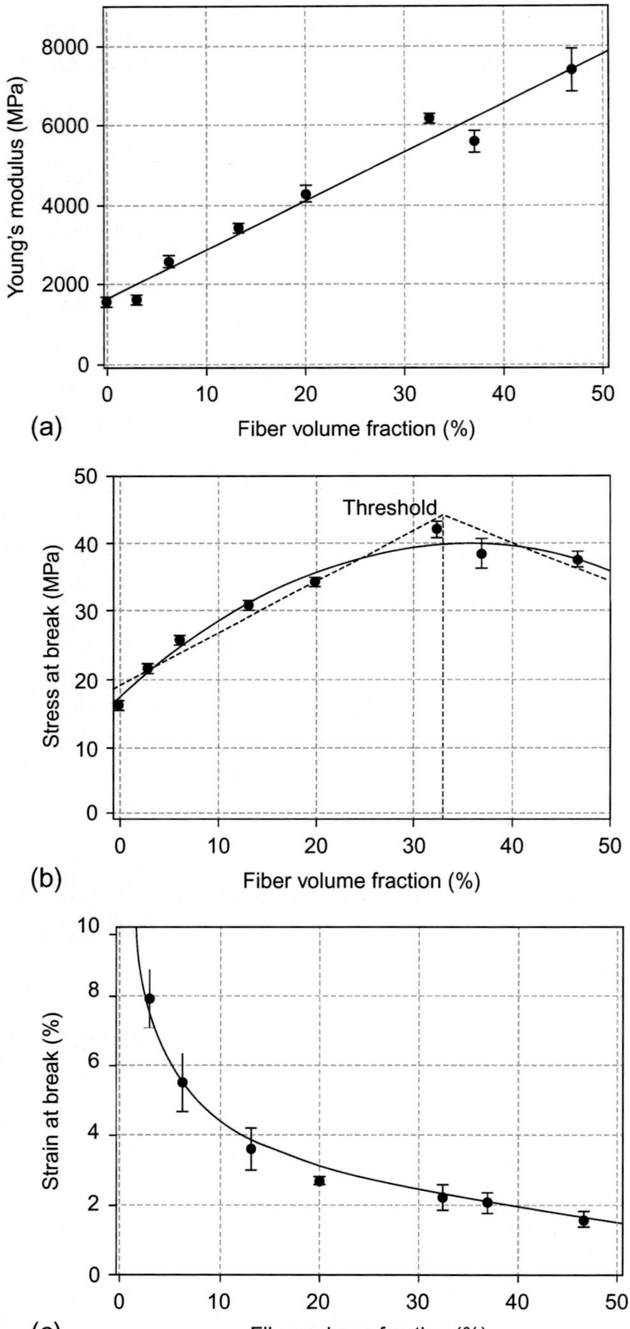

Figure 14.7 Evolution of (a) the Young's modulus, (b) stress, and (c) strain at break of PP–flax composites with fiber volume fraction (Ausias et al., 2013).

With high fiber content, the value of resistance decreases linearly (Figure 14.7b), suggesting a drop in load transfer between matrix and fibers.

This behavior has also been shown by other authors for polypropylene composites reinforced with lignocellulosic fibers (Elsabbagh et al., 2009; Vlasveld et al., 2005; Arbelaiz et al., 2005). This trend can also be explained by a strong decrease of fiber aspect ratio for volume fraction above the threshold. Moreover, the elongation at break of PP–flax fiber composite decreases significantly compared to the value recorded for the pure matrix and stabilized from a mass rate of 30% (Figure 14.7c). This drop is generally explained by strain concentration between fibers.

The composite-reinforced polypropylene with 13.4% by volume of fiberglass has a modulus of elasticity of 6200 MPa (Ausias et al., 2013). The value of the polypropylene composite/flax being equivalent volume of 4260 MPa, the PP/glass is twice as efficient as the PP/flax. Properties of flax fiber-reinforced PP have been compared to a composite made with glass fiber for the same fiber weight fraction. Young's modulus of composite with glass fiber is more than twice that obtained with flax fiber. However, there are still the issues of fiber matrix bound and fiber orientation and dispersion, which will be discussed in the next section.

14.4.2 Impact of the fiber geometry on the composite performances

Bledzki et al. (2007) shows that the load transfer between fibers and matrix does not depend only on the intrinsic properties of fiber and matrix but also on parameters of geometry (aspect ratio) and arrangement of fibers in the matrix as the fiber distribution. In an injection process, the viscosity of the fluid is strongly linked to the microstructure induced by the flow conditions, affecting the physical properties of the final product (Djalili-Mohagddam and Toll, 2006). Thus, the mechanical properties can be explained by the dispersion and orientation of fibers in the polypropylene matrix.

These results show clearly the impact of the fiber volume fraction on the fiber morphology and on the composites' mechanical properties. A limit threshold could be defined around 32% volume (Ausias et al., 2013); from these values, the stress at break and the aspect ratio begin to decrease inducing a lower reinforcement. This fiber volume fraction could be assimilated to an optimal value for plant fiber composites.

As underlined by Figure 14.8, a threshold could be noticed into the aspect ratio above this fiber fraction value. The high quantity of fibers induces the formation of bundles and consequently a decrease into the average length/diameter. Consequently, the composite mechanical performances, and especially the stress at break, are affected. This result will be highlighted in the following section dedicated to the impact of the processing method on the composite performances.

14.4.3 Influence of the process (injection molding and film stacking) on the PLLA–flax composite performances

Several manufacturing methods can be applied to biocomposites in order to obtain a wide range of mechanical properties. Injection molding is widely used in the

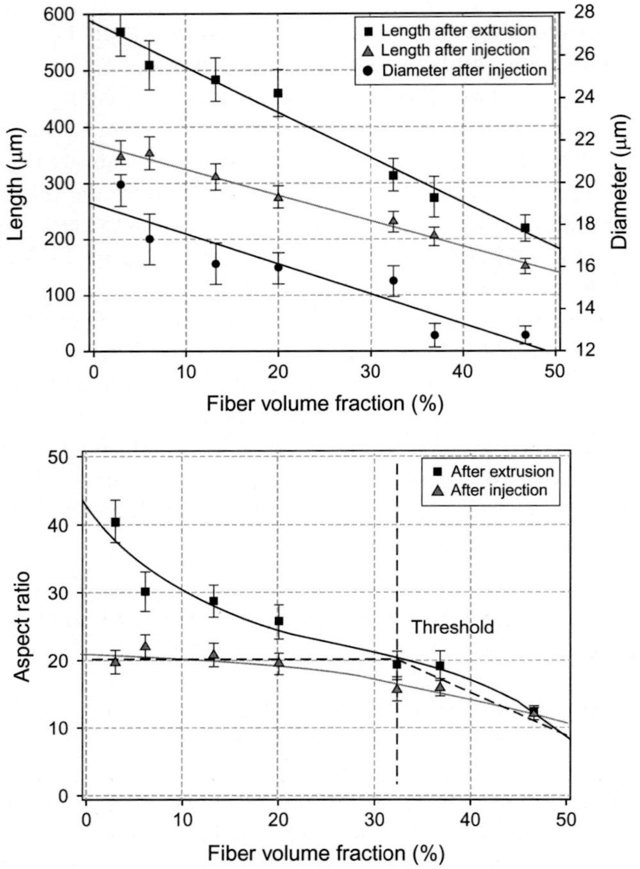

Figure 14.8 Evolution of the (a) length and diameter and (b) aspect ratio of fibers with fiber volume fraction after compounding and injection step (Ausias et al., 2013).

automotive industry due high production rate with complex geometry. Film stacking with heating plate is also a well-known process with higher properties. Vacuum film stacking is basically used for thermoset resin composites. However, some authors (Lystrup, 2006) used it to manufacture thermoplastics composite for wind turbine blades. The influence of the process is evaluated on a similar flax/PLLA biocomposite with randomly dispersed flax fiber (volume fraction of 26.5%) (Table 14.5) (Le Duigou et al., 2009).

Mechanical properties of injected biocomposites are lower both for Young's modulus and tensile strength. Film stacking assisted by vacuum biocomposites exhibit intermediate properties, while film stacking with heating plate appear to be the most competitive manufacturing method. Indeed, unlike film stacking with heating plate, which requires only one manufacturing step, injection molding should be achieved after the compounding process responsible for reduction of fiber length from 4 to 2.3 mm (Table 14.6), fiber diameters, and aspect ratio.

Table 14.5 Mechanical properties of flax/PLLA biocomposites as a function of processing method (injection molding and film stacking) (Le Duigou et al., 2009)

Material	Process	V_f (%)	Young modulus (MPa)	Tensile strength (MPa)	Strain at break (%)	References
PLLA/flax	Film-stacking heating plate	26.5	9525 ± 122	89 ± 2	1.2 ± 0.1	Le Duigou et al. (2009)
	Film-stacking vacuum	26.5	8042 ± 653	56.1 ± 2.8	1.4 ± 0.1	Le Duigou et al. (2010, 2011)
	Injection molding	26.5	7320 ± 380	53.1 ± 2.8	1.2 ± 0.1	Le Duigou et al. (2008)
PLLA/flax	Film stacking heating plate	26.5	8856 ± 297	81.3 ± 2.8	1.2 ± 0.1	Bodros et al. (2007)

Table 14.6 **Geometry of reinforcement manufactured by injection, film stacking with heating plates, and vacuum (Le Duigou et al., 2009)**

Manufacturing		Volumic content (V_f) (%)	Length (μm)	Diameter (μm)	Aspect ratio (*L*/*D*)
Injection	Without extrusion	–	4064 ± 670	182 ± 76	25.8 ± 11.1
	After extrusion	26.5	2319 ± 1242	133 ± 60	19.8 ± 13.8
	After injection	26.5	334 ± 277	44 ± 42	9.1 ± 5.8
Film-stacking heating plates		26.5	10,000 ± 1000	24 ± 19	438.4 ± 52.6
Film-stacking vacuum		26.5			

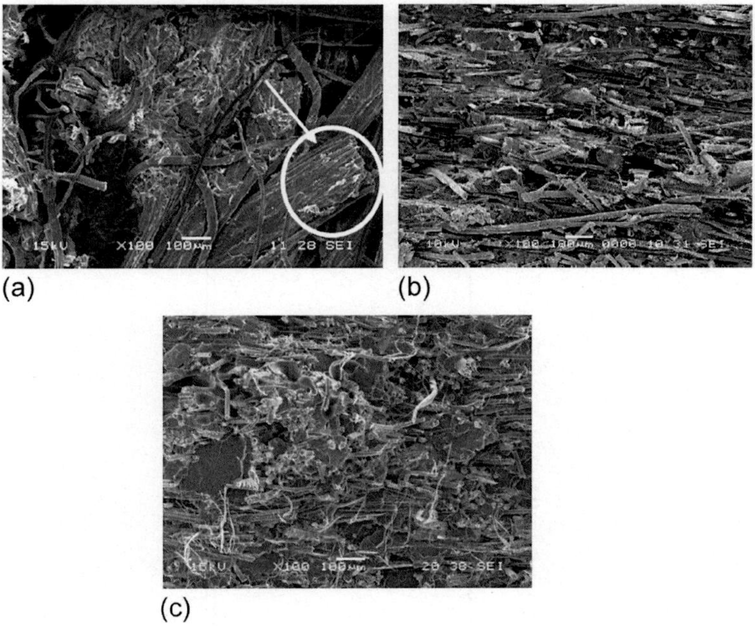

Figure 14.9 SEM micrograph of the fracture surface of PLLA–flax 20 wt.% manufactured by (a) injection, (b) film stacking with heating plates, and (c) vacuum (Le Duigou et al., 2009).

Film-stacked biocomposite exhibits higher aspect ratio as fibers do not undergo shearing (Table 14.6 and Figure 14.9) as they are based on a papermaking process to manufacture flax mat. As interfacial evaluation of bond strength between flax and PLA gives a critical fiber length around 1 mm (Le Duigou et al., 2010), mechanical properties between different processing routes could be first explained by a difference in fiber geometry and heterogeneity.

Besides mechanical degradation, plant fiber reinforcement can undergo thermal degradation. Film stacking with vacuum stand around 60 min against 8 min with heating plates for temperature set at 185 °C. Indeed, for vacuum process, the wetting process is mainly driven by capillarity forces that require low viscosity. Heating plate can generate enough mechanical stress that induces impregnation (Ouagne and Bréard, 2010). Processing time can explain the difference between performances of both film-stacking processes.

The injection step induces high shearing stress (screw rotation, pressure), which reduces again length and fiber diameters but bundles may be still present (Figure 14.9).

Polymer matrix can also be degraded during biocomposite manufacturing. For example, injected samples undergo high reduction of their molecular weight evidenced by GPC and DSC (Table 14.7) (Le Duigou et al., 2009). Thermal properties during first and second heating (irreversible degradation) are presented in Table 14.7.

Injected and film-stacked PLLA exhibit metastable crystallization process occurring during the heating step (Le Duigou et al., 2009). Adding flax fibers in injected samples involves cold crystallization as well as an increasing melting enthalpy due to degradation. In addition, molecular weight decreases by 45%.

Table 14.7 Thermal properties and molar weight of PLLA and flax/PLLA as a function of manufacturing method (Le Duigou et al., 2009)

		First cooling		Second heating				Molar weight
Manufacturing	T_g (°C)	T_c (°C)	ΔH_c (J g^{-1})	T_c (°C)	ΔH_c (J g^{-1})	T_m (°C)	ΔH_m (J g^{-1})	(M_w) (g mol^{-1})
PLLA injection molding	66.5	–	–	123.3	36.5	175.1	36.3	182,637
PLLA heating plate	65.4	–	–	120.9	33	172.3	27.6	195,000
Injection 30%	62.2	114.5	45.6	–	–	170.4	44	101,000
Film-stacking heating plate 30%	64.3	–	–	121.1	30	173	29.5	196,000
Film-stacking vacuum 30%	61.8	–	–	120.2	31	172	28	–

For film-stacked biocomposites, no crystallization appears at cooling while melting enthalpy does not vary so much. PLLA matrix within film-stacked biocomposites crystallizes at the second heating step; therefore, degradation is low. Injected biocomposites exhibit a crystallization enthalpy at cooling as well as melting enthalpy at heating higher than neat PLLA due to higher chain scission during extrusion and injection. Indeed, viscosity of PLLA/flax is 100 times higher than neat PLLA ($\eta_{0PLLA}=510\,Pa\,s$; $\eta_{0Biocomposite}=52{,}300\,Pa\,s$) (Le Duigou et al., 2009).

For structural application, film stacking must be used to minimize degradation. Using vacuum makes it possible to manufacture large parts but degradation and mechanical properties will be lower than using heating plate.

14.4.4 Impact of the process on the composite water uptake

There is not a strong influence of the manufacturing route on water diffusion kinetics, though the injected samples do not show a clear saturation plateau, so a diffusion coefficient has not been estimated for these. The film-stacked samples, with similar overall fiber content to the injected samples, pick up weight and stabilize more quickly than the injected samples, but this may simply reflect the fact that their edges were machined, so there are exposed fibers along the sides of the specimen (Le Duigou et al., 2009). Water ingress is then possible by capillarity. The injected samples are completely covered by a thin layer of PLLA, which acts as a barrier to water and protects the fibers.

As for the unreinforced PLLA (Le Duigou et al., 2009), water plasticizes the biocomposites. There is an influence of the manufacturing method; the injected specimens tend to retain their linear elastic behavior, while the film-stacked specimens show a more significant increase in strain during aging.

Figure 14.10 (Le Duigou et al., 2009) shows how maximum stress of the two biocomposites evolves with immersion time and temperature. In a similar way to modulus, there is a large drop at short times followed by a stabilization. The strength of

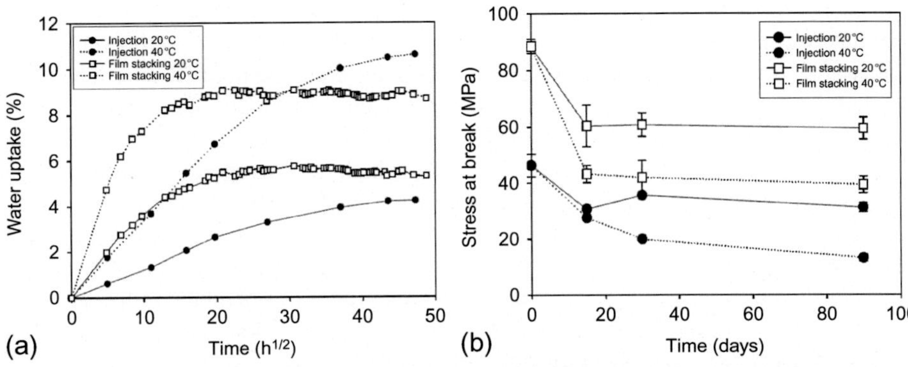

Figure 14.10 (a) Absorption behavior of injected and film-stacked samples at 20 and 40 °C. (b) Evolution of biocomposite modulus and stress at break versus immersion time (Le Duigou et al., 2009).

the film-stacked specimens remains superior to that of the injected specimens in all cases, due to the longer reinforcement length and higher molecular weight. Changes in properties can be related to water uptake, but for the film-stacked samples the changes occur too fast here and cannot be plotted.

14.5 Impact of the process on the plant fiber composite microstructure

14.5.1 *Injection molding: skin-core effect*

If the microstructure and the orientation of fiberglass-reinforced injected composites are well known, it is not the case for biocomposites. Indeed, the orientation of short glass fibers in a center-gated disk and the existence of a skin-core effect on injected parts have been demonstrated in numerous publications. By experimental work on composite disks, Bay and Tucker (1992a,b) showed that the injected part can be divided into several layers of preferential orientation of fibers, symmetrical about the mid-thickness of the sample. They noticed a layer of biaxially oriented fibers from the edge toward the center, a layer of radially oriented fibers in the flow, a transition layer with random orientation and, finally, a core layer where the fibers are oriented perpendicular to the flow.

Some authors studied the orientation of vegetal fibers in thermoplastics composites in more usual ways. Bledzki and Faruk (2006) studied the elaboration of injection-molded microcellular wood fiber–PP composites. They highlighted a three-layer sandwich structure with various fiber orientations similar to a skin-core effect. Correa et al. (1996) investigated the anisotropy of injected cellulose fibers composites and evidenced by microscopic observations that fiber alignment varies, depending on the position inside the mold, influenced by the flow induced at the injection point. A paper by Vilaseca et al. (2010) is dedicated to the evaluation of the structure and mechanical properties of injected abaca/PP composites. They proceeded to orientation measurements through optical microphotograph analysis, and showed the existence of a skin-core effect with perpendicular orientation at the flow direction of the injected part.

The PP–flax composite (Bourmaud et al., 2013a) shows a good dispersion with the presence of many elementary fibers. However, despite two process cycles (an extrusion and an injection), the presence of a few bundles could be noticed. These bundles are due to the flax stems morphology. In the stems, the elementary fibers are located at the periphery, and assembled by bundles of several dozen. These elementary fibers are linked by pectic cements (Morvan et al., 2003), which are damaged during the retting phase in order to separate the fibers; this separation process could be carried on during the extrusion or injection steps thanks to the important shear rate suffered by fibers (Bourmaud and Baley, 2007). The polished sections here do not reveal 3-D effects in the fiber distribution, which is particularly interesting for out-of-plane properties.

Figure 14.11 presents the scan and micrograph of (a) the skin and (b) the core of a PP/flax 20 wt.% fiber sample (Bourmaud et al., 2013a).

In the case of PP/flax fibers (Bourmaud et al., 2013a), the skin-core effect is present, and a good alignment of the flax fiber with the flow direction can be noticed in the

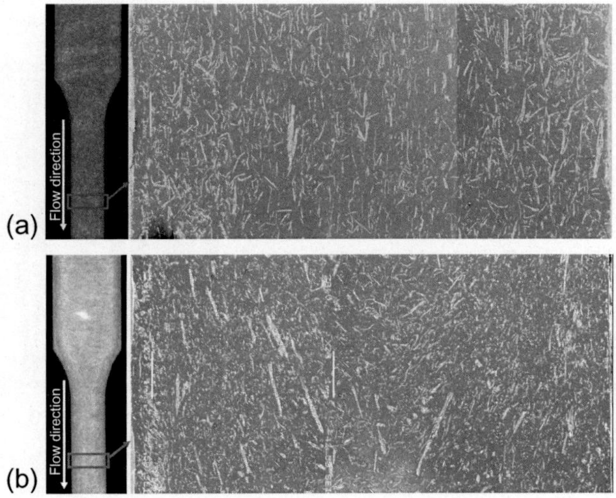

Figure 14.11 Scan and micrograph of (a) skin and (b) core of PP/flax sample (scale bar 1 mm) (Bourmaud et al., 2013a).

skin layer. The skin layer seems to be larger than those of PP/glass fiber (Bourmaud et al., 2013a), and the orientation is less obvious than for the PP/glass fibers. The difference in longitudinal stiffness of flax fibers (54 GPa) and glass (72 GPa) may partly explain this observation; another explanation could be fiber morphology and the presence of bundles, as the fibers with high aspect ratio are prone to bending. In the skin layer, the fibers seem to be less aligned in the center of the specimen. As already noticed in PP/glass fibers, this phenomenon is less conspicuous on the edge probably because of the decrease in speed in these areas.

In the core of the specimen, the majority of fibers seem to appear as circles or ellipses, which highlights a low orientation. Some fibers exhibit perpendicular or in flow orientation, which shows the presence of a divergent flow (Vincent et al., 2005) in the core of the sample. As observed in PP/glass fibers (Bourmaud et al., 2013a), in the core of the sample, a progressive orientation from the edges to the center of the specimen could be noticed; this phenomenon could be correlated with the speed profile, close to zero at the edge, and maximal in the center of the part.

14.5.2 Impact of individualization on plant fiber composites mechanical performances

Biocomposites could exhibit fiber bundles due to the plant structure, the fiber loading, or the processing method. Figure 14.12 shows the stress–strain curves of the (a) PP/flax core and (b) skin specimen (with clusters) and the corresponding scans that show the fiber bundles (Bourmaud et al., 2013a).

Tensile experiments have been carried out, and the characterized specimens have been scanned before and after the tensile tests. Thanks to the scan micrographs, we can have an interesting representation of the fiber dispersion in the composites that

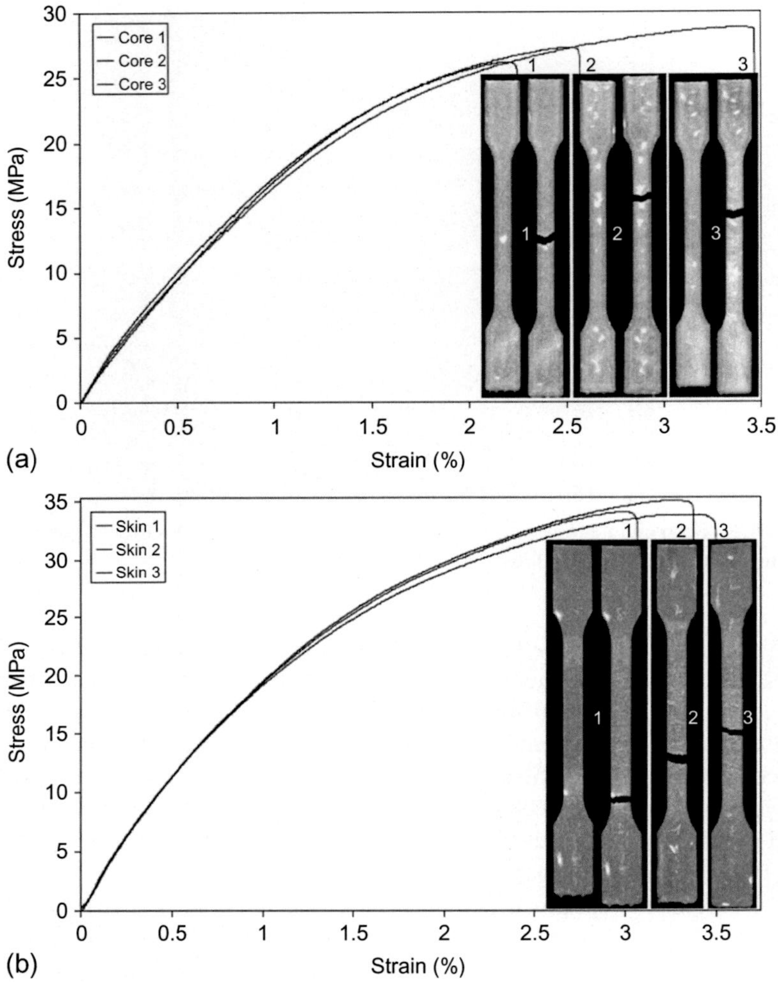

Figure 14.12 Stress–strain curves of the (a) PP/flax core and (b) skin specimen (with clusters) and the corresponding scans (Bourmaud et al., 2013a).

highlight the bundles localization (Figure 14.12). We can notice that the core samples feature a lot of bundles in the useful cross-section of the specimens, and that the stress–strain curves show that the more important their size is, the earlier the composite break occurs. Thus, samples 1 and 2 (Figure 14.12a), which have more important fiber bundles than sample 3, reveal premature breaking.

Contrary to the core samples (Figure 14.12b), scans of the skin show a lower quantity of bundles; the fibers seem to be better dispersed, probably due to the more important shear rate at the wall of the part. Among the three scanned samples, only one of them (number 1) has a bundle in the effective cross-section and this sample has an elongation of 3.0%, as opposed to 3.4% and 3.5% for the others.

However, whether the sample is taken from core or skin, no significant variation is induced on the Young's modulus values, regardless of the presence or absence of fiber bundles. Thus, the presence of bundles, which could appear as a major defect, has no influence on the stiffness of the composite. On the other hand, a preliminary breaking has an influence on the other mechanical properties; the more precocious the rupture is, the lower the strength and strain at break. As demonstrated by Thomason and Vlug (1996, 1997; Thomason et al., 1996), who studied the properties of glass fiber-reinforced PP, from 40% (w/w), the improvement of mechanical properties is reduced, probably because of fiber packing problems, which induce an increase in void content and out-of-plane fiber orientation.

14.5.3 Link between fiber bundles and damage initiation

The impact of the fiber bundles during a tensile experiment could be highlighted by *in situ* SEM observations. Thus, Figure 14.13 (Bourmaud et al., 2013a) presents the SEM micrograph of a PP/flax core sample under loading, and shows the existence of many flow or transverse-orientated elementary fibers, as well as perpendicular to the flow-orientated bundles.

As explained previously, these bundles come from the structure of the flax stem in which they are assembled in bundles of several tens of fibers by pectic cements (Morvan et al., 2003). During the tensile solicitation, we can notice that microcracks appear, mainly through the transverse-orientated fiber bundles. As evidenced by Baley et al. (2006) in a previous study, due to the presence of the fiber bundles, the flax fiber composite becomes weaker and more sensitive to breaking. Figure 14.13 shows that the fibers or bundles that are perpendicular to the flow-orientated bundles create preferential ways for the propagation of cracks.

In the breaking areas that include bundles, we can observe that the cracks induce a dislocation of the bundle between the flax fibers (indicated by arrows on Figure 14.13)

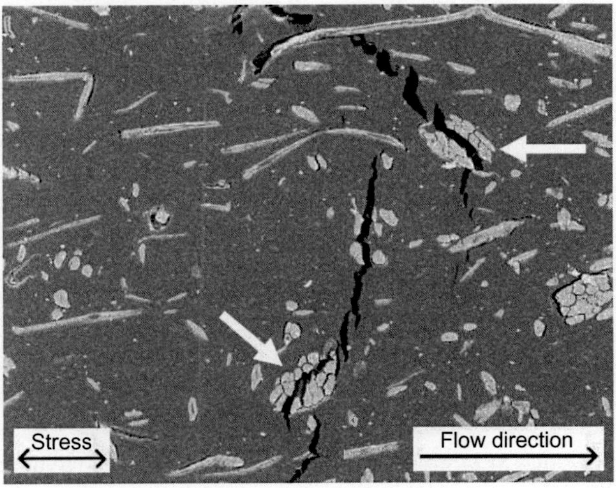

Figure 14.13 SEM micrograph of a core PP/flax sample under loading (Bourmaud et al., 2013a).

(Bourmaud et al., 2013a). The adhesion between the matrix and the fiber doesn't seem to be affected, which shows that the middle lamella interface is weaker than that of the fiber matrix. These observations confirm the importance of the fiber dispersion in order to obtain a composite material that has the best possible properties and the minimal quantity of bundles.

14.5.4 Interest of the fiber retting on plant fiber composites properties

To reduce the amount of bundles, it is necessary to carry out an efficient retting (Bourmaud et al., 2010), and to use a suitable process for the elaboration of compounds (Shon and White, 1999) in order to divide the bundles without cell walls alteration.

The influence of retting on the properties of single flax fibers and short flax fibers/polypropylene composites has been studied in a recent work (Martin et al., 2013). The mechanical properties of short fibers-reinforced composites were increased with a high degree of retting. Intrinsic tensile properties, surface, and aspect ratio of fibers play a key role. A high degree of retting enables easier splitting of the fiber bundles during fiber/matrix extrusion and injection molding. The division of composite reinforcements is more important for a high degree of retting, which results in smaller fiber bundle diameters and a higher aspect ratio.

In another work dedicated to the study of epoxy–flax composites, the influence of elementary fibers properties, their individualization and the fiber/matrix bond strength on the tensile behavior of associated unidirectional composites were studied (Coroller et al., 2013). Morphological analyses highlighted the importance of a hackling step for fiber dispersion; this process reduces the number of bundles in the final composite (Figure 14.14). The analysis of the UD composites' mechanical properties confirms

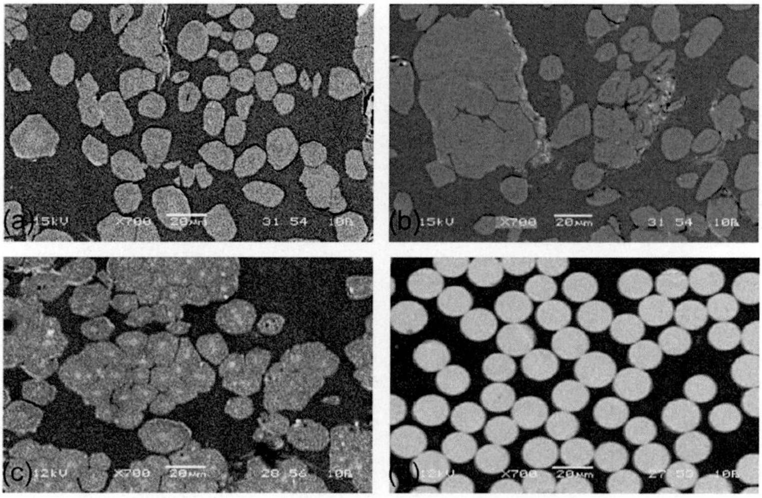

Figure 14.14 Composites SEM images: (a) Hermes; (b) Andrea; (c) Marylin; and (d) Glass (Coroller et al., 2013).

this point; although the Young's modulus values are similar to the mixture rule prediction whatever the fiber dispersion, this is not the case for the tensile strength, which is mainly influenced by the fiber dispersion. There is a major stake in elaborating composites with great fiber individualization, especially for the materials strength. In this way, it could be interesting to use optimized performs, enabling easier handling, as well as a controlled orientation and loading rate.

14.6 Conclusion

In this chapter, we were interested in the mechanical properties of composite materials reinforced by plant fibers. The elaboration of these materials must be done by taking account of specificities of the plant cell walls. Indeed, their functions in nature are multiple and we have an interest in selecting fibers that play a structural role. For example, a flax stem is reinforced by fibers that are support tissues and assembled in cohesive bundles. Consequently, their extraction requires a first step of retting in order to facilitate it and to individualize the bundles to the maximum.

The vegetal fibers have a microstructure similar to a composite material, which it is necessary to control and to understand. The different walls and layers are reinforced by cellulose microfibrils. Thus, their mechanical properties and their behavior during a solicitation depend on various parameters such as the rate on cellulose, the microfibril angle, the biochemical composition, the organization within the walls, or the fraction of absorbed water. Thus, the mechanical properties of plant fibers are not comparable and even for a given plant, major differences in behaviors can exist.

Various parameters could influence the properties of vegetal fibers such as the variety, the pedoclimatic conditions, agricultural practices (work of the soil, treatments, plant development), the retting, and the scutching conditions. During a transformation cycle with a polymeric matrix, the fibers are subjected to various solicitations by thermohydromechanic couplings. The water contained in the plants, which is a cell wall plasticizer, is greatly influenced by temperature and particularly sensitive to the process. Moreover, in the case of extrusion and of injection, dimensions of fibers will evolve while decreasing length but also by diameter thanks to the fibers bundles individualization. Thus, processing induces at the same time as an evolution of properties but also of the geometry of the reinforcements. By local techniques such as nanoindentation, it is possible to follow the evolution of the properties of the walls after a cycle of processing. We can note that after a thermal cycle, the parameters of surface of fiber are also modified. The plant fibers-reinforced composite materials properties are influenced by the stages of transformation but also by the fiber volume fraction, the aspect ratio, the temperature, or by time. Thus, for two similar materials, the comparison of the properties obtained between two processing methods illustrates the importance of the process parameters. If the behavior in temperature of the vegetable walls is limited, we can note the conjugated importance of time (necessary to the fibers impregnation) and of the temperature.

The orientation, the distribution of fibers within the matrix, their aspect ratio, and their individualization are parameters to be taken into account to analyze the laminate properties and the mechanisms of initiation and development of the damages. Moreover, retting as well as scutching are very important stages because they influence the cohesion of the beams and, thus, the individualization of fibers within the polymeric matrices.

The optimization of the performances of the biocomposites imposes an analysis and a knowledge of used fibers as well as an adaptation of the processing parameters in order to respect the integrity of the vegetable walls as well as possible.

References

AFNOR, 2009. Determination of tensile properties of elementary flax fibres. XPT 25-501-2.

Alix, S., Philippe, E., Morvan, C., Baley, C., 2008. Putative role of pectins in the tensile properties of flax fibres: a comparison between linseed and flax fibers varieties. In: Pectins and Pectinases. Wageningen, The Netherlands.

Altaner, C.M., Jarvis, M.C., 2008. Modelling polymer interactions of the 'molecular Velcro' type in wood under mechanical stress. J. Theoret. Biol. 253, 434–445.

Amada, S., Ichikawa, Y., Munekata, T., Nagase, Y., Shimizu, H., 1997. Fiber texture and mechanical graded structure of bamboo. Compos. B Eng. 28, 13–20.

Amada, S., Untao, S., 2001. Fracture properties of bamboo. Compos. Part B Eng. 32, 451–459.

Andersons, J., Joffe, R., 2011. Estimation of the tensile strength of an oriented flax fiber-reinforced polymer composite. Compos. A Appl. Sci. Manufactur. 42, 1229–1235. http://dx.doi.org/10.1016/j.compositesa.2011.05.005.

Angelini, L.G., Lazzeri, A., Levita, G., Fontanelli, D., Bozzi, C., 2000. Ramie (Boehmeria nivea (L.) Gaud.) and Spanish Broom (Spartium junceum L.) fibres for composite materials: agronomical aspects, morphology and mechanical properties. Ind. Crops Prod. 11, 145–161.

Antich, P., Vazquez, A., Mondragon, I., Bernal, C., 2006. Mechanical behavior of high impact polystyrene reinforced with short sisal fibers. Compos. A Appl. Sci. Manufactur. 37, 139–150.

Arbelaiz, A., Fernández, B., Ramos, J.A., Retegi, A., Llano-Ponte, R., Mondragon, I., 2005. Mechanical properties of short flax fibre bundle/polypropylene composites: influence of matrix/fibre modification, fibre content, water uptake and recycling. Compos. Sci. Technol. 65, 1582–1592.

Arbelaiz, A., Fernández, B., Valea, A., Mondragon, I., 2006. Mechanical properties of short flax fibre bundle/poly(ε-caprolactone) composites: influence of matrix modification and fibre content. Carbohyd. Polym. 64, 224–232.

Ausias, G., Bourmaud, A., Coroller, G., Baley, C., 2013. Study of the fibre morphology stability in polypropylene–flax composites. Polym. Degrad. Stab. 98, 1216–1224. http://dx.doi.org/10.1016/j.polymdegradstab.2013.03.006.

Baiardo, M., Zini, E., Scandola, M., 2004. Flax fibre–polyester composites. Compos. A Appl. Sci. Manufactur. 35, 703–710.

Baley, C., 2002. Analysis of the flax fibres tensile behaviour and analysis of the tensile stiffness increase. Compos. A Appl. Sci. Manufactur. 33, 939–948.

Baley, C., Bourmaud, A., 2014. Average tensile properties of French elementary flax fibers. Mater. Lett. 122, 159–161. http://dx.doi.org/10.1016/j.matlet.2014.02.030.

Baley, C., Grohens, Y., Pillin, I., 2005a. Etat de l'art sur les matériaux composites biodégradables. Rev. Compos. Matér. Av. 14, 135–166.

Baley, C., Morvan, C., Grohens, Y., 2005b. Influence of the absorbed water on the tensile strength of flax fibers. Macromol. Sympos. 222, 195–202.

Baley, C., Perrot, Y., Busnel, F., Guezenoc, H., Davies, P., 2006. Transverse tensile behaviour of unidirectional plies reinforced with flax fibres. Mater. Lett. 60, 2984–2987.

Baley, C., Le Duigou, A., Bourmaud, A., Davies, P., 2012. Influence of drying on the mechanical behaviour of flax fibres and their unidirectional composites. Compos. A Appl. Sci. Manufactur. 43, 1226–1233. http://dx.doi.org/10.1016/j.compositesa.2012.03.005.

Basu, S., Shivhare, U., Muley, S., 2011. Moisture adsorption isotherms and glass transition temperature of pectin. J. Food Sci. Technol. 10, 1–5.

Batra, S.K., 1998. Handbook of Fibre Science and Technology. Fibre Chemistry, vol. 4. Marcel Dekker, New York.

Bay, R.S., Tucker, C.L., 1992a. Fiber orientation in simple injection moldings. Part I: Theory and numerical methods. Polym. Compos. 13, 317–331. http://dx.doi.org/10.1002/pc.750130409.

Bay, R.S., Tucker, C.L., 1992b. Fiber orientation in simple injection moldings. Part II: Experimental results. Polym. Compos. 13, 332–341. http://dx.doi.org/10.1002/pc.750130410.

Beaugrand, J., Berzin, F., 2013. Lignocellulosic fiber reinforced composites: influence of compounding conditions on defibrization and mechanical properties. J. Appl. Polym. Sci. 128, 1227–1238. http://dx.doi.org/10.1002/app.38468.

Beckermann, G.W., Pickering, K.L., 2008. Engineering and evaluation of hemp fibre reinforced polypropylene composites: Fibre treatment and matrix modification. Compos. A Appl. Sci. Manuf. 39, 979–988.

Bert, F., 2013. In: Arvalis (Ed.), Lin Fibre: Culture et Transformation. Institut du Végétal, Paris.

Bisanda, E.T.N., Ansell, M.P., 1992. Properties of sisal/CNSL composites. J. Mater. Sci. 27, 1690–1700.

Bismark, A., Aranberri-Askargorta, I., Springer, J., 2002. Surface characterization of flax, hemp and cellulose fibers – Surface properties and the water uptake behaviour. Polym. Compos. 23, 872–894.

Bledzki, A.K., Faruk, O., 2006. Injection moulded microcellular wood fibre–polypropylene composites. Compos. A Appl. Sci. Manufactur. 37, 1358–1367.

Bledzki, A.K., Mamun, A.A., Faruk, O., 2007. Abaca fibre reinforced PP composites and comparison with jute and flax fibre PP composites. Expr. Polym. Lett. 11, 755–762.

Bodros, E., Baley, C., 2008. Study of the tensile properties of stinging nettle fibres (Urtica dioica). Mater. Lett. 62, 2143–2145.

Bodros, E., Pillin, I., Montrelay, N., Baley, C., 2007. Could biopolymers reinforced by randomly scattered flax fibre be used in structural applications? Compos. Sci. Technol. 67, 462–470.

Bos, H.L., 2004. The potential of flax fibres as reinforcement for composite materials. Thesis report, Eindhoven.

Bos, H., Van Den Oever, M., Peters, O., 2002. Tensile and compressive properties of flax fibres for natural fibre reinforced composites. J. Mater. Sci. 37, 1683–1692.

Bos, H.L., Molenveld, K., Teunissen, W., van Wingerde, A.M., van Delft, D.R.V., 2004. Compressive behaviour of unidirectional flax fibre reinforced composites. J. Mater. Sci. 39, 2159–2168. http://dx.doi.org/10.1023/b:jmsc.0000017779.08041.49.

Bourmaud, A., 2011. Contribution à l'étude multi échelles de fibres végétales et de biocomposites. Université de Bretagne Sud, Paris.

Bourmaud, A., Baley, C., 2007. Investigations on the recycling of hemp and sisal fibre reinforced polypropylene composites. Polym. Degrad. Stab. 92, 1034–1045.

Bourmaud, A., Baley, C., 2009. Rigidity analysis of polypropylene/vegetal fibre composites after recycling. Polym. Degrad. Stab. 94, 297–305.

Bourmaud, A., Baley, C., 2010. Effects of thermo mechanical processing on the mechanical properties of biocomposite flax fibers evaluated by nanoindentation. Polym. Degrad. Stab. 95, 1488–1494.

Bourmaud, A., Pimbert, S., 2008. Investigations on mechanical properties of poly(propylene) and poly (lactic acid) reinforced by miscanthus fibers. Compos. A Appl. Sci. Manufactur. 39, 1444–1454.

Bourmaud, A., Morvan, C., Baley, C., 2010. Importance of fiber preparation to optimize the surface and mechanical properties of unitary flax fiber. Ind. Crops Prod. 32, 662–667.

Bourmaud, A., Ausias, G., Lebrun, G., Tachon, M.L., Baley, C., 2013a. Observation of the structure of a composite polypropylene/flax and damage mechanisms under stress. Ind. Crops Prod. 43, 225–236.

Bourmaud, A., Morvan, C., Bouali, A., Placet, V., Perré, P., Baley, C., 2013b. Relationships between micro-fibrillar angle, mechanical properties and biochemical composition of flax fibers. Ind. Crops Prod. 44, 343–351. http://dx.doi.org/10.1016/j.indcrop.2012.11.031.

Brito, J., Silva, F., Leao, M., Almeida, G., 2008. Chemical composition changes in eucalyptus and pinus woods submitted to heat treatment. Bioresour. Technol. 99, 8545–8548.

Buschle-Diller, G., Zeronian, S.H., 1993. Photodegradation of oxidized cotton cellulose. J. Appl. Polym. Sci. 47, 1319–1328. http://dx.doi.org/10.1002/app.1993.070470802.

Burgert, I., 2006. Exploring the micromechanical design of plant cell walls. Am. J. Bot. 93, 1391–1401. http://dx.doi.org/10.3732/ajb.93.10.1391.

Chand, N., Tiwary, R.K., Rohatgi, P.K., 1988. Bibliography Resource structure properties of natural cellulosic fibres — an annotated bibliography. J. Mater. Sci. 23, 381–387. http://dx.doi.org/10.1007/bf01174659.

Charlet, K., Baley, C., Morvan, C., Jernot, J.P., Gomina, M., Bréard, J., 2007. Characteristics of Hermès flax fibres as a function of their location in the stem and properties of the derived unidirectional composites. Compos. A Appl. Sci. Manufactur. 38, 1912–1921.

Charlet, K., Eve, S., Jernot, J.P., Gomina, M., Breard, J., 2009a. Tensile deformation of a flax fiber. Procedia Eng. 1, 233–236.

Charlet, K., Jernot, J., Gomina, M., Bréard, J., Morvan, C., Baley, C., 2009b. Influence of an Agatha flax fibre location in a stem on its mechanical, chemical and morphological properties. Compos. Sci. Technol. 69, 1399–1403.

Charlet, K., Eve, S., Jernot, J.P., Gomina, M., Breard, J., 2009c. Tensile deformation of a flax fiber. Procedia Eng. 1, 233–236.

Charlet, K., Jernot, J.P., Eve, S., Gomina, M., Bréard, J., 2010. Multi-scale morphological characterisation of flax: from the stem to the fibrils. Carbohyd. Polym. 82, 54–61.

Christiansen, A., 1990. How overdrying reduces its bonding to phenol formaldehyde adhesives: a critical review of the literature. Part I. Physical responses. Wood Fiber Sci. 22, 441–449.

Christiansen, A., 1991. How overdrying reduces its bonding to phenol formaldehyde adhesives: a critical review of the literature. Part II. Chemical reactions. Wood Fiber Sci. 23, 69–84.

Cichocki Jr., F.R., Thomason, J.L., 2002. Thermoelastic anisotropy of a natural fiber. Compos. Sci. Technol. 62, 669–678.

Coroller, G., Lefeuvre, A., Le Duigou, A., Bourmaud, A., Ausias, G., Gaudry, T., Baley, C., 2013. Effect of flax fibres individualisation on tensile failure of flax/epoxy unidirectional composite. Compos. A Appl. Sci. Manufactur. 51, 62–70. http://dx.doi.org/10.1016/j.compositesa.2013.03.018.

Correa, R.A., Nunes, R.C.R., Filho, W.Z.F., 1996. Effect of injections on the orientation of short fibre composites. An optical microscopic analysis. Polym. Test. 15, 467–475.

Cosgrove, D.J., 2005. Growth of the plant cell wall. Nat. Rev. Mol. Cell Biol. 6, 850–861. http://www.nature.com/nrm/journal/v6/n11/suppinfo/nrm1746_S1.html.

Davies, G.C., Bruce, D.M., 1998. Effect of environmental relative humidity and damage on the tensile properties of flax and nettle fibers. Text. Res. J. 68, 623–629.

Djalili-Mohagddam, M., Toll, S., 2006. Fibre suspension rheology: effect of concentration, aspect ratio and fibre size. Rheol. Acta 45, 315–320.

Duval, A., Bourmaud, A., Augier, L., Baley, C., 2011. Influence of the sampling area of the stem on the mechanical properties of hemp fibers. Mater. Lett. 65, 797–800.

Elsabbagh, A., Steuernagel, L., Ziegmann, G., 2009. Effect of fiber/matrix chemical modification on the mechanical properties and water absorption of extruded flax/polypropylene composite. J. Appl. Polym. Sci. 111, 2279–2289.

Fratzl, P., Burgert, I., Gupta, H.S., 2004. On the role of interface polymers for the mechanics of natural polymeric composites. Phys. Chem. Chem. Phys. 6, 5575–5579.

Gassan, J., Bledzki, A., 2001. Thermal degradation of flax and jute fibers. J. Appl. Polym. Sci. 82, 1417–1422.

Ganster, J., Fink, H., 1999. Physical constants of cellulose. In: Grulke, E.A. (Ed.), Polymer Handbook. Wiley, New York.

Gérardin, P., Petric, M., Petrissans, M., Lambert, J., Ehrhrardt, J.J., 2007. Evolution of wood surface free energy after heat treatment. Polym. Degrad. Stab. 92, 653–657.

Gindl, W., Schöberl, T., 2004. The significance of the elastic modulus of wood cell walls obtained from nanoindentation measurements. Compos. A Appl. Sci. Manufactur. 35, 1345–1349.

Gindl, W., Reifferscheid, M., Adusumalli, R.-B., Weber, H., Röder, T., Sixta, H., Schöberl, T., 2008. Anisotropy of the modulus of elasticity in regenerated cellulose fibres related to molecular orientation. Polymer 49, 792–799.

Girault, R., Bert, F., Rihouey, C., Jauneau, A., Morvan, C., Jarvis, M., 1997. Galactans and cellulose in flax fibres: putative contributions to the tensile strength. Int. J. Biol. Macromol. 21, 179–188.

Goda, K., Sreekala, M.S., Gomes, A., Kaji, T., Ohgi, J., 2006. Improvement of plant based natural fibers for toughening green composites–Effect of load application during mercerization of ramie fibers. Compos. A Appl. Sci. Manuf. 37, 2213–2220.

Gorshkova, T., Morvan, C., 2006. Secondary cell-wall assembly in flax phloem fibres: role of galactans. Planta 223, 149–158.

Gorshkova, T.A., Salnikov, V.V., Pogodina, N.M., Chemikosova, S.B., Yablokova, E.V., Ulanov, A.V., et al., 2000. Composition and distribution of cell wall phenolic compounds in flax (*Linum usitatissimum* L.) stem tissues. Ann. Bot. 85, 477–486.

Gorshkova, T.A., Wyatt, S.E., Salnikov, V.V., Gibeaut, D.M., Ibragimov, M.R., Lozovaya, V.V., Carpita, N.C., 1996. Cell-wall polysaccharides of developing flax plants. Plant Physiol. 110, 721–729. http://dx.doi.org/10.1104/pp. 110.3.721.

Gourier, C., Le Duigou, A., Bourmaud, A., Baley, C., 2014. Mechanical analysis of elementary flax fibre tensile properties after different thermal cycles. Compos. A Appl. Sci. Manufactur 64, 159–166. http://dx.doi.org/10.1016/j.compositesa.2014.05.006.

Hearle, J.W.S., 1963. The fine structure of fibers and crystalline polymers. III. Interpretation of the mechanical properties of fibers. J. Appl. Polym. Sci. 7, 1207–1223.

His, I., Andème-Onzighi, C., Morvan, C., Driouich, A., 2001. Microscopic studies on mature flax fibers embedded in LR white: immunogold localization of cell wall matrix polysaccharides. J. Histochem. Cytochem. 49, 1525–1535.

Ho, T.N., Ngo, A.D., 2005. Influence of temperatures and humidity on the tensile strength and stiffness of hemp and coir fibers. In: 5th Canadian International Composites Conference, 2005, Vancouver.

Hughes, M., Carpenter, J., Hill, C., 2007. Deformation and fracture behaviour of flax fibre reinforced thermosetting polymer matrix composites. J. Mater. Sci. 42, 2499–2511.

Idicula, M., Malhotra, S.K., Joseph, K., Thomas, S., 2005. Dynamic mechanical analysis of randomly oriented intimately mixed short banana/sisal hybrid fibre reinforced polyester composites. Compos. Sci. Technol. 65, 1077–1087.

Jain, S., Kumar, R., Jindal, U.C., 1992. Mechanical behaviour of bamboo and bamboo composite. J. Mater. Sci. 27, 4598–4604. http://dx.doi.org/10.1007/bf01165993.

John, M.J., Anandjiwala, R.D., 2008. Recent developments in chemical modification and characterization of natural fiber-reinforced composites. Polym. Compos. 29, 187–207. http://dx.doi.org/10.1002/pc.20461.

John, M.J., Anandjiwala, R.D., 2009. Chemical modification of flax reinforced polypropylene composites. Compos. A Appl. Sci. Manufactur. 40, 442–448.

Joseph, P.V., Joseph, K., Thomas, S., 1999. Effect of processing variables on the mechanical properties of sisal-fiber-reinforced polypropylene composites. Compos. Sci Technol. 59, 1625–1640.

Joshi, S.V., Drzal, L.T., Mohanty, A.K., Arora, S., 2004. Are natural fiber composites environmentally superior to glass fiber reinforced composites? Compos. A Appl. Sci. Manufactur. 35, 371–376.

Keckes, J., Burgert, I., Fruhmann, K., Muller, M., Kolln, K., Hamilton, M., 2003. Cell-wall recovery after irreversible deformation of wood. Nat. Mater. 2, 810–813.

Kelly, A., Tyson, W.R., 1965. Tensile properties of fibre-reinforced metals: copper/tungsten and copper/molybdenum. J. Mech. Phys. Solids 13, 329–338.

Khalil, H., Rozman, H., Ahmad, M., Ismail, H., 2000. Acetylated plant-fiber reinforced polyester composites: a study of mechanical, hygrothermal, and aging characteristics. Polym. Plast. Technol. Eng. 39, 757–781.

Kumar, R.P., Amma, M.L.G., Thomas, S., 1995. Short sisal fiber reinforced styrene-butadiene rubber composites. J. Appl. Polym. Sci. 58, 597–612. http://dx.doi.org/10.1002/app.1995.070580315.

Lapierre, I., Belmont, P., 1994. Bilan écologique du lin de la graine au tissu. Ecole Nationale Supérieure des Arts et Industries Textiles de Roubaix.

Le Duigou, A., Pillin, I., Bourmaud, A., Davies, P., Baley, C., 2008. Effect of recycling on mechanical behaviour of biocompostable flax/poly(L-lactide) composites. Compos. Part A 39, 1471–1478.

Le Duigou, A., Davies, P., Baley, C., 2009. Seawater ageing of flax/poly(lactic acid) biocomposites. Polym. Degrad. Stab. 94, 1151–1162.

Le Duigou, A., Deux, J.-M., Davies, P., Baley, C., 2010. PLLA/Flax Mat/Balsa Bio-Sandwich Manufacture and Mechanical Properties. Appl. Compos. Mater. 1–18. http://dx.doi.org/10.1007/s10443-010-9173-8.

Le Duigou, A., Davies, P., Baley, C., 2011. Environmental impact analysis of the production of flax fibres to be used as composite material reinforcement. J. Biobased Mater. Bioenerg. 5, 1–13.

Le Duigou, A., Bourmaud, A., Balnois, E., Davies, P., Baley, C., 2012. Improving the interfacial properties between flax fibres and PLLA by a water fibre treatment and drying cycle. Ind. Crops Prod. 39, 31–39. http://dx.doi.org/10.1016/j.indcrop.2012.02.001.

Le Troedec, M., Sedan, D., Peyratout, C., Bonnet, J.P., Smith, A., Guinebretiere, R., et al., 2008. Influence of various chemical treatments on the composition and structure of hemp fibres. Compos. A Appl. Sci. Manuf. 39, 514–522.

Lefeuvre, A., Bourmaud, A., Lebrun, L., Morvan, C., Baley, C., 2013. A study of the yearly reproducibility of flax fiber tensile properties. Ind. Crops Prod. 50, 400–407. http://dx.doi.org/10.1016/j.indcrop.2013.07.035.

Lefeuvre, A., Bourmaud, A., Morvan, C., Baley, C., 2014. Elementary flax fibre tensile properties: correlation between stress–strain behaviour and fibre composition. Ind. Crops Prod. 52, 762–769. http://dx.doi.org/10.1016/j.indcrop.2013.11.043.

Li, X., Tabil, L., Panigrahi, S., Tabil, L.-G., 2009. A study on flax fibre reinforced polyethylene biocomposites by injection moulding. Appl. Eng. Agric. 25, 525–531.

Liu, L., Yu, J., Cheng, L., Qu, W., 2009. Mechanical properties of poly(butylene succinate) (PBS) biocomposites reinforced with surface modified jute fibre. Compos. A Appl. Sci. Manuf. 40, 669–674. http://dx.doi.org/10.1016/j.compositesa.2009.03.002.

Lystrup, A., 2006. Vacuum consolidation of thermoplastic composites for wind turbine rotor blades. In: Liholt, H., Madsen, B., Andersen, T.L., Mikkelsen, L.P., Thygesen, A. (Eds.), 27th Riso International Symposium on Material Science: Polymer Composites Materials for Wind Power Turbines. Riso National Laboratory, Roskilde, Denmark.

Mano, B., Araújo, J.R., Spinacé, M.A.S., De Paoli, M.A., 2010. Polyolefin composites with curaua fibres: effect of the processing conditions on mechanical properties, morphology and fibres dimensions. Compos. Sci. Technol. 70, 29–35.

Martin, N., Mouret, N., Davies, P., Baley, C., 2013. Influence of the degree of retting of flax fibers on the tensile properties of single fibers and short fiber/polypropylene composites. Ind. Crops Prod. 49, 755–767. http://dx.doi.org/10.1016/j.indcrop.2013.06.012.

Martin, N., Davies, P., Baley, C., 2014. Comparison of the properties of scutched flax and flax tow for composite material reinforcement. Ind. Crops Prod. 61, 284–292.

Mohanty, A., Misra, M., Hinrichsen, G., 2000. Biofibres, biodegradable polymers and biocomposites: an overview. Macromol. Mater. Eng. 1, 1–24.

Morvan, C., 2010. Le Lin, Les Fibres Cellulosiques: Leurs Parois: Variabilité et Défauts. Ecole d'été, Lorient.

Morvan, C., Andème-Onzighi, C., Girault, R., Himmelsbach, D.S., Driouich, A., Akin, D.E., 2003. Building flax fibres: more than one brick in the walls. Plant Physiol. Biochem. 41, 935–944.

Nussbaum, R., 1999. Natural surface inactivation of Scots pine and Norway spruce evaluated by contact angle measurements. Holz als Roh Werkstoff 57, 419–424.

Ouagne, P., Bréard, J., 2010. Continuous transverse permeability of fibrous media. Compos. A: Appl. Sci. Manuf. 41 (1), 22–28.

Paiva Júnior, C.Z., de Carvalho, L.H., Fonseca, V.M., Monteiro, S.N., d'Almeida, J.R.M., 2004. Analysis of the tensile strength of polyester/hybrid ramie-cotton fabric composites. Polym. Test 23, 131–135.

Pervaiz, M., Sain, M.M., 2003. Carbon storage potential in natural fiber composites. Resour. Conserv. Recyc. 39, 325–340.

Pickering, K.L., Beckermann, G.W., Alam, S.N., Foreman, N.J., 2007. Optimising industrial hemp fibre for composites. Compos. A Appl. Sci. Manuf. 38, 461–468.

Placet, V., 2009. Characterization of the thermo-mechanical behaviour of Hemp fibres intended for the manufacturing of high performance composites. Compos. A Appl. Sci. Manufactur. 40, 1111–1118.

Placet, V., Cissé, O., Lamine Boubakar, M., 2014. Nonlinear tensile behaviour of elementary hemp fibres. Part I: Investigation of the possible origins using repeated progressive loading with in situ microscopic observations. Compos. A Appl. Sci. Manufactur. 56, 319–327. http://dx.doi.org/10.1016/j.compositesa.2012.11.019.

Rask, M., Madsen, B., Sørensen, B.F., Fife, J.L., Martyniuk, K., Lauridsen, E.M., 2012. In situ observations of microscale damage evolution in unidirectional natural fibre composites. Compos. A Appl. Sci. Manufactur. 43, 1639–1649. http://dx.doi.org/10.1016/j.compositesa.2012.02.007.

Rao, K.M.M., Rao, K.M., 2007. Extraction and tensile properties of natural fibers: Vakka, date and bamboo. Compos. Struct. 77, 288–295.

Roe, P.J., Ansell, M.P., 1985. Jute-reinforced polyester composites. J. Mater. Sci. 20, 4015–4020.

Saheb, D.N., Jog, J.P., 1999. Natural fiber polymer composites: A review. Adv. Polym. Technol. 18, 351–363. http://dx.doi.org/10.1002/(sici)1098-2329(199924)18:4<351::aid-adv6>3.0.co;2-x.

Salmén, L., 2004. Micromechanical understanding of the cell-wall structure. C. R. Biol. 327, 873–880.

Satyanarayana, K.G., Sukumaran, K., Mukherjee, P.S., Pavithran, C., Pillai, S.G.K., 1990. Natural fibre-polymer composites. Cem. Concr. Compos. 12, 117–136.

Sedan, D., Pagnoux, C., Smith, A., Chotard, T., 2007. Propriétés mécaniques de matériaux enchevêtrés à base de fibre de chanvre et matrice cimentaire. 18ème Congrès Français de Mécanique.

Sernek, M., 2002. Comparative Analysis of Inactivated Wood Surfaces. http://scholar.lib.vt.edu/theses/available/etd-04292002-083427/unrestricted/WSInactivation.pdf.

Shah, A.N., Lakkad, S.C., 1981. Mechanical properties of jute-reinforced plastics. Fibre Sci. Technol. 15, 41–46.

Shon, K., White, J.L., 1999. A comparative study of fiber breakage in compounding glass fiber-reinforced thermoplastics in a buss kneader, modular co-rotating and counter-rotating twin screw extruders. Polym. Eng. Sci. 39, 1757–1768. http://dx.doi.org/10.1002/pen.11570.

Siaotong, B., Tabil, L., Panigrahi, S., Crerar, W., 2006. Determination of optimum extrusion parameters in compounding flax fibres reinforced polyethylene composites. Am. Soc. Agric. Biol. Eng., 1–12.

Stamboulis, A., Baillie, C.A., Peijs, T., 2001. Effects of environmental conditions on mechanical and physical properties of flax fibers. Compos. A Appl. Sci. Manufactur. 32, 1105–1115. http://dx.doi.org/10.1016/S1359-835X(01)00032-X.

Thomason, J.L., Vlug, M.A., 1996. Influence of fibre length and concentration on the properties of glass fibre-reinforced polypropylene: 1. Tensile and flexural modulus. Compos. A Appl. Sci. Manufactur. 27, 477–484.

Thomason, J.L., Vlug, M.A., 1997. Influence of fibre length and concentration on the properties of glass fibre-reinforced polypropylene: 4. Impact properties. Compos. A Appl. Sci. Manufactur. 28, 277–288.

Thomason, J.L., Vlug, M.A., Schippr, G., Krikor, H.G., 1996. Influence of fibre length and concentration on the properties of glass fibre-reinforced polypropylene: Part 3. Strength and strain at failure. Compos. A Appl. Sci. Manufactur. 27, 1075–1084.

Tissaoui, J., 1996. Effects of Long-Term Creep on the Integrity of Modern Wood Structures. http://scholar.lib.vt.edu/theses/public/etd-12803097520/etd.pdf.

Trivaudey, F., Placet, V., 2010. Influence de l'Angle des MicroFibrilles (AMF) sur le module d'élasticité longitudinal des fibres végétales – Approche expérimentale et numérique. J. Sci. Tech.

Tröger, F., Wegener, G., Seemann, C., 1998. Miscanthus and flax as raw material for reinforced particleboards. Ind. Crops Prod. 8, 113–121.

Varma, D.S., Varma, M., Varma, I.K., 1984. Coir fibers. Text Res. J. 54, 827–832. http://dx.doi.org/10.1177/004051758405401206.

Velde, K.V.D., Baetens, E., 2001. Thermal and mechanical properties of flax fibres as potential composite reinforcement. Macromol. Mater. Eng. 286, 342–349.

Vilaseca, F., Valadez-Gonzalez, A., Herrera-Franco, P.J., Pèlach, M.À., López, J.P., Mutjé, P., 2010. Biocomposites from abaca strands and polypropylene. Part I: Evaluation of the tensile properties. Bioresour. Technol. 101, 387–395.

Vincent, M., Giroud, T., Clarke, A., Eberhardt, C., 2005. Description and modeling of fiber orientation in injection molding of fiber reinforced thermoplastics. Polymer 46, 6719–6725.

Vlasveld, D.P.N., Parlevliet, P.P., Bersee, H.E.N., Picken, S.J., 2005. Fibre-matrix adhesion in glass-fibre reinforced polyamide-6 silicate nanocomposites. Compos. A Appl. Sci. Manufactur. 36, 1–11.

Wardrop, A.B., Bland, D.E., 1959. The process of lignification in woody plants. 4ème congrès Int. Biochim, vol. 2. Pergamon Press, New York, pp. 76–81.

Zhou, Q., Greffe, L., Baumann, M., Malmstrom, E., Teeri, T., Brumer, H., 2005. Use of xyloglucan as a molecular anchor for the elaboration of polymers from cellulose surfaces: a general route for the design of biocomposites. Macromolecules 38, 3547–3549.

Zykwinska, A., Thibault, J.-F., Ralet, M.-C., 2007. Organization of pectic arabinan and galactan side chains in association with cellulose microfibrils in primary cell walls and related models envisaged. J. Exp. Bot. 58, 1795–1802. http://dx.doi.org/10.1093/jxb/erm037.

Zykwinska, A., Thibault, J.-F., Ralet, M.-C., 2008. Competitive binding of pectin and xyloglucan with primary cell wall cellulose. Carbohyd. Polym. 74, 957–961. http://dx.doi.org/10.1016/j.carbpol.2008.05.004.

Mechanical properties of oil palm biocomposites enhanced with micro to nanobiofillers

15

H.P.S. Abdul Khalil, R. Dungani*,†, M.S. Hossain*, N.L.M. Suraya*, S. Aprilia‡, A.A. Astimar#, Z. Nahrul Hayawin#, Y. Davoudpour**

*Universiti Sains Malaysia, Penang, Malaysia; †School of Life Sciences and Technology, Institut Teknologi Bandung, Bandung, Indonesia; ‡Syiah Kuala University, Banda Aceh, Indonesia; #Malaysian Palm Oil Board, Kuala Lumpur, Malaysia

15.1 Introduction

Wood-based industries are now facing a problem with the supply of raw materials, not only from natural forests, but also from other wood plantations. The utilization of oil palm biomass for raw material wood-based industries will provide a strategic role in promoting oil palm and wood alternative sources of raw materials that are environmentally friendly, as well as the barrier of tropical forest destruction. Oil palm plantations are established primarily in tropical regions, where oil palm is produced in 42 countries worldwide on about 27 million acres (Abdul Khalil et al., 2010a). Distribution of oil palm is restricted to the tropics and it is mainly cultivated in Malaysia, Indonesia, and Thailand in Southeast Asia; Nigeria in Africa; Colombia and Ecuador in South America; and Papua New Guinea in Oceania (UNEP, 2011). Oil palm production has nearly doubled in the last decade, and oil palm has been the world's foremost fruit crop, in terms of production, for almost 20 years.

The current status of oil palm biomass in Asia shows that the two largest palm oil producer countries are Malaysia, with a total of 4.85 million acres of plantations in 2010 (Joshua, 2011), and Indonesia, with 7.80 million acres in 2010 (Slette and Wiyono, 2011). The two countries combined account for 85% of the palm oil production worldwide. Despite this enormous production, the oil consists of only a minor fraction of the total biomass produced in the plantation. The remainder consists of a huge amount of lignocellulosic materials in the form of fronds, trunks, empty fruit bunches (EFB), palm pressed fibre (PPF), oil palm ash (OPA), and so forth. The amount of biomass produced by an oil palm tree include oil and lignocellulosic materials that average 231.5 kg dry weight per year (Sumathi et al., 2008). Therefore, there is an urgency to find alternative sources for local raw material and oil palm biomass appears to be the most viable alternative, which is possible to be utilized as a value-added product as well as a future wood-based industry (Mohamad et al., 2005). Palm lumber or conventional composites such as plywood (Abdul Khalil et al., 2010b), oriented strainboard (Mullen and Kurniawan, 2010), medium-density fibreboard (Ibrahim et al., 2013), polymer composite (thermoset

Biocomposites: Design and Mechanical Performance. http://dx.doi.org/10.1016/B978-1-78242-373-7.00026-3

and thermoplastic) (Abdul Khalil et al., 2008a; Abdullah et al., 2012), and pulp and paper (Yusoff et al., 1984) can be produced from the parts of oil palm biomass.

Generally, biomass wastes have attracted worldwide attention as potential raw materials for the lumber and composites industries due to their viability, easy processability, light weight, nonhazardous, recyclable, and biofriendly properties. Utilization of biomass leads to several advantages such as low density, greater deformability, less abrasiveness to equipment, biodegradability, and low cost (Abdul Khalil et al., 2008a,b). Oil palm biomass in various forms such as lumber, strands, particle, fibre, filler, and other forms can be used as raw material to prepare many processing products, as shown in Figure 15.1.

Recently, biocomposites enhanced with oil palm biofiller ranging from micro- to nanosize has started to gain interest among researchers (Abdul Khalil et al., 2008c; Paul et al., 2007; Shafigh et al., 2011; Arami-Niya et al., 2010). Nanomaterials have attracted scientific and industrial interest due to their low density, high mechanical properties, and many other improved properties. As compared to microsize filler, nanofilled biocomposites exhibit superior property enhancements (Zhang et al., 2011). These properties enhancement can be explained by the fact that the smaller diameter of nanofiller with a higher surface-to-volume ratio increases interaction with matrix and thereby improves the interfacial properties to a larger extent (Dagani, 1999).

Mechanical properties play a major role during the process of selecting a material for engineering applications. The mechanical properties of filled composites depend strongly on the filler size, filler–matrix interface adhesion, and filler loading (Bhat and Abdul Khalil, 2011; Ibrahim et al., 2012). A relatively strong interfacial bond is needed for an effective transfer of the applied load, as a weak interface will probably lead to a premature failure of the composite (De Rosa et al., 2009). As reported by a few previous researchers, the suitability of EFB, OPS, and OPA as filler has proven to be a good biofiller for biocomposites. Figure 15.2 shows the illustration of micro- to nanostructure of oil palm biomass.

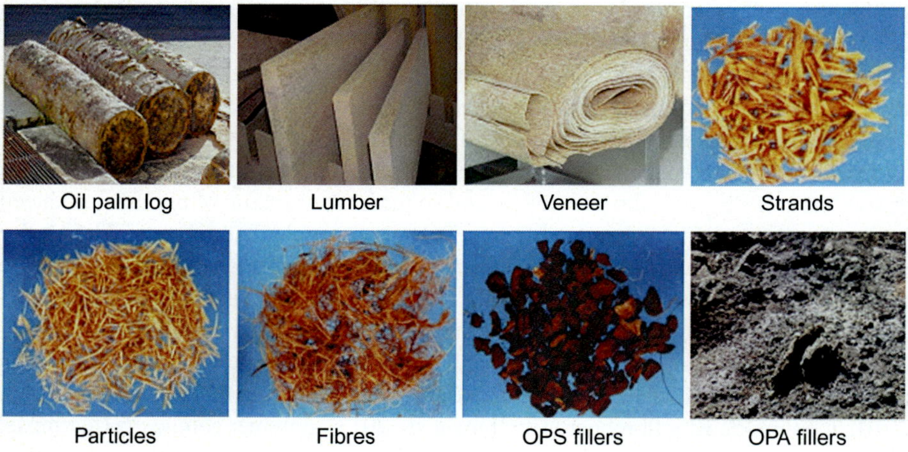

Figure 15.1 Various types of raw materials from oil palm biomass for biocomposites.

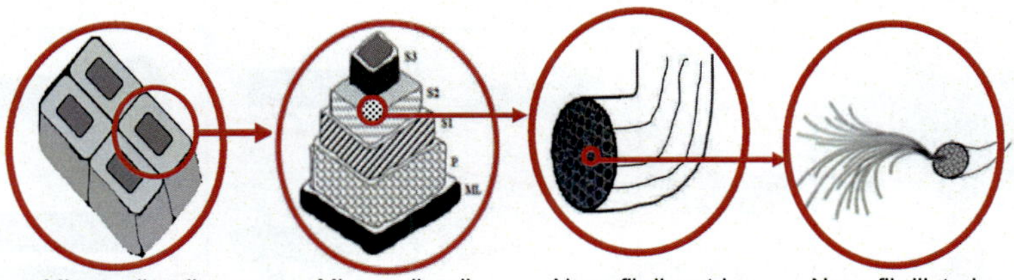

Figure 15.2 Illustration of micro- to nanostructure of oil palm biomass.

15.2 Oil palm biomass: an alternative to wood lumber and wood composite products

15.2.1 Overview of oil palm biomass

Biomass material is one of the important sources of alternative material for the production of biocomposite products. An increasing global awareness about environmental issues is acting as the driving force behind the utilization of biomass material as valuable products. A greater emphasis is on the promotion of the biomass material in the industrialized as well as the developed world to counter environmental issues. The "waste to wealth" ecofriendly concept is becoming increasingly important as a means to exploit and use biomass as a raw material for value-added biocomposite products. Biomass fibre (natural fibres, fibres from agricultural waste, industrial waste, and other wood) has various advantages of a techno-economical nature than synthetic fibres such as glass fibres, carbon fibres, and so forth. History has shown, for example, how in 1938 Henry Ford's use of soybean residue could be the main raw material for the production of car components of the inner frame (70% cellulose fibres) (United Soybean Board, 2013).

Biomass can play an important supplementary role, especially in the form of fibre or filler. The nonwood materials were identified as suitable for the production of paper, composites, and engineered material. Fibre-producing materials such as oil palm trunk (OPT), oil palm frond (OPF), EFB, flax, hemp, ramie, kenaf, jute, abaca, sisal, and coir are seen as promising lignocellulosic raw material for the manufacturing of natural fibre-reinforced/filled composites for different applications. Availability, price, performance, and biodegradable properties were among the factors that acted as a catalyst to promote the use of lignocellulosic fibres of biomass as value-added products.

In line with the development of crude palm oil production, the oil palm industry also produces biowaste (biomass waste) in two forms: namely, waste from the mill and waste from the field, as presented in Figure 15.3. The oil palm sector produces large amounts of biomass that are categorized as agricultural waste and now only 10% is used as an alternative raw material for biocomposite-based industries, industrial raw materials, fertilizer, animal feed, chemical derivatives, and others. Globally, oil palm biomass is produced and unutilised in a million metric tonnes annually. The anticipated higher fresh fruit bunch yields and increase in planted areas in the world

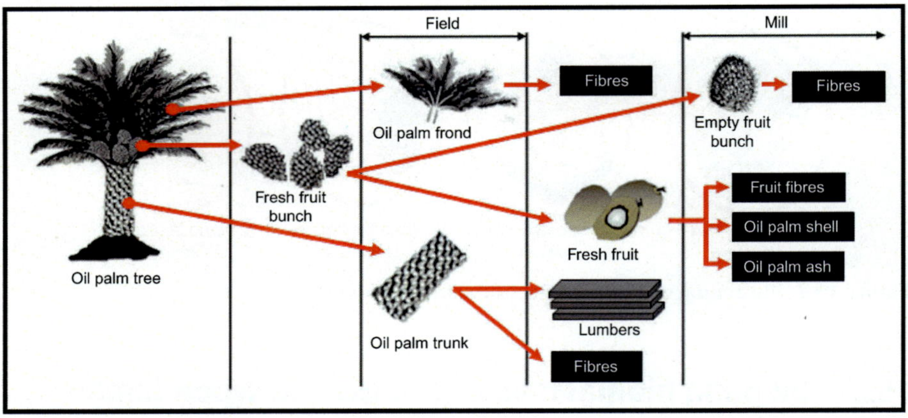

Figure 15.3 Biomass wastes from oil palm.

Table 15.1 The availability of oil palm biomass in Malaysia in a year

Oil palm biomass	Amount (million tonnes)
Oil palm frond (OPF)	25
Oil palm trunk (OPT)	39
Oil palm ash (OPA)	5
Fresh fruit bunch (FFB)	
Palm oil mill effluent (POME)	39
Empty fruit bunch (EFB)	14
Fibres	8
Palm kernel shell	5

are expected to produce more than 150 million metric tonnes of biomass annually. Malaysia scenario oil palm biomass are expected to produce between 87 and 96 million metric tonnes annually. Table 15.1 shows the estimated amount of oil palm biomass by category (Abdul Khalil, 2012).

15.2.2 Morphology, structure, physical, and mechanical properties of oil palm biomass

15.2.2.1 Morphology and structure of oil palm biomass

The morphology and characteristic, so-called fundamental characterization and properties, will not only help open up a new avenue for this fibre, but also emphasize the importance of this biomass as future material. Oil palm trees are included in the monocotyledon class of plants. As a monocotyledonous species, the OPT structure is quite different compared to wood. This monocotyledonous species does not have a lateral meristem, thus the monocotyledon growth is determined by only the apical meristem.

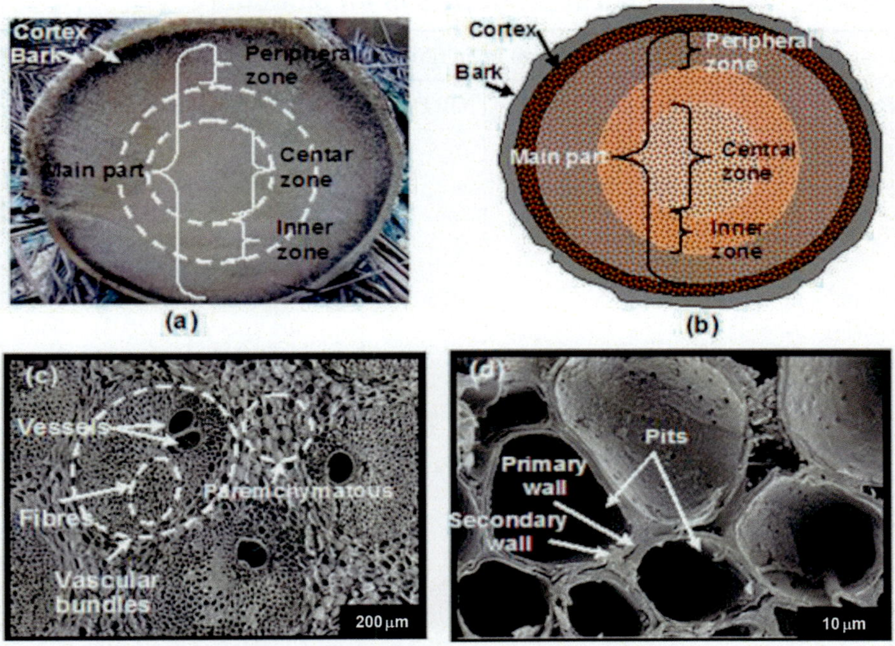

Figure 15.4 Morphology of OPT (a) Actual end crosscut of OPT (Dungani et al., 2012); and (b) schematic drawing and structure of oil palm fibres (Dungani et al., 2012) (c) Vascular bundle structure; and (d) Cell wall structure (Abdul Khalil et al., 2012b).

The structure of OPT consists of fibres in the form of a vascular bundle system and ground parenchymatous cells (Figure 15.4a). It is divided into inner, central, and peripheral zones, and the distribution of fibrous strands depends on the number of bundles present (Killmann and Choon, 1985; Abdul Khalil et al., 2006, 2008a) (Figure 15.4b).

Vascular bundles are the main component of OPT that supports the structural features of the trunk. The bundle also contains three wide metaxylem vessels, or the two metaxylem vessels. These metaxylem vessels are very large, isolated from each other, and each vessel is surrounded by parenchyma cells. The xylem fibre in the vascular bundle of OPT fibre contained one or two vessels in the peripheral region and two or three vessels in the core (Figure 15.4c). Fibres of oil palm wood spread or scattered over the vascular bundles and filled the area surrounded by parenchyma cells. The fibres are comprised of lumen, cell wall, and pits. A transverse sectional view of oil palm fibres is presented in Figure 15.4d, where the fibres attached one to the others in a very compact formation and various sizes and shapes were distinguished. The presence of pits was also identified at the fibre wall as well as companion cells. Meanwhile, the structure of the cell wall of all palm fibre consists of the primary layer (P) and secondary layers (S1, S2, and S3), which is similar to wood structure. Generally, the result of analysis of OPF morphology indicated that the OPF surface is porous and the fibre is constituted of vessel elements. Table 15.2 shows morphological properties of oil palm fibre in different references.

Table 15.2 **Morphological properties of oil palm fibre**

Properties	Type of oil palm fibre				References
	EFB	OPF	OPT	PPF	
Fibre length (mm)	0.33–17.5	0.28–1.03	0.66–1.37	30–50	Amar et al. (2005), Law et al. (2007), Abdul Khalil et al. (2008a), Hassan et al. (1996), and Lau et al. (2005)
Width of fibre (μm)	8.00–12.50	15.10	20.50	–	Amar et al. (2005), Law et al. (2007), Abdul Khalil et al. (2008a), and Hassan et al. (1996)
Width of lumen (μm)	7.90–8.50	8.20	17.60	100	Amar et al. (2005) and Sreekala et al. (1997)
Wall-thickness (μm)	2.00–3.38	1.80	8.00	–	Amar et al. (2005), Law et al. (2007), Abdul Khalil et al. (2008a), and Hassan et al. (1996)
Runkel ratio	0.47–0.59	0.57–0.84	0.26–1.37	–	Amar et al. (2005), Law et al. (2007), and Hassan et al. (1996)
Area of fibre (μm^2)	75.60	126.20	86.70	–	Abdul Khalil et al. (2008a) and Hassan et al. (1996)

Layers S1, S2, and S3 are bound strongly and form a sandwich-like structure in which the microfibril angle S1 and S3 are parallel to the S2 layer. This sandwich structure provides additional strength to the fibres for resistance to water stress, buckling resistance of the compression force, and bending stiffness of the bending force. Primary walls of all palm fibre looks like a thin layer, as shown in Figure 15.5. Some primary wall can be clearly distinguished between the middle lamella next to each other. The thickness of the layer S2, OPT is estimated to have the highest strength as the strength depends on the microfibril cellulose fibres, which are positioned parallel to the fibre axis in the layer S2 (Abdul Khalil et al., 2007a,b, 2010c; Bhat et al., 2010).

EFB fibre is multicellular fibre, hard and strong with the middle of the so-called lacuna. The porous surface morphology is important to provide a better mechanical link together with the resin matrix for composite fabrication. The fibre cross-section is shaped like a polygon with vascular bundle or package that is simple and is surrounded by a thickened layer of cells. Vascular bundles in monocot plants are usually surrounded by a thick cell wall layer that serves to provide buckling resistance against lateral compression force. OPF fibre consists of various sizes of vascular bundles. Vascular bundles are widely used in thin-walled parenchyma tissue. Each file consists of fibrous sheath, vessels, fibres, and phloem tissue parenchyma. Xylem and phloem

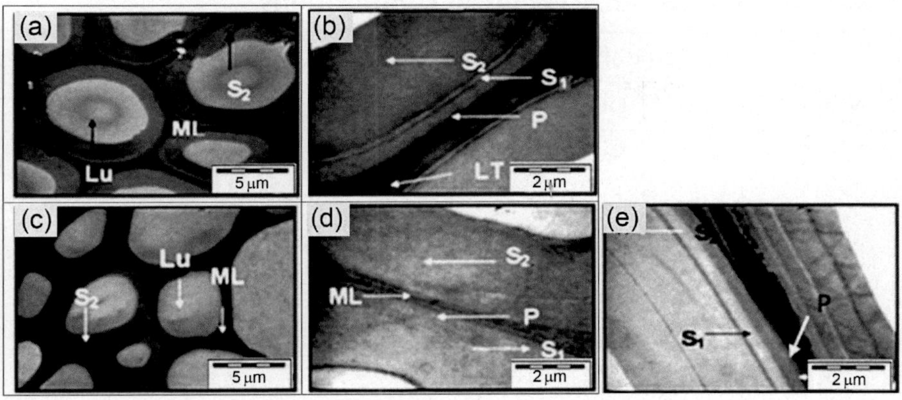

Figure 15.5 Cross section of (a) and (b) the EFB fibre, (c) and (d) OPF fibre, (e) OPT fibres. Lu (lumens), ML (middle lamella); S1, S2 (secondary wall layer) P (primary wall).
Source: Abdul Khalil et al. (2006).

tissue is clearly distinguishable in the phloem divided into two separate parts in each file (Abdul Khalil et al., 2008a).

15.2.2.2 Chemical composition of oil palm fibres

The three-dimensional polymeric components, principally composed of macromolecules of cellulose, hemicelluloses, and lignin, determine most of the properties of lignocellulosics (Ndazi et al., 2006). Chemical analysis of oil palm fibres shows that the principal component is cellulose. OPT fibre is also believed to have a high dimensional stability, because its high extractive content (especially oil and wax) will reduce the hygroscopic fibre property. The amount of cellulose in the fibres will influence the economical aspects in the production of fibre properties. Table 15.3 shows the difference in chemical composition between different types of oil palm fibre.

15.2.2.3 Physical and mechanical aspects of oil palm biomass

Lignocellulosic, in this context, is that of oil palm fibres that have different properties due to differences in chemical composition and morphology of the matrix. Polymer molecules of cellulose, hemicellulose, lignin, extractives, and pectin are responsible for the physical and mechanical properties of lignocellulosic. The diameter and density of these fibres will affect the properties of the fibre. This may be due to the firmly bound chemical structure of the fibre. Lignin binds the three-dimensional cellulose network as well as the fibrils (Sreekala et al., 1997). Properties of lignocellulosic fibres depend mainly on the cellulose content and microfibrillar angle. Therefore, strength properties of the fibres are dependent mainly on the fibrillar structure, microfibrillar angle, and cellulose content. The fact is that, the strength and Young's modulus of the OPEFB fibre are greater than those of oil palm mesocarp fibre. But the mesocarp fibre shows a higher percentage of elongation. As depicted in Table 15.2, OPEFB fibre is higher in cellulose content than mesocarp fibre. The microfibrrial angle of OPEFB

Table 15.3 **Oil palm fibre chemical compositions**

Constituent	Type of oil palm fibre (wt.%)				References
	EFB	OPF	OPT	PPF	
Cellulose	42–65	40–50	29–37	13.97–40	Law et al. (2007), Khoo and Lee (1991), Law and Jiang (2001), Sreekala et al. (1997), Abdul Khalil et al. (2006), Astimar et al. (2002), Abdul Khalil et al. (2008a), and Teoh and Mashitah (2010)
Hemicellulose	17–33	34–38	12–17	26.1	Khoo and Lee (1991) and Astimar et al. (2002)
Holocellulose	65–87	80–84	42–45	24	Law and Jiang (2001), Abdul Khalil et al. (2006), Astimar et al. (2002), and Abdul Khalil et al. (2008a)
Lignin	17–21	20–21	18–23	24.90–25.60	Sreekala et al. (1997), Astimar et al. (2002), Abdul Khalil et al. (2008a), and Teoh and Mashitah (2010)
Extractive	0.9–4.1	4.5	4.1	3	Khoo and Lee (1991), Law and Jiang (2001), Abdul Khalil et al. (2006), Abdul Khalil et al. (2008a)

fibre is 42° and mesocarp fibre shows about 46° (Sreekala et al., 1997). The variation of value for mechanical properties shows that results reported by researchers are not identical because of variation in the kind of oil palm fibre used and also the irregular sectional area, which fluctuates along the length of OPF (Virk et al., 2010). The important mechanical properties of the oil palm fibre are given in Table 15.4.

15.2.3 Properties enhancement of oil palm biomass

The various modification treatments will increase performance of the fibres for use in biocomposite products. Chemical modification of fibres is a chemical reaction between some reactive parts of wood components and a simple chemical reagent to form a covalent bond between both wood and chemical (Rowell, 1975). Figure 15.6, shows an illustration of the principle of chemical modification.

The formation of covalent bonds between wood polymers and reagent by the chemical modification of ether and acetals is depicted in Figure 15.6 (Matsuda, 1996). Chemical agents diffuse those into lumina, without altering the cell walls; meanwhile, bulking in the cell wall, where the cavities in the cell wall are filled and thereby the pathways for water are blocked. Furthermore, the cell wall polymers

Table 15.4 Mechanical properties of selected oil palm fibres

Fibres	Mechanical properties				References
	Tensile strength (MPa)	Elongation at break (%)	Young's modulus (GPa)	Toughness (MPa)	
OPEFB	248–580	14	2.0	2000	Sreekala et al. (1997) and Mukherjee and Satyanaraya (1986)
	100–400	4–18	1.7–9.0	–	Sreekala et al. (2002), Jawaid et al. (2010), Jawaid et al. (2011), and Joseph et al. (2010)
OPF	20–200	3–16	2.0–8.0	–	Joseph et al. (2010) and Zuhri et al. (2010)
OPT	300–600	5–25	8.0–45.0	–	Ahmad et al. (2010) and Killman and Hong (1989)
PPF	80	17	500	500	Sreekala et al. (1997) and Mukherjee and Satyanaraya (1986)

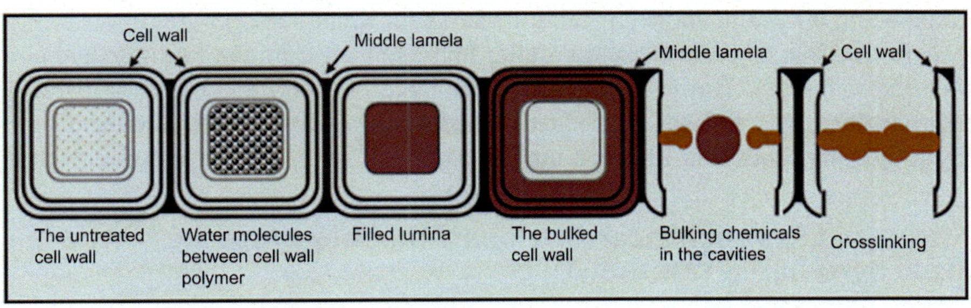

Figure 15.6 Principle of chemical modification for enhancement of oil palm fibre.

are altered and this may lead to cross-linking, reduction of OH-groups and/or (undesired) cleavage of the chains.

There are a number of weaknesses of the use of oil palm biomass (especially in the case of OPT) in composite materials, particularly, in the case of dimensional stability, strength, and durability. These weaknesses can limit the performance of OPT. The various modification treatments can be used to enhance performance of OPT to be comparable with the wood fibres. Among the existing modification methods, impregnation is the most popular and effective (Dungani et al., 2012; Furuno et al., 2004). It is found that impregnation by phenol-formaldehyde (PF) resin or nonresin chemical agents can improve the quality of the OPT (Dungani et al., 2012).

The impregnation of oil palm lumber using a variety of fillers (resin and nonresin) is the most versatile material for building, light construction, furniture, and other uses. Chemical modification increases the quality, for example, aesthetic texture, dimensional stability, resistance to termite attack, water resistance, hardness, stiffness, and so on (Deka et al., 2002). Additionally, Abdul Khalil et al. (2010a) reported that the use of modified OPT core as reinforcement in matrix improve the mechanical, physical, chemical resistant, aesthetic, and decay properties, compared with ordinary wooden lumber. Impregnated oil palm lumber has also been used to slow the UV degradation of the surface.

15.3 Designing of various biocomposites from oil palm biomass

15.3.1 Chronological development of biocomposites

The development of biocomposites and wood-based products began in the 1870s. The evolution expanded from a very basic conventional wood composite to advanced agro-based biocomposites. The mechanized manufacturing of furniture from wood was established during the Renaissance in the first half of fifteenth century. Solid wood was the raw material of choice in producing conventional biocomposites and the use of wood has been growing. The development evolved from plywood, hardboard, particleboard, medium-density fibre board (MDF), laminated veneer lumber (LVL), oriented strand board (OSB), as shown in Table 15.5 (Abdul Khalil, 2012).

A further big step ahead occurred after high demand from the furniture and construction industries. However, due to the shortage of wood and concerns over the environment, the wood-based product started to revolutionize into advanced agro-based biocomposites that involved the utilization of agricultural waste to substitute

Table 15.5 Choronological development of biocomposites from agro-based raw materials

Year	Agro-based biocomposites
1950	Plywood
1960–1970	Hardboard and particleboard
1970–1980	Waferboard, laminated veneer lumber (LVL), medium-density fibreboard (MDF), fireproof panel, wood plastic composite (WPC)
1980–1990	Oriented strand board (OSB), composite gypsum fibreboard, joineryboard structure with radius
1990–2005	Recycled wood fibre products, composite wood/nonwood, furniture component manufacturing, high-performance product (resistance to fire, moisture, and deterioration), thermoset and thermoplastic composite
2005–2010	Hybrid composites, agro-composite lumber, activated carbon composite, composite thermoplastics, thermosets and elastomer
2010–present	Bionanocomposites

for the role of wood and timber. Legislative pressures for greener technologies as well as customer demands for more environmentally friendly consumer goods also forced manufacturers to consider the environmental impact of their products at all stages of their life cycle, including materials selection processing, recycling, and final disposal (Bismarck et al., 2005).

Advanced agro-based polymer biocomposites such as wood plastic composites (WPC), thermoset composites, and thermoplastic composites are the major development in the history of the composites industry. The history of WPC production began in the United States several decades ago (1970s) and since then it has become a common material in various forms from state-of-the-art technology facilities to individual homes (Pritchard, 2004). It has been used for large-scale applications such as cladding, roofing, and structural construction material. Composites have encompassed almost all material domains (house furnishing, packaging, automobiles, aerospace, transportation, sport, leisure, etc.). All synthetic polymers (thermoplastics, thermosets, and elastomers) can be used as matrices. Thermoset and thermoplastic reinforced agro-based fibre has gain an extremely good interest among biocomposites manufacturers. These innovative uses provide an opportunity to greatly expand the purpose and market of composites.

The worldwide availability of plant fibres and other abundantly accessible agro waste is responsible for this research interest in the field of polymer science, engineering, and the urge for a sustainable technology. Various renewable and abundant agricultural wastes offer the potential to act as a biodegradable reinforcing material alternative for the use of synthetic fibre. The resurrection of interest in using natural fibres as reinforcement in polymer composites is mainly due to their techno-ecological advantages (Jawaid et al., 2012). The polymer matrix in advanced biocomposites is required to fulfil several functions, most of which are vital to the performance of the material. The roles of the matrix in advanced fibre-reinforced composites perform a variety of other functions to determine the mechanical behaviour of a produced composite.

While we are in the early decades of the development of agro-based polymer composites, researchers around the globe are constantly trying to come out with new ideas to produce more advanced and sustainable biocomposites. This research effort has introduced hybrid composites from various type of natural fibre combinations that give much higher performance, especially in terms of mechanical properties. In hybrid composites, the advantages of one type of fibre could complement the disadvantages of another type. Hybrid composites fabrication by proper material design could help to achieve a balance in cost and performance. Various researchers have tried blending two fibres in order to achieve the best utilization of the positive attributes of one fibre or to reduce its negative attributes (Abdul Khalil et al., 2011b, 2012a).

Recently, the biocomposites industry has entered another level of advanced reinforced composites that involved a nanoscale of reinforcement. Nanotechnology has become a rapidly growing field with many potential applications. It includes the production and application of biological, chemical, and physical systems at scales ranging from individual atoms or molecules to the submicron level, as well as integration of the resulting nanostructure to a larger system. Nanobiocomposites are a special class of composites, in which one of the components contains nanoscaled dimension. Nanomaterial is defined as the material having a size of roughly 1 to 100 nanometer (nm) or at least one dimension in the nanometer range (Abdul Khalil et al., 2011c). The development of nanomaterial in biocomposites will give strong reinforcement effects, which result in a tremendous interest from the scientific researcher as well as from industry.

15.3.2 Conventional

Conventional wood-based composites manufacture products made primarily from wood with only a few percent of resin and other additives. A useful way to classify conventional wood-based composites is based on specific gravity, density, raw material, and processing methods (Stark et al., 2010). Conventional wood composite materials fall into a few main categories based on the physical configuration of the wood used to make the products: plywood, oriented strandboard (OSB), particleboard, and fibreboard. Within limits, the performance of a conventional type of composite can be tailored to the end-use application of the product (Youngquist, 1999).

15.3.2.1 Plywood

This product is a flat panel made of veneer layers from either a soft or hardwood and these layers are arranged perpendicularly using odd numbers and resin is spread on these layers. It is stacked together by using a clamp or pressing machine to form a panel. Each layer is identified with a different name; for example, the outer layer is the face, the inner layer is the middle or the core, and the wood grain layer that is at a right angle with both the face and back side is known as the cross layer (Abdul Khalil and Hashim, 2004). The properties of plywood depend on the quality of the different layers of veneer, order of layer placement, adhesive used, and control of bonding conditions. The grade of the panel depends upon the quality of the veneers used, particularly of the face and back. The type of panel refers to the durability of the adhesive-to-wood bond and depends upon the adhesive-bonded joint, particularly its water resistance, and upon veneer grades used.

15.3.2.2 Oriented strandboard

OSB is an engineered structural-use panel manufactured from thin wood strands bonded together with water-resistant resin. The wood strands typically have an aspect ratio (strand length divided by width) of at least 3. OSB panels are usually made up of three layers of strands, the outer faces having longer strands aligned in the long-direction of the panel and a core layer that is counteraligned or laid randomly using the smaller strands or fines. The orientation of different layers of aligned strands gives OSB its unique characteristics, including greater bending strength and stiffness in the oriented or aligned direction. Control of strand size, orientation, and layered construction allows OSB to be engineered to suit different uses (Stark et al., 2010).

15.3.2.3 Particleboard

Particleboard is a panel material composed of small discrete pieces of wood or other lignocellulosic materials that are bonded together in the presence of heat and pressure by a synthetic resin adhesive. Particleboard is normally used to produce products suitable for specific geometric resin levels, board densities, and manufacturing processes. Additives can be incorporated during manufacturing to enhance its specific properties, such as improving its dimensional stability and increasing both its fire and moisture resistance (Abdul Khalil and Hashim, 2004). The way particleboard furnishes product is dependent on the raw material. In order to manufacture a board of adequate strength the particles must be compressed to at least 5% above their natural density. In practice, the raw material is usually compressed to nearer 50% of its natural density; so if a raw material of about 400 kg/m^3 is used, then the finished board will have a density of approximately 600 kg/m^3. This degree of compression is needed to achieve good chip-to-chip contact. The vast majority of particleboard is made from wood. Wood is obtainable in three forms: as round wood, which is being used less due to costs; as residues from other processes; and as recovered wood (normally processed in the form of large chips) (Irle and Barbu, 2010).

15.3.2.4 Fibreboard

Fibreboard includes hardboard, MDF, and cellulosic fibreboard. Several things differentiate fibreboard from particleboard, most notably the physical configuration of the wood element. Because wood is fibrous by nature, fibreboard exploits the inherent strength of wood to a greater extent than does particleboard. Fibreboard is normally classified by density and can be made by either dry or wet processes. Dry processes are applicable to boards with high density (hardboard) and medium density (MDF). Wet processes are applicable to both high-density hardboard and low-density cellulosic fibreboard (Stark et al., 2010).

15.3.3 Polymer composites

Composite is defined as a combination of two or more materials (reinforcing elements, fillers, and composite matrix binder), differing in the form or composition. The individual materials that make up composites are called constituents. Most composites have two constituent materials: a binder or matrix, and reinforcement. The reinforcement is usually much stronger and stiffer than the matrix, and gives the composite its good properties. The matrix holds the reinforcements in an orderly pattern. The reinforcements are usually discontinuous, and thus the matrix helps to transfer load among the reinforcements. Numerous studies have utilized polymer matrix in the oil palm fibre-reinforced biocomposites, as presented in Table 15.6 (Abdul Khalil, 2012).

15.3.3.1 Without biofiller

Major constituents in a natural fibre-reinforced composite are the reinforcing fibres and matrix, which acts as a binder for the fibres. Fibrous composites consist of two types, which are categorized based on fibre length. The reinforcement of this type of composites is in the form of fibres that are either continuous (long fibres) or discontinuous (chopped fibres, short fibres, etc.) (Berthelot, 1999). Growing global environmental concern, the high rate of depletion of petroleum and mineral resources, as

Table 15.6 Reported works on oil palm fibre-reinforced biocomposites

Types of polymer matrix	Biocomposites	References
Thermoset	EFB/polyester	Abdul Khalil et al. (2008b) and Abdul Khalil et al. (2007b)
	EFB/polyurethane	Badri et al. (2006) and Barra et al. (2006)
	EFB/phenol formaldehyde	Chai et al. (2009) and Sreekala et al. (2000)
	OPF/phenol formaldehyde	Sreekala et al. (2002)
Thermoplastic	EFB/polyethylene	Roshafima and Wan Aizan (2008) and Ishak et al. (1998)
	EFB/polypropylene	Rozman et al. (1998)
	EFB/poly(vinyl chloride)	Abu Bakar et al. (2005)

well as new environmental regulations have forced the search for new fibre-reinforced composite materials that are compatible with the environment. Oil palm fibres are natural fibres and represent an environmentally friendly alternative to conventional reinforcing fibres. These fibres are available in abundance, renewable, nontoxic, and their low cost are of industrial economic interest.

The oil palm EFB is one of the biomass materials, which is a by-product from the palm oil industry. This oil palm EFB has high cellulose content and has potential as natural fibre resources. Many studies showed that oil palm fibres have the potential to be an effective reinforcement in thermoplastics and thermosetting materials (Abdul Khalil et al., 2012a). Oil palm EFB fibres are lignocellulosic fibres where the cellulose and hemicellulose are reinforced in a lignin matrix similar to that of other natural fibres. These oil palm EFB consist of high cellulose content and are potential natural fibre resources, but applications account for only a small percentage of total biomass production. High cellulose content and high toughness value of oil palm EFB fibres make it suitable for application in polymer composites (Sreekala et al., 2004; John et al., 2008). Fibres from oil palm biomass are in the form of thread-like bundles. However, after processing, EFB fibres are available physically in long as well as in short lengths of about 50–60 and 10–30 μm, respectively. Average fibre diameter is around 200 μm.

The performance of a fibre composite is judged by its length, shape, orientation, composition of the fibres, and the mechanical properties of the matrix. The orientation of the fibre in the matrix is an indication of the strength of the composite, and the strength is greatest along the longitudinal directional of the fibre. This does not mean the longitudinal fibres can take the same quantum of load irrespective of the direction in which it is applied. Optimum performance from longitudinal fibres can be obtained if the load is applied along its direction. The slightest shift in the angle of loading may drastically reduce the strength of the composite.

15.3.3.2 Biofiller impregnated

Filler particles can be defined as the inorganic or organic particles designed to strengthen composites, decrease thermal expansion, minimize polymerization shrinkage, and reduce the amount of swelling caused by water absorption. The filler strengthens the matrices by their hardness, relative to the matrix. Filler particle size varies from 100 μm to very small size 0.1–100 nm (Abdul Khalil et al., 2011c; Beun et al., 2007). The strength of the composite usually depends on the diameter of the filler, the interparticle spacing, and the volume fraction of the composite (Srinivasan, 2010).

Biocomposites enhanced with biofiller have been used extensively in various fields because this filler behaves isotropically to the matrix. The mechanical properties of filled composites depend strongly on the filler size, filler–matrix interface adhesion and filler loading (Bhat and Abdul Khalil, 2011; Ibrahim et al., 2012). Several studies have been done by utilizing oil palm biomass as filler in polymer composites from microsized to the current nanosized, as shown in Table 15.7. For instance, EFB and OPA have stimulated much interest among researchers in developing advanced and quality biocomposites enhanced with biofiller. University Sains Malaysia has developed hybrid composite technology with nanobiofiller from oil palm shell in matrix EFB fibres.

Figure 15.7 shows the filler from various oil palm biomasses ranging from micro- to nanosize under the scanning electron microscope (SEM) and the transmission electron microscope (TEM). The suitability of EFB as filler is proven as a good biofiller for biocomposites. For example, a study on mechanical properties of EFB-filled polypropylene composites showed that flexural, modulus, and toughness strength were increased. According to researchers, filler size is one of the major factors in determining the properties of biocomposites produced. It was found that the smaller the filler size, the higher impact strength resulted. On the other hand, the tendency of filler agglomeration or nonuniformity and poor adhesion can be reduced to a greater extent with the help of coupling agents (Azman et al., 2010).

OPA, an abundantly available throwaway waste from fired-boiler furnaces, has currently emerged to be an ideal material used as biofiller. The ash is produced after the combustion of oil palm fibre and shell as boiler fuel to produce steam for palm oil mill consumption (Bhat and Abdul Khalil, 2011). Combinations of the unsaturated polyester with natural fillers have been reported to improve the mechanical and thermal properties of composites. Utilization of OPA as filler in the manufacture of polymer composites was found to cut the consumption of more binders or matrix of the composite materials (Ibrahim et al., 2012). Filler loading is one of the important parameters that should be considered in producing biocomposites enhanced with biofiller. According to Bhat and Abdul Khalil (2011), favourable interfacial properties and the

Table 15.7 Reported works on oil palm biomass as filler in biocomposites

Biocomposite	References
Oil palm ash/epoxy	Abdul Khalil et al. (2013)
Oil palm ash/polyester	Ibrahim et al. (2012)
Oil palm ash/polypropylene	Bhat and Abdul Khalil (2011)
Oil palm shell/oil palm trunk lumber	Dungani et al. (2013)
EFB (carbon black)/epoxy	Abdul Khalil et al. (2010c)
EFB/polypropylene	Basuki et al. (2004)

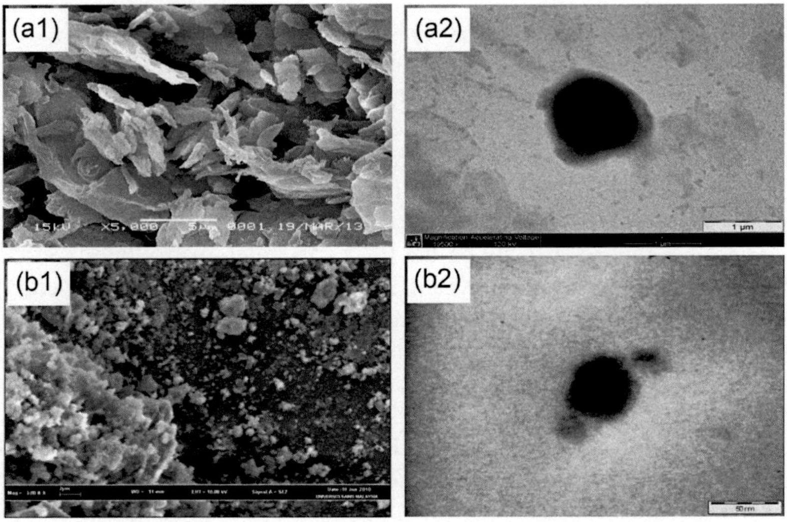

Figure 15.7 Filler from various oil palm biomasses under SEM and TEM. (a1) oil palm shell under SEM; (a2) oil palm shell under TEM; (b1) oil palm ash under SEM; (b2) oil palm ash under TEM.

lack of agglomeration at low OPA loading will give the high values of the mechanical properties such as tensile strength.

15.3.4 Hybrid composites

Hybrid composites are those systems in which one kind of reinforcing material is incorporated in a mixture of different matrices (blends), or two or more reinforcing and filling materials are present in a single matrix, or both approaches are combined. The incorporation of two or more lignocellulosic fibres into a single matrix has led to the development of hybrid composites. The behaviour of hybrid composites is a weighed sum of the individual components in which there is more favourable balance between the inherent advantages and disadvantages. While using a hybrid composite that contains two or more types of fibres, the advantages of one type of fibre could complement with the disadvantages in the other type (Jawaid and Abdul Khalil, 2011a).

The strength of hybrid composites is dependent on the properties of fibres, aspect ratio, fibre content, length of an individual fibre, orientation of fibre, extent of intermingling of fibres, fibre–matrix interface bonding, arrangement of both fibres and also on the failure strain of individual fibres (Sreekala et al., 2002). Several studies have been conducted of hybrid composites combining oil palm fibres with thermoset and thermoplastic resin. Jawaid and Abdul Khalil (2011b) has reported several studies on oil palm fibre-based hybrid composites, as shown in Table 15.8. Hybrid composites made by using oil palm EFB and woven jute fibre were studied, and the composites produced showed the improvement in physical properties and also in impact strength, which is evident from the scanning electron micrograph in Figure 15.8 (Jawaid et al., 2011).

Table 15.8 **Reported works on oil palm fibre-based polymer hybrid composites**

Hybrid	Matrix	References
Oil palm/glass	Polyester	Kumar et al. (1997), Agrawal et al. (2000), Abdul Khalil et al. (2007b), Karina et al. (2008), and Wong et al. (2010)
	Phenol formaldehyde(PF)	Sreekala et al. (2002), Sreekala et al. (2004), and Sreekala et al. (2005)
	Epoxy	Abu Bakar and Abdul Khalil (2004), Abu Bakar and Abdul Khalil (2005), and Mridha et al. (2007)
	Vinyl ester	Abdul Khalil et al. (2009)
	Natural rubber	Anuar et al. (2006)
Oil palm/sisal	Natural rubber	Jacob et al. (2005), Jacob et al. (2006, 2007), and John et al. (2008)
Oil palm/jute	Epoxy	Jawaid et al. (2010)

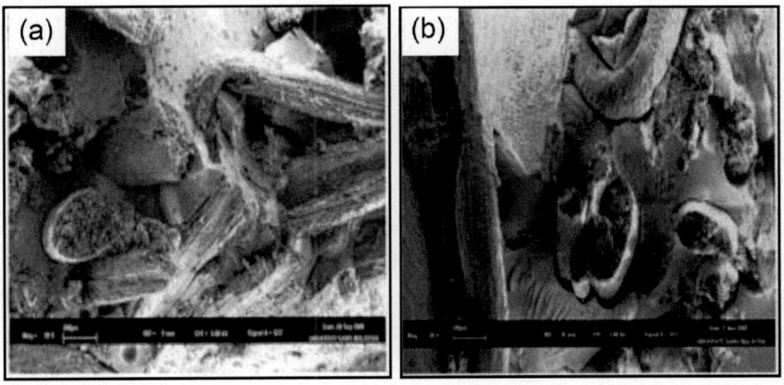

Figure 15.8 (a) SEM micrograph of impact fracture sample of pure EFB, (b) SEM micrograph of impact fracture sample of EFB/woven jute/EFB composite.
Source: Jawaid et al. (2011).

15.4 Properties of oil palm nanobiocomposites

The term polymer nanocomposite broadly describes any number of multicomponent systems, where the primary component is the polymer and the filler material has at least one dimension below 100 nm (Koo, 2006). Polymer nanocomposites are generally lightweight, require low filler loading, are often easy to process with good technique, and provide property enhancements extending orders of magnitude beyond those realized with conventional composites or microcomposites (Chisholm et al., 2005). Rapid advancements in nanocomposite technologies have been realized as new classes of nanoscale fillers continue to emerge.

15.4.1 Structural and morphology

Morphologic studies of oil palm biocomposites were carried out using spectra infrared (FT-IR), SEM, and X-ray diffractometry (X-RD). The chemical structure in the panel was shown in wave numbers and also explained the functional presence in different parts of oil palm raw materials and after biocomposites were made (Siyamak et al., 2012).

The morphology of biocomposite can be observed by SEM. According to Bhat et al. (2010), in SEM images the observation showed that the porous morphology of dried OPT has been filled with the resin. PF and urea formaldehyde (UF) resins also penetrated into the cell wall structures of oil palm trunk lumber (OPTL) and covered the parenchymatous cells. Meanwhile, Abdul Khalil et al (2011b) used SEM to conduct morphological analysis of the surface of nanostructured OPA-filled epoxy composites where it showed the neat epoxy/filler loading of nanostructured OPA-filled epoxy composites. Similarly, the study by Bhat and Abdul Khalil (2011) showed OPA in polypropylene composite where it is clear that good interfacial adhesion is formed between the PP matrix and OPA. However, 100 g and 3% OPA and 100 g and 5% OPA showed poor morphology as compared to the rest of the composite materials, as given in Figure 15.9.

15.4.2 Physical properties

The physical properties for the production of oil palm nanobiocomposite include density, water absorption, and thickness swelling. Abdul Khalil et al. (2010a) studied the density of OPTCL composite materials that increased with the increase in the resin loading percentage and subsequently the density decreased with the higher nanoloading percentage of the resin. This is due to the better interfacial adhesion between the OPTCL fibres and the resin materials in case of critical loading. The density of a composite depends on the relative proportion of matrix and reinforcing materials. Many studies have been conducted on the characterization of composite physical properties, as presented in Table 15.9. The composite physical properties is one of the most important factors determining the quality of composites.

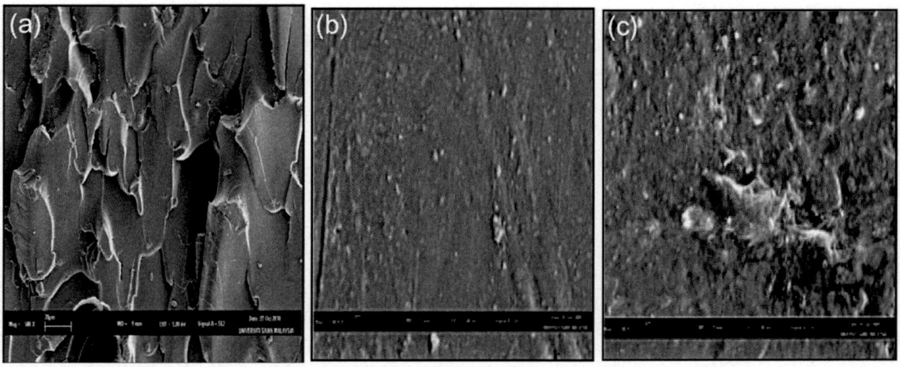

Figure 15.9 SEM micrograph of oil palm biomass nanocomposites (a) 3% filler loading of nanostructured OPA-filled epoxy composites (Fizree, 2012); (b) 100 g polypropylene + 3% OPA (Bhat and Abdul Khalil, 2011); (c) 100 g polypropylene + 5% OPA (Bhat and Abdul Khalil, 2011).

Table 15.9 Physical properties of some products based on oil palm biocomposites

Biocomposites	Material bending/resin	Physical properties	References
Laminated veneer lumber from OPT	Urea formaldehyde	• Density: 441.67–596.77 kg/m^3 • Water absorption: 52–68% • Thickness swelling: 79–89%	Razak et al. (2008)
Polyester–Glass–EFB	–	• Density: 1.24–1.36 g/cc • Water absorption: 1.4–3.1% • Thickness swelling: 1.74–2.03%	Karina et al. (2008)
Veneer laminated-OPT	Phenol formaldehyde	• Density: 0.64–0.66 g/cm^3 • Water absorption: 56.66–58.26% • Thickness swelling: 11.18–12.83%	Abdul Khalil et al. (2010a)
Veneer laminated-OPT-OFB	Phenol formaldehyde	• Density: 0.72–0.74 g/cm^3 • Water absorption: 43.61–50.49% • Thickness swelling: 7.41–9.85%	Abdul Khalil et al. (2011a)
Oil palm stem plywood	Phenol formaldehyde resin	• Density: 607–749 kg/m^3 (untreated) • Density: 668–879 kg/m^3 (treated)	Loh et al. (2010)
Oil palm trunk lumber	Phenol formaldehyde resin	• Density: 440–860 kg/m^3 • Water absorption: 65.67–54.70% • Thickness swelling: 5.59–3.65%	Bhat et al. (2010)
	Urea formaldehyde	• Density: 430–810 kg/m^3 • Water absorption: 70.49–58.10% • Thickness swelling: 6.53–4.40%	
Oil palm trunk lumber	Phenol formaldehyde resin	• Density: 400–650 kg/m^3 • Water absorption: 21–30.2%	Abdul Khalil et al. (2010a)
Particleboard panel (rubber wood)	Starch from oil palm trunk	• Density: 0.57–0.74 g/cm^3 • Water absorption: 6.97–17.58% (24 h)	Sulaiman et al. (2013)
Palm oil-based rigid polyurethane nanocomposite foams	Polyurethane from oil palm polyol	Density: 40.0–46.6 kg/cm^3	Chuayjuljit et al. (2010)

Water absorption in palm oil nanobiocomposites is dependent on temperature, fibre loading, orientation of fibres, permeability of the fibre, surface protection, area of the exposed surfaces, diffusivity, and so on (Sreekala et al., 2002). As the cell walls are fully embedded with the resin and the water content replaced with resin, OPT with 120 min impregnation time has the least water absorption compared with other impregnated samples (Abdul Khalil et al., 2010a). Therefore, by increasing the impregnation time and the resin content, the water absorption decreases. It is clear that the more resin replaces in the OPT structure, the less cavities are available to absorb water.

15.4.3 Mechanical

Mechanical properties of composites depend on several factors, such as the properties of constituent reinforcement and matrix; their relative volume fraction; and the shape, size, and architecture of the reinforcement phase. A relatively strong interfacial bond is needed for an effective transfer of the applied load, as a weak interface will probably lead to a premature failure of the composite (De Rosa et al., 2009). Mechanical properties are commonly taken into consideration in the general usage areas of product composited based on oil palm.

Several research works have been carried out to identify the parameters that govern mechanical behaviour of particulate oil palm composites. Several studies on other lignocellulosics such as EFB-, OPT-, and OPF-filled thermoplastic and thermoset composites have indicated that the mixing procedures could play an important role in determining the properties of composites. However, one property (namely, the impact strength) is often listed among the major disadvantages of oil palm biocomposites. In recent years, the development of new fibre manufacturing techniques and improved composite processing methods along with enhancement of fibre/matrix adhesion has improved the current situation fairly.

Some studies, incorporation of coupling agents and compatibilizers in EFB composites have also been reported to impart some improvement in properties. The properties of the composites were believed to be predominantly influenced by the type of bonding produced. In general, the composites with isocyanate treated fibres showed superior tensile and flexural properties than those without treatment. Karina et al. (2008) has studied EFB in polyester to replace glass in many applications that do not require very high bearing capabilities. They have compared the glass and EFB 40–70%. Meanwhile, the reinforcing effect of EFB in polyester composites at various fibre loadings in composite, overall fibre content was (by weight) 20% and 40%. From the analysis were shown each composition with different percentage and size shows different types of behaviour and properties (Abdul Khalil et al., 2012a). Bhat et al. (2010), has studied OPT impregnated with PF and UF resin as a matrix using a high-pressure vacuum impregnation chamber. Many researchers have investigated the mechanical properties in oil palm-reinforced thermoplastic and thermoset biocomposites that are shown in Table 15.10.

15.4.4 Thermal

Several researchers who studied thermal stability, such as Abdul Khalil et al. (2012b), reported thermal stability for the EFB-reinforced laminated biocomposites, and noted

Table 15.10 Mechanical properties for thermoplastics and thermoset polymer-reinforced oil palm biocomposites

Biocomposite	Tensile strength (MPa)	Flexural strength (MPa)	Tensile modulus (MPa)	Flexural modulus (GPa)	Impact strength (J/m)	References
Polypropylene–EFB	8–23	15–40	500–3500	1.25–2.25	75–94	Rozman et al. (2000)
Polypropylene–EFB effect maleic anhydride	–	11–50	–	1.5–4.9	30–130	Rozman et al. (2003)
Polyurethane–EFB Isocyanate	9–23	28–78	500–1300	0.4–1.55	7–20	Rozman et al. (2004)
Polyurethane–EFB mat	9–32	17–80	400–2100	0.25–1.8	–	Rozman et al. (2007)
Polyester–glass–EFB	–	36.8–165.9 33.9–163.3	–	–	–	Karina et al. (2008)
Polypropylene–oil palm wood flour	31–25	–	–	1.9–2.9	31–25	Zaini et al. (1996)
Polypropylene–EFB pulp, with black liquor NaOH content	9.5–14.5	16–27	1580–2400	1.53–3.1	–	Tay et al. (2010)
Polypropylene–EFB fibre	22–35 (MAPP) 20–27 (TMPTA)	–	–	2.5–2.7 (MAPP) 2.2–2.7 (TMPTA)	25–33 (MAPP) 25–38 (TMPTA)	Khalid et al. (2008)
Poly(butylene adipate-co-terephthalate)–EFB	6–13	6–11	20–250	100–600	–	Siyamak et al. (2012)
Poly (vinyl chloride)–EFB effect acrylic	–	0 phr acrylic 80–75 9 phr acrylic 69–63	–	0 phr acrylic 3.5–4.5 9 phr acrylic 2.7–3.2	–	Abu Bakar et al. (2005)
Polylactic acid/epoxidized palm oil/fatty nitrogen compounds modified clay nanocomposites	30–17	–	600–300	–	–	Jaffar Al-Mulla (2011)

that the initial degradation temperature occurs at a higher temperature compared to the OPT laminated biocomposite materials. Siyamak et al. (2012) reported that thermal stability of composite declined as the fibre content in composites improved. It is also obvious that the devolatilization of moisture of EFB fillers occurred in the temperature range of 70–100 °C even after the EFB fillers incorporated into the poly(butylene adipate-co-terephthalate) matrix. Such moisture content might play a significant role in the degradation processes due to the fact that the OH group in water is more reactive than the OH group available in the EFB filler. Bhat et al. (2010) also investigated OPTL impregnated with PF and UP resins, and found that the weight lost below 100 °C can be attributed to the evaporation of moisture. The presence of PF and UF resins in OPTL showed better thermal stability compared with dried OPT, while PF resin in OPTL showed more thermal stability than UF in OPTL. Rihayat et al. (2006) studied thermal properties of polyurethane-nanoclay based on palm oil polyol, and the result indicated an improvement in thermal stability of polyurethane because the organic material can prevent the heat from expanding quickly and limit further degradation. Thermal resistance of nanocomposites improved above 62% as compared to pure polymer.

15.5 Product designing and application of oil palm biocomposites

15.5.1 Current application

Traditionally, the main uses of oil palm fibre bundles are in ropes and twines, mainly produced through cottage industries. Oil palm biomass is converted into yarn, string, ropes, floor mats, bags, floor and wall coverings, and different handicrafts. The oil palm fibre can also be used in the manufacture of cellulose pulp to manufacture craft paper of high resistance and other types of thin paper, as shown in Figure 15.10. It can also be used in manufacturing reinforced composite for application in the automotive industry, sporting goods, appliances, building construction, and medical equipment.

The bulk of the fibrous waste is not used and even contributes to severe environmental problems when left at processing plants and farms alone. Previous studies

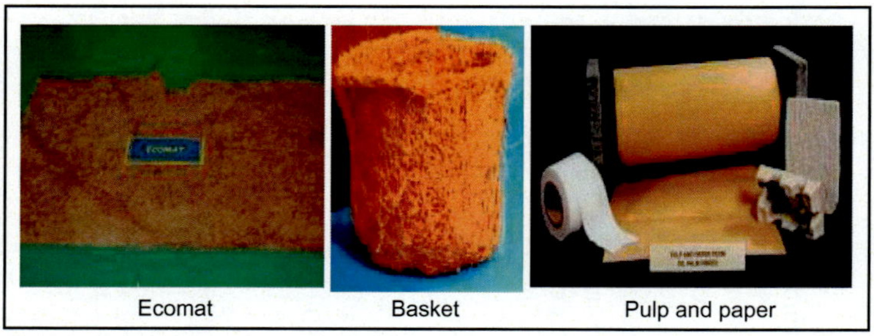

Figure 15.10 The traditional products from oil palm fibre bundles.

and the latest on biomass waste and other agricultural residues have shown potential in its use for the production of various types of value-added products such as medium-density panels, block board, LVL, mineral-bonded particleboard, plywood, chipboard, thermoset and thermoplastic composites, nanobiocomposites, and pulp and paper manufacturing (Sulaiman et al., 2008; Laemsak and Okuma, 2000; Chew and Ong, 1985; Noorbaini, 2009; Abdul Khalil and Bhat, 2010; Shuit et al., 2009; Sreekala et al., 2002).

In the past decades, composites materials have been utilised to solve technological problems for a long time but only in the 1960s did these materials start capturing the attention of industries with the introduction of biomass-based composites. Since then, biocomposite materials have become common engineering materials and are designed and manufactured for various applications including automotive components, sporting goods, aerospace parts, consumer goods, and in the marine and oil industries.

Some examples of the variety of products and applications for oil palm biocomposite industry are booming in the world, as shown in Figure 15.11. Oil palm biocomposite products are divided into conventional and advanced biocomposites (polymer). Many applications of biomass-based biocomposite products of oil palm biomass have been and will be in the market and have contributed to the national economy.

15.5.1.1 Construction

There are many uses of potential value-added products made from oil palm biomass such as construction materials. EFB can also be used as a major component of specialized construction materials. Significant research studies have been conducted on the development of new construction materials using EFB. However, the application of these construction materials in real construction is limited. Abdul Khalil et al. (2012a) investigated conversion of OPT and oil palm EFB into new plywood. Regarding the utilization of EFB, other researchers such as Zaidon et al. (2007) and Deraman et al. (1999) studied mixing EFB and rubber wood into a particleboard.

15.5.1.2 Furniture

With increasing developments of newer biocomposites in the furniture industry, the use of oil palm biomass particleboard in furniture making could support the concept of reducing waste. Blockboard can be manufactured from OPT as the centre core. It is lighter than blackboard manufactured from normal wood. The physical and mechanical properties indicate that blockboard from OPT is suitable for non-load bearing applications, such as paneling and furniture applications. The EFB can be used for the manufacture of MDF (Norul Izani et al., 2012). In another study on MDF, the board was produced from OPT. MDF manufactured from trunk with its light colour will compare better than that manufactured from fronds, which have dark spots on the surface.

OPT and fronds have also been used as raw materials in moulded particleboard manufacture. Several items, such as table tops of various sizes and shapes and chair seats, have been successfully tried in industrial-scale plants.

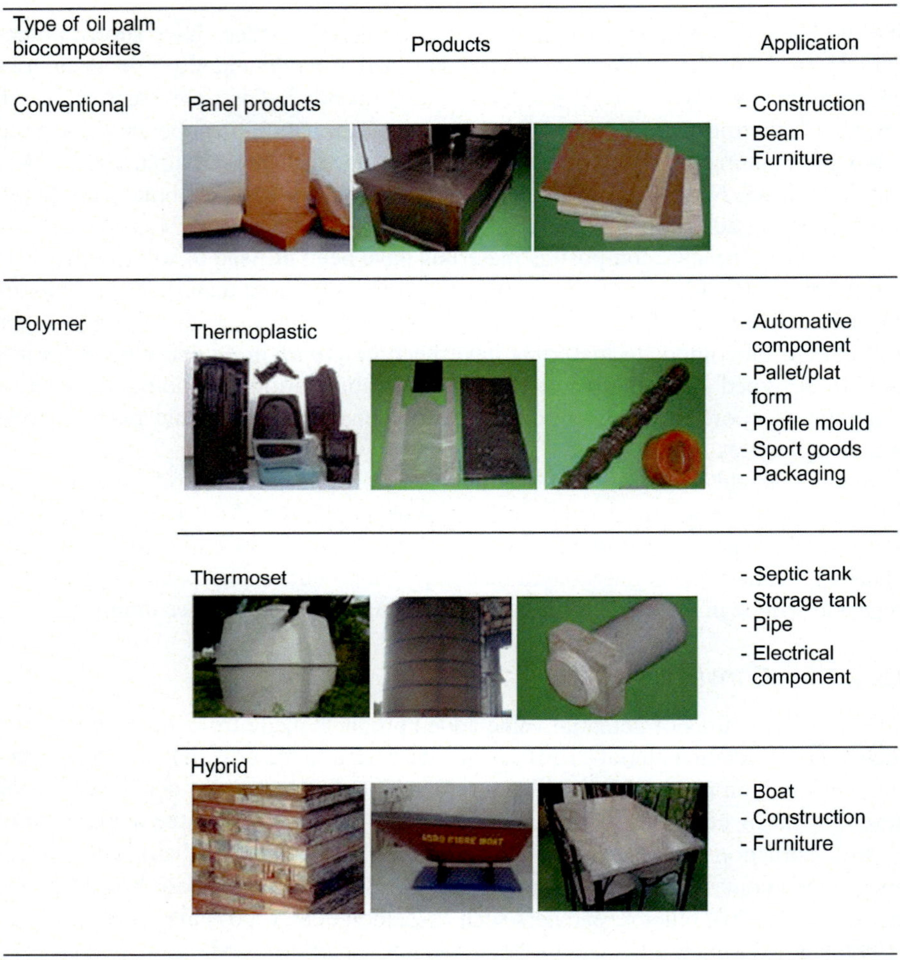

Figure 15.11 The oil palm biocomposite products and its applications in various fields.

15.5.1.3 Automotive

The application of polymer–matrix composites in the automotive field is very important for reduce operating costs. Most common of composites in automotive applications are the exterior body panel and various components under the hood (Aström, 1997). The use of fibre-reinforced plastic composites in the automotive industry has grown significantly in recent years because of their low weight, design flexibility, corrosion resistance, and cost effectiveness (Rudd, 2001). Natural fibres have already found applications in the automotive industry, primarily in nonstructural parts such as interior panels, parcel shelves, and so on (Hoydonckx, 2007). The oil palm fibre can be used as filler in thermoplastic and thermoset composites. This composite has wide application in automotive components. In Malaysia, research and development in this field finally reached the commercialization stage when Proton (Malaysian National Carmaker) signed an agreement with the Malaysian Palm Oil Board (MPOB) to develop the field of thermoplastic composites, thermosets, and elastomers for use in the manufacture of car components protons (Shuit et al., 2009).

15.5.1.4 Sporting goods

Biocomposite materials have been used in some sporting goods. A flax-based bicycle and tennis racquet are currently available on the market (François, 2013). Advanced composites are now found in products used for 7 of the top 10 most popular outdoor sports and recreational activities (Lee, 2010). Carbon-reinforced composites (hybrids with other fibres) continue to successfully replace wood and metal in high-volume components such as fishing rods, tennis racquets, golf clubs, spars/shafts for kayak paddles, windsurfing masts, kites, bicycle handlebars, and ski equipment.

15.5.1.5 Medical equipment

An advance in the field of composite technology has led to the development of various applications in medicine, especially modern orthopedic and prosthetic devices. The composites made from various combinations of metals, ceramics, and polymers used in biomedical applications. However, not all materials are suitable for biomedical applications (Hench, 1991). Composites of fibre-reinforced polymers or biocomposites in recent years has been widely used for biomedical purposes (Scholz et al., 2011). This natural fibre-reinforced composite can then be used to make a variety of implants such as intramedullary rods and artificial joints. The applications include spinal fusion, hard tissue applications, including prosthetic socket, dental post, external fixator, bone plate, orthodontic archwire, orthodontic bracket, total hip replacement, and composite screws and pins. Meanwhile, composite materials have been extensively used in dentistry and prosthesis designers are now incorporating these materials into other applications.

15.5.2 Potential applications

Oil palm biomass can be used almost entirely, not only in biocomposites but also in other applications. As shown in Table 15.11, researchers have developed ruminen food/silage, filter and adsorbent from oil palm shell; sugar production from oil palm trunk; organic mulch, electricity generation, biofuel and hybrid composite from EFB; adsorbent for sulphur gas and cement replacement from OPA; filter and carbon active from oil palm shell; and also animal feedstock from palm kernel cake. The combination of mechanical and physical properties along with the ecofriendly nature of belonging has sparked a lot of research and industrial sectors, particularly the construction, furniture, and automotive industries to consider these fibres as a potential material to replace man-made fibres in the production of products that are safe and environmentally friendly.

Table 15.11 Other application of oil palm biomass

Oil palm biomass	Products of application	References
Oil palm fronds	Ruminen food/silage	Hassan et al. (1996)
Oil palm trunk	Sugar production	Rahman et al. (2006)
Empty fruit bunches	Organic mulch	Husin et al. (2005)
	Electricity generation	Leete (2007)
	Biofuel	Shuit et al. (2009)
	Hybrid composite	Abdul Khalil et al. (2008c)
Oil palm ash	Adsorbent for sulfur gas	Zainudin et al. (2005)
	Cement replacement	Tay and Show (1995)
Oil palm shell	Filter	Jusoh et al. (1995)
	Carbon active	Arami-Niya et al. (2010) and Lua and Guo (2001)

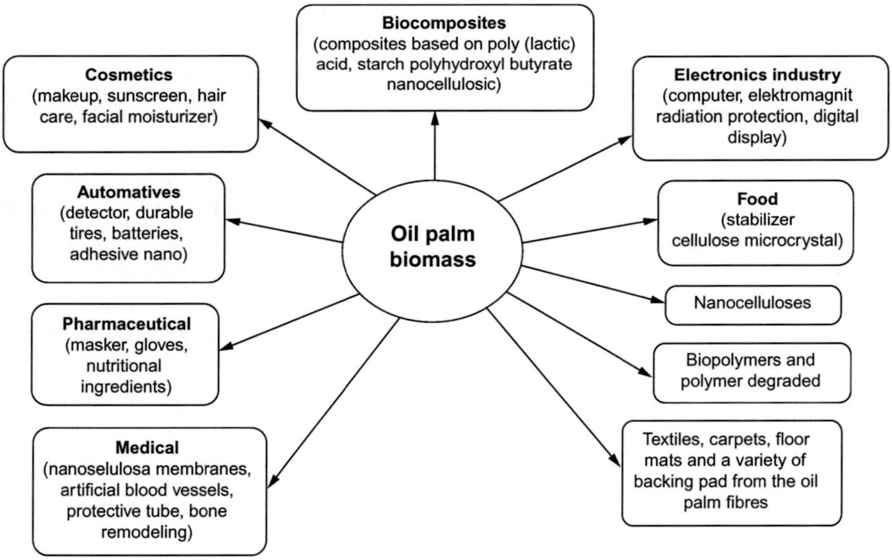

Figure 15.12 Applications and potentials of oil palm biomass in various fields.

Oil palm biomass can play an important supplementary role, especially in the form of fibre. Natural fibre materials were identified as suitable for the production of paper, composites, and engineered material. Fibre-producing materials, such as oil palm trunk, OPF, EFB, and PPF, are seen as promising lignocellulosic raw material for the manufacturing of natural fibre-reinforced/filled composites for different applications. Products produced using natural fibre improved their homogeneity and quality. Innovative products created from ecofriendly new technology to produce competitive products in several fields in the development of oil palm biomass applications are shown in Figure 15.12.

15.6 Conclusions

With the increasing area of the oil palm plantation, the industry will proportionately generate more residues. Therefore, the industry must be ready to deal with the residues wisely. The availability of oil palm biomass amounts to seven times the total wood used in world. Therefore, the wood-based industry should consider using more sustainably produced oil palm biomass in their traditional products. The conversion of oil palm fibre into fibre-reinforced composite is now technically feasible. The products include paper and paperboard, mattresses, and reconstituted panel board, such as particleboard and fibreboard. Oil palm biomass are natural fibres that have excellent physical and mechanical properties and can be used in producing composite products for various end uses such as WPC, fibre-filled thermoplastic and thermoset composites, automotive components, and other value-added biocomposite products. Oil palm fibres incorporated in a polymer matrix increase the stiffness, fatigue life, and other properties. Fibres are mechanically more effective in achieving a stiff, strong composite than particles. Oil palm fibres such as graphite are stiff (Young's modulus is 1700–9000 MPa) and strong (the tensile strength is 100–580 MPa). In terms of strength and stiffness per unit weight, oil palm fibres have strength comparable with man-made fibres. Hence, nanobiocomposites made from them can be as strong as steel but much lighter.

The development of nanotechnology in the last decade has produced nanocomposite materials. This material, consisting of a polymeric matrix material and nanomaterial/nanofillers, has many uses ranging from high-barrier packaging for food and electronics (Ahmadi et al., 2004) to automotive and aerospace applications (Njuguna and Pielichowski, 2003) due to their improved properties. At low-filler contents as compared with their conventional micro and macro or neat counterparts, they exhibit superior property enhancements (improved mechanical and thermal properties, gas permeability, resistance and fire retardancy) that are related to the nanostructure achieved in processing these materials.

Biocomposite applications are wide and contribute to the growth of many other industries. It is acknowledged that the biocomposite industry is an important contributor to the economy of a country. Reduction in the supply of raw materials has sparked concern and in this context, biomass (palm oil) has been seen as an alternative material for the biocomposite industry to produce value-added products. As a result, plant operators are encouraged to explore the potential of the available natural resources for the production of new value-added products that will enhance growth, competitiveness, and sustainability. This clearly shows the oil palm biomass is a raw material to be exploited because the cost of energy and raw materials it uses are cheaper and more environmentally friendly than man-made fibres.

Acknowledgement

The authors would like to thank the Universiti Sains Malaysia (USM), Penang, Malaysia, for providing Research Grant, No.: RUI-1001/PTEKIND/814255 as financial support.

References

Abdul Khalil, H.P.S., 2012. Biomass and biocomposite: potential and future of green technology in Malaysia. In: Public Lecture Series of Professorship. Universiti Sains Malaysia, Penang, Malaysia, p. 22.

Abdul Khalil, H.P.S., Bhat, A.H., 2010. Oil Palm Biomass: Fiber Cultivation, Production and its Varied Applications. Nova Science Publishers, New York.

Abdul Khalil, H.P.S., Hashim, R., 2004. Komposit panel berasaskan sumber kayu. Universiti Sains Malaysia Publisher, Penang, Malaysia.

Abdul Khalil, H.P.S., Siti Alwani, M., Mohd Omar, A.K., 2006. Chemical composition, anatomy, lignin distribution and cell wall structure of Malaysian plant waste fibers. BioResources 1 (2), 220–232.

Abdul Khalil, H.P.S., Siti Alwani, M., Mohd Omar, A.K., 2007a. Cell wall structure of various tropical plant waste fibers. Mokchae Konghak 35, 9–15.

Abdul Khalil, H.P.S., Kumar, R.N., Asri, S.M., Nik Fuaad, N.A., Ahmad, M.N., 2007b. Hybrid thermoplastic pre-preg oil palm frond fibers (OPF) reinforced in polyester composites. Polym.-Plast. Technol. Eng. 46, 43–50.

Abdul Khalil, H.P.S., Siti Alwani, M., Ridzuan, R., Kamarudin, H., Khairul, A., 2008a. Chemical composition, morphological characteristics, and cell wall structure of Malaysian oil palm fibers. Polym.-Plast. Technol. Eng. 47, 273–280.

Abdul Khalil, H.P.S., Kang, C.W., Khairul, A., Ridzuan, R., Mohd Adawi, T.O., 2008b. The effect of different laminations on mechanical and physical properties of hybrid composite. J. Reinf. Plast. Compos. 30, 24–30.

Abdul Khalil, H.P.S., Noorshashillawati Azura, M., Issam, A.M., Said, M.R., Mohd Adawi, T.O., 2008c. Oil palm empty fruit bunches (OPEFB) reinforced in new unsaturated polymer composites. J. Reinf. Plast. Compos. 27, 1817–1826.

Abdul Khalil, H.P.S., Kang, C.W., Khairul, A., Ridzuan, R., Adawi, T.O., 2009. The effect of different laminations on mechanical and physical properties of hybrid composites. J. Reinf. Plast. Compos. 28 (9), 1123–1137.

Abdul Khalil, H.P.S., Bhat, A.H., Jawaid, M., Amouzgar, P., Ridzuan, R., Said, M.R., 2010a. Agro-wastes: mechanical and physical properties of resin impregnated oil palm trunk core lumber. Polym. Compos. 3 (4), 638–644.

Abdul Khalil, H.P.S., Nurul Fazita, M.R., Bhat, A.H., Jawaid, M., Nik Fuad, N.A., 2010b. Development and material properties of new hybrid plywood from oil palm biomass. Mater. Des. 31, 417–424.

Abdul Khalil, H.P.S., Firoozian, P., Bakare, I.O., Hazizan, M.A., Ahmad, M.N., 2010c. Exploring biomass based carbon black as filler in epoxy composites: flexural and thermal properties. Mater. Des. 31, 3419–3425.

Abdul Khalil, H.P.S., Nurul Fazita, M.R., Jawaid, M., Bhat, A.H., Abdullah, C.K., 2011a. Empty fruit bunches as a reinforcement in laminated bio-composites. J. Compos. Mater. 45 (2), 219–236.

Abdul Khalil, H.P.S., Bhat, A.H., Ireana Yusra, A.F., 2011b. Green composites from sustainable cellulose nanofibrils: a review. Carbohydr. Polym. 87, 963–979.

Abdul Khalil, H.P.S., Fizree, H.M., Jawaid, M., Omar, S.A., 2011c. Preparation and characterization of nano-structured materials from oil palm ash: a bio-agricultural waste from oil palm mill. BioResources 6 (4), 4537–4546.

Abdul Khalil, H.P.S., Amouzgar, P., Jawaid, M., Hassan, A., Ahmad, F., Hadiyana, A., Dungani, R., 2012a. New approach to oil palm trunk core lumber material properties enhancement via resin impregnation. J. Biobased Mater. Bio. 6, 1–10.

Abdul Khalil, H.P.S., Jawaid, M., Hassan, A., Paridah, M.T., Zaidon, A., 2012b. Oil palm biomass fibres and recent advancement in oil palm biomass fibres based hybrid biocomposites. In: Composites and Their Applications. InTech, Croatia, pp. 187–220 (Chapter 8).

Abdul Khalil, H.P.S., Fizree, H.M., Bhat, A.H., Jawaid, M., Abdullah, C.K., 2013. Development and characterization of epoxy nanocomposites based on nano-structured oil palm ash. Compos. Part B 53, 324–333.

Abdullah, C.K., Jawaid, M., Abdul Khalil, H.P.S., Zaidon, A., Hadiyane, A., 2012. Oil palm trunk polymer composite: morphology, water absorption and thickness swelling behaviours. BioResources 7 (3), 2948–2959.

Abu Bakar, A., Abdul Khalil, H.P.S., 2004. Influence of oil palm fibre loading on the mechanical and physical properties of glass fibre reinforced epoxy bi-layer hybrid laminated composite. In: Paper Read at Proceeding of 3rd USM-JIRCAS Joint International Symposium, 9–11 March, Penang, Malaysia.

Abu Bakar, A., Abdul Khalil, H.P.S., 2005. Lignocellulose-based hybrid bilayer laminate composite: Part I – studies on tensile and impact behavior of oil palm fiber-glass fiber-reinforced epoxy resin. J. Compos. Mater. 39 (8), 663–684.

Abu Bakar, A., Hassan, A., Mohd Yusof, A.F., 2005. Effect of oil palm empty fruit bunch and acrylic impact modifier on mechanical properties and processability of unplasticized poly(vinyl chloride) composites. Polym.-Plast. Technol. Eng. 44, 1125–1137.

Agrawal, R., Saxena, N.S., Sreekala, M.S., Thomas, S., 2000. Non-isothermal crystallization kinetics of glass oil palm fibre reinforced phenolformaldehyde composites. J. Sci. Ind. Res. 59 (2), 136–139.

Ahmad, Z., Saman, H.M., Tahir, P.M., 2010. Oil palm trunk fibers as a bio-waste resource for concrete reinforcement. Int. J. Mech. Mater. Eng. 5 (2), 199–207.

Ahmadi, S.J., Huang, Y.D., Li, W., 2004. Synthetic routes, properties and future applications of polymer-layered silicate nanocomposites. J. Mater. Sci. 39, 1919–1925.

Amar, K.M., Manjusri, M., Lawrence, T.D., 2005. Natural Fibers, Biopolymers, and Biocomposites. CRC Press, Taylor & Francis, Boca Raton, FL.

Anuar, H., Ahmad, S.H., Rasid, R., Nik Daud, N.S., 2006. Tensile and impact properties of thermoplastic natural rubber reinforced short glass fiber and empty fruit bunch hybrid composites. Polym.-Plast. Technol. Eng. 45 (9), 1059–1063.

Arami-Niya, A., Wan Daud, W.M.A., Mjalil, F.S., 2010. Using granular activated prepared from oil palm shell by $ZnCl_2$ and physical activation for methane adsorption. J. Anal. Appl. Pyrolysis 89, 197–203.

Astimar, A.A., Das, K., Husin, M., Anis, M., 2002. Preparation of cellulose from oil palm empty fruit bunches via ethanol digestion: effect of acid and alkali catalysts. J. Oil Palm Res. 14 (1), 10–17.

Aström, B.T., 1997. Manufacturing of the polymer–matrix composites. http://www.eolss.net/Eolss-sampleAllChapter.aspx (accessed 18 October 2013).

Azman, H., Salema, A.A., Ani, F.N., Bakar, A.A., 2010. A review on oil palm empty fruit bunch fiber-reinforced polymer composite materials. Polym. Compos. 31(12), 2079–2101. http://dx.doi.org/10.1002/pc.

Badri, K.H., Mat Amin, K.A., Othman, Z., Manaf, H.A., Khalid, N.K., 2006. Effect of filler-to-matrix blending ratio on the mechanical strength of palm-based biocomposite boards. Polym. Int. 55, 190–195.

Barra, G.M.O., Fredel, M.C., Qureshi, H.A.A., Taylor, A.W., Clemenceau, C., 2006. Properties of chemically treated natural amorphous silica fibers as polyurethane reinforcement. Polym. Compos. 27, 582.

Basuki, W., Guritnob, P., Ismail, H., 2004. Oil palm empty fruit bunch filled polypropylene composites. Int. J. Polym. Mater. 53, 295–306.

Berthelot, J.M., 1999. Composite Materials: Mechanical Behavior and Structural Analysis. Springer-Verlag, New York.

Beun, S., Glorieux, T., Devaux, J., Vreven, J., Leloup, G., 2007. Characterization of nanofilled compared to universal and microfilled composites. Dent. Mater. 23 (1), 51–59.

Bhat, I.H., Abdul Khalil, H.P.S., 2011. Exploring “nano filler” based on oil palm ash in polypropylene composites. BioResources 6 (2), 1286–1297.

Bhat, I.H., Abdullah, C.K., Abdul Khalil, H.P.S., Ibrahim, M.H., Nurul Fazita, M.R., 2010. Properties enhancement of resin impregnated agro waste: oil palm trunk lumber. J. Reinf. Plast. Compos. 29 (22), 3301–3308.

Bismarck, A., Mishra, S., Thomas, L., 2005. Chapter 2: plant fibers as reinforcement for green composites. In: Book of Natural Fibres, Biopolymers and Biocomposites. Taylor and Francis Publication, Boca Raton, FL.

Chai, L., Zakaria, S., Chia, C., Nabihah, S., Rasid, R., 2009. Physico-mechanical properties of PF composite board from EFB fibers using liquefaction technique. Iran. Polym. J. 18, 917–923.

Chew, L.T., Ong, C.L., 1985. Particleboard from oil palm trunk. In: Proceedings of the National Symposium on Oil Palm By-Products for Agro-Based Industries, Kuala Lumpur. PORIM Bulletin, vol. 11, pp. 99–108.

Chisholm, N., Mahfuz, H., Rangari, V.K., Ashfaq, A., Jeelani, S., 2005. Fabrication and mechanical characterization of carbon/SiC-epoxy nanocomposites. Compos. Struct. 67, 115–124.

Chuayjuljit, S., Maungchareon, A., Saravari, O., 2010. Preparation and properties of palm oil-based rigid polyurethane nanocomposite foams. J. Reinf. Plast. Compos. 29 (2), 132–141.

Dagani, R.O.N., 1999. Putting the nano into composites. Chem. Eng. News 77, 25–37.

De Rosa, I.M., Santulli, C., Sarasini, F., 2009. Acoustic emission for monitoring the mechanical behaviour of natural fibre composites: a literature review. Compos. Part A 40, 1456–1469.

Deka, M., Saikia, C.N., Baruah, K.K., 2002. Studies of thermal degradation and termite resistance properties of chemically modified wood. Bioresour. Technol. 84 (2), 151–157.

Deraman, M., Zakaria, S., Husin, M., Aziz, A.A., Ramli, R., Mokhtar, A., Sahri, M.H., 1999. X-ray diffraction studies on fiber of oil palm empty fruit bunch and rubberwood for medium-density fiberboard. J. Mater. Sci. Lett. 18, 249–253.

Dungani, R., Jawaid, M., Abdul Khalil, H.P.S., Jasni, J., Aprilia, S., Hakeem, K.R., Hartati, S., Islam, M.N., 2012. A review on quality enhancement of oil palm trunk waste by resin impregnation: future materials. BioResources 8 (2), 3136–3156.

Dungani, R., Nazrul, M.D., Abdul Khalil, H.P.S., Hartati, S., Abdullah, C.K., Dewi, M., Hadiyane, A., 2013. Termite resistance study of oil palm trunk lumber (OPTL) impregnated with oil palm shell meal and phenol formaldehyde resin. BioResources 8 (4), 4937–4950.

Fizree, H.M., 2012. Characterization of nano structured oil palm ash and its influence as filler in epoxy composites. M.Sc. Thesis, Universiti Sains Malaysia, Penang, Malaysia.

François, V., 2013. Lineo – flax fibres impregnation. Available from: www.lineo.eu (accessed 10 October 2013).

Furuno, T., Imamura, Y., Kajita, H., 2004. The modification of wood by treatment with low molecular weight phenol-formaldehyde resin: a properties enhancement with neutralized phenolic-resin and resin penetration into wood cell walls. Wood Sci. Technol. 37 (5), 349–361.

Hassan, O.A., Ishida, M., Shukri, M.I., Tajuddin, Z.A., 1996. Oil-palm fronds as a roughage feed source for ruminants in Malaysia. In: Extension Bulletin, vol. 420. ASPAC, Food & Fertilizer Technology Center, p. 8.

Hench, L.L., 1991. Bioceramic: from concept to clinic. J. Am. Ceram. Soc. 1 (74), 1487–1510.

Hoydonckx, H., 2007. Renewable furan resins in composite applications. composites. Innovation 2007 Conference Proceedings. www.netcomposites.com (accessed 13 October 2013).
Husin, M., Ridzuan, R., Mokhtar, A., Wan, H.W.H., Kamaruddin, H., Ropandi, M., Astimar, A.A., 2005. Research and development of oil palm biomass in wood based industries. In: Malaysian Palm Oil Board Bulletin/Palm Oil Developments. MPOB Information Series, vol. 36.
Ibrahim, M.S., Sapuan, S.M., Faieza, A.A., 2012. Mechanical and thermal properties of composites from unsaturated polyester filled with oil palm ash. J. Mech. Eng. Sci. 2, 133–147.
Ibrahim, Z., Astimar, A.A., Ridzuan, R., Anis, M., Lee, S., 2013. Effect of refining parameters on medium density fibreboard (MDF) properties from oil palm trunk (*Elaeis guineensis*). Open J. Compos. Mater. 3, 127–131.
Irle, M., Barbu, M.C., 2010. Chapter 1: wood-based panel technology. In: Book of Wood-Based Panels: An Introduction for Specialists. Brunel University Press, London, pp. 1–90.
Ishak, M.Z.A., Aminullah, A., Ismail, H., Rozman, H.D., 1998. Effect of silane based coupling agents and acrylic acid based compatibilizers on mechanical properties of oil palm empty fruit bunch filled high density polyethylene composites. J. Appl. Polym. Sci. 68, 2189–2203.
Jacob, M., Varughese, K.T., Thomas, S., 2005. Water sorption studies of hybrid biofiber reinforced natural rubber biocomposites. Biomacromolecules 6 (6), 2969–2979.
Jacob, M., Varughese, K.T., Thomas, S., 2006. Dielectric characteristics of sisal-oil palm hybrid biofibre reinforced natural rubber biocomposites. J. Mater. Sci. 41 (17), 5538–5547.
Jacob, M., Sabu, T., Varughese, K.T., 2007. Biodegradability and aging studies of hybrid biofiber reinforced natural rubber biocomposites. J. Biobased Mater. Bioenergy 1, 118–126.
Jaffar Al-Mulla, E.A., 2011. Polylactic acid/epoxidized palm oil/fatty nitrogen compounds modified clay nanocomposites: preparation and characterization. Korean J. Chem. Eng. 28 (2), 620–626. http://dx.doi.org/10.1007/s11814-010-0373-6.
Jawaid, M., Abdul Khalil, H.P.S., 2011a. Cellulosic/synthetic fibre reinforced polymer hybrid composites: a review. Carbohydr. Polym. 86, 1–18.
Jawaid, M., Abdul Khalil, H.P.S., 2011b. Effect of layering pattern on the dynamic mechanical properties and thermal degradation of oil palm-jute fibers reinforced epoxy hybrid composite. BioResources 63, 2309–2322.
Jawaid, M., Abdul Khalil, H.P.S., Bakar, A.A., 2010. Mechanical performance of oil palm empty fruit bunches/jute fibres reinforced epoxy hybrid composites. Mater. Sci. Eng. A 527 (29–30), 7944–7949.
Jawaid, M., Abdul Khalil, H.P.S., Bakar, A.A., 2011. Hybrid composites of oil palm empty fruit bunches/woven jute fiber: chemical resistance, physical, and impact properties. J. Compos. Mater. 0(0), 1–8. http://dx.doi.org/10.1177/0021998311401102.
Jawaid, M., Abdul Khalil, H.P.S., Hassan, A., Abdallah, E., 2012. Bi-layer hybrid composites biocomposites: chemical resistant and physical properties. BioResources 7 (2), 2344–2355.
John, M.J., Francis, B., Varughese, K.T., Thomas, S., 2008. Effect of chemical modification on properties of hybrid fiber biocomposites. Compos. A Appl. Sci. Manuf. 39 (2), 352–363.
Joseph, S., Appukuttan, S.P., Kenny, J.M., Puglia, D., 2010. Dynamic mechanical properties of oil palm microfibril-reinforced natural rubber composites. J. Appl. Polym. Sci. 117 (3), 1298–1308.
Joshua, B., 2011. Mixed palm oil results for 2010, 2011 growth supported by NKEA projects. Economic Transformation Programme.
Jusoh, A., Mohd Noor, M.J.M., Ghazali, A.H., 1995. Potential of burnt oil palm shell (BOPS) granules in deep bed filtration. J. Islamic Acad. Sci. 8, 143–148.

Karina, M., Onggo, H., Abdullah, A.H.D., Syampurwadi, A., 2008. Effect of oil palm empty fruit bunch fiber on the physical and mechanical properties of fiber glass reinforced polyester resin. J. Biol. Sci. 8 (1), 101–106.

Khalid, M., Salmiaton, A., Chuah, T.G., Ratman, C.T., Choong, S.Y.T., 2008. Effect of MAPP and TMPTA as compatibilizer on the mechanical properties of cellulose and oil palm fiber empty fruit bunch–polypropylene biocomposites. Compos. Interfaces 15 (2–3), 251–262.

Khoo, K.C., Lee, T.W., 1991. Pulp and paper from oil palm. Appita J. 44 (6), 385–388.

Killman, W., Hong, L.T., 1989. Anatomy and properties of oil palm stem. In: Proceeding of the National Symposium of Oil Palm By-Products for Agro-Based Industries, Kaula Lumpur, 1842.

Killmann, W., Choon, L.S., 1985. Anatomy and properties of oil palm stem. PORIM Bull. 11, 18–42.

Koo, J.H., 2006. Polymer Nanocomposites: Processing, Characterization, and Applications. McGraw-Hill, New York.

Kumar, R.N., Wei, L.M., Rozman, H.D., Abusamah, A., 1997. Fire resistant sheet moulding composites from hybrid reinforcements of oil palm-fibres and glass fibre. Int. J. Polym. Mater. 37 (1–2), 43–52.

Laemsak, N., Okuma, M., 2000. Development of boards made from oil palm frond properties of binderless boards from steam exploded fibers of oil palm frond. J. Wood Sci. 46, 322–326.

Lau, H.L.N., Choo, Y.M., Ma, A.N., Chuah, C.H., 2005. Identification of water-soluble components from palm-pressed fiber. In: Proceedings of the 2005 MPOB International Congress (2005 PIPOC), September 25–29. Sunway Pyramid Convention Centre, Selangor, Malaysia, p. 23.

Law, K.N., Jiang, X., 2001. Comparative papermaking properties of oil palm empty fruit bunch. Tappi J. 84 (1), 1–13.

Law, K.N., Wan Rosli, W.D., Arniza, G., 2007. Morphological and chemical nature of fiber strands of oil palm empty-fruit-bunch (OPEFB). BioResources 2 (3), 351–360.

Lee, S., 2010. An overview of advanced composite material and their industrial applications. www.industryhk.org/english/news/news_fed/…/presentation_acm.pdf (accessed 17 October 2013).

Leete, R., 2007. 50 years of economic and social development. In: Dalam Malaysia: From Kampung to Twin Towers. Oxford Fajar, Legal Research Board, Kuala Lumpur.

Loh, Y.F., Paridah, M.T., Hoong, Y.B., Bakar, E.S., Hamdan, H., Anis, M., 2010. Properties enhancement of oil palm plywood through veneer pretreatment with low molecular weight phenol-formaldehyde resin. J. Adhes. Sci. Technol. 24, 1729–1738.

Lua, A.Ch., Guo, J., 2001. Microporous oil-palm-shell activated carbon prepared by physical activation for gas-phase adsorption. Lagmuir 17, 7112–7117.

Matsuda, H., 1996. Chemical modification of solid wood. In: Hon, D.N.S. (Ed.), Chemical Modification of Lignocellulosic Materials. Marcel Dekker, Inc, New York, Basel, Hong Kong, pp. 159–183.

Mohamad, H., Anis, M., Wan Hasamudin, W.H., 2005. Energizing the wood-based industry in Malaysia. In: Proceedings of the 6th National Seminar on the Utilisation of Oil Palm Tree. Oil Palm Tree Utilisation Committee (OPTUC), Malaysia, pp. 6–13.

Mridha, S., Keng, S.B., Ahmad, Z., 2007. The effect of OPWF filler on impact strength of glass-fiber reinforced epoxy composite. J. Mech. Sci. Technol. 21 (10), 1663–1670.

Mukherjee, P.S., Satyanaraya, K.G., 1986. An empirical evaluation of structure-property relationships in natural fibers and their fracture behaviour. J. Mater. Sci. 21, 4162–4170.

Mullen, R.L., Kurniawan, A., 2010. Chipboard. U.S. Patent No. 7,819,147, 26 October.

Ndazi, B., Tesha, J.V., Bisanda, E.T.N., 2006. Some opportunities and challenges of producing bio-composites from non-wood residues. J. Mater. Sci. 41, 6984–6990.

Njuguna, J., Pielichowski, K., 2003. Polymer nanocomposites for aerospace applications: properties. Adv. Eng. Mater. 5, 769–778.

Noorbaini, S., 2009. Effect of Cold Setting Adhesives and Glue Spread on Properties of Oil Palm Trunk Laminated Veneer. Thesis, Universiti Sains Malaysia.

Norul Izani, M.A., Paridah, M.T., Astimar, A.A., Mohd Nor, M.Y., Anwar, U.M.K., 2012. Mechanical and dimensional stability properties of medium density fiberboard produced from treated oil palm empty fruit bunch. J. Appl. Sci. 12 (6), 561–567.

Paul, K.T., Satphaty, S.K., Manna, L., Chakraborty, K.K., Nando, G.B., 2007. Preparation and characterization of nano structured materials from fly ash: a waste from thermal power stations, by high energy ball milling. Nanoscale Res. Lett. 2, 397–404.

Pritchard, G., 2004. Two technologies merge: wood plastic composites. Reinf. Plast. 48 (6), 26–29.

Rahman, S.H.A., Choudhury, J.P., Ahmed, A.L., 2006. Production of xylose from oil palm empty fruit bunch fiber using sulfuric acid. Biochem. Eng. J. 30, 97–103.

Razak, W., Samsi, H.W., Mohamed, A., Sulaiman, O., 2008. Utilization potential of 30year-old oil palm trunks laminated veneer lumbers for non-structural purposes. J. Sustain. Dev. 1 (3), 212–224.

Rihayat, T., Saari, M., Suraya, A.R., Mahmood, M.H., Dahlan, K.Z.H.M., Wan Yunus, W.M.Z., Sapuan, S.M., 2006. Synthesis and thermal characterization of polyurethane/clay nanocomposites based on palm oil polyol. Polym.-Plast. Technol. Eng. 45, 1323–1326.

Roshafima, R.A., Wan Aizan, W.A.R., 2008. Low density polyethylene/tapioca starch biofilm with palm oil as processing aid for food packaging. In: International Conference on Advances Materials and Processing Technologies. Gulf International Convention Centre, Gulf Hotel Kingdom of Bahrain.

Rowell, R.M., 1975. Chemical modification of wood: advantages and disadvantages. In: Proceedings Am. Wood Preservers' Association. pp. 1–10.

Rozman, H.D., Peng, G.B., Mohd Ishak, Z.A., 1998. The effect of compounding techniques on the mechanical properties of oil palm empty fruit bunch polypropylene composites. J. Appl. Polym. Sci. 70, 2647–2655.

Rozman, H.D., Lai, C.Y., Ismail, H., Mohd Ishak, Z.A., 2000. The effect of coupling agents on the mechanical and physical properties of oil palm empty fruit bunch–polypropylene composites. Polym. Int. 49, 1273–1278.

Rozman, H.D., Saad, M.J., Mohd Ishak, Z.A., 2003. Material properties flexural and impact properties of oil palm empty fruit bunch (EFB)–polypropylene composites—the effect of maleic anhydride chemical modification of EFB. Polym. Test. 22, 335–341.

Rozman, H.D., Ahmadhilmi, K.R., Abubakar, A., 2004. Material properties: polyurethane (PU)—oil palm empty fruit bunch (EFB) composites: the effect of EFBG reinforcement in mat form and isocyanate treatment on the mechanical properties. Polym. Test. 23, 559–565.

Rozman, H.D., Ahmad Hilme, K.R., Abubakar, A., 2007. Polyurethane composites based on oil palm empty fruit bunches: effect of isocyanate/hydroxyl ratio and chemical modification of empty fruit bunches with toluene diisocyanate and hexamethylene diisocyanate on mechanical properties. J. Appl. Polym. Sci. 106, 2290–2297.

Rudd, C., 2001. Composite for Automotive Applications. Rapra Technology Ltd, iSmithers Rapra Publishing, UK. ISBN: 1-85957-265-0.

Scholz, M.S., Blanchfield, J.P., Bloom, L.D., Coburn, B.H., Elkington, M., Fuller, J.D., Gilbert, M.E., Muflahi, S.A., Pernice, M.F., Rae, S.I., Trevarthen, J.A., White, S.C., Weaver, P.M., Bond, I.P., 2011. The use of composite materials in modern orthopaedic medicine and prosthetic devices: a review. Compos. Sci. Technol. 7 (16), 1791–1803.

Shafigh, P., Jumaat, M.Z., Mahmud, H., 2011. Oil palm shell as a lightweight aggregate for production high strength lightweight concrete. Construct. Build Mater. 25, 1848–1853.

Shuit, S.H., Tan, K.T., Lee, K.T., Kamaruddin, A.H., 2009. Oil palm biomass as a sustainable energy source: a Malaysian case study. Energy 34, 1225–1235.

Siyamak, S., Ibrahim, N.A., Abdolmohammadi, S., Wan Yunus, W.M.Z., Rahman, M.Z.A.B., 2012. Effect of fiber esterification on fundamental properties of oil palm empty fruit bunch fiber/poly(butylene adipate-*co*-terephthalate) biocomposites. Int. J. Mol. Sci. 13, 1327–1346.

Slette, J.P., Wiyono, I.E., 2011. Oilseeds and products update 2011. USDA Foreign Agricultural Service, Washington, D.C., USA. [online] URL: http://www.usdaindonesia.org/public/uploaded/Oilseeds%20and%20Products%20Update_Jakarta_Indonesia.pdf (accessed 28 January 2013).

Sreekala, M.S., Kumaran, M.G., Thomas, S., 1997. Oil palm fibers: morphology, chemical composition, surface modification and mechanical properties. J. Appl. Polym. Sci. 66, 821–835.

Sreekala, M.S., Kumaran, M.G., Joseph, S., Jacob, M., Thomas, S., 2000. Oil palm fibre reinforced phenol formaldehyde composites: influence of fibre surface modifications on the mechanical performance. Appl. Compos. Mater. 7, 295.

Sreekala, M.S., George, J., Kumaran, M.G., Thomas, S., 2002. The mechanical performance of hybrid phenol – formaldehyde-based composites reinforced with glass and oil palm fibers. Compos. Sci. Technol. 62, 339–353.

Sreekala, M.S., Kumaran, M.G., Geethakumariamma, M.L., Thomas, S., 2004. Environmental effects in oil palm fiber reinforced phenol formaldehyde composites: Studies on thermal, biological, moisture and high energy radiation effects. Adv. Compos. Mater. 13 (34), 171–197.

Sreekala, M.S., Thomas, S., Groeninckx, G., 2005. Dynamic mechanical properties of oil palm fiber/phenol formaldehyde and oil palm fiber/glass hybrid phenol formaldehyde composites. Polym. Compos. 26 (3), 388–400.

Srinivasan, R., 2010. Engineering Materials and Metallurgy. Tata McGraw Hill, New Delhi.

Stark, N.M., Cai, Z., Carll, C., 2010. Chapter 11: Panel products, glued-laminated timber, structural composite lumber and wood–nonwood composite materials. In: Book of Wood Handbook. U.S. Department of Agriculture, Forest Service, Forest Products Laboratory, Madison, WI, pp. 11-1–11-28, General Technical Report FPL-GTR-190.

Sulaiman, O., Hashim, R., Raza, K.W., Hashim, W.S., Azmy, M., 2008. Evaluation on some finishing properties of oil palm plywood. Holz Roh Werkst 66, 5–10.

Sulaiman, N.S., Hashim, R., Amini, M.H.M., Sulaiman, O., Hiziroglu, S., 2013. Evaluation of the properties of particleboard made using oil palm starch modified with epichlorohydrin. BioResources 8 (1), 283–301.

Sumathi, S., Chai, S.P., Mohamed, A.R., 2008. Utilization of oil palm as a source of renewable energy in Malaysia. Renew. Sustain. Energy Rev. 12 (9), 2404–2421.

Tay, J.H., Show, K.Y., 1995. Use of ash derived from oil-palm waste incineration as a cement replacement material. Resour. Conserv. Recycl. 13, 27–36.

Tay, G.S., Mohd. Zaim, J., Rozman, H.D., 2010. Mechanical properties of polypropylene composite reinforced with oil palm empty fruit bunch pulp. J. Appl. Polym. Sci. 116, 1867–1872.

Teoh, Y.P., Mashitah, M.D., 2010. Cellulose production by *Pycnoporus sanguineus* on oil palm residues through pretreatment and optimization study. J. Appl. Sci. 10 (12), 1036–1043.

United Nations Environment Programme (UNEP), 2011. Oil palm plantation: threats and opportunities for tropical ecosystem. [online] Available from world wide web: www.unep.org/GEAS (accessed 22 June 2012).

United Soybean Board, 2013. Soy-based adhesives provide new greener options for North American construction professionals. Wood Design and Building (Spring),41–44, Technical Solution.

Virk, A.S., Hall, W., Summerscales, J., 2010. Failure strain as the key design criterion for fracture of natural fibre composites. Compos. Sci. Technol. 70 (6), 995–999.

Wong, K.J., Nirmal, U., Lim, B.K., 2010. Impact behavior of short and continuous fiber reinforced polyester composites. J. Reinf. Plast. Compos. 29 (23), 3463–3474.

Youngquist, J.A., 1999. Chapter 10: wood-based composites and panel products. In: Book of Wood Handbook—Wood as an Engineering Material. U.S. Department of Agriculture, Forest Service, Forest Products Laboratory, Madison, WI, pp. 10-1–10-31, General Technical Report FPL-GTR-113.

Yusoff, M.N.B.M., Khoo, K.C., Lee, T.W., 1984. Preliminary characterization of oil palm stem as a raw material for pulp and paper. Malaysian Forester 47 (1/2), 28–41.

Zaidon, A., Nizam, A.M.N., Noor, M.Y.M., Abood, F., Paridah, M.T., Yuziah, M.Y.N., Jalaludin, H., 2007. Properties of particleboard made from pretreated particles of rubberwood, EFB and rubberwood-EFB blend. J. Appl. Sci. 7 (8), 1145–1151.

Zaini, W.I.J., Fuad, M.Y.A., Ismail, Z., Mansor, M.S., Mustafah, J., 1996. The effect of filler content and size on the mechanical properties of polypropylene/oil palm wood flour composites. Polym. Int. 40, 51–55.

Zainudin, N.F., Lee, K.T., Kamaruddin, A.H., Bhatia, S., Mohamed, A.R., 2005. Study of adsorbent prepared from oil palm ash (OPA) for flue gas desulfurization. Sep. Purif. Technol. 45, 50–60.

Zhang, C., Tjiu, W.W., Liu, T., Lui, W.Y., Phang, I.Y., Zhang, W.D., 2011. Dramatically enhanced mechanical performance of nylon-6 magnetic composites with nanostructured hybrid one-dimensional carbon nanotube – two-dimensional clay nanoplatelet heterostructures. J. Phys. Chem. B 115, 3392–3399.

Zuhri, M., Yusoff, M., Sapuan, S.M., Ismail, N., Wirawan, R., 2010. Mechanical properties of short random oil palm fibre reinforced epoxy composites. Sains Malaysiana 39 (1), 87–92.

Design, processing, and characterization of triaxially braided natural fiber epoxy-based composites

16

I.I. Qamhia, R.F. El-Hajjar
University of Wisconsin, Milwaukee, WI, USA

16.1 Introduction

Natural fibers and biobased polymers such as poly-(lactic acid), natural oils, and various epoxies with large biobased sources have been previously used to manufacture biobased composites. Natural fibers are considered an attractive reinforcement in composite materials for several product applications due to their renewable and biodegradable nature (Cheng et al., 2009; Graupner and Mussig, 2011; Soykeabkaew et al., 2009). Both thermosetting and thermoplastic resins have been used to manufacture composites with natural fibers. The most commonly used thermosetting matrices are phenolic, epoxy, and polyester resins (Malkapuram et al., 2009; Ku et al., 2011). The use of natural fiber reinforcement with epoxy resins is extensively reported in the literature. Yan et al. (2012) investigated the effect of alkali treatment with 5% by weight NaOH on epoxy-based natural fiber composite systems with flax, linen, and bamboo fabrics. The tensile fractured surfaces were seen to exhibit an improved fiber/epoxy interfacial adhesion and an increase in the tensile and flexural properties, like flexural strength. The treatment, however, had a negative effect on the tensile strength and modulus of the single-strand yarns of the three fibers. Pinto et al. (2014) produced untreated alkali and silane surface-treated plain weaved jute fibers/epoxy composites by vacuum infusion. The treated jute fiber/epoxy composites showed a 300% increase in the elastic modulus compared to neat epoxy resin. The silane treated composites had significant improvements in moisture resistance ability. On the other hand, silane-treated bamboo fiber/epoxy composites had lower elastic modulus and comparable tensile strength with the untreated fiber composites (Kushwaha and Kumar, 2010). Shih et al. (2012) prepared novel epoxy-based composites with plant and waste fibers made from bananas fiber, banana leaf fiber, and recycled chopstick fibers. The reinforced epoxy composites had an 80–117% higher tensile strength compared to neat epoxy. The chemically modified fiber-reinforced composites with 3-aminopropyltriethoxysilane also had higher decomposition temperature compared to the untreated ones. The molecular deformation of regenerated cellulose fibers in epoxy and polyester, studied using Raman spectroscopy analysis, has shown the presence of large interfacial shear stress between the fibers and the resin (Mottershead and Eichhorn, 2007). Higher

Biocomposites: Design and Mechanical Performance. http://dx.doi.org/10.1016/B978-1-78242-373-7.00007-X

properties have been found when biocompatible resins were used with regenerated cellulose fibers (Wollerdorfer and Bader, 1998).

Compared to other natural fiber systems, regenerated cellulose fibers can offer less variability in the properties of the fiber (Gindl and Keckes, 2006). Meredith et al. investigated natural fiber/epoxy composites for high performance structural applications (Meredith et al., 2013). Test laminates of preimpregnated woven flax and cordenka (regenerated cellulose) were prepared in an autoclave and tested for their static and energy absorption properties. The tensile strengths ranged from 63 to 92.6 MPa and interlaminar shear strength from 10.7 to 23.3 MPa. Specific energy absorption varied between 21.2 and 34.2 kJ/kg. El-Hajjar and Qamhia investigated the tensile mechanical properties of unidirectional Lyocell/epoxy composites under wet and dry conditions using a multiscale analysis approach (El-Hajjar and Qamhia, 2013a). A unit cell finite element model was successfully proposed to relate micro- to macromechanical behavior for predicting the 3-D orthotropic elastic–plastic constants of regenerated cellulose/epoxy composites. The acoustic properties, dynamic and static modulus of elasticity, and the bending strength and density of regenerated cellulose fibers in an epoxy matrix were investigated by Buksnowitz et al. (2010). The static and the dynamic modulus were found to be relatively high compared to glass/epoxy reference samples when the low density of the composites is taken into account.

Braiding fibers, including regenerated cellulose fibers, can produce a unique fiber preform that can be impregnated with a thermosetting matrix using various techniques like resin transfer molding, vacuum assisted casting, or in preimpregnated forms. The advantages of a braided system is that it increases the through-the-thickness yarn interlacing properties, thus reducing the susceptibility to delamination (Miravete et al., 2000). Braided composites have a wide range of possible applications due to the out-of-plane resistance improvement relative to uniaxial fabrics and filament wound composites (Poe et al., 1997). Complex shapes at high rates through automation techniques can be formed using braided preforms (Scholz et al., 1997). Triaxially braided architectures are designed by using three fiber orientations for the axial and the two-braider yarns. The amount of fibers in distinct amounts and orientations can be used to achieve target properties. Advanced braids can be constructed such that quasi-isotropic properties may be obtained through optimization of the fiber preform. The most effective parameters are the axial yarn size, braider yarn size, braider angle, and the axial yarn spacing. The stiffness and strength properties are typically determined from the fiber, matrix, interface mechanical properties, and fabric architecture. Challenges in direct application of classical lamination plate theories rise from the presence of the tow undulations in the bias yarns. Several approaches in energy methods, finite element, and classical lamination plate theory methods have been proposed to model the elastic response (Byun, 2000; Yang et al., 1986; Masters et al., 1993; Naik, 1994; Quek et al., 2003). In one of these approaches, Quek et al. (2003) proposed a representative unit cell (RUC) of the braided architecture that accounted for the undulation in the tows and the bias tow angle by separating the axial and bias yarns into separate layers with subsequent averaging. Shokrieh and Mazloomi (2010) proposed a modified equation to better capture the shear properties. The three-layer analytical representations were evaluated by El-Hajjar et al. (2013)

for triaxially braided glass fiber/epoxy composites, and by Qamhia et al. (2015) on triaxially braided regenerated cellulose fiber/epoxy composites.

In this study, a micro- to macromechanics approach using a three-layer discrete analytical model is used to predict the mechanical properties in a quasi-isotropic triaxially braided regenerated cellulose composite with a high biocontent epoxy resin. The results are validated by investigating the tensile behavior of composites prepared by a resin infusion technique. Both longitudinal and transverse test coupons are studied. The load redistribution in the composites is investigated by a full-field study using a digital image correlation (DIC) technique. Damage initiation and propagation is examined using a passive acoustic emission (AE) technique. The analytical model was examined for the implications of changing the braider angle, aspect ratios, crimp angle, and the yarn spacing on the elastic properties of the braided composite. In the second part of the study, the viscoelastic properties of the quasi-isotropic composite are investigated. The manufactured composites are evaluated using dynamic–mechanical analysis for wet and dry conditions and for transverse and axial directions using a three-point bending test setup. Two different studies were carried out: in the first study, specimens are cyclically loaded and exposed to a temperature ramp; in the second, the specimens are ramped under progressively increasing temperature and frequency loadings.

16.2 Processing of triaxially braided cellulose and bioepoxy composites

In the experimental investigation, composite panels are prepared from triaxially braided regenerated cellulose fibers (BioMid; GSC and ENC, Inc., Burnaby, BC, Canada) and a high-biocontent epoxy resin system (Super Sap 100/1000; Entropy Bio-Resins, Gardena, California, USA) using a resin infusion technique. The mechanical properties of the fibers and the resin are shown in Table 16.1. The regenerated cellulose fibers are reported to have a 95% degree of crystallinity and stability at high temperatures up to 360 °C (680 °F) compared to other natural fibers that start to degrade at 150 °C (302 °F) (Composites_Technology, 2012). The fibers used in the study are braided into a quasi-isotropic sleeve consisting of axial and +60/−60 degrees bias yarns in a single ply. The braiding was especially performed by A&P Technology (Cincinnati, Ohio, USA). The triaxial braid fiber architecture is produced by having the bias yarns alternating two over and two under the axial yarns. The quasi-isotropic properties are

Table 16.1 Regenerated cellulose fiber and epoxy matrix properties

	Regenerated cellulose fiber	Epoxy
E_x GPa (Msi)	46 (6.67)	1.45 (0.21)
E_y GPa (Msi)	46 (6.67)	1.45 (0.21)
G_{xy} GPa (Msi)	18 (2.61)	0.53 (0.076)
Poisson's ratio	0.23	0.35

Table 16.2 Triaxial braiding parameters for regenerated cellulose quasi-isotropic ply

Braided angle (degree)	60
Undulation amplitude (mm)	0.0005
Yarn aspect ratio	0.18
Fiber volume fraction (%)	60.5
Raw fiber yield—Braider yarns (yds/lb)	2653
Raw fiber yield—axial yarns (yds/lb)	2653
Material content in axial yarns (% by volume)	33.4
Material content in braider yarns (% by volume)	66.6

obtained by controlling the bias angle and the spacing of the axial and bias yarns to yield similar properties in both axial and transverse directions. The spacing between the bias yarns was at 2.6 mm, whereas the center-to-center spacing between the axial yarns was kept at 5.2 mm and the width of the axial bundle was approximately 2.5 mm between the bias yarns. The percentage by weight in each bias yarn and in the axial yarn is equal (Table 16.2) to obtain quasi-isotropic properties. Following the braiding process, the braided sleeve was subsequently slit into a flat sheet to allow for further processing. The biobased epoxy resin is reported by the supplier to have a 37% bio-content, mainly from coproducts of other green industries such as wood pulp and bio-fuels production. The resin is classified as a U.S. Department of Agriculture (USDA) biopreferred product using ASTM D6866 (ASTM, D6866, 2007). The mixed resin has initial viscosity of 725 cps and a pot life of 25 min at 25 °C (77 °F).

The regenerated fibers were stored at room temperature and a humidity level of 30%. A vacuum-assisted resin infusion method was used to manufacture the composite panels with minimal levels of porosity (Figure 16.1). The process resulted in an average fiber volume fraction of 0.60 ± 0.02, which was calculated using volume measurements of the matrix and fiber constituents (ASTM, D2734, 2009). A wet layup

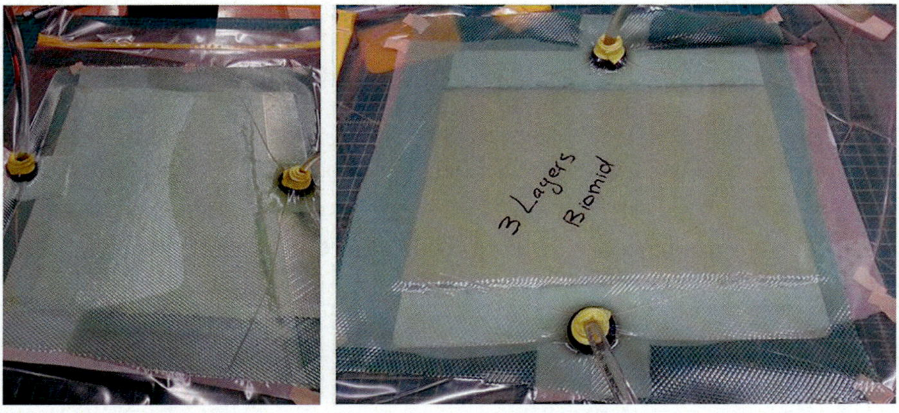

Figure 16.1 Fabrication of triaxially braided regenerated cellulose/epoxy composites using a vacuum-assisted resin infusion method.

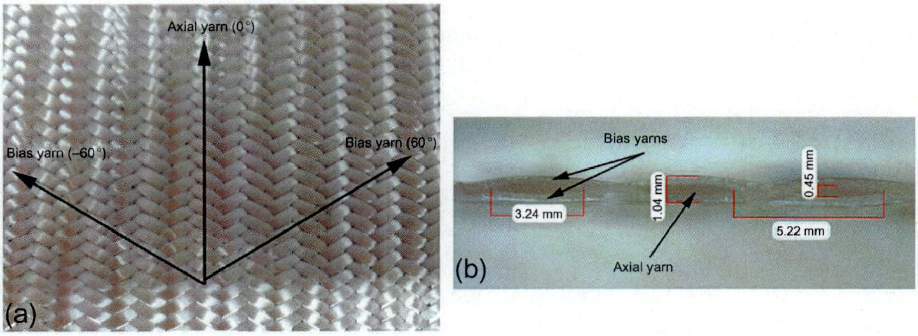

Figure 16.2 (a) A quasi-isotropic triaxially braided BioMid preform and (b) a cross-section micrograph of the cured composite.

technique was used to prepare the composites in the initial trials, but the fibers weren't fully infused; the axial yarns and porosity levels, especially, were relatively high. The increased difficulty in processing may be attributed to the swelling of the fibers as they are wetted with resin. The modified process used silicon connectors for the resin introduction and flow mesh to improve the wetting process. The fibers were infused under room temperature conditions. A cross-section micrograph of the final cured composite and the triaxially braided preform architecture are shown in Figure 16.2. Panels were cut into coupons in transverse and longitudinal directions.

16.3 Analytical model

Several factors affect the strength and stiffness properties of triaxially braided composites; these factors are typically the fibers, matrix, interface mechanical properties, and fabric architecture. The direct application of classical lamination plate theories to these composites can raise some challenges due to tow undulations in the bias yarns. Several methods have been proposed to model the elastic response of the braided composites. These methods included macro- and micromechanical approaches energy methods, finite element methods, as well as classical lamination plate theory methods (Byun, 2000; Yang et al., 1986; Masters et al., 1993; Naik, 1994; Quek et al., 2003). The RUC of the braided architecture proposed by Quek et al. (2003) took into account the tow undulation and the bias tow angle through its constituents by separating the axial and bias yarns into separate layers with subsequent averaging. A modified equation was proposed by Shokrieh and Mazloomi (2010) to better capture the shear properties. In both models reported, the thicknesses of all three layers were assumed to be identical. The application of the 3-layer analytical representations was evaluated by El-Hajjar et al. (2013) for application on triaxially braided models of glass fiber composites. The 3-layer representation approach yielded consistent results that showed good correlation with the experimental results.

For the triaxially braided regenerated cellulose architecture, the modified 3-layer discrete analytical model is evaluated for prediction of the mechanical properties using a micro- to macromechanics approach. The crimp effect of the braider yarns is taken into account by applying the following sinusoidal function to the bias layers (Quek et al., 2003; Shokrieh and Mazloomi, 2010; Qamhia et al., 2015):

$$\tan(\beta) = \frac{\pi A}{L}\cos\left(\frac{\pi x'}{L}\right) \tag{7.1}$$

The local coordinates of the undulating yarn along a ply is defined by the 1–2–3 local coordinates (Quek et al., 2003). Two sets of rotation transformation matrices are used around the *y*-axis and *z*-axis for the braider yarns for alignment in the global axes (Figure 16.2). The first two rotations are used for considering the crimp effect and the second set is for the bias angles that are shown in Equations (7.2)–(7.6), respectively.

$$\hat{T}_1 = \begin{pmatrix} \cos(\beta)^2 & 0 & \sin(\beta)^2 & 0 & 2\cos(\beta)\sin(\beta) & 0 \\ 0 & 1 & 0 & 0 & 0 & 0 \\ \sin(\beta)^2 & 0 & \cos(\beta)^2 & 0 & -2\cos(\beta)\sin(\beta) & 0 \\ 0 & 0 & 0 & \cos(\beta) & 0 & \sin(\beta) \\ -\cos(\beta)\sin(\beta) & 1 & \cos(\beta)\sin(\beta) & 0 & \cos(\beta)^2 - \sin(\beta)^2 & 0 \\ 0 & 0 & 0 & \sin(\beta) & -2\cos(\beta)\sin(\beta) & \cos(\beta) \end{pmatrix} \tag{7.2}$$

$$\hat{T}_2 = \begin{pmatrix} \cos(\beta)^2 & 0 & \sin(\beta)^2 & 0 & \cos(\beta)\sin(\beta) & 0 \\ 0 & 1 & 0 & 0 & 0 & 0 \\ \sin(\beta)^2 & 0 & \cos(\beta)^2 & 0 & -\cos(\beta)\sin(\beta) & 0 \\ 0 & 0 & 0 & \cos(\beta) & 0 & -\sin(\beta) \\ -2\cos(\beta)\sin(\beta) & 1 & 2\cos(\beta)\sin(\beta) & 0 & \cos(\beta)^2 - \sin(\beta)^2 & 0 \\ 0 & 0 & 0 & \sin(\beta) & 0 & \cos(\beta) \end{pmatrix} \tag{7.3}$$

The stress–strain relation can be obtained by using the averaged transformed local stiffness calculated over one wave-length (Shokrieh and Mazloomi, 2010):

$$\{\sigma\} = \frac{1}{2l}\int\left[\hat{T}_1\right]^{-1}[C]\left[\hat{T}_2\right]\mathrm{d}x\{\varepsilon\} \tag{7.4}$$

And the second rotation, as seen in Figure 16.2, for the braider angle transformation is given by (Quek et al., 2003)

$$T_1 = \begin{pmatrix} \cos(\theta)^2 & \sin(\theta)^2 & 0 & 0 & 0 & 2\cos(\theta)\sin(\theta) \\ \sin(\theta)^2 & \cos(\theta)^2 & 0 & 0 & 0 & -2\cos(\theta)\sin(\theta) \\ 0 & 0 & 1 & 0 & 0 & 0 \\ 0 & 0 & 0 & \cos(\theta) & -\sin(\theta) & 0 \\ 0 & 0 & 0 & \sin(\theta) & \cos(\theta) & 0 \\ -\cos(\theta)\sin(\theta) & \cos(\theta)\sin(\theta) & 0 & 0 & 0 & \cos(\theta)^2 - \sin(\theta)^2 \end{pmatrix} \tag{7.5}$$

$$T_2 = \begin{pmatrix} \cos(\theta)^2 & \sin(\theta)^2 & 0 & 0 & 0 & \cos(\theta)\sin(\theta) \\ \sin(\theta)^2 & \cos(\theta)^2 & 0 & 0 & 0 & -\cos(\theta)\sin(\theta) \\ 0 & 0 & 1 & 0 & 0 & 0 \\ 0 & 0 & 0 & \cos(\theta) & -\sin(\theta) & 0 \\ 0 & 0 & 0 & \sin(\theta) & \cos(\theta) & 0 \\ -2\cos(\theta)\sin(\theta) & 2\cos(\theta)\sin(\theta) & 0 & 0 & 0 & \cos(\theta)^2 - \sin(\theta)^2 \end{pmatrix} \quad (7.6)$$

The stress–strain relation in the global stiffness axes is given as (Shokrieh and Mazloomi, 2010)

$$\{\sigma\} = [\mathrm{T}_1]^{-1}[\bar{\mathrm{C}}][\mathrm{T}_2]\{\varepsilon\} \quad (7.7)$$

where $[\bar{\mathrm{C}}]$ is the stiffness matrix obtained from rotation around the *y*-axis. After calculating the stiffness matrix of each layer, the stiffness matrix of the unit cell can be determined as (Shokrieh and Mazloomi, 2010)

$$[C^{\mathrm{RUC}}] = t^{+\theta}[C_{\mathrm{global}}]^{+\theta} + t^{-\theta}[C_{\mathrm{global}}]^{-\theta} + t^{0}[C_{\mathrm{global}}]^{0} \quad (7.8)$$

where $t^{+\theta}$, $t^{-\theta}$, and t^0 are specified as the thickness of each layer to the thickness of the laminate. Previous studies have considered these thicknesses to be identical (Shokrieh and Mazloomi, 2010). The actual thickness values obtained from microstructural examinations are used in the current study.

The predictions provided by the analytical model for the effect of changing the fiber volume fraction and the aspect ratio on the mechanical properties such as the elastic modulus and the Poisson's ratio in *x* and *y* directions are shown in Figure 16.3. A high correlation is seen between the analytical model predictions and the experimental results. Differences in stiffness between experiments and model results were found to be less than 4% in both the longitudinal and transverse directions. The results also show how the in-plane fiber architecture of the quasi-isotropic triaxially braided composite yields close stiffness in both the axial and transverse directions. The variation of longitudinal and transverse elastic modulus with respect to the fiber volume fraction shows good correlation with the experimental data. The same is also observed for the Poisson ratio predictions by the model in the longitudinal direction (ν_{xy}), where the difference between the experimental value obtained from DIC results and the analytical model is around 2%. For the Poisson effect in the transverse direction (ν_{yx}), however, the results from experimental and analytical approaches vary by around 15%. The same trend is also seen for the variation of Poisson ratio with the aspect ratio. The effects of the crimp angle on the mechanical constants were also investigated. A relatively high sensitivity of the transverse modulus of elasticity to the crimp is seen. However, there is no significant variation in the predicted longitudinal stiffness with the crimp angle as the behavior in that direction is dominated

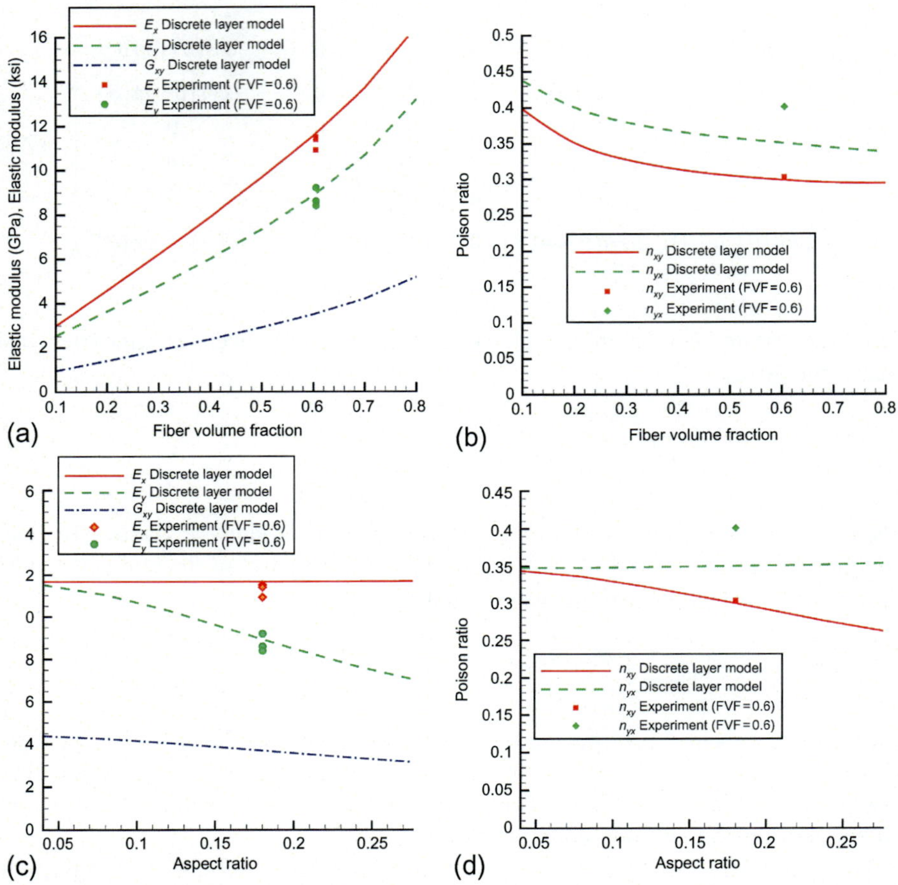

Figure 16.3 Results of the analytical discrete 3-layers model for (a) fiber volume fraction versus the elastic modulus, (b) fiber volume fraction versus Poisson's ratio, (c) aspect ratio versus the elastic modulus, and (d) aspect ratio versus Poisson's ratio.

by the longitudinal fibers that have no undulations. In contrast with E_x, E_y decreases by increasing the aspect ratio because this parameter affects the bias yarns that have a larger contribution in the transverse direction. Note that there is a physical limitation on the aspect ratio as this is a function of the yarn spacing and the fiber volume fraction of the composite. In addition, a comparison of the effects of changes in the aspect ratio shows less effect on the predicted properties of the quasi-isotropic material compared with other braided systems (Quek et al., 2003). Similar results are shown for the effect of crimp angle on Poisson ratio, where the Poisson ratio in the transverse direction is more sensitive to changes in the aspect ratio than in the longitudinal direction. The test results show that the microstructure effects are pronounced in the transverse specimen largely due to the relatively large spacing between the axial yarns.

16.4 Mechanical characterization of regenerated cellulose/epoxy composites

16.4.1 Static characterization

Tensile testing using ASTM ID: D3039 (ASTM, D3039, 2007) standard for tensile properties of polymer reinforced fiber composites was used to obtain the elastic properties of the triaxially braided composite specimens. A displacement-controlled mode at a displacement rate of 1.3 mm/min was used. The tensile properties for both unnotched tensile coupons and notched specimens were evaluated in the longitudinal and transverse directions. The tensile coupons were 178 mm long by 38 mm wide and notched specimens, while the notched specimens were 152 mm long by 76 mm wide. Two notches were machined at each side leaving an uncracked ligament of 36 mm. Tensile load was applied using an electromechanical test system with a 97.8 kN (22.0 kips) load capacity (Figure 16.4). In order to record accurate strain measurements, an extensometer was used to record the elastic deformation along a 25 mm length at the middle of the specimens. The maximum error of the recorded load was within ±0.022 kN (5.0 lb). Three replicates were used for each specimen/orientation investigated. Simultaneous mechanical testing, 3-D DIC, and AE tests were carried out on a subset of the specimens. The full field strains on the surface of the composite specimen were captured using DIC in order to confirm the loading curves obtained by simple mechanical loading. The DIC technique uses a random speckle pattern applied

Figure 16.4 Tensile testing of (a) unnotched and (b) notched braided regenerated cellulose/epoxy specimens.

to the specimen that is captured by using a couple-charged device (CCD) camera. These images are then processed by using correlation algorithms to compare strained maps with a reference image taken before any loading is applied. The procedure has been previously applied by the authors by comparing strain results obtained from 3-D DIC with finite element predictions for carbon/epoxy laminates with through thickness cutouts (El-Hajjar and Petersen, 2011). The DIC tests were performed by using the Dantec Dynamics Q-400 system (Figure 16.5). In order to investigate damage initiation and damage propagation as the composite specimens are loaded, AE technique was used to capture the cumulative energy of the recorded emissions. A Mistras 2001 data acquisition system was used for recording the data, including waveforms. A Physical Acoustics WS-alpha transducer was used for measuring the AE signals. The sensor has an operating frequency of 100–1000 kHz with a 63.74 dB peak sensitivity occurring at 523.23 kHz. A preamplifier coupled with a band-pass filter adjusted to pass frequencies in the range of 10–1200 kHz was used to improve the quality of the captured signals.

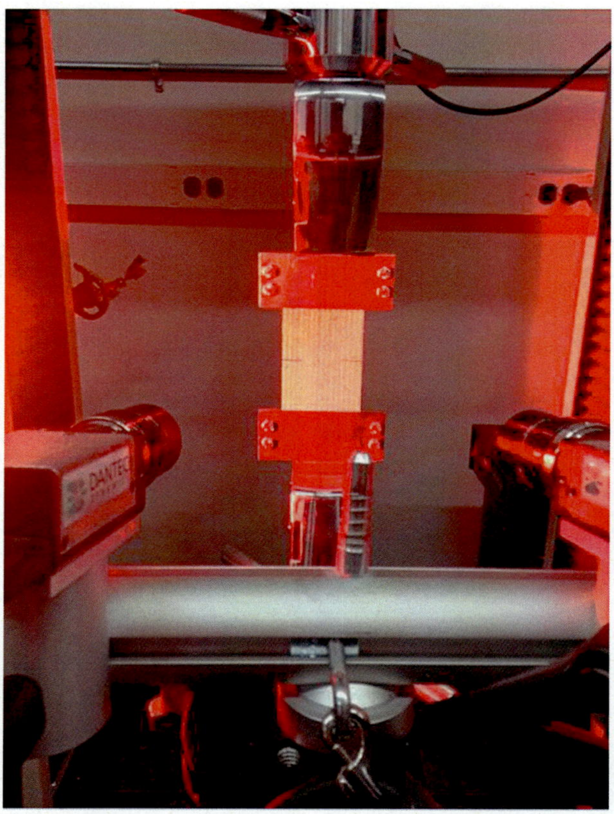

Figure 16.5 DIC test setup for a notched triaxially braided regenerated cellulose specimen tested under uniaxial tension.

16.4.1.1 Experimental results and discussion

The tensile stress–strain curves for longitudinal and transverse specimens are shown in Figure 16.6. Two regions can be immediately identified for the loading curve, in which the first is linear elastic followed by a region with a lower stiffness as the loading proceeds. Similar loading properties have been seen for other regenerated cellulose composites such as Lyocell/epoxy composites (El-Hajjar and Qamhia, 2013b). The stiffness properties are extracted within the 0–0.6% of the strain range from extensometer readings. Good correlation can be observed between the analytical model prediction and the experimental results for both the transverse and axial directions. The results of cumulative AEs are superimposed with the stress–strain curves showing that damage initiation starts at a much lower stress level for samples loaded in the transverse direction. The initial emissions are likely due to the matrix cracking. For samples loaded in the longitudinal direction, higher energy emissions before failure are observed, where the final cumulative energy at failure is 35% higher than the transverse direction. This is believed to be associated with the fracture of the axial fibers oriented in the direction of loading. While a distinction between initial and final ply failure cannot be directly made, the incidence of high AEs can be used as an indication of the point of detrimental irreversible damage.

The results for 3-D DIC show a high correlation between the loading curves obtained by mechanical testing and the full-field strain measurements. The strain fields from DIC are also plotted in Figure 16.6 to show the correlation with the extensometer measurements. The Poisson ratios, ν_{xy} and ν_{yx} were obtained from DIC results by averaging the strain fields over the area of the specimen (Table 16.3). Figure 16.7 shows the failure progression in longitudinal and transverse coupons from DIC results. Figure 16.7a shows that the load in the longitudinal tension test is uniformly distributed. Consequently, the interlaminar shear stresses as a result of increasing the shearing stresses in the axial and bias yarn interface are believed to be causes of the interfacial matrix cracking (Littell et al., 2009). A significant effect of the

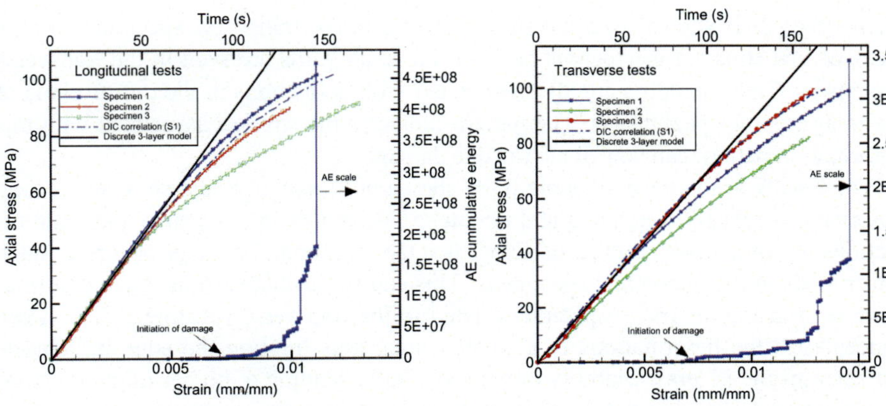

Figure 16.6 Stress versus strain and acoustic emission cumulative energy plots of RF/epoxy quasi-isotropic composites (a) longitudinal and (b) transverse.

Table 16.3 Elastic properties of triaxially braided regenerated cellulose/epoxy composite system

Mechanical property	Analytical	Coupons	DMA[a]
E_x GPa (Msi)	11.67 (1.69)	11.22 (1.63)	3.5 (0.51)
E_y GPa (Msi)	8.74 (1.27)	9.06 (1.31)	3.0 (0.44)
V_{xy}	0.296	0.303	–
V_{yx}	0.351	0.401	–
G_{xy} GPa (Msi)	3.46 (0.52)	–	–

[a]DMA values are for the storage modulus.

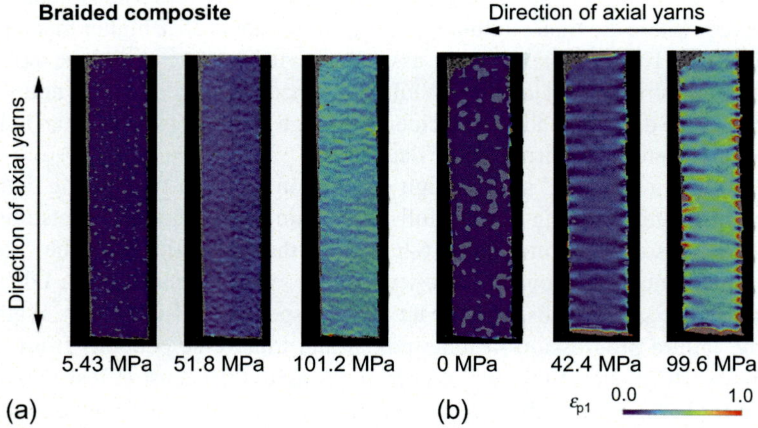

Figure 16.7 Progression of principal strain distribution in (a) longitudinal specimen and (b) transverse specimen.

microstructure is shown from the strain patterns in the transverse specimens. Large areas of low stress in the regions between the axial yarns are seen in the transverse specimens. These patterns are also associated with deflections in the out-of-plane of the image as the specimen is. The pattern indicates the stresses associated with this deformation pattern can contribute to edge damage.

The results of the tension strength are shown in Figure 16.8 and show interesting observations related to notched and unnotched behaviors in transverse and axial directions. A comparison of the AEs shows that the initiation of damage occurs at much lower loads in the unnotched specimens. This can be attributed to the edge deformations seen in Figure 16.7, especially related to the transverse specimens. The notch concentrates the deformations in a smaller area, thus limiting the edge influences, and also results in less variability in the test results, leading to higher initiation damage load levels. However, the slightly reduced ultimate strength in both orientations in the notched specimens is most likely caused by the increased stress concentrations at the

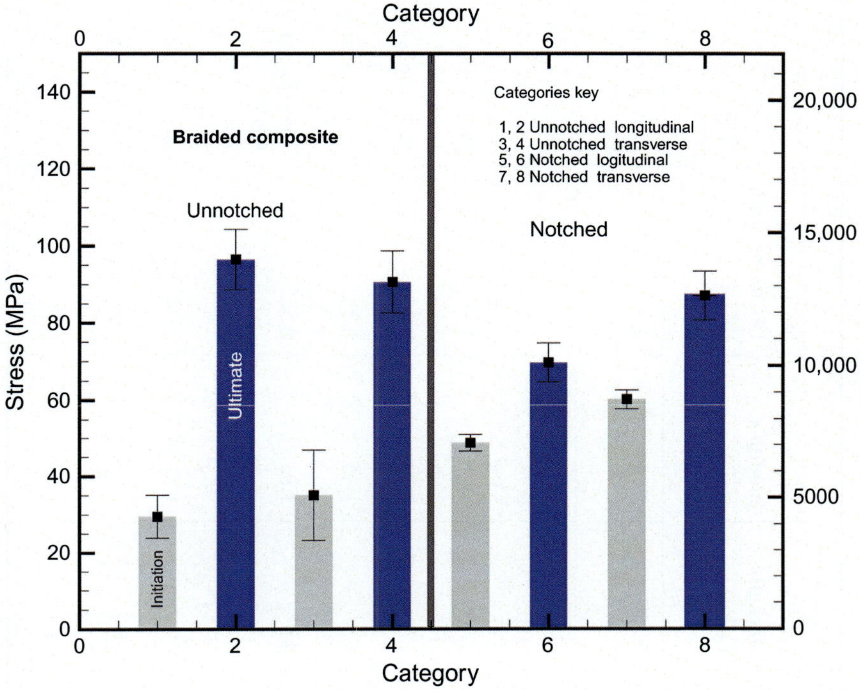

Figure 16.8 Initiation and ultimate failure of unnotched and notched RF/epoxy specimens.

notches. A comparison of the BioMid triaxially braided composites system to other natural fiber-based composites systems is shown in Table 16.4.

16.4.2 Viscoelastic behaviors of triaxially braided composites

The use of viscoelastic techniques such as DMA to characterize cellulose and regenerated cellulose composites have been previously reported. Bledzki et al. studied the effect of fibrous reinforcement incorporation of regenerated cellulose fibers produced in chemical–technical viscose processes and natural abaca fibers on the dynamic–mechanical performance of native polylactic acid (PLA) and its composites (Bledzki et al., 2009). They found out that the incorporation of fibers enhances the T_g to a higher temperature due to the altered mobility of the polymer chains. Soykeabkaew et al. used dynamic–mechanical analysis to investigate the effect of temperature on the storage modulus of three different types of regenerated cellulose composites; namely, low-draw ratio Lyocell, high-draw ratio Lyocell, and Bocell fibers (Soykeabkaew et al., 2009). Tajvidi et al. (2006) studied the dynamic–mechanical properties of natural fiber/polypropylene (PP) composites. The natural fibers used included wood flour, Kenaf fibers, rice hulls, and newsprint. A decrease in the mechanical loss factor was observed for all the composites as compared to the pure PP resin, indicating a more elastic behavior of the composites. The latter two studies show the suitability of DMA to characterize regenerated cellulose composites and its efficiency as a

Table 16.4 Comparison of the elastic properties of different natural fiber/polymer systems

Fiber	Resin	Volume fraction	Resin modulus (GPa)	Resin tensile strength (MPa)	Composite modulus (GPa)	Composite tensile strength (MPa)	Normalized modulus (GPa)	Normalized tensile strength (MPa)	Reference
This study									
BioMid	Epoxy	0.59	1.45	20.68	11.22	96.5	9.84	4.67	Qamhia et al. (n.d.)
Other studies									
Kenaf	PP	0.4[a]	1.36[b]	33.7[b]	6.8	28	5.00	0.83	Wambua et al. (2003)
Coir	PP	0.4[a]	1.36[b]	33.7[b]	1.3	10.6	0.96	0.31	Wambua et al. (2003)
Sisal	PP	0.4[a]	1.36[b]	33.7[b]	5.3	33.8	3.90	1.00	Wambua et al. (2003)
Hemp	PP	0.4[a]	1.36[b]	33.7[b]	6.9	52.2	5.07	1.55	Wambua et al. (2003)
Jute	PP	0.4[a]	1.36[b]	33.7[b]	3.6	26.5	2.65	0.79	Wambua et al. (2003)
Jute[c]	Vinyl ester	0.16	3.45[b]	76[b]	4.8	27	1.39	0.36	Khondker et al. (2005)
Jute[d]	Vinyl ester	0.17	3.45[b]	76[b]	5.5	34	1.59	0.45	Khondker et al. (2005)
Jute[e]	Vinyl ester	0.20	3.45[b]	76[b]	6.3	44	1.83	0.58	Khondker et al. (2005)
Jute[f]	Vinyl ester	0.19	3.45[b]	76[b]	5.7	87	1.65	1.14	Khondker et al. (2005)
Jute[g]	Vinyl ester	0.18	3.45[b]	76[b]	5.2	79	1.51	1.04	Khondker et al. (2005)
Chorn chaff	PP	0.09	1.42	29.8	1.64	25.4	1.15	0.85	Ku et al., (2011)
Coir[h]	PP	0.3	0.64	25.1	0.93	20.2	1.45	0.80	Ku et al. (2011)
Coir[i]	PP	0.3	0.64	25.1	1.05	21.7	1.64	0.86	Ku et al., (2011)
Jute	PBS	0.2	0.68	16.9	1.29	22.2	1.90	1.31	Ku et al. (2011)

Fiber	Resin	Volume fraction	Resin modulus (GPa)	Resin tensile strength (MPa)	Composite modulus (GPa)	Composite tensile strength (MPa)	Normalized modulus (GPa)	Normalized tensile strength (MPa)	Reference
Jute	PBS	0.3	0.68	16.9	1.15	18.1	1.69	1.07	Ku et al. (2011)
Hemp	HDPE	0.4	1.11	–	6.53	–	5.88	–	Ku et al. (2011)
Hardwood A	HDPE	0.36	1.28	–	4.91	–	3.84	–	Ku et al. (2011)
Hardwood B	HDPE	0.43	1.26	–	5.19	–	4.12	–	Ku et al. (2011)
Rice Husk	HDPE	0.44	1	–	3.06	–	3.06	–	Ku et al. (2011)
Arctic flax[j]	Epoxy	0.47	4.5[b]	67[b]	39	279	8.67	4.16	Yan et al. (2014)
UD-flax[k]	Epoxy	0.32	4.5[b]	67[b]	15	132	3.33	1.97	Yan et al. (2014)
Flax	PP	0.4	1.3	28.17	7.61	29.41	5.85	1.04	Yan et al. (2014)
Flax	PLA	0.4	3.51	50.53	7.35	44.32	2.09	0.88	Yan et al. (2014)

[a]By weight.
[b]Not reported/estimated based on average value.
[c]Woven, half-bleached.
[d]Woven, unbleached.
[e]Woven, full-bleached.
[f]Flat braided, uncoated yarn.
[g]Flat braided, coated yarn.
[h]Untreated.
[i]Alkaline treated.
[j]Unidirectional.
[k]Unidirectional.

materials characterization comparison tool between similar composites. Jacob et al. (2006) investigated the effect of fiber weight fraction and alkali treatment on the thermomechanical properties of a natural fiber-reinforced natural rubber composite by dynamic–mechanical analysis.

Dynamic–mechanical analysis is a thermomechanical technique that monitors property changes due to a temperature and/or a frequency or a time change. The technique measures the total energy stored and dissipated in the material due to dynamic stimuli. DMA separates the viscoelastic properties into an elastic part and a viscous part. The elastic part is measured by the storage modulus (E'), which accounts for the energy stored in the material due to the dynamic response. The loss modulus (E''), on the other hand, measures the energy dissipated in the material due to the viscous part of the dynamic response. This energy is lost due to friction and internal motions. The equations of the storage and loss modulus are given by Menard (2008) as follows:

$$E' = \frac{\sigma_0}{\varepsilon_0}\cos(\delta) = \frac{f_0}{bk}\cos(\delta) \tag{7.9}$$

$$E'' = \frac{\sigma_0}{\varepsilon_0}\sin(\delta) = \frac{f_0}{bk}\sin(\delta) \tag{7.10}$$

where f_0 is the force applied at the peak of the sinusoidal wave; k is the sample displacement at the peak; b is a term for sample geometry; σ_0 and ε_0 are the maximum stress and the strain at maximum stress, respectively; and δ is the phase angle. The phase angle (δ) is the primary DMA output. The phase lag between a sinusoidal applied stress (e.g., $\sigma = \sigma_0 \sin \omega t$) and a measured strain (or vice versa) is due to the viscoelastic behavior of most materials. A fully elastic response produces an in-phase response while a fully viscous response produces an out-of-phase response. The tangent of the phase angle (tan δ) is called damping and is defined as the ratio of the loss to storage modulus (Equation (7.3)). Damping is a measure of how efficiently a material loses energy to molecular rearrangements and internal friction as a response to the dynamic stimuli.

$$\tan(\delta) = \frac{E'}{E''} \tag{7.11}$$

For a three-point bending clamp, the model equation for a rectangular cross-section sample is given as

$$E = K_s \frac{L^3}{6I}\left[1 + \frac{6}{10}(1+\nu)\left(\frac{t}{L}\right)^2\right] \tag{7.12}$$

where E is the elastic modulus; K_s is the measured stiffness; I is the moment of inertia of the rectangular cross-section; L and t are the length and the thickness of the specimen, respectively; and ν is the Poisson's ratio. The equation assumes that the local deformations in the region of the supports are negligibly small, and that there is no correction factor between the measured stiffness (K_s) and the model stiffness (K) for a rectangular beam under three-point flexure.

16.4.3 Experimental approach

Similar to the composite panels used for the validation of the analytical models discussed in the previous sections, the composite samples used for this study were also prepared from triaxially braided regenerated cellulose fibers (BioMid; GSC and ENC, Inc., Burnaby, BC, Canada) and a high-biocontent epoxy resin system (Super Sap 100/1000; Entropy Bio-Resins, Gardena, California, USA). The regenerated fibers were stored at room temperature and a humidity level of around 30%. The panels from which the samples were cut were prepared by a vacuum-assisted resin infusion technique. For comparison purposes, pure epoxy samples were prepared and tested under the same dynamic conditions. The pure resin samples were prepared by a wet layup technique from the excess resin that was used in preparing the composites panels. Using the same resin batch ensures consistency in the composition and sets an accurate control point to compare the effect of the triaxially braided regenerated cellulose fibers on the mechanical behavior of the composites. In addition, all prepared samples were postcured in a closed oven for 240 ± 5 min at a fixed temperature of 75 °C (167 °F) to ensure full cure of the resin.

The effect of triaxially braided regenerated cellulose reinforcement on the viscoelastic properties was examined using dynamic mechanical analysis. The study was carried out on a DMA system (Q800; TA instruments, New Castle, Delaware, USA). A three-point bending mode with a controlled strain of 15 μm (displacement) was used. For the first part of the study, pure epoxy samples and samples of triaxially braided regenerated cellulose reinforced composites were tested at a constant frequency of 1.0 Hz. A temperature range of 0–120 °C (32–248 °F) and a heating rate of 3 °C/min (5.4 °F/min) were chosen. The supported span width was 20 mm (0.8 in). The test parameters were chosen to comply with the general recommendations of the ASTM D4065-12 (ASTM, 2012) and ASTM D5023-07 (ASTM, 2007) standards. For the regenerated cellulose-reinforced composites, the effect of moisture on the behavior of the composites was investigated by soaking samples in water at a room temperature of 22 °C (72 °F) for a period of 120 ± 5 min. The change in mass for the surface dry composites was recorded and the wetted samples were tested in DMA to investigate the effect of soaking on the viscoelastic properties. In the second series of tests conducted for the frequency effects, the same DMA loading and running parameters discussed above were used except that frequency values of 1, 10, and 40 Hz were selected. The temperature and frequency ramping were performed simultaneously. The frequency effects study was carried out on a temperature range of 20–135 °C (68–275 °F).

16.4.4 Results and discussion

A cross-section micrograph for the triaxially braided regenerated cellulose composites was shown in Figure 16.2. The waviness in the bias yarns makes the thickness variable across the specimens. In our calculations, the average thickness was used for the DMA calculations of storage modulus. The results of the first part of the study by DMA are shown in Figures 16.9–16.13. Figure 16.9 shows the variation in the storage modulus with temperature. For the three sets of samples, a nonlinear degradation

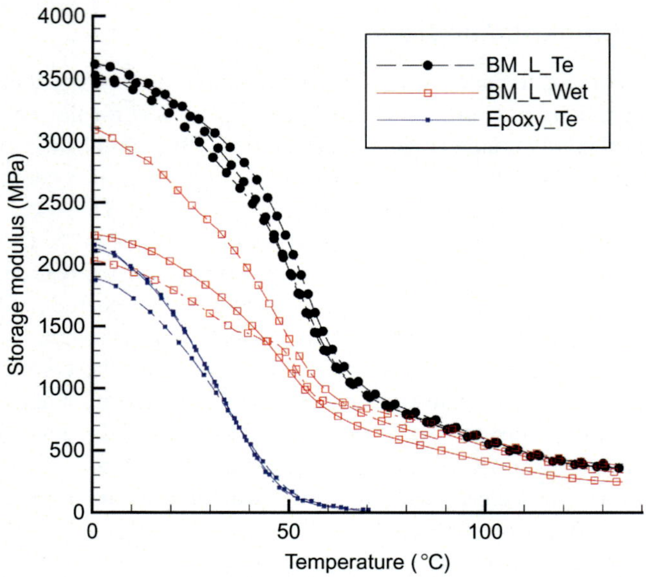

Figure 16.9 Changes in the storage modulus of neat epoxy and wet and dry triaxially braided cellulose composite with temperature ramping.

in the storage modulus is observed as the temperature increases. As expected, the neat epoxy specimens show the highest degradation in modulus as these were held as controls. The wet specimens showed the second largest drop. The epoxy specimens were found to lose all of their elastic properties at a temperature of 70 °C (158 °F), while the dry and wet regenerated composites retained 11% and 15%, respectively, of their storage modulus for a temperature up to 140 °C (284 °F); double that for epoxy. Due to the degradation of the mechanical properties of cellulose as it absorbs water, soaking the samples for 2 h is seen to reduce the storage modulus significantly at the lower temperatures. At higher temperatures, the storage properties of the wet and dry composites are similar. One explanation might be that the fibers recover some of their mechanical strength as the water evaporates. One general observation was that for the same testing parameters, the wet composite samples were slightly warped after testing while the dry samples didn't show any obvious curvature. The curvature results from the uneven swelling of the cellulose fibers. Note that the values of the storage modulus at room temperature obtained for the triaxially braided regenerated cellulose/epoxy composites are much lower than the values of the elastic modulus obtained by the mechanical testing of uniaxial tension specimens shown in Table 16.3. This can be attributed to the nonlinearity in the load displacement response and the data-reduction approach used in the analysis of the DMA results, which is based on a point-measurement approach.

For the loss modulus results (Figure 16.10), a shift in the peak is seen between the neat resin and the wet and dry regenerated cellulose/epoxy specimens. The same trend in peak shifts is also shown in the tan δ plots (Figure 16.11), where the glass transition

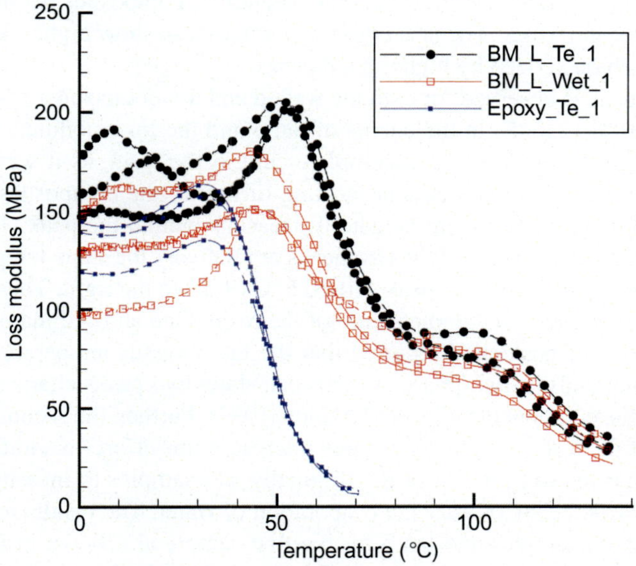

Figure 16.10 Changes in the loss modulus of neat epoxy and wet and dry triaxially braided cellulose composite with temperature ramping.

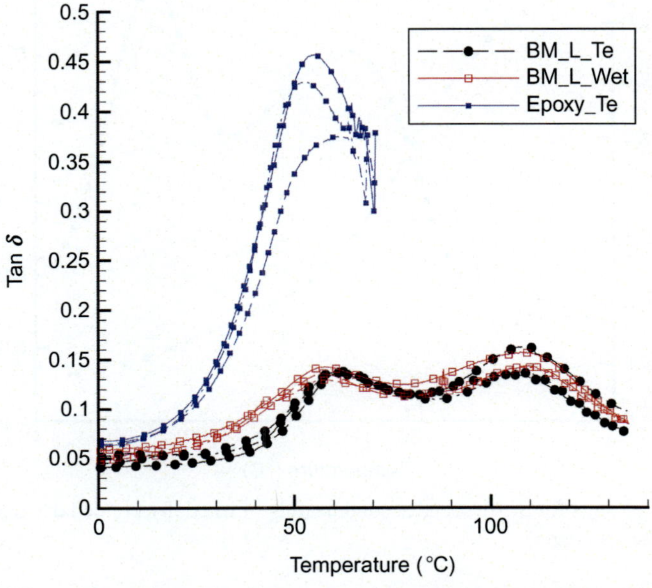

Figure 16.11 Changes in tan δ of neat epoxy resin, wet and dry triaxially braided cellulose composite with temperature ramping.

temperature (T_g) is higher for the regenerated cellulose composites but drops with the wetting of the composites. The neat epoxy specimens also show higher damping characteristics as characterized by higher tan δ peaks.

A second peak is observed for both the wetted and dry composites at a temperature of around 110 °C (230 °F) in the curves of tan δ and the loss modulus. Initially, this peak was thought to indicate an incomplete cure in the resin, as it was reported by Margem et al. (2010); thus, a second heating of one of the composites sample was carried out (Figure 16.12) to investigate the reasons behind the peak. The curve for the second heating still showed the two peaks at relatively the same temperatures and damping rates; with the second peak shifted 5 °C (9 °F) to the right. Thus, this second peak is likely not due to incomplete cure of the resin. Guo and Ashida observed similar duplications of peaks and reported that the tan δ versus temperature curves for composites materials of two-phase systems can show two peaks characteristic of the T_g of each of the components (Guo and Ashida, 1993). Further, the second peak occurs after the T_g of the resin, so it is not of main interest in this effort. In addition, a second heating was carried out for two of the originally wet samples to investigate changes in the viscoelastic behavior after the evaporation of water. The results for the storage modulus for both samples and tan δ for the first sample are shown in Figure 16.13. A slight enhancement in the storage modulus and shifts in the tan δ peaks are seen

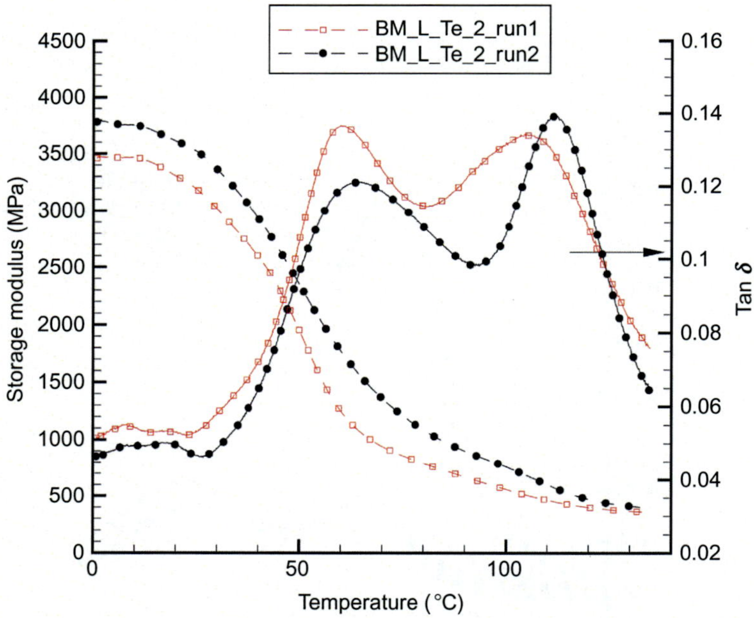

Figure 16.12 Results for the first and second heating of triaxially braided dry regenerated cellulose/epoxy specimen.

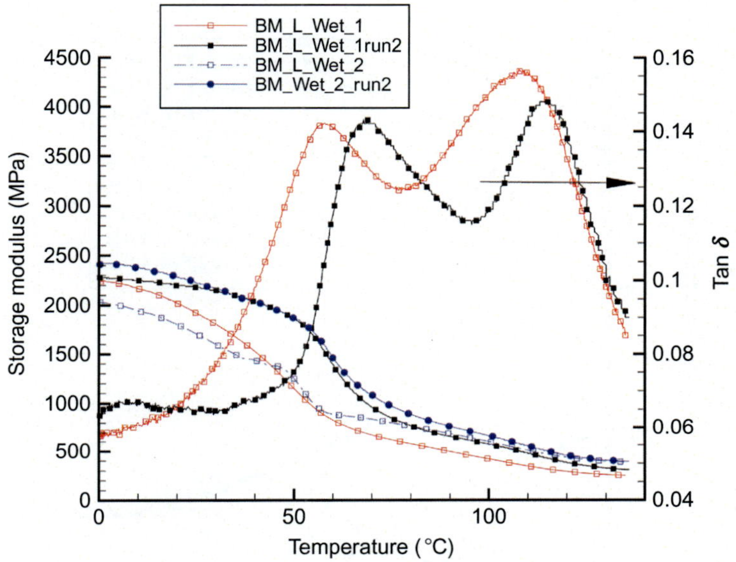

Figure 16.13 Results for the first and second heating of wet regenerated triaxially braided cellulose/epoxy specimens.

for the second heating. The storage modulus values however, do not recover to the values reported for the dry composites (Figure 16.9). Thus, part of the deterioration in the mechanical properties that is caused by exposure to water is irreversible if the composites are dried.

The second part of the study involved the investigation of the effect of the frequency of loading on the viscoelastic behavior of the triaxially braided regenerated cellulose composites. For this part of the study, triaxially braided regenerated cellulose composites were tested in the axial and transverse directions. The results are shown in Figures 16.14–16.16. For the storage modulus results (Figure 16.14), a 5% increase in the storage modulus of the longitudinal and transverse braided BioMid composites at room temperature is observed as the frequency increases from 1 to 40 Hz. The increase is more significant at higher temperatures and is around 50% for the highest test temperature of 135 °C (275 °F). Overall, the same trend in the degradation of the storage modulus for both epoxy and the longitudinal composites is seen at different frequencies. At room temperature and normal humidity levels, the storage modulus of the transverse composites is 12–15% lower for the transverse composites. This result is analogous to that seen by the authors in a previous study (El-Hajjar and Qamhia, 2013a), where the average elastic modulus was 24% higher in the longitudinal direction due to the undulations of the bias yarns that dominate the behavior in the transverse direction. The results for the loss modulus and tan δ plots are shown in Figures 16.15 and 16.16.

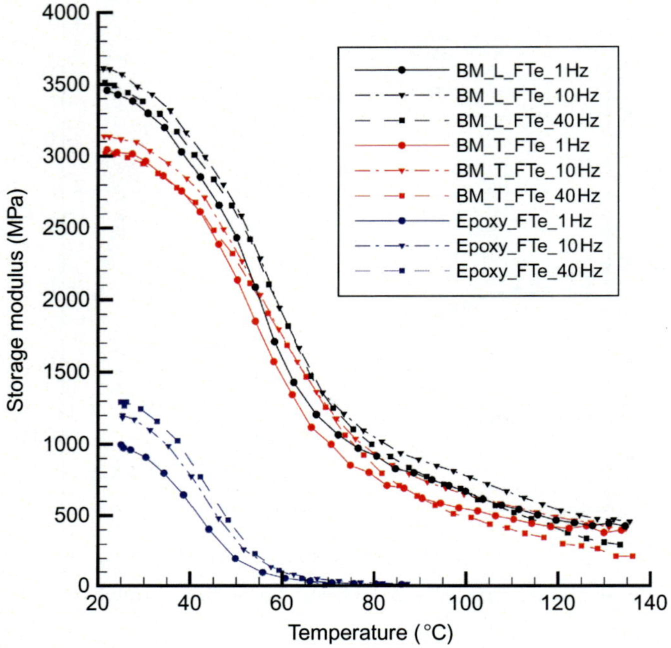

Figure 16.14 Change in storage modulus of neat epoxy, longitudinal and transverse dry cellulose composite with temperature ramping at different frequencies.

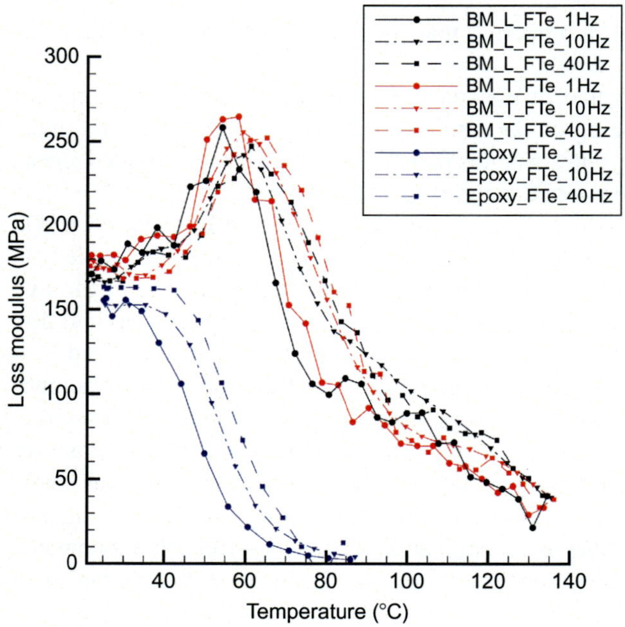

Figure 16.15 Change in loss modulus of neat epoxy, longitudinal and transverse dry cellulose composite with temperature ramping at different frequencies.

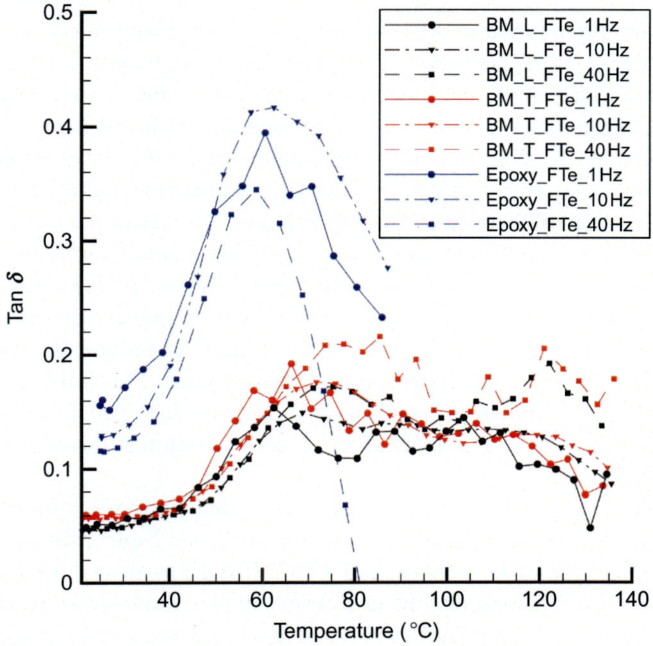

Figure 16.16 Change in tan δ plots of neat epoxy, longitudinal and transverse dry cellulose composite with temperature ramping at different frequencies.

16.5 Conclusions

In this study, a micro- to macromechanics approach using a 3-layer discrete analytical model is used to predict the mechanical properties in a quasi-isotropic triaxially braided regenerated cellulose composite with a high-biocontent epoxy resin. The results are validated by investigating the tensile behavior of composites prepared by a resin infusion technique. Both longitudinal and transverse test coupons are studied. The load redistribution in the composites is investigated by a full-field study using a DIC technique. Damage initiation and propagation is examined using a passive AE technique. The analytical model was examined for the implications of changing the braider angle, aspect ratios, crimp angle, and the yarn spacing on the elastic properties of the braided composite. In the second part of the study, the viscoelastic properties of the quasi-isotropic composite are investigated. The manufactured composites are evaluated using dynamic mechanical analysis for wet and dry conditions and for transverse and axial directions using a three-point bending test setup. Two different studies were carried out: in the first study, specimens are cyclically loaded and exposed to a temperature ramp; in the second, the specimens are ramped under progressively increasing temperature and frequency loadings.

In the first part of the study, the behavior of quasi-isotropic triaxially braided regenerated cellulose composites with a high-biocontent epoxy resin is investigated. The resin infusion setup is found to be the most successful in fabricating the composites with minimal porosity levels and a high fiber volume fraction. The initial elastic stiffness results are approximately 20% higher in the longitudinal direction. The 3-layer discrete analytical model is evaluated for the quasi-isotropic braided composite. The effects of the braider angle, the aspect ratios, the crimp angle, and the yarn spacing on the elastic properties of the braided composite can be investigated. The model is successfully calibrated to show reasonable predictions for the elastic properties of the composites and to study the effects of fiber volume fraction and aspect ratio (crimp angle) on the elastic modulus and Poisson ratio. For changes in the aspect ratio of the bias yarns, the sensitivity of the transverse properties is significantly higher. The AE results show that although more energy is dissipated due to damage in the longitudinal direction due to fiber breaks, damage initiation starts at lower stresses in the transverse direction. The AEs can be used to establish limiting failure stresses and strains corresponding to the onset of significant internal damage. The full-field strain measurements show the behavior of the transverse coupons that results in significant out-of-plane deformation, creating areas of stress concentration at the edges from which failure can propagate.

For the DMA study, the addition of the regenerated cellulose is found to shift the glass transition to higher temperatures. The peak in the *tan* δ curve for a constant loading frequency of 1 Hz shifts from around 38 °C (100 °F) to around 55 °C (131 °F) for the dry composites. Additionally, the neat resin samples show better energy damping, which is characterized by the higher values of tan δ for the neat epoxy samples. When the cellulose is exposed to a humid environment, the glass transition temperature of the composite shifts to a lower temperature, but this temperature is still higher than that of the neat resin. The mechanical properties also drop drastically, which appears in the DMA results as a drop in the elastic modulus. On the other hand, the wet specimens recover their storage modulus as temperature is increased approaching that of the dry specimens at high temperatures exceeding 70 °C (158 °F). When the water evaporates and the sample is reheated, part of the mechanical strength is recovered, as indicated by an increased storage modulus. However, some of the elastic properties are permanently lost. Finally, the DMA approach is able to capture the variation in the modulus in the axial and transverse directions of the triaxially braided composite. This difference is due to the undulations in the bias yarns that are braided above and below the main axial yarns leading to some loss in the quasi-isotropic properties in the transverse direction.

16.6 Future challenges and opportunities

Future challenges limiting the growth of the use of these materials will be in pairing the correct applications not only with the suitable natural fiber system but also with the manufacturing process. Hybridizing the process with multiple fibers can be easily achieved with the braiding process to achieve several design objectives. The issues with mechanical property degradation due to moisture exposure will continue to be an area of interest, and urgent methods to improve or mitigate the loss of properties will be necessary. The behavior of the composite with a polymer binder with realistically conditioned accelerated tests can be used to help certify the use of these materials in a variety of structural applications. The nonmechanical properties are also areas of opportunity for future research. The ability to structure the material with advancing forming can potentially result in important benefits such as in acoustic properties or vibrational response.

References

ASTM, 2007. D5023-07 Standard Test method for plastics: dynamic mechanical properties: in flexure (three-point bending).

ASTM, 2012. D4065-12 standard practice for plastics: dynamic mechanical properties: determination and report of procedures.

ASTM, D2734, 2009. Standard Test Methods for Void Content of Reinforced Plastics. ASTM International, West Conshohocken, PA.

ASTM, D3039, 2007. Standard Test Method for Tensile Properties of Polymer Matrix Composite Materials. ASTM International, West Conshohocken, PA.

ASTM, D6866, 2007. Standard Test Methods for Determining the Biobased Content of Solid, Liquid, and Gaseous Samples Using Radiocarbon Analysis. ASTM International, West Conshohocken, PA.

Bledzki, A.K., Jaszkiewicz, A., Scherzer, D., 2009. Mechanical properties of PLA composites with man-made cellulose and abaca fibres. Compos. A Appl. Sci. Manuf. 40 (4), 404–412.

Buksnowitz, C., Adusumalli, R., Pahler, A., Sixta, H., Gindl, W., 2010. Acoustical properties of Lyocell, hemp, and flax composites. J. Reinf. Plast. Compos. 29, 3149–3154.

Byun, J.-H., 2000. The analytical characterization of 2-D braided textile composites. Compos. Sci. Technol. 60 (5), 705–716.

Cheng, Q., Wang, S., Harper, D.P., 2009. Effects of process and source on elastic modulus of single cellulose fibrils evaluated by atomic force microscopy. Compos. A Appl. Sci. Manuf. 40 (5), 583–588.

Composites_Technology, 2012. New cellulose-based fiber. Composites Technology. Gardner Business Media, Cincinnati, Ohio, USA.

El-Hajjar, R., Petersen, D.R., 2011. Adhesive polyvinyl chloride coatings for quantitative strain measurement in composite materials. Compos. Part B 42 (7), 1929–1936.

El-Hajjar, R., Qamhia, I., 2013a. A kinematic and acoustic emission methodology for investigation of progressive damage behavior of triaxially braided regenerated cellulose composites. In: 18th International Nondestructive Testing and Evaluation of Wood Symposium: Madison, Wisconsin.

El-Hajjar, R.F., Qamhia, I.I., 2013b. Modeling and characterization of the moisture-dependent bilinear behavior of regenerated cellulose composites. J. Wood Sci. 59 (4), 331–336.

El-Hajjar, R.F., Shams, S.S., Kehrl, D.J., 2013. Closed form solutions for predicting the elastic behavior of quasi-isotropic triaxially braided composites. Compos. Struct. 101, 1–8.

Gindl, W., Keckes, J., 2006. Strain hardening in regenerated cellulose fibres. Compos. Sci. Technol. 66 (13), 2049–2053.

Graupner, N., Mussig, J., 2011. A comparison of the mechanical characteristics of kenaf and lyocell fibre reinforced poly(lactic acid) (PLA) and poly(3-hydroxybutyrate) (PHB) composites. Compos. A Appl. Sci. Manuf. 42 (12), 2010–2019.

Guo, W., Ashida, M., 1993. Dynamic viscoelasticities for short fiber-thermoplastic elastomer composites. J. Appl. Polym. Sci. 50 (8), 1435–1443.

Jacob, M., et al., 2006. Dynamical mechanical analysis of sisal/oil palm hybrid fiber-reinforced natural rubber composites. Polym. Compos. 27 (6), 671–680.

Khondker, O.A., et al., 2005. Tensile, flexural and impact properties of jute fibre-based thermosetting composites. Plast. Rubber Composit. 34, 450–462.

Ku, H., et al., 2011. A review on the tensile properties of natural fiber reinforced polymer composites. Compos. Part B 42 (4), 856–873.

Kushwaha, P.K., Kumar, R., 2010. Effect of silanes on mechanical properties of bamboo fiber-epoxy composites. J. Reinf. Plast. Compos. 29 (5), 718–724.

Littell, J.D., et al., 2009. Characterization of damage in triaxial braided composites under tensile loading. J. Aerosp. Eng. 22 (3), 270–279.

Malkapuram, R., Kumar, V., Singh Negi, Y., 2009. Recent development in natural fiber reinforced polypropylene composites. J. Reinf. Plast. Compos. 28, 1169–1189.

Margem, F.M., et al., 2010. The dynamic-mechanical behavior of epoxy matrix composites reinforced with ramie fibers. Revista Matéria 15, 164–171.

Masters, J.E., et al., 1993. Mechanical Properties of Triaxially Braided Composites: Experimental and Analytical Results, Vol. 15. American Society for Testing and Materials, Philadelphia, PA.

Menard, K.P., 2008. Dynamic Mechanical Analysis: A Practical Introduction, second ed. CRC Press, Boca Raton, FL.

Meredith, J., et al., 2013. On the static and dynamic properties of flax and Cordenka epoxy composites. Compos. Sci. Technol. 80, 31–38.

Miravete, A., Clemente, R., Castejon, L., 2000. 3-D Textile Reinforcements in Composite Materials. CRC Press LLC, Corporate Blvd, NW, Boca Raton, FL 33431, USA.

Mottershead, B., Eichhorn, S.J., 2007. Deformation micromechanics of model regenerated cellulose fibre-epoxy/polyester composites. Compos. Sci. Technol. 67 (10), 2150–2159.

Naik, R.A., 1994. Analysis of woven and braided fabric reinforced composites. NASA CR-194930.

Pinto, M., et al., 2014. Fabrication and mechanical characterization of jute fiber/epoxy laminar composites. In: Barthelat, F., et al. (Eds.), Mechanics of Biological Systems and Materials, vol. 4. Springer International Publishing, Cham, Switzerland, pp. 173–177.

Poe, C., Dexter,H.B., Raju, I.S., 1997. A review of the NASA textile composites research. NASA Langley Technical Report.

Qamhia, I., Shams, S.M., El-Hajjar, R., 2015. Analytical and experimental investigation of quasi-isotropic triaxially braided regenerated cellulose reinforced composites. Mech. Adv. Mater. Struct. 22 (12). http://dx.doi.org/10.1016/j.compositesa.2008.10.021.

Quek, S.C., et al., 2003. Analysis of 2D triaxial flat braided textile composites. Int. J. Mech. Sci. 45 (6–7), 1077–1096.

Scholz, D.B., et al., 1997. Advanced Technology Composite Fuselage – Materials and Processes. Langley Research Center, Hampton, Virginia.

Shih, Y.-F., et al., 2012. Plant fibers and wasted fiber/epoxy green composites. Compos. Part B 43 (7), 2817–2821.

Shokrieh, M.M., Mazloomi, M.S., 2010. An analytical method for calculating stiffness of two-dimensional tri-axial braided composites. Compos. Struct. 92 (12), 2901–2905.

Soykeabkaew, N., Nishino, T., Peijs, T., 2009. All-cellulose composites of regenerated cellulose fibres by surface selective dissolution. Compos. A: Appl. Sci. Manuf. 40 (4), 321–328.

Tajvidi, M., Falk, R.H., Hermanson, J.C., 2006. Effect of natural fibers on thermal and mechanical properties of natural fiber polypropylene composites studied by dynamic mechanical analysis. J. Appl. Polym. Sci. 101 (6), 4341–4349.

Wambua, P., Ivens, J., Verpoest, I., 2003. Natural fibres: can they replace glass in fibre reinforced plastics? Compos. Sci. Technol. 63, 1259–1264.

Wollerdorfer, M., Bader, H., 1998. Influence of natural fibres on the mechanical properties of biodegradable polymers. Ind. Crop. Prod. 8 (2), 105–112.

Yan, L., Chouw, N., Yuan, X., 2012. Improving the mechanical properties of natural fibre fabric reinforced epoxy composites by alkali treatment. J. Reinf. Plast. Compos. 31 (6), 425–437.

Yan, L., Chouw, N., Jayaraman, K., 2014. Flax fibre and its composites – a review. Compos. Part B 56, 296–317.

Yang, J.-M., Ma, C.-L., Chou, T.-W., 1986. Fiber inclination model of three-dimensional textile structural composites. J. Compos. Mater. 20 (5), 472–484.

Mechanical performance of polyurethane (PU)-based biocomposites

17

M.I. Aranguren, N.E. Marcovich, M.A. Mosiewicki
Institute of Research in Materials Science and Technology (INTEMA), National University of Mar del Plata, Mar del Plata, Argentina

17.1 Introduction

Polyurethanes (PUs) are a family of polymers that has grown based on their versatility. They can be formulated from varied raw materials and processing methods, leading to polymers with properties designed to satisfy specific conditions: from structural to functional and responsive materials, which can be manufactured into very different applications; from insulating or cushioning foams to biomedical devices, as well as nonstructural applications such as adhesives or coatings.

To obtain a PU, a monomer containing at least two hydroxyl groups (diol) must be reacted with another one containing at least two isocyanate groups (diisocyanate), to produce urethane bonds, as shown below in a very simplified scheme (Scheme 17.1).

Incorporation of molecules containing a larger number of reactive groups (polyols and/or polyisocyanates) leads to network structures that can be rigid or flexible depending on the length of the original chains between reactive groups but also on the nature of the chosen reactives (e.g., linear or aromatic isocyanates).

Depending on the formulation chosen, cross-linked elastomers with a wide range of modulus and extensibility can be obtained, as well as thermoplastic PUs that are linear, phase-separated, structured polymers. Besides, these materials can be prepared as compact polymers, foams, or fibers, which explain their extensive utilization in so many different areas.

To further tune their behavior as well as prices, fillers and reinforcements can also be added to formulate composite materials. In particular, from the last decade of the twentieth century to now, increasing interest has arisen in the use of vegetable fibers and particles as reinforcements or fillers of polymer composites. While this endeavor has required the use of modifiers to be used in most thermoplastics matrices of high-volume production (Marcovich et al., 2014; Reboredo et al., 2008; Aranguren, 2006), in the case of PU the compatibility with polar vegetable materials is high and, thus, composites are generally prepared without modifiers or compatibilizers with usually good results.

Although other natural fibers are possible, such as animal fibers, like wool or silk, in which proteins are the main components, in this chapter we will focus on the use of vegetable materials, which have been more frequently analyzed in the open literature and have already found a commercial niche.

Biocomposites: Design and Mechanical Performance. http://dx.doi.org/10.1016/B978-1-78242-373-7.00010-X

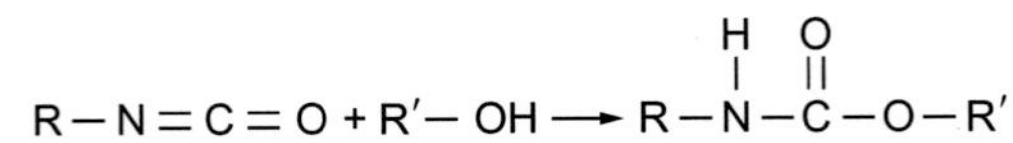

Scheme 17.1 Urethane bond formation from reaction of isocyanate and hydroxyl groups.

Vegetable fibers and particles offer the advantages of large worldwide availability, reduced wearing of the processing equipment, neutral carbon balance, and high specific properties, providing additional benefits in their use for acoustic and thermal insulation because of their natural porous structure (Peng et al., 2011).

17.2 Vegetable particles/fibers and synthetic PUs

17.2.1 Cross-linked PU elastomers

Given the relative low price of the fillers (which frequently are wastes from agricultural- or forestry-derived activities), it is logical to consider matrices of low price. This was the case in the work of Rácz et al. (2009), who used a recycled polyol to produce lightweight reinforced PU for acoustic panels in construction. The experimental polyol was obtained from PU scraps from the upholstery and automotive industries. The highly viscous polyol required the addition of a viscosity modifier to prepare filled PU foams. The cross-linking agent was a polymeric aromatic isocyanate (pMDI) and the filler was pine wood flour (WF) (average size of 100 μm, concentration range 0–20 wt.%). The polyol had an average humidity content of 5 wt.%, which acted as a foaming agent, contributing to reduce total weight and improve insulation. The density of the microfoamed composites was in the range of 0.55–0.76 kg m^{-3} and, as could be expected, the tensile properties (modulus and strength) of the composites increased with the concentration of WF (234 ± 28 and 4.8 ± 0.3 MPa for the neat PU, and 604 ± 21 and 9.2 ± 0.6 MPa for the 20 wt.% WF composite). Flexural properties of the composites showed a similar trend, with the modulus of the 20 wt.% WF composite being more than twice that of the neat PU. On the other hand, the more rigid composite showed a more fragile behavior and had a reduced impact resistance (8.60 ± 1.14 and 4.70 ± 1.01 kJ m^{-2}, for the neat PU and 20 wt.% WF composite, respectively).

Although WF is hygroscopic, the composites absorbed less water than the neat PU. The interfacial interactions that most probably involve covalent bonding between particles and polymer prevent the swelling of the particles and matrix by incoming moist during water immersion tests (Rácz et al., 2009). This result is important in maintaining good insulation properties.

Bledzki et al. (2001) compared jute (bidimensional mat) and flax incorporated in PU composites and found that flax composites presented higher strength and stiffness. As usual, the increase in fiber concentration resulted in increased Young's modulus and flexural strength.

Batouli et al. (2014) mixed a low molecular weight polyester-diol with methylene diphenyl diisocyanate, MDI (at NCO/OH = 1.3) at high speed and allowed it to freely foam. Kenaf core ground and dried (70 °C for 4 h under vacuum) was added to the formulation to achieve weight percentages of 5, 10, and 15 wt.%. In this case, the addition of fibers was not successful in improving PU properties, and 15 wt.% of the powdered fibers were necessary to just reach properties similar to that of the unfilled matrix. Actually, maximum stress under flexural tests was 0.178 ± 0.03 MPa for the pure PU, but it was as low as 0.084 ± 0.06 MPa for the 5 wt.% filled foam. The authors explained this result by proposing that kenaf located inside the foam cells, filling the pores, instead of reinforcing the wall cells. Although not mentioned by the authors, it is also possible that the relatively large size of the fibers have compromised the integrity of the cell walls, reducing the strength of the structures, so that higher fiber concentrations are needed to produce high strength materials.

An elastomeric cross-linked reinforced PU with good damping properties, as bearing pad for acoustic vibration isolation in railway/underground lines, was prepared from polyethylene adipate diol 4,4′diphenylmethane diisocyanate with 1,4-butane diol or 1,6-hexanediol, glycerin (Gil, 2009; Oprea, 2008) and 1–15 wt.% cork. As in most cases, filler addition increased the Young's modulus of the materials and decreased the elongation at break.

Rials and Wolcott (1998) prepared composites by spraying a PU based on a polyether polyol and an isocyanate prepolymer on two cellulosic pulps, one obtained from recycled paper and the other being a virgin thermomechanical pulp of southern pine. Dynamic–mechanical studies showed that the glass transition temperature, T_g (assigned to the temperature of the maximum in tan δ), shifted from −40 to −26 °C, while the peak intensity was reduced from 1.5 to 0.31, a larger change than expected from dilution effect. As typically reported, the storage modulus at high temperature was dramatically increased with respect to the neat PU.

Fibers from oil palm empty fruit bunch were ground to different particle sizes (EFB), dried, and used as filler in a PU made from MDI and polyethylenglycol (PEG). Particles were incorporated in the isocyanate component; the PEG was added and mechanically stirred. The mixtures were cured under pressure (125 °C, 5 h, 500 kg/cm^2) and then postcured at 125 °C for 24 h (Rozman et al., 2001). The contribution of the EFB to the concentration of OH was considered and the total OH content (from EFB and PEG) remained constant. A maximum in tensile strength and modulus was observed at about 50 wt.% EFB for all particle sizes, with the exception of the largest particles (35–60 mesh), for which the maximum occurs at 60 wt.%. Moreover, samples made with larger particles appear to be tougher under tensile tests, which was explained by introducing the idea that smaller particles have more OH per unit weight and some of these groups could not react with the isocyanate, which led to weaker interface and lower toughness. An alternative explanation could be that smaller particles lead to more rigid but more fragile materials, due to the reduced extensibility and mobility of the polymer that interacts with the particles.

17.2.2 Thermoplastic PUs

El-Shekeil and coworkers (2011, 2012a, 2013) prepared composites from a polyester-based thermoplastic polyurethane (TPU) and kenaf bast fibers extracted by mechanical decortication (KF). Fibers were ground and then separated into three different sizes (300–425, 125–300, and <125 μm) by sieving. Samples containing 30 wt.% of fibers were prepared by melt blending followed by compression molding (El-Shekeil et al., 2011). The authors found that the tensile strength decreases with increasing mixing time (from 11 to 15 min), shows a maximum with mixing rate (i.e., 33.5 MPa at 40 rpm), and with increasing fiber size (24.2 MPa for fibers <125 μm; 33.5 MPa for fibers between 125–300 μm and 30 MPa for the 300–325 μm range). Additionally, the impact strength of the composites ranged from 2.76 to 2.97 J/m^2 for increasing fiber lengths (El-Shekeil et al., 2012a). The authors concluded that the 30 wt.% fiber composite exhibited the best tensile strength (i.e., 33.5 MPa), while tensile and flexural modulus and flexural strength increased and strain and impact strength decreased at increasing fiber content. The hardness (Shore D) of the 30 wt.% composite was greater than that corresponding to the neat TPU (66 vs. 55), while abrasion resistance was lower (0.074 vs. 0.012 wt.% loss) (El-Shekeil et al., 2012a). Treatment of the kenaf fibers with 4% pMDI or with 2% NaOH + 4% pMDI (El-Shekeil et al., 2012b) showed that the first treatment did not significantly affect composite properties, but the second one increased the tensile properties with respect to the nontreated fiber composite (i.e., 30% and 42% increase of the tensile strength and modulus, respectively).

17.3 Biopolyurethane composites

The efforts to replace petroleum-based polymers with other ones produced from renewable resources have also been applied to formulate biobased PUs, being the hydroxyl-containing component the one that is replaced partially or completely by biodiols or biopolyols, as exemplified in the following sections.

17.3.1 Biopolyurethane composites made from vegetable short fibers/particles

Mosiewicki and coworkers investigated the behavior of solid and foamed composites based on a PU obtained from the reaction of a polyol derived from tung oil with pMDI and WF (Mosiewicki et al., 2009b; Casado et al., 2009). Mechanical test results showed that increasing the WF content in the PU matrix induced an increase of the tensile modulus (from 0.91 ± 0.12 to 3.03 ± 0.40 GPa, for the 0 and 30 wt.% WF samples, respectively) and strength, as well as the impact strength (about 70% increase at 30 wt.% of WF). Tensile results were explained by the higher modulus and strength of the wood fibers compared with those of the matrix, in addition to the good adhesion wood-polymer due to physical interactions (H-bonds) and chemical interfacial bonding through reactions of pMDI with –OH from the WF and from the PU matrix, that act as extra crosslinks. Impact behavior was also related to the strong interface and the generation of a tortuous fracture path through the WF-reinforced material. The propagation energy and the total energy absorbed during the test increase significantly compared to that of the unreinforced PU because of new mechanisms of energy dissipation resulting from the presence of the filler. Thus, the ductile index also increases with WF content, indicating that the composites behave in a more ductile manner as compared to the matrix.

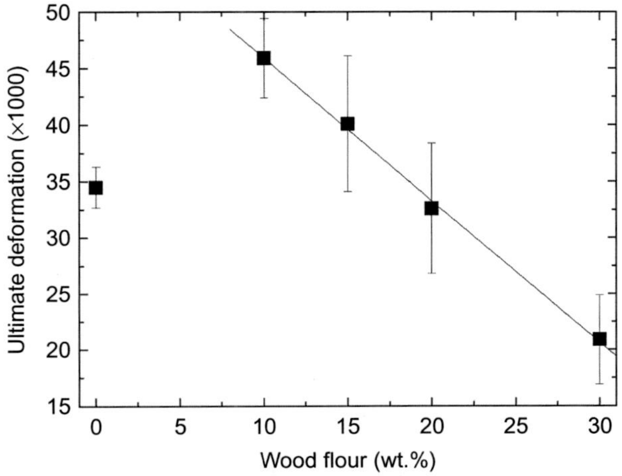

Figure 17.1 Ultimate deformation versus WF content for biocomposites based on tung oil-PU.
Reprinted from Casado et al. (2009), with permission from Wiley.

This study showed very unusual results for microcomposites; the addition of moderate amounts of WF (10–15 wt.%) led also to increased elongation at break (Figure 17.1). In this system, the initiation of the failure by interfacial debonding at multiple sites, which is followed by the coalescence of the cracks and final catastrophic crack growth (typical failure mechanism of microcomposites at low strains (Hull, 1993)) was not present because of the excellent interfacial adhesion, which largely delayed the final breakage of the material. Besides, the WF added to the composites and coreacted with the PU could act as chain extender. However, as more WF is added, the increased cross-linking density and rigidity of the particles produced the overall typical effect of increasing the composite modulus and reducing deformability of the material. Additionally, simple mechanical models were used to analyze the results, finding that the Hirsch and Halpin–Tsai models (Casado et al., 2009) were adequate for modeling the dependence of the tensile modulus with WF concentration. More interesting, they applied a model for the ultimate tensile stress proposed by Pukanszky et al. (1989) that allows quantifying the reinforcing effect of a filler through calculation of a fitting parameter, "*B*," related to the quality of the filler–matrix inteface. According to D'Almeida and De Carvalho (1998), for $B<3$ no reinforcing effect is obtained. The solid curve in Figure 17.2 was fitted to the experimental results corresponding to different WF concentrations, obtaining $B=7.1$ (for the 0–15 wt.% range), which indicates that the strong interface led to a remarkable reinforcing effect. The samples with 20 and 30 wt.% of WF deviate from the previous trend, because there was not enough PU to wet the filler. Even so, using all experimental values in the fitting process (dashed line, Figure 17.2) gives a B parameter higher than 3 ($B=5.6$), confirming the reinforcing effect of WF. SEM images of the interface also supported this feature.

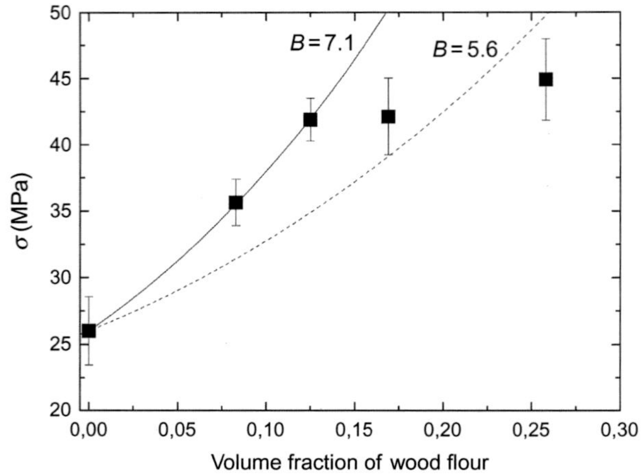

Figure 17.2 Tensile strength and Pukanszky model (solid line) versus WF content for biocomposites based on tung oil-PU.
Reprinted from Casado et al. (2009), with permission from Wiley.

Transparent composite sheets were prepared by the reaction of a mixture of soy oil-based polyol and petrochemical polyol with pMDI in the presence of different contents of microcrystalline cellulose (MCC) (0–1 wt.% with respect to the total polyol mass) by compression molding (Luo et al., 2013a). The flexural strength and flexural modulus of the PU were increased by 60.3% and 103.3%, respectively, by the addition of only 1 wt.% MCC, suggesting a very effective reinforcement. At this concentration, the notched Izod impact strength also increased by 74%. The crack propagates freely in the neat biobased PU, but in the composites, the crack occurs first and then peeling takes place from the cracked surface along the MCC that separates from the matrix, thus increasing the absorbed energy in the composites (Latere Dwan'isa et al., 2004). The storage modulus of the composite sheets was higher than that of the neat PU at temperatures below T_g. Additionally, the T_g of the biobased PU, 149 °C, was shifted to 153 °C at 1 wt.% MCC loading. These results agreed with the high mechanical strength of MCC and the additional cross-linking contributed by it, resulting in the improved mechanical properties of the composites.

Bioreinforced foams of high density (300 kg/m^3) were reported by Kuranska et al. (2013). In this case the reference rigid PU foams were prepared using pMDI, polyether polyols, an amine catalyst, silicone surfactant, and water as a chemical blowing agent. This reference petrochemical formulation was modified with rapeseed oil-based polyols and flax fibers with lengths in the range of 0–0.5 mm. Two different rapeseed oil-based polyols were prepared: the first one by epoxidation and further opening of the oxirane rings (P1, hydroxyl = 256 mg KOH/g) and the second one by esterification with triethanolamine (P2, hydroxyl number = 365 mg KOH/g). Composites prepared from P1 and flax fibers (five parts of fibers per hundred parts of the polyols, php) increased the compressive strength about 40% in comparison to the reference neat PU. Composites containing 10 php of the natural fillers showed the worsening of the compressive strength, which was explained by a fraction of the fibers not being incorporated in the struts between the cells, causing disorder in the cell structure. In another work, Kuranska and Prociak (2012) studied rigid PU foams of about 40 kg/m^3 density, modified with flax and hemp fibers of different lengths and rapeseed oil-based polyol (30 wt.% as a replacement for petrochemical polyol). The addition of 5 php flax fibers with a length of 0.5 mm led to the highest compressive strength of the series, up to about 18% for the strength determined in a direction perpendicular to the foam rise in comparison to the reference material. Compressive strength in all three directions was also affected by the apparent density of foamed composites. The materials with long fibers had a lower value of compressive strength. Based on the structure analysis it was concluded that shorter fibers are more adequate to interact with the building of the cells without causing their damage.

Merlini et al. (2011) reported the influence of the fiber volume fraction, fiber length, and alkaline treatment on the mechanical and thermal properties of short random banana fiber-reinforced PU obtained from a commercial castor oil (CO)-based polyol and MDI. Polymeric composites were prepared by hand lay up followed by compression molding at room temperature. The tensile strength and Young's modulus of the composites increased with increasing fiber volume fraction and length (Table 17.1), with treated banana fiber composites displaying higher values than the untreated ones, due to the stronger interfacial interactions between the treated fibers and the PU matrix. The treated fibers presented higher surface roughness and free OH groups, promoting mechanical and chemical adhesion between fiber and matrix (Vilaseca et al., 2007).

Dynamic–mechanical analysis of the above samples showed the typical characteristics of the largest reinforcing effect above the glass transition temperature of the composites, where the differences of rigidity between matrix and fibers are maxima. The stiffness of the composites increased due to the presence of fibers, which restricted the mobility of the polymer chains, leading also to the T_g increase. As expected, the intensity of the tan δ peak decreases with the amount of banana fiber.

Miléo et al. (2011) worked on castor oil PU reinforced with cellulose fibers from sugarcane straw (5–20 wt.%). Cellulose was extracted by steam explosion of the straw, followed by delignification with NaOH. The incorporation of cellulose fibers led to the increase of the composites' stiffness; the flexural modulus was 163.7 ± 25.1 MPa

Table 17.1 Tensile properties of banana fiber-reinforced PU

	Banana fiber (vol%)	Fiber length (mm)	Tensile modulus (MPa)	Tensile stress (MPa)	Tensile strain (%)
	0	–	5.89 ± 1.06	1.96 ± 0.43	31.35 ± 1.80
Untreated fibers	10	20	15.77 ± 2.24	2.62 ± 0.35	18.00 ± 1.73
	10	30	22.21 ± 2.99	4.39 ± 0.32	17.00 ± 1.53
Treated fibers	10	20	39.99 ± 1.55	5.73 ± 0.25	16.00 ± 0.50
	10	30	31.32 ± 3.40	6.07 ± 0.42	20.00 ± 1.01

Values extracted from Merlini et al. (2011), with permission from Elsevier.

for the pure PU, while the value raised up to 171.0 ± 9.7 and 354.5 ± 25.1 for the 5 and 20 wt.% composites, respectively. On the other hand, fracture surfaces showed a poor dispersion of the fibers in the matrix, although with good fiber–matrix adhesion, without fiber pull out.

17.3.2 Biopolyurethane composites made from vegetable long fibers

Bakare et al. (2010), used a modified rubber seed oil to synthesize a PU resin and prepare composites incorporating long sisal fibers (10–30 wt.%). The tensile and flexural strength and modulus of the unidirectional fiber-reinforced composites increased with fiber loading (69–103 MPa and 2.28–3.25 GPa for the flexural strength and modulus, respectively).

Ramires and coworkers (2013) prepared PUs using sodium lignosulfonate (NaLS), castor oil, and two diols of different chain length. Randomly oriented fiber composites, containing 30 wt.% of curaua (average diameter of 116 μm, 3 cm length) and coir fibers (average diameter of 230 μm, 3 cm length) were then obtained. In neat lignopolyurethanes, the Izod impact strength increased with increasing chain length of the polyol used. Longer chain polyols lead to a lower degree of cross-linking and, thus, to materials with elastomeric characteristics and high molecular mobility, resulting in high impact strength. On the other hand, composites made with curaua fiber had higher tensile strength (485 MPa) than that made from coir (120 MPa), due to the higher cellulose content and higher crystallinity index of the curaua fibers (67.7% and 65%, respectively) compared with those of the coir fibers (50% and 41%, respectively). Moreover, curaua composites exhibited higher impact strength than coir ones and increased Izod impact strength with respect to that of the neat lignopolyurethane, which was the result of the high efficiency of the load transfer from the matrix to the fibers. The aspect ratio of the fibers (~259 for curaua and ~130 for coir) also influenced the impact strength of the composites. Generally, fibers with high aspect ratios are more efficient reinforcements of composites, leading to materials with better mechanical properties.

17.3.3 Foamed biopolyurethanes

Chemically modified tung oil was also used as the main polyol component in the formulation of filled viscoelastic PU foams (Ribeiro da Silva et al., 2013ab). A small part of the modified tung oil was replaced by a low molecular weight diol (ethylene glycol) leading to reactive liquid mixtures of adequate viscosity that facilitated the mechanical stirring of reactants. The chosen filler was rice husk ash (from combustion of rice husk) because of its high silica content. The dynamic–mechanical tests revealed that the foams exhibited two different and broad thermal transitions, the temperature of their maxima depending on filler concentration.

Figure 17.3 shows the compression stress–strain curves for selected specimens. As expected for rigid foams, the compressive stress–strain curves show a linear elastic region, followed by a plateau (Latere Dwan'isa et al., 2004; Gibson and Ashby, 1997; Ashby, 2006) and then a further increase. At the end of the elastic region, the stress

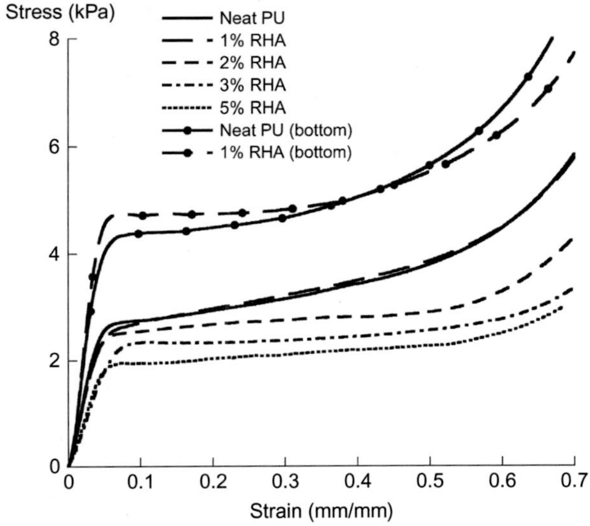

Figure 17.3 Compression stress–strain curves for rice husk ash filled viscoelastic polyurethane foams.
Reprinted from Ribeiro da Silva et al. (2013b), with permission from Elsevier.

reaches a peak value that corresponds to the compressive strength. This maximum stress is sometimes followed by a decrease before the stress plateau is reached (Tondi et al., 2009; Ashby, 2006). Cracks initiate at the peak stress value and the material tends to generate fragments as a result of these cracks. The plateau originates in the coexistence of collapsed and uncollapsed zones, and it is typical of rigid foams undergoing successive cell wall fractures. Beyond the plateau, densification takes place and the stress rises sharply as complete densification begins. When total cell collapse is completed, the compacted material begins to behave like a nonporous solid (Tu et al., 2001). Overall, compression modulus, compressive strength, and storage modulus increased as foam density increases but decreased as rice husk ash concentration increases due to the detrimental changes induced by the filler in the foam cellular structure (severe disruption of the foam morphology). The same is true for the stress value at the plateau. However, densification strain exhibits the opposite behavior, indicating that reinforced foams can sustain slightly higher deformations without collapsing, probably due to a reduced reactivity of the components induced by the filler.

Microcellular PU composites were investigated by Aranguren et al. (2007), who used long hemp fibers as reinforcement. The PU matrix was synthesized from a CO-based polyol and pMDI, while microfoamed composites with preferentially orientated hemp fibers were prepared; the threads were long enough to behave as continuous fibers, extending over the whole length of the sample specimen. SEM micrographs of the unfilled foam and the hemp-foamed composite are shown in Figure 17.4. In general, hemp fiber composites showed relatively high bending properties, which were related to the large aspect ratio of the hemp fibers. In this respect, one-directional hemp composites offered the best bending behavior. Dynamic–mechanical studies of

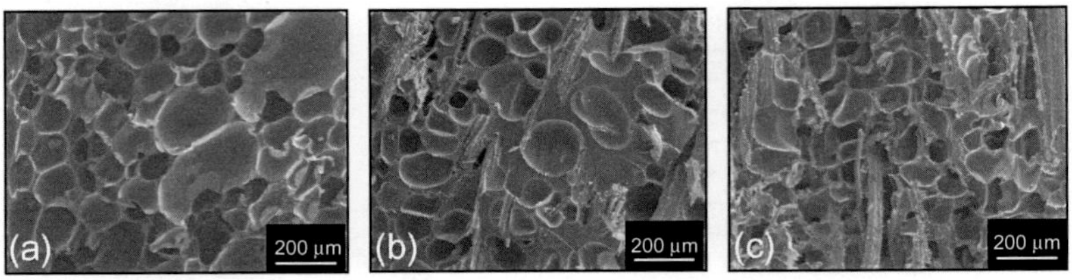

Figure 17.4 Scanning electronic micrographs: (a) unfilled foam; (b) long unidirectional hemp fibers reinforced foams; (c) long unidirectional hemp fibers reinforced foams with added water.
Reprinted from Aranguren et al. (2007), with permission from Wiley.

the composites showed very wide temperature transitions. The storage modulus and tan δ curves for the unreinforced PU and the composites showed that the peak, which corresponds to the main relaxation of the matrix and is related to its glass transition, is in the range of 90–118 °C for the composites, while it is around 70–73 °C for the unreinforced sample. As usual, it was observed that the difference between glass and rubber modulus is much higher in the matrix than in the composites. The differences in properties between the unfilled and filled PU were attributed to the combination of the hydrodynamic effects of the particles embedded in a viscoelastic medium and to the mechanical restraint introduced by the filler at high concentrations. As was discussed in other examples reviewed in this chapter, the interfacial restricted mobility of the polymer chains is further reduced by covalent interfacial bonding, a characteristic property of composites made from vegetable fibers and in situ polymerized PU matrices. The static flexural properties of the hemp-reinforced PU composites indicated that the best properties were obtained for the aligned one-directional fiber composite. The lowest properties corresponded to the two directional-fiber composite, while the four directional materials had intermediate properties. An interesting observation was that these materials suffered important aging due to a secondary reaction of the unsaturations present in the vegetable oil-based polyol. As a result, the modulus and strength of the materials increased with aging, but the deformation to rupture did not decrease, and for this reason aged materials showed improved modulus and toughness compared to those of the original composites, under the test conditions reported.

Gu et al. (2013) studied composite PU foams prepared by the free rise pouring method, using a commercial soy-based polyol of hydroxyl number = 55.67 mg KOH/g. Commercial steam explosion pulp of maple with particle size in the range of 180–850 μm was added in 13.3 parts by weight to prepare the filled foams. Although the incorporation of wood pulp fibers in the PU did not significantly alter the compressive strength, it caused a slight reduction in tensile strength, but contributed to the resistance to deformation (up to about 6%).

Comparable results were obtained by Mosiewicki et al. (2009a) that prepared rigid PU foams using a polyol of low viscosity synthesized by alcoholysis of CO with triethanolamine. The modified oil was used as the polyol component in the formulation of rigid PU foams filled up to 15 wt.% of WF (particle size <64 μm). The results showed that the compression modulus and yield strength decreased as WF content increased, which was attributed to the cell disruptions introduced by a microsized filler. Moreover, similar results obtained with a commercial rigid PU system led to the conclusion that these new materials can be a valuable alternative to replace petroleum-derived formulations, and could be used in semistructural applications, where low weight is desirable.

An improvement in the mechanical properties with the incorporation of natural fibers into foams was reported by Luo et al. (2013b). In this work, lignin/soy-based PU biofoams were prepared by a free-rising method in the presence of water as blowing agent, while lignin powder (from bioethanol production) was used as a reactive reinforcing filler (15 wt.%, 75 μm). The PU foam was formulated using a commercial soybean phosphate ester polyol (OH value = 272 mg KOH/g), a petrochemical polyol (OH value = 635 mg KOH/g) and pMDI. Mechanical properties of the samples were improved by the incorporation of lignin, with an optimum at 10 wt.%. Similarly, the glass transition temperature increased with the addition of lignin (172.8–182.1 °C for the neat PU, and 15 wt.% composite, respectively), a behavior that is typically associated with increasing cross-link density. The authors proposed that lignin was acting as a chain extender and cross-linking agent. The specific compressive modulus (compressive modulus/density) also increased with lignin content (109.09 and 153.45 MPa/g/ml for the 5 and 10 wt.% composites, respectively) up to about 15 wt.%, decreasing at higher concentrations. The flexural modulus presented the same trend: increased from 195.80 to 354.86 MPa with lignin content from 5 to 15 wt.%, with the optimum at 10 wt.% lignin. It was proposed that the lignin–polyurethane interactions determine the incorporation of urethane–urethane and ester–urethane ordered structures from matrix in the lignin amorphous phase. Consequently, at high lignin content the urethane structures could not crystallize further and the thermomechanical properties of the corresponding blends are highly reduced (Ciobanu et al., 2004). On the other hand, the notched Izod impact strength was not significantly affected by the presence of lignin, their values being 1.33, 1.29, 1.28, and 1.32 J/m for biofoams containing 0, 5, 10, and 15 wt.% of lignin, respectively.

17.3.4 Water sorption and biodegradability of composites based on biopolyurethanes

Because vegetable fibers and particles are hygroscopic, it is important to monitor changes of the mechanical properties of the composites due to the action of humid environments. This effect was investigated by Mosiewicki et al. (2012) on tung oil-based PU reinforced with WF. The storage modulus (DMA), as well as the tensile modulus and the tensile ultimate stress, decreased as the moisture content of neat matrix and composites increased. The comparison between the mechanical properties of the recently prepared dry samples and the wet samples showed the plasticization effect of the water molecules in the PU and its composites. Although increasing amounts of WF led to increasing water absorption, the incorporation of low percentages of WF resulted in lowering the water sorption with respect to the neat PU. This unexpected result was due to the very good compatibility of the system and the changes generated in the polymer as a consequence of the coreaction with WF.

The tensile and flexural properties of wet samples (30 days of water immersion) of long sisal fiber composites made with a rubber seed oil PU were investigated by Bakare et al. (2010). A slight loss in tensile strength after water immersion was observed at 25–30 wt.% sisal. The decrease in mechanical properties was attributed to the formation of hydrogen bonding between the water molecules and the hydroxyl groups of the cellulosic component of the sisal fiber. However, the flexural strength and modulus were not adversely affected by moisture absorption.

The biodegradation is also important for this type of materials that intend to replace synthetic polymers and composites. Thus, the biodegradation of tung oil-based PU reinforced with WF was followed during 383 days of exposure to soil or vermiculite media (Aranguren et al., 2012). The hydrolytic degradation was the most important mechanism of deterioration in all cases. A shift of the glass transition toward higher temperatures was observed (from around 65 to above 100 °C), which was explained as the result of the preferential attack and removal of free or dangling-pendant chains that plasticize the original material. The effect on the storage modulus was an increase in the whole range of temperatures observed, with the composites biodegraded for one year almost doubling the value of the storage modulus of the recently prepared samples.

17.4 PU nanocomposites based on vegetable-derived nanofibers

The previous sections illustrated the versatility of the PU composites and the wide range of possibilities that derive from the use of lignocellulosic materials as fillers/reinforcements for these polymers. It is not surprising then, that the idea of using PU matrices with biobased nanoparticles and nanofibers has been developed during the last years. The following section is dedicated to the bionanocomposites made from PUs and plant derived nanoparticles; in particular, composites containing nanocellulose will be considered.

Vegetable material is composed of cellulose, hemicellulose, lignin, and other minor components. Cellulose is present as fibers, bundles of associated microfibrils, which are in turn formed by cellulose chains physically linked to one another by strong hydrogen bonds. The fibrils are formed by crystalline well-ordered regions that alternate with intermediate amorphous regions. These fibrils can also associate to one another through extensive hydrogen bonds. By preferentially degrading the amorphous regions by acid or enzimatic methods, crystals can be obtained with 5–20 nm of diameter and hundreds to thousands nm in length. These nanoelements are usually called cellulose nanocrystals or nanowhiskers, a term that was first used for crystals obtained from tunicates, which are longer than wood-derived crystals. Cellulose nanocrystals (NCC) show modulus in the order of the 150 GPa, and, as is the case with other nanoparticles, if a low concentration of NCC is well dispersed in a polymeric matrix, the extensibility of the polymer can be preserved, while the modulus can be largely increased.

Cellulose nanocrystals obtained by sulfuric acid hydrolysis of MCC were incorporated into an elastomeric PU by Marcovich et al. (2006). They used a polymeric

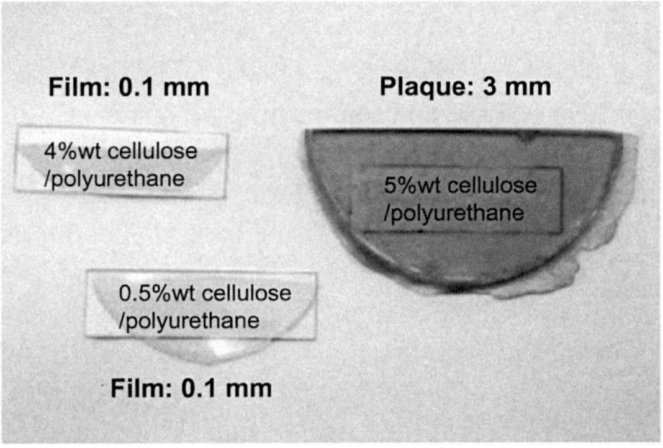

Figure 17.5 Photographs of elastomeric polyurethane materials reinforced with NCCs.

aromatic isocyanate (pMDI) and a mixture of a diol (PEG) and a polyol (functionality >2.0) for the PU synthesis. The NCC was added to the OH-components after redispersion of the freeze-dried nanocellulose in dimethylformamide (DMF), followed by solvent evaporation. Very good dispersion was obtained by this method and the samples were transparent even after the addition of 5 wt.% of NCC, as shown in Figure 17.5. A yellow-brown tint was observed as a result of the original brown dark color of the polymeric isocyanate.

Dynamic–mechanical analysis of these samples revealed large upward shifts in the glass transition temperature (T_g) of the cellulose nanocomposites as compared to the unfilled PU (about 16 °C, with 5 wt.% nanocellulose). The T_g increase resulted not just from the restricted mobility of the polymer in the presence of the nanocrystals, but also from the interfacial reaction between the cellulose and the isocyanate that covalently bonded matrix and nanocrystals; the latter acting as multifunctional cross-linkers, which caused even a larger increase in T_g. Notice that these same reasons were used to explain the behavior of traditional microcomposites, although the changes are observed at much lower concentrations of the reinforcing elements in the case of the nanocomposites.

Although the low-frequency viscosity of the unreacted mixtures was several decades higher than that of the unfilled PU, the sample containing 5 wt.% of NCC had a tensile modulus just 2.44 times that of the unfilled sample, which was explained by the indirect interaction cellulose–PU–cellulose that is formed after cross-linking. Thus, the same reason that contributes to excellent nanocellulose dispersion also reduces its reinforcement effect. This suggests that if high modulus is desired, an optimum degree of dispersion must be achieved that still allows for the formation of direct cellulose contacts.

Li et al. (2013) obtained nanocrystalline cellulose from hydrolysis of medical absorbent cotton with anhydrous phosphoric acid. The system was neutralized with ammonia and the whole mixture was used to modify a PU foam. The nanocrystalline cellulose performed as reinforcing material, the phosphates as flame retardants, and the hydrolyzed saccharides contributed as partial replacement of polyols. It was observed that the addition of the modifier improved the compression strength of the foams up to an optimum of 6 wt.%, but then dropped for larger additions. At the optimum formulation the compression strength was 107.5 kPa, which is 4.29 times that of the unfilled foam, but because of the retained phosphates the flame performance was also improved (lower heat release and longer time for ignition). The cell structure of the modified foam was more regular than that corresponding to the neat PU, which was related to the improved mechanical properties of the foams.

Juntaro et al. (2012) prepared sheets from bacterial cellulose (BC) that were impregnated by a PU-based resin containing UV polymerizable groups to produce transparent nanocomposites with excellent compatibility and, consequently, no gaps between cellulose and matrix, which otherwise would lead to light diffraction and opaque samples. The authors compared the tensile properties of the composites with those of a sheet of BC and those of the unfilled PU. While the PU showed a Young's modulus of 0.19 GPa, the cellulose sheet reached the 5.09 GPa. Interestingly, the composites reached modulus ranging from 5.4 to 11.6 GPa depending on the processing conditions; the highest modulus achieved by the nanocomposite made with the BC dried at the highest imposed pressure. A similar trend was observed for the tensile strength (PU: 16.3 MPa, BC: 113 MPa, and nanocomposites: 72–151 MPa). Figure 17.6 illustrates the tensile response of these nanocomposites. The image shows the behavior of composites made from aqueous sheets dried under a pressure of 400 kPa (Type A) or 6.9 kPa (Type B), or by solvent exchange with ethanol and drying under low pressure (Type C). The composites had higher modulus and lower extensibility than the neat PU. The interesting fact is that the results for the nanocomposites were even better than for the cellulose sheet. The explanation given by

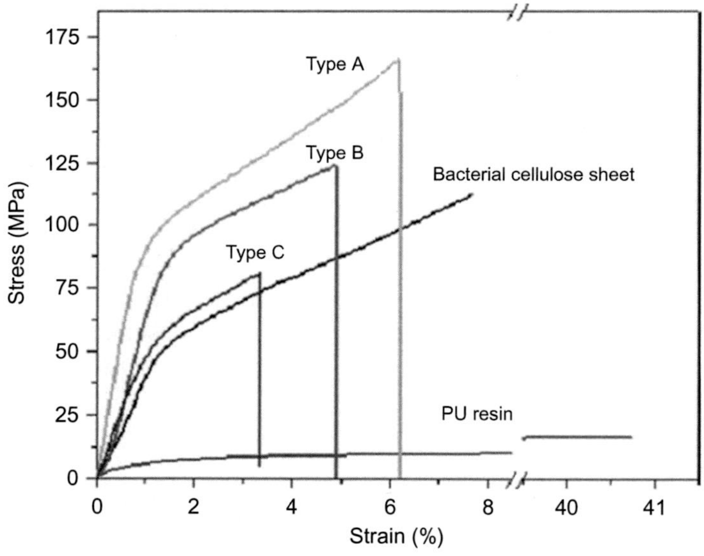

Figure 17.6 Stress–strain curves of bacterial cellulose sheets impregnated by a PU-based resin.
Reprinted from Juntaro et al. (2012), with permission from Elsevier.

the authors is that in BC sheet the stress transfer occurs only through the skeletal 3-D structure formed by the fibers, but in the nanocomposites the interstices instead of being air-filled are occupied by the PU. Thus, stress transfer occurs through the fibers and also from fiber to polymer to fiber, which improves the mechanical properties.

Other authors compared materials prepared as micro- or nanocomposites. Wu et al. (2007) prepared a segmented PU using polytetramethylene glycol, 1–4 butanodiol as diol and chain extender, respectively, and MDI as isocyanate component, which were polymerized in the usual two-step method. Microcomposites were prepared in a similar way, but cellulose fibers (1–2 mm length), diol, and MDI were mixed with DMF and reacted; in the second step, the chain extender was added and reacted. For the nanocomposites, MCC was first dispersed in DMF with traces of LiCl (80 °C for 6 h), and then mixed with diol and reacted with MDI to be chain extended in a second step. TEM images allow confirming that the second type of composites contained nanocellulose. Comparison of the effect of the two types of fibers showed that microcomposites (5 wt.% of cellulose) had the typical reported behavior of higher modulus and lower extensibility than the unfilled PU. On the other hand, nanocomposites showed higher modulus (twice that of the microcomposite with the same cellulose concentration and 2.6 times that of the neat PU), but largely increased extensibility (1.5 times that of the PU). The example illustrates the important role of the size of the reinforcing elements, because interfacial compatibility, strong physical interaction, and covalent bonding are the same in both types of composites.

Regarding thermoplastic PU-based nanocomposites, Cherian and coworkers (2011) synthesized a degradable-segmented PU using approximately 2:1:1 of hard segment (diisocyanate)/soft segment (polycaprolactone-diol)/chain extender (1,4-butanediol). PU films were prepared by a solvent (tetrahydrofuran) casting process. The authors used pineapple leaf fibers as the raw material to obtain nonwoven nanocellulose mats by combining high-pressure defibrillation and chemical purification procedures. Microscopy studies showed that the obtained mats were composed of bundles of cellulose nanofibers with diameters ranging between 5 and 15 nm and estimated lengths of several micrometers. The nanocomposites were prepared by compression molding, which involved stacking the nanocellulose fiber mats between PU films. The results showed that nanofibrils reinforced the PU efficiently. The addition of 5 wt.% of cellulose nanofibrils to PU increased the strength nearly 300% and the stiffness by 2600%. This composite also exhibited the highest strain to fracture. Moreover, the developed composites were utilized to fabricate various versatile medical implants, such as cardiovascular ones, and have the potential for being used in scaffolds for tissue engineering, repair of articular cartilage, vascular grafts, urethral catheters, mammary prostheses, penile prostheses, adhesion barriers, and artificial skin.

Rueda et al. (2013) extracted cellulose nanocrystals, with an aspect ratio of 17 from MCC and then incorporated them in a polycaprolactone-based PU by a casting/evaporation method using DMF as solvent. Thermal, mechanical, and morphological properties indicated, in general, favorable matrix–nanocrystals interactions arising from efficient dispersion of NCC in the thermoplastic PU. Low NCC quantity in PU nanocomposites led to a tough material without loss in ductility, whereas an increase in NCC content enhanced the soft and hard segment crystallization, which resulted in an increase in the material stiffness and stability versus temperature.

Besides mechanical performance, nanoreinforcements have also been added to tune the functional properties of PU. Let's just briefly mention that nanocellulose has been used to improve the recovery force of shape memory PUs that have the capability of autonomously recovering a previously "memorized" shape as a response to a change in the surrounding environment; for example, temperature (Auad et al., 2008, 2010, 2012; Han et al., 2012) or humidity (Zhu et al., 2012; Luo et al., 2011, 2012) changes, among others. The mechanical characterization of these materials is a useful tool for tailoring their properties, as dramatic changes in their rigidity can be displayed as a response to sensed external changes.

Floros et al. (2012) prepared nanocrystalline cellulose by a process of several steps involving alkali treatment of the fibers, steam explosion, bleaching, and finally oxalic acid treatment followed again by steam explosion (NCC length ~200–250 nm and diameter ~4–5 nm). Samples with different concentrations of the NCC were prepared from a biobased TPU derived entirely from oleic acid. The best results were obtained with 0.5% of NCC in the TPU, with an elongation at break that improved from 178% to 269% and stress at break that increased from 29.3 to 40.5 MPa.

Lin et al. (2013) developed new nanocomposites consisting of a CO-based PU matrix filled with acetylated cellulose nanocrystals (ACNs). The ACN exhibited improved dispersion in tetrahydrofuran as a blending medium, and reduced polarity as compared with unmodified NCC, resulting in a high loading level of 25 wt.% in the nanocomposite. As the ACN loading level increased from 0% to 25%, the tensile strength and Young's modulus of the nanocomposites increased from 2.79 to 10.41 MPa and from 0.98 to 42.61 MPa, respectively. When the ACN loading level was 10 wt.%, the breaking elongation of the nanocomposites reached the maximum value of more than twice that of the PU.

Seydibeyoglu and Oksman (2008) considered the effects of cellulose from the micro- to the nanoscale as reinforcement of a thermoplastic PU. In this case, nanofibrillated cellulose from hard wood fibers was obtained by means of a high-pressure homogenizer. Again, composite materials were prepared using a compression molding technique, which involved stacking the cellulose fiber mats between thermoplastic PU films. The results showed that both microfibers and nanofibrils reinforced the PU, nanofibrils being a much more effective reinforcement than the microsized cellulose fibers; for example, the addition of 16.5 wt.% cellulose nanofibrils to PU increased the strength nearly 500% and the stiffness by 3000% while the use of 18.7 wt.% microfibers lead to increases of 200% and 500%, respectively. The authors attributed this behavior to the fact that cellulose nanofibrils integrate within the polymer matrix much better due to the smaller size but also to the better properties of nanosized fibrils compared with the microsized fibers.

Waterborne PUs are fully reacted urethane polymers dispersed in water, and they show many excellent features compared to conventional organic solvent-based PUs, including higher rates of biodegradation (Santiago de Oliveira Patricio et al., 2013). However, they show low mechanical strength and thermal stability, restricting some of their potential applications. In this way, reinforcements from renewable biomass are viable supplements for these polymers, as was demonstrated by Liu and coworkers (2013), who used cellulose nanocrystals to improve properties of rosin-based waterborne PUs. Tensile strength of the composite films increased from 28.2 to 52.3 MPa while the tensile modulus was improved from 316.2 to 1045.4 MPa with increasing cellulose amount from 0 to 20 wt.%, respectively. However, elongation at break was sharply reduced from 267% to 22% for the same filler concentration range.

Santiago de Oliveira Patricio et al. (2013) prepared nanocomposites based on cellulose nanocrystals and synthesized waterborne PU using a two-step method commonly referred to as a prepolymer method. They developed nanocomposites with different properties by altering the mode and step in which the nanofillers were incorporated during the PU formation. The process utilized is as follows: WPU/NCC nanocomposites were obtained by physical mixtures of waterborne PU and cellulose nanocrystals in an aqueous suspension; WPU/NCC-P samples were obtained by adding the NCC suspension during the dispersion step of the prepolymer (before adding the hydrazine hydrate used to react with the remaining free −NCO groups); and WPU/NCC-PP composites were produced by adding the cellulose nanocrystals, dispersed in the polyols, at the beginning of the prepolymer synthesis. The authors demonstrated that the degree of interaction between the nanofillers and the WPU through hydrogen bonds could be controlled by the method of incorporation of the NCC to the polymer network and proposed a chemical reaction between the −NCO groups and the −OH groups of the cellulose crystals in WPU/NCC-PP. In this case, it was noticed that a high degree of interaction between the hard segments and the NCC affected the morphology, reducing phase separation, so that strength was decreased. WPU/NCC-P samples exhibited the typical behavior of composites reinforced by agents with a high aspect ratio and good adhesion, whereas WPU/NCC samples performed as composites with poor adhesion or a less efficient dispersion.

Gao et al. (2012) filled a biocompatible waterborne polyurethane (WPU) based on castor oil (CO)/polyethylene glycol (PEG) with low-level loadings of Eucalyptus globulus NCC obtained by sulfuric hydrolysis (average length and diameter of 518.0 ± 183.4 and 21.7 ± 13.0 nm, respectively). The nanocomposites showed significant enhancement in tensile properties: 1 wt.% NCC increased the tensile strength from 5.43 to 12.22 MPa, while the Young's modulus reached the maximum of 4.83 MPa (416% higher than the neat PU) at a loading of 4 wt.% NCC.

17.5 Final Remarks

PUs have proven to be very compatible with vegetable fibers, particles, and nanoreinforcements. Clearly, the method of processing the composite is of importance, especially if the polymerization is carried out in the presence of the vegetable material, in which case this can coreact with the polymeric matrix. This interfacial reaction has been considered to be the reason for reduced water absorption even when the components are hygroscopic, the increase of glass transition temperature, and in some cases the improved toughness and impact properties.

The incorporation of vegetable fibers into foams have been shown to have opposing effects, because although the fibers contribute with their highest rigidity, they also disturb the integrity and regularity of the cells, which can reduce the performance of composite foams.

Vegetable-based nanoelements proved to lead to composites of improved properties, while their effect is observed at much lower concentrations that in traditional composites.

Although traditional composites will continue to find their place in different applications, the biobased and WPUs will receive more attention as they offer new ways to reach more environmentally friendly materials. On the other hand, PU nanocomposites and bio-PU nanocomposites will continue to receive attention, and it is to be expected that they will find their market as functional special applications.

References

Aranguren, M.I., 2006. Polymer composites. In: Somasundaran, P., Hubbard, A. (Eds.), Encyclopedia of Surface and Colloid Science, second ed., vol. 6. Taylor & Francis/CRC Press, New York, pp. 4796–4810.

Aranguren, M.I., Rácz, I., Marcovich, N.E., 2007. Microfoams based on castor oil polyurethanes and vegetable fibers. J. Appl. Polym. Sci. 105 (5), 2791–2800.

Aranguren, M.I., González, J.F., Mosiewicki, M.A., 2012. Biodegradation of a vegetable oil based polyurethane and wood flour composites. Polym. Test. 31 (1), 7–15.

Ashby, M.F., 2006. The properties of foams and lattices. Philos. Trans. R. Soc. A Math. Phys. Eng. Sci. 364, 15–30.

Auad, M.L., Contos, V.S., Nutt, S., Aranguren, M.I., Marcovich, N.E., 2008. Characterization of nano-cellulose reinforced shape memory polyurethanes. Polym. Int. 57, 651–659.

Auad, M.L., Mosiewicki, M.A., Richardson, T., Aranguren, M.I., Marcovich, N.E., 2010. Nanocomposites made from cellulose nanocrystals and tailored segmented polyurethanes. J. Appl. Polym. Sci. 115, 1215–1225.

Auad, M.L., Richardson, T., Hicks, M., Mosiewicki, M.A., Aranguren, M.I., Marcovich, N.E., 2012. Shape memory-segmented polyurethanes: behavior dependence on nanocellulose addition and testing conditions. Polym. Int. 61, 321–327.

Bakare, I.O., Okieimen, F.E., Pavithran, C., Abdul Khalil, H.P.S., Brahmakumar, M., 2010. Mechanical and thermal properties of sisal fiber-reinforced rubber seed oil-based polyurethane composites. Mater. Des. 31, 4274–4280.

Batouli, S.M., Zhu, Y., Nar, M., D'Souza, N.A., 2014. Environmental performance of kenaf-fiber reinforced polyurethane: a life cycle assessment approach. J. Clean. Prod. 66, 164–173.

Bledzki, A.K., Zhang, W., Chate, A., 2001. Natural-fibre-reinforced polyurethane microfoams. Combust. Sci. Technol. 61, 2405–2411.

Casado, U., Marcovich, N.E., Aranguren, M.I., Mosiewicki, M.A., 2009. High strength composites based on tung oil polyurethane and wood flour: effect of the filler concentration on the mechanical properties. Polym. Eng. Sci. 49, 713–721.

Cherian, B.M., Lopes Leão, A., Ferreira de Souza, S., Manzine Costa, L.M., Molina de Olyveira, G., Kottaisamy, M., Nagarajand, E.R., Thomas, S., 2011. Cellulose nanocomposites with nanofibres isolated from pineapple leaf fibers for medical applications. Carbohydr. Polym. 86, 1790–1798.

Ciobanu, C., Ungureanu, M., Ignat, L., Ungureanu, D., Popa, V.I., 2004. Properties of lignin–polyurethane films prepared by casting method. Ind. Crop. Prod. 20, 231–241.

D'Almeida, J.R.M., De Carvalho, L.H., 1998. An investigation on the tensile strength of particulate filled polymeric composites. J. Mater. Sci. 33, 2215–2219.

El-Shekeil, Y.A., Sapuan, S.M., Azbdan, K., Zainundin, E.S., 2011. Development of a new kenaf bast fiber-reinforced thermoplastic polyurethane composite. BioResources 6 (4), 4662–4672.

El-Shekeil, Y.A., Sapuan, S.M., Abdan, K., Zainudin, E.S., 2012a. Influence of fiber content on the mechanical and thermal properties of Kenaf fiber reinforced thermoplastic polyurethane composites. Short Commun. Mater. Des. 40, 299–303.

El-Shekeil, Y.A., Sapuan, S.M., Khalina, A., Zainudin, E.S., Al-Shuja'a, O.M., 2012b. Influence of chemical treatment on the tensile properties of kenaf fiber reinforced thermoplastic polyurethane composite. Express Polym. Lett. 6 (12), 1032–1040.

El-Shekeil, Y.A., Sapuan, S.M., Azaman, M.D., Jawaid, M., 2013. Optimization of blending parameters and fiber size of kenaf-bast-fiber-reinforced the thermoplastic polyurethane composites by Taguchi method. Adv. Mater. Sci. Eng. 2013, Article ID 686452, 5 pages.

Floros, M., Hojabri, L., Abrahama, E., Jose, J., Thomas, S., Pothan, L., Lopes Leao, A., Narine, S., 2012. Enhancement of thermal stability, strength and extensibility of lipid-based polyurethanes with cellulose-based nanofibers. Polym. Degrad. Stab. 97, 1970–1978.

Gao, Z., Peng, J., Zhong, T., Sun, J., Wang, X., Yue, C., 2012. Biocompatible elastomer of waterborne polyurethane based on castor oil and polyethylene glycol with cellulose nanocrystals. Carbohydr. Polym. 87, 2068–2075.

Gibson, L.J., Ashby, M.F., 1997. Cellular Solids: Structure and Properties. Cambridge University Press, Cambridge.

Gil, L., 2009. Cork composites: a review. Materials 2, 776–789.

Gu, R., Sain, M.M., Konar, S.K., 2013. A feasibility study of polyurethane composite foam with added hardwood pulp. Ind. Crop Prod. 42, 273–279.

Han, J., Zhu, Y., Hu, J., Luo, H., Yeung, L.-Y., Li, W., Meng, Q., Ye, G., Zhang, S., Fan, Y., 2012. Morphology, reversible phase crystallization, and thermal sensitive shape memory effect of cellulose whisker/SMPU nano-composites. J. Appl. Polym. Sci. 123 (2), 749–762.

Hull, D., 1993. An introduction to composite materials. In: Cambridge Solid State Science Series, first ed., vol. 150. Cambridge University Press, Cambridge, pp. 87.

Juntaro, J., Ummartyotin, S., Sain, M., Manuspiya, H., 2012. Bacterial cellulose reinforced polyurethane-based resin nanocomposite: a study of how ethanol and processing pressure affect physical, mechanical and dielectric properties. Carbohydr. Polym. 87, 2464–2469.

Kuranska, M., Prociak, A., 2012. Porous polyurethane composites with natural fibres. Compos. Sci. Technol. 72, 299–304.

Kuranska, M., Aleksander, P., Mikelis, K., Ugis, C., 2013. Porous polyurethane composites based on bio-components. Compos. Sci. Technol. 75, 70–76.

Latere Dwan'isa, J.P., Mohanty, A.K., Misra, M., Drzal, L.T., Kazemizadeh, M.J., 2004. Biobased polyurethane and its composite with glass fiber. J. Mater. Sci. 39, 2081–2087.

Li, S., Li, C., Li, C., Yan, M., Wu, Y., Cao, J., He, S., 2013. Fabrication of nano-crystalline cellulose with phosphoric acid and its full application in a modified polyurethane foam. Polym. Degrad. Stab. 98, 1940–1944.

Lin, S., Huang, J., Chang, P.R., Wei, S., Xu, Y., Zhang, Q., 2013. Structure and mechanical properties of new biomass-based nanocomposite: castor oil-based polyurethane reinforced with acetylated cellulose nanocrystal. Carbohydr. Polym. 95, 91–99.

Liu, H., Cui, S., Shanga, S., Wang, D., Song, J., 2013. Properties of rosin-based waterborne polyurethanes/cellulose nanocrystals composites. Carbohydr. Polym. 96, 510–515.

Luo, H., Hu, J., Zhu, Y., 2011. Polymeric shape memory nanocomposites with heterogeneous twin switches. Macromol. Chem. Phys. 212, 1981–1986.

Luo, H., Hu, J., Zhu, Y., 2012. Path-dependent and selective multi-shape recovery of a polyurethane/cellulose-whisker nanocomposite. Mater. Lett. 89, 172–175.

Luo, X., Mohanty, A., Misra, M., 2013a. Green composites from soy-based biopolyurethane with microcrystalline cellulose. Macromol. Mater. Eng. 298, 412–418.

Luo, X., Mohantya, A., Misra, M., 2013b. Lignin as a reactive reinforcing filler for water-blown rigid biofoam composites from soy oil-based polyurethane. Ind. Crop Prod. 47, 13–19.

Marcovich, N.E., Bellesi, N.E., Auad, M.L., Nutt, S.R., Aranguren, M.I., 2006. Cellulose micro/nanocrystals reinforced polyurethane. J. Mater. Res. 21 (4), 870–881.

Marcovich, N.E., Reboredo, M.M., Aranguren, M.I., 2014. Natural fiber thermoplastic composites. In: Olagoke, O. (Ed.), Handbook of Thermoplastics, second ed. Taylor and Francis Books, Inc, Florida, USA (Chapter 29).

Merlini, C., Soldi, V., Barra, G.M.O., 2011. Influence of fiber surface treatment and length on physico-chemical properties of short random banana fiber-reinforced castor oil polyurethane composites. Polym. Test. 30, 833–840.

Miléo, P.C., Mulinari, D.R., Baptista, C.A.R.P., Rocha, G.J.M., Gonçalves, A.R., 2011. Mechanical behaviour of polyurethane from castor oil reinforced sugarcane straw cellulose composites. Procedia Eng. 10, 2068–2073.

Mosiewicki, M.A., Dell'Arciprete, G.A., Aranguren, M.I., Marcovich, N.E., 2009a. Polyurethanes foams obtained from castor oil based polyol and filled with wood flour. J. Compos. Mater. 43 (25), 3057–3072.

Mosiewicki, M.A., Casado, U., Marcovich, N.E., Aranguren, M.I., 2009b. Polyurethanes from tung oil: polymer characterization and composites. Polym. Eng. Sci. 49, 685–692.

Mosiewicki, M.A., Casado, U., Marcovich, N.E., Aranguren, M.I., 2012. Moisture dependence of the properties of composites made from tung oil based polyurethane and wood flour. J. Polym. Res. 19 (2), 1–8.

Oprea, S., 2008. Effects of fillers, namely cork, on polyurethane resin based polyurethane bearing materials for passive isolation. J. Compos. Mater. 42, 2673–2685.

Peng, X., Fan, M., Hartley, J., Al-Zubaidy, M., 2011. Properties of natural fiber composites made by pultrusion process. J. Compos. Mater. 46 (2), 237–246.

Pukanszky, B., Tudos, F., Jancar, J., Kolarik, J., 1989. Possible mechanisms of polymer–filler interaction in polypropylene-$CaCO_3$ composites. J. Mater. Sci. Lett. 8, 1040.

Rácz, I., Andersen, E., Aranguren, M.I., Marcovich, N.E., 2009. Wood flour recycled polyol based polyurethane lightweight composites. J. Compos. Mater. 43, 2871–2884.

Ramires, E.C., de Oliveira, F., Frollini, E., 2013. Composites based on renewable materials: polyurethane-type matrices from forest byproduct/vegetable oil and reinforced with lignocellulosic fibers. J. Appl. Polym. Sci. 129 (4), 2224–2233.

Reboredo, M.M., Aranguren, M.I., Marcovich, N.E., 2008. Selected topics in wood flour–polypropylene composites: thermal, mechanical and time dependent response. In: Nwabunma, D., Kyu, T. (Eds.), Polyolefin Composites. John Wiley and Sons, Inc., Hoboken, NJ.

Rials, T.G., Wolcott, M.P., 1998. Morphology-property relationships in wood-fibre-based polyurethanes. J. Mater. Sci. Lett. 17, 317–319.

Ribeiro da Silva, V., Mosiewicki, M.A., Yoshida, M.I., Coelho da Silva, M., Stefani, P.M., Marcovich, N.E., 2013a. Polyurethane foams based on modified tung oil and reinforced with rice rusk ashes I: synthesis and physical chemical characterization. Polym. Test. 32 (2), 438–445.

Ribeiro da Silva, V., Mosiewicki, M.A., Yoshida, M.I., Coelho da Silva, M., Stefani, P.M., Marcovich, N.E., 2013b. Polyurethane foams based on modified tung oil and reinforced with rice husk ash II: mechanical characterization. Polym. Test. 32 (4), 665–672.

Rozman, H.D., Tay, G.S., Abubakar, A., Kumar, R.N., 2001. Tensile properties of oil palm empty fruit bunch-polyurethane composites. Eur. Polym. J. 27, 1759–1765.

Rueda, L., Saralegui, A., Fernández d'Arlas, B., Zhou, Q., Berglund, L.A., Corcuera, M.A., Mondragon, I., Eceiza, A., 2013. Cellulose nanocrystals/polyurethane nanocomposites. Study from the viewpoint of microphase separated structure. Carbohydr. Polym. 92, 751–757.

Santiago de Oliveira Patricio, P., Pereira, I.M., Ferreira da Silva, N.C., Ayres, E., Vargas Pereira, F., Lambert Oréfice, R., 2013. Tailoring the morphology and properties of waterborne polyurethanes by the procedure of cellulose nanocrystal incorporation. Eur. Polym. J. 49, 3761–3769.

Seydibeyoglu, M.O., Oksman, K., 2008. Novel nanocomposites based on polyurethane and micro fibrillated cellulose. Compos. Sci. Technol. 68, 908–914.

Tondi, G., Fierro, V., Pizzi, A., Celzard, A., 2009. Tannin-based carbon foams. Carbon 47, 1480–1492.

Tu, Z.H., Shim, V.P.W., Lim, C.T., 2001. Plastic deformation modes in rigid polyurethane foam under static loading. Int. J. Solids Struct. 38, 9267–9279.

Vilaseca, F., Mendéz, J.A., Pèlach, A., Llop, M., Cañigueral, N., Gironès, J., Turon, X., Mutjè, P., 2007. Composite materials derived from biodegradable starch polymer and jute strands. Process Biochem. 42, 329–334.

Wu, Q., Henriksson, M., Liu, X., Berglund, L.A., 2007. A high strength nanocomposite based on microcrystalline cellulose and polyurethane. Biomacromolecules 8 (12), 3687–3692.

Zhu, Y., Hu, J., Luo, H., Young, R.J., Deng, L., Zhang, S., Fan, Y., Ye, G., 2012. Rapidly switchable water-sensitive shape-memory cellulose/elastomer nano-composites. Soft Matter 8, 2509–2517.

Index

Note: Page numbers followed by *f* indicate figures and *t* indicate tables.

C

D

T

X

Y

Z

Printed in the United States
By Bookmasters